Nano-size Polymers

Stoyko Fakirov
Editor

Nano-size Polymers

Preparation, Properties, Applications

Springer

Editor
Stoyko Fakirov
Department of Mechanical Engineering
The University of Auckland
Auckland
New Zealand

ISBN 978-3-319-39713-9 ISBN 978-3-319-39715-3 (eBook)
DOI 10.1007/978-3-319-39715-3

Library of Congress Control Number: 2016940802

Printed on acid-free paper

This Springer imprint is published by Springer Nature
The registered company is Springer International Publishing AG Switzerland

Preface

Nanomaterials and nanotechnologies are part of the most attractive areas of scientific research and commercial interest because they offer new opportunities for improving human life. Nanomaterials are materials with morphological features on the nano-range, which usually are defined as smaller than one-tenth of a micrometer in at least one direction, with special properties that stem from their nanoscale dimensions.

A common characteristic of all nano-size materials, regardless of their chemical composition and method of preparation, is the extremely large ratio of surface area to volume. The natural tendency to reduce this free surface is the driving force for agglomeration of nanoparticles into larger formations.

Nanomaterials are not new—they have been used for many centuries, but it is only in recent decades that their unique properties were demonstrated at the nano-scale, where quantum effects are possible. A revolutionary discovery in nanoscience was the preparation of carbon nanotubes. It turned out that this all-carbon material is the ideal material—it has the highest heat conductivity, electrical conductivity, mechanical properties, etc.—and thus attracted the attention of researchers from various fields, including polymer science and technology. For example, with tensile moduli in the terapascal range and lengths exceeding 10 μ of carbon nanotubes, simple composite models predict order-of-magnitude enhancement in modulus at loadings less than 1 %. For this reason, a decade ago, it was believed that the most common polymer composites comprising about 30 % glass fibers will be replaced by the nanocomposites having only 2–5 % nano-size filler as reinforcement! Unfortunatelly, this has proved to be elusive.

The case of nanocomposites is one example of polymer nanotechnology, in which the polymer remains in a bulk state. Completely different is the situation when the polymer itself has to be converted into nano-size material, as, for example, in electrospinning. This technique exists since the beginning of the last century and nowadays it is used in hundreds of labs worldwide because of its elegancy, economic setup, and the fascinating electron microscopic images of the nanofibers obtained, which can be seen in some 2000 papers published per year.

It is important to mention a terminological peculiarity related with nanomaterials. In the literature, the terms “nanomaterials”, “nano-structured materials” (and even “nano-size materials”) are considered synonymous and used interchangeably, which is hardly correct. A good example in this respect, are the polymer materials, whose structural elements (lamellae, microfibrils, etc.) are nano-sized. Their common characteristic feature is the fact that they are **interconnected** via *tie molecules* and cannot be isolated as separate particles. In contrast, the nanomaterials, or even better, the nano-size materials are of the same sizes but they represent **individual**, not interconnected particles.

In addition to electrospinning, bulk polymers can be converted into nano-size materials by the recently developed concept of micro- or nanofibrilar composites starting from polymer blends. Nano-size polymers can also be obtained directly during synthesis. In the pharmaceutical industry, nano-size polymers are prepared using their solution. Many more applications also exist.

Because of the variety of potential applications (including the industrial and military ones), governments have invested billions of dollars in nanotechnology research. Through its National Nanotechnology Initiative, the USA has invested 3.7 billion dollars. The European Union has invested 1.2 billion, and Japan 750 million dollars.

This book represents an attempt to collect all available techniques used for converting the bulk polymers into nano-sized materials, as well as to discuss the properties of these materials and the potential for their applications. The chapters are prepared by leading worldwide specialists working in the respective areas.

Finally, the editor would like to thank all the contributors for their valuable contributions to this book as well as for being patient during the compilation work. Thanks are also extended to the Centre for Advanced Composite Materials of the Auckland University, Auckland, where this project was realized, as well as to the Ministry of Science and Innovations, New Zealand, for their financial support.

Auckland
May 2016

Stoyko Fakirov

Contents

Part I
Introduction

Chapter 1
Nanostructured and Nano-size Polymer Materials: How to Generate Them and Do We Need Them?

Manfred Stamm

The world is facing a lot of global challenges, which are of eminent importance for our living. This includes increasing demand of energy, lack of drinking water for a large part of our growing population, environmental problems connected with global heating, health problems with fast spreading of new diseases, and increasing resistance against existing medication, and many more [1–3]. Science and society is hardly capable to tackle those problems properly, which is connected with increasing complexity, limitation of resources, and lack of acceptance of necessary and sometimes painful demands. These challenges have been recognized by politicians and the scientific community, however, and there are many efforts to tackle these problems. In many (if not most) cases also materials aspects are important, where efficient insulation or lightweight construction can save energy, polymer membranes are used for drinking water generation or solar cells with new organic semiconductors are developed for cheap and flexible solar cells. One should not forget that also in the past the progress of mankind was largely connected with materials development, ranging from stone, bronze, and iron age to modern times, where then many other materials became important.

So materials research is highly contributing to developments in many areas, while in most cases it is little recognized, because often individual achievements are only marginal. They sum up however, and materials properties now can be tailored in a much better way to increasing needs. One of the developments with many facets are "*nanomaterials*," where according to the definition at least one dimension in the material is smaller than 1000 nm, in a more rigorous and commonly accepted definition smaller than 100 nm (see, e.g., wikipedia.org/nanomaterials). Polymer

M. Stamm (✉)
Leibniz-Institut für Polymerforschung Dresden, Hohe Strasse 6, 01069 Dresden, Germany
e-mail: stamm@ipfdd.de

M. Stamm
Technische Universität Dresden, Physical Cemistry of Polymer Materials, 01069 Dresden, Germany

S. Fakirov (ed.), *Nano-size Polymers*,
DOI 10.1007/978-3-319-39715-3_1

nanomaterials then contain at least one polymer as a component. Although this terminology is relatively new, nanomaterials exist already since a long time, but have experienced a lot of new developments over the last years.

Following the definition given above, nanomaterials cover a wide range of different challenging materials [4–7]. They include for instance the so-called *nanostructured materials*. The preparation of a single material or a blend of materials can be performed in such a way that crystal domains or phases are formed at nanoscale which as a function of their size can alter materials properties significantly. Domain walls, interfaces, interphases or defects may play an important role. So for instance just the introduction of nanoholes into a homopolymer can significantly change mechanical properties of this material. Block-copolymers, where micro-phase segregation occurs at molecular level, should also be considered here. If newly formed phases are anisotropic at nano, micro, or macro scale, this again will modify properties in a particular way.

Nanocomposites are another important class of nanomaterials [8–10]. The addition or in situ synthesis of nanoparticles, which often are formed from inorganic components, can introduce new functionalities. In many cases already at very low concentrations of those particles a significant change of properties is observed. Those nanocomposites offer plenty of possibilities, ranging from conductive or magnetic materials to optically active or mechanically reinforced materials for instance. Some nanocomposites are already known since a long time, which includes automotive tires filled with carbon black or colored glasses containing metal nanoparticles. New multifunctional formulations are now on the way, which include self-repairing composites or materials with sensing properties.

Another class of nanomaterials uses *nano-sized and isolated nanoscale objects* which also can exhibit very special properties [7, 11, 12]. They constitute the main topic of this book. This could be nanoscopicaly thin platelets or nanofibers for instance. Ultra-thin mats of electro-spun polymer nanofibers used for pollution filters are an example as well as nanowhiskers of cellulose. Also platelets containing nanocoatings or colloidal particles with a nanoshell fall into this class, where properties are determined to a large extend also by the surface properties. Those surface properties can be tuned by a thin nanoscopic coating. Smart surfaces then even may change properties in response to an external trigger like temperature, pH, or light [13], and the properties of the material become responsive or adaptive to environmental conditions for instance.

This survey, although by far not complete, already shows the wide range of nanomaterials—and their huge potential. Indeed future materials development will surely progress also with the further development of nanomaterials. Polymers also have their valuable place there, mostly because of their low price, but also because of mechanical properties, easy processability, versatility, low weight, biocompatibility, etc. With nanomaterials their range of application can be significantly widened. Raw materials are presently easily available from oil, but also increasingly from renewable resources and recycling. Thus, polymeric nanomaterials should have a bright future. Environmental properties are undoubtedly there, but they could be largely reduced for by proper handling of polymer waste. Similarly, safety aspects in particular for nanoparticles have to be considered [14, 15], but can be taken to be less severe for nanomaterials in the bulk.

In this volume many aspects of polymer nanomaterials are covered, with emphasis on nano-sized polymers. This first includes the several routes for their preparation, in particular:

Extraction of nano-size fibers from biosources
Direct synthesis of nano-size polymers
Precipitation of nanoparticles from solution
Preparation of nanostructures via self-assembly
Nanofibers by electrospinning
Nanostructured materials by deformation of polymer blends
Spin-coating of nanoscopic polymer films

Those nano-size polymer objects and materials can be used in different applications ranging from mechanically reinforced nanocomposites, membranes, scaffolds, nonwovens to photonic, nanosensing or nanoelectronic devices as well as drug delivery and catalytically active systems. The applications are based on the special properties of the nanostructured polymers, which in many cases rely on the improved mechanical properties and the low weight. But also composite or hybrid materials are of particular interest because of their tunable and possibly multi-functional properties as well as the relative cost efficient way to prepare nanoscopic materials with designed functionalities. Going to the nanoscale offers many opportunities and challenges for new and exciting materials and properties, which are difficult to achieve otherwise. Several examples will be discussed in this book.

Nanocharacterization

An important aspect for the development of nano-size materials is that it now routinely has become possible to characterize materials at nanoscopic level [7, 12, 16]. The big advancement in the development of nano-size materials thus is also a consequence of the advancement of appropriate characterization techniques. Using techniques like transmission electron microscopy (TEM), scanning electron microscopy (SEM) or atomic force microscopy (AFM) it is possible to resolve structures in the nanometer and in some cases even sub-nanometer range. The proper sample preparation, however, is still a delicate and important step.

AFM is a widespread technique to visualize topography and surface properties of thin films and nanoobjects adsorbed to a flat surface or dispersed in the bulk. The lateral resolution is typically some nanometers depending on tip radius and operation mode, while the height resolution may go down to 0.1 nm. It is mostly limited by surface roughness of the sample, since most "flat" surfaces have a relatively high roughness (e.g., standard silicon wafer typically 0.5–1 nm, polymer film on silicon wafer prepared by spin-coating typically 1–2 nm, commercial polymer film

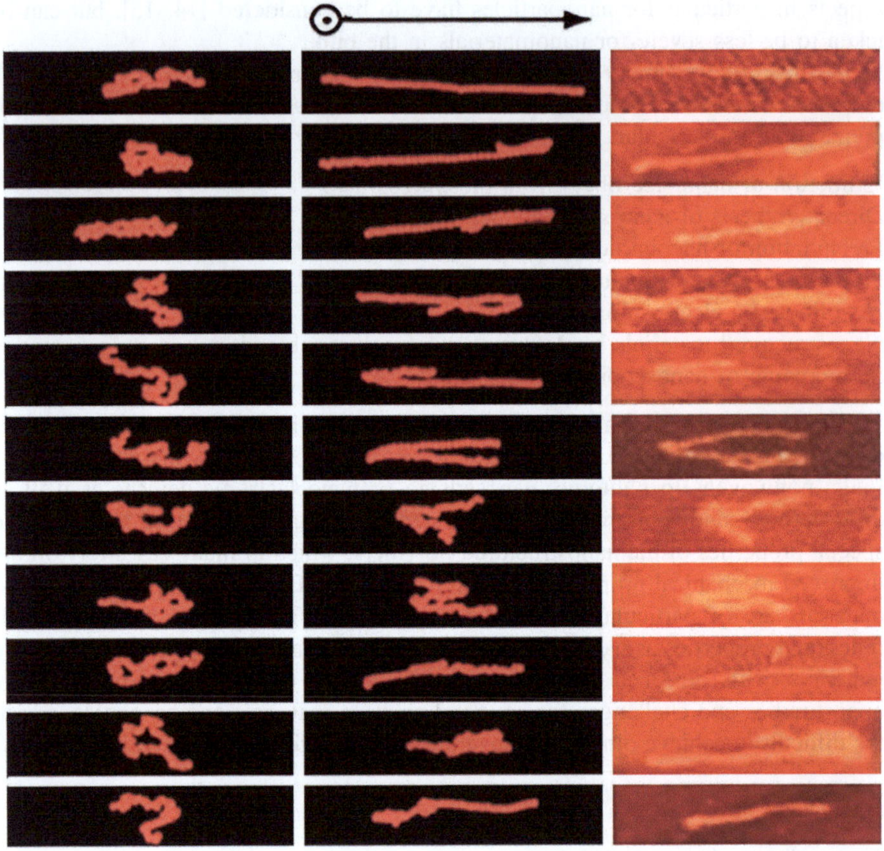

Fig. 1.1 Individual polymer molecules under shear: comparison of chain conformations from simulations and AFM images, linear shear flow is in x-direction. The *left columns* show simulations of conformations at lower and higher shear rates; *right column* shows AFM images of 2D-conformations of individual chains adsorbed from solution on mica surface at high shear rate during spin-coating. Scale of AFM images in x-direction is approx. 500 nm (reproduced from Ref. [17] with permission from the Royal Society of Chemistry)

typically 5–50 nm). By adsorption on an atomically flat mica surface, however, (local roughness typically 0.05 nm) it is possible to visualize by AFM individual adsorbed polymer molecules (Fig. 1.1) [17, 18]. In that way their chain conformation and size can be directly measured. It has to be considered, however that only the 2D-conformation after adsorption is measured and that during adsorption and subsequent relaxation on the surface changes with respect to the 3D-conformation in solution are possible. Also other nano-sized objects might rearrange or deform during the adsorption process and therefore AFM measurements of soft 3D-objects after adsorption have to be taken with care. With nanostructured thin films AFM in tapping mode can provide valuable information on lateral phase segregation where contrast between phases is achieved by viscoelastic differences between phases. So also mechanical information of nanostructured thin films can be obtained by AFM and local mechanical data can be determined by nanoindentation also quantitatively [16].

TEM has also been developed significantly, and offers many imaging possibilities down to sub-nanometer resolution. The preparation of the 50–100 nm thick slices, needed for transmission experiments, is in many cases a challenge. Microtome cuts, sometimes at liquid nitrogen temperature, are most commonly used for polymers, but for cuts of soft/hard material combinations also focused ion beam (FIB) techniques are applied, while for nanoparticles in dispersion cryo-quenching of a thin liquid film is used. Alternatively, the particles might be adsorbed on a thin carbon film which then is dried. The experiments are performed under ultra-high vacuum, and differences of scattering from areas of different electron densities in the sample are visualized.

With the advancement of synchrotron radiation sources and lab X-ray instrumentation also scattering experiments from ultra-thin films and from individual nanoobjects are possible. Ultra-thin films are analyzed in grazing incidence geometry and small or wide angle X-ray scattering is detected typically with an area detector. In that way, lateral nanostructures, but also the structure within a film can be obtained. Typical examples are measurements on the lateral nanostructures in diblock copolymer thin films [19] or the determination of molecular orientation and degree of crystallization in thin films of organic conductors [20]. To analyze individual nano-size objects the X-ray beam is focused by mirrors or refractive optics to a spot, which can be in the range between 50 nm and several micrometers [21]. The scattering is detected by an area detector, but radiation damage in the beam is a serious limitation. Nanoparticles in dispersion are commonly investigated by light scattering which provides information on size and shape.

Chemical analysis [16] is similarly possible at the nanometer scale by electron energy loss spectroscopy (EELS), energy-dispersive X-ray spectroscopy (EDX), or spatially resolved X-ray photoelectron spectroscopy (XPS, XPEEM). The first two of those techniques EELS and EDX can be performed at specially equipped electron microscopes. They detect the characteristic electron energy loss from different elements or the characteristic element specific X-ray emission, respectively, and can be applied in the electron microscopes with nanoscopic spatial resolution. So element specific images for different elements can be determined, which provide locally resolved chemical information. The last technique XPEEM needs preferentially the high flux of a synchrotron source to obtain nanoscopic spatial resolution. The X-ray beam illuminates the sample at incident different X-ray energies and the respective photoelectron image of the sample is recorded. This again provides information of distribution of elements in the sample, but only electrons from the region close to the surface are detected.

Many other characterization techniques can be used with nano-size objects, but application has to be checked in each individual case. Sample preparation is often critical, and much better information of nano-size objects can be obtained by combination of different techniques.

Nanobiomaterials

Nature uses nanostructured components in many examples [22–24]. A living cell for instance is built with plenty nanoscopic functional elements. Up to now, this complex element cannot be copied synthetically, although one is trying to construct micro- or nanocompartments which mimic specific aspects of living cells for instance for biotechnological applications like drug production. So the lipid membrane from a living cell is a highly sophisticated nano-size element which acts as a separation device with many functionalities mediated by proteins and polypeptides which are built in. Those proteins and polypeptides are again nano-size functional elements and can be taken as the basis of our life, by using nano-size objects we can deal in many aspects of mimic nature.

A fascinating biomaterial is wood which consists of lightweight hierarchical structural elements containing cellulose, lignin, and other components at different length scales [24, 25]. Trees can reach a height of more than 100 m and withstand many external influences and stresses. Of course one is and has been using wood as a lightweight construction material itself. Lightweight constructions in automotive or other applications can only partly mimic those concepts from nature.

By decomposition of wood one can separate the nano-sized elements called nanowhiskers. They consist mainly of thin fibers of cellulose and are nano-size building blocks. Depending on the starting material and on the chemical and mechanical treatment, the nanowhiskers can be macroscopically long, while their diameter is in the nanorange. They can be used as thin mats forming nanomembranes. By blending with synthetic polymers one obtains nanocomposites where the high uniaxial strength of the fibers shall reinforce the synthetic polymers. This concept does not work out in all cases because of poor dispersion of the nanofibers and insufficient contact with the polymer matrix material. Cellulose fibers on the other hand are environmental friendly since they are biodegradable, can be obtained from many different natural resources and are available in large quantities. Several aspects of cellulose nanowhiskers and their nanocomposites will be described in one chapter of the book.

Nanocomposites

The use of biomaterials in nanocomposites is limited because of limited inherent physical properties of biocomponents in some cases. To achieve high performance and multifunctional materials one therefore uses nanocomposites of synthetic polymers with inorganic materials [8–10], an area which has emerged as a challenging and largely growing field. So glass and carbon fibers are used for mechanical reinforcement, but also oxydic, metallic, or semiconducting nanoparticles are applied for optical, electrical, magnetic, catalytic, or sensing applications.

Again dispersion and contact of nanoparticles with the matrix are important. The field is very wide and we will only cover some special aspects of nanocomposites in this book.

Polymer Nanoparticles and Individual Polymer Chains

To obtain polymer nano-size objects [11, 12], direct synthesis is a possible way. One could perform polymerization at very high dilution, which however is difficult because of contaminations, or use micro- or nanoemulsions with small compartments to perform polymerization. In this way nano-sized polymer materials can be produced, which are either composed out of small polymer nanoparticles or contain nano-size holes. One chapter of the book describes procedures and properties.

Nanoprecipitation is another technique to obtain nano-sized polymer particles. An organic polymer solution is rapidly given to a nonsolvent (typically water), and nanoparticles are generated during precipitation of the polymer at the interfaces. The process depends critical on all components, and surfactant can be added to further control nanoparticle size. In this way one can produce polymer nanoparticles at large scale to be used in biomedical, biotechnological, or environmental applications. Also here the chapter of the book describes further details.

To obtain nano-sized objects with individual polymer chains, one can use highly diluted polymer solutions where individual polymer chains can be sprayed in the drops of an aerosol or adsorbed from solution onto a substrate. In the first case, one obtains after the evaporation of the solvent nano-size particles from collapsed single chains, which relax very fast to their equilibrium conformation after heating in bulk. In the second case chains are adsorbed on the substrate and adopt 2D-chain conformations [17, 18]. Depending on the strength of adsorption and the adsorption process those chains might be used to obtain information on the chain conformation in solution prior to the adsorption process. So in this way, chain conformations in solution can be visualized using AFM techniques, and for instance conformational changes in polyelectrolyte solutions may be resolved. So with increasing pH the chain conformation of the weak polyelectrolyte PVP changes from expanded Gaussian chain to pearl-necklace conformation and collapsed globule which can be described as a secondorder phase transition [26]. The adsorbed single chains on the other hand can also be taken as nano-size objects which can be manipulated further. So oppositely charged nanoparticles adsorb on top of an adsorbed polyelectrolyte molecule, and conducting nanowires can be produced based on a single polymer molecule by further functionalization. There is a recent review on this topic [18], and of course plenty of applications deal with nanoelectronics and sensorics of individual carbon nanotubes CNTs and graphene [11], which also might be considered as nanoobjects from single molecules.

In those so-called bottom-up processes one tries to use nanoscopic objects like polymer molecules, nanoparticles, etc. and possibly their self-assembly to generate nanostructured devices. As mentioned above individual nanoobjects like single polymer molecules, carbon nanotubes, graphene sheets, silicon nanorods, DNA nanoorigami, etc. are employed to build electronic nanodevices [7, 11, 27]. This is still an open field of research, which offers many opportunities and challenges.

Nanofibers

Nanofibers of most polymer materials can be obtained by electrospinning from solution [28]. Typically, a solution of the polymer is exposed to a high electric field which results in the formation of nanofibers. Parameters can be varied to generate fibers of thickness from several nanometers to micrometers. All sorts of fibers can be generated ranging for composite fibers to core-shell structures and porous or hollow fibers. Production at larger scale can be performed by modification of the technique or by parallel arrangement. Those fibers may be used as mats, filters, membranes, biodegradable scaffolds, drug release agents, or fillers in nanocomposites. Their production and use is described in several chapters of the book.

Nanostructured templates and nanoobjects including nanofibers can also be produced by self-assembly of diblock copolymers in bulk or in solution [29]. Here the segregation of the two immiscible blocks of the copolymers is used, and phase segregation occurs at the scale of the copolymer molecules themselves. Depending on the molecular weight of the blocks typical structures of micro-phase-segregation in bulk are between 10 and 50 nm. Different morphologies ranging from spherical, cylindrical to gyroid, and lamellar are formed by self-assembly depending on relative molecular weight and volume fraction of the components. So for instance a regular hexagonal pattern of standing cylinders can be generated in a thin copolymer film. At removal of one component by selective UV-mediated decomposition empty holes or arrays of standing cylinders can be achieved. With concepts from supramolecular chemistry a third low molecular weight component can be added which preferentially is located in one of the phases. It then is easily removed by dissolution, to again result in empty cylindrical holes or cylinder nanostructures of the copolymers. If the same concept is used with bulk copolymer materials, nanofibers are obtained (Fig. 1.2) [29, 30]. At different copolymer compositions also nanospheres or nanosheets may be produced. Since the gyroid phase is a co-continuous structure, the removal of one component results here in a nanoporous morphology [31], which for instance may be used as a cathode for high energy-density LiS-battery cathodes. Much more complex nano-size objects may be obtained from tri or multiblock copolymers. Micelles, which form by phase segregation in solution, offer very similar possibilities for production of nano-size objects.

A very elegant and more economical way to produce nanofibers proceeds by deformation of phase segregated blends. For instance, a phase segregated blend of two immiscible polymers may be highly oriented by shear to produce fibrillar morphologies of the phases. At a temperature between the melting points of the two phases one component is then melted to produce an isotropic matrix for the second

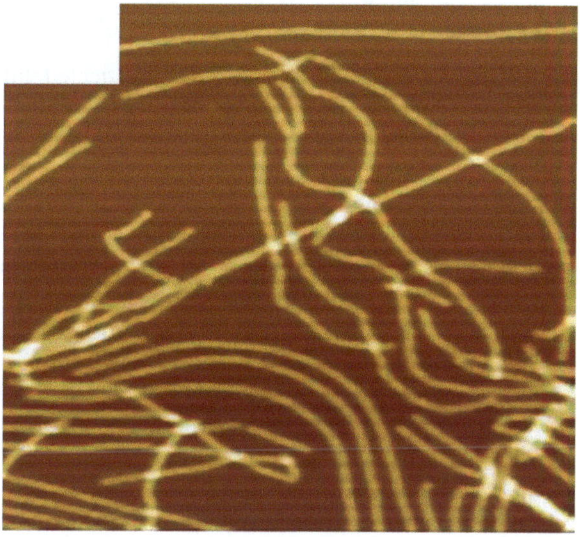

Fig. 1.2 Thin polymer nanofibers produced by supramolecular assembly of polystyrene-polyvinyl pyridine diblock copolymer PS-*b*-P4VP with 4-pentadecylphenol (PDP) after solvent extraction of PDP visualized by AFM. Shown area is 2.5 × 2.5 μm^2 (reproduced from Ref. [38] with permission from Wiley-VCH)

nanofibrillar component. This process is described in detail in one of the contributions of this book and interesting mechanically reinforced nanofibrillar composite polymer materials can be produced.

Nanofilms and Coatings

Very smooth and uniform thin films of polymers can be produced by spin-coating of polymer solutions [16, 32]. This process is often performed with large and smoothly polished silicon wafers, and by optimization of parameters like concentration of the polymer solution, molecular weight of the polymers, type of solvent, and rotation speed very homogeneous films in a thickness range from some nanometers to micrometers may be produced. The situation becomes complex when mixtures of many components are used. Different competing processes might happen like phase segregation of components, enhancement of components at surface or interface to substrate, orientation of chains and possibly crystallization of components. This happens in an environment, where strong shear forces are acting and the solvent usually evaporates very fast during spinning. Depending on diffusion and relaxation times as well as forces and interactions involved, different morphologies and layers might be formed. This on the other hand offers the possibility to use spin-coating for the production of nanostructured films. Also spin-coating is the theme of one of the contributions in the book.

Spin-coated homogeneous films of photoresists are used in microelectronics industry to produce lateral nanostructures by lithography [33]. Although optical illumination systems with inherent limitations given by the wavelength of light are applied to crosslink (or decompose) the photoresist, where nonilluminated (or illuminated) regions are dissolved away, it is possible to generate nanostructures in the range of presently <14 nm by sophisticated lithographic illumination procedures. 22–28 nm structures are in the production line of current processors [34], and complex 3D-structures are generated by multiple lithographic processes. So both film thicknesses as well as lateral structures are in the range of several nanometers, and are in the range of the size of the polymer molecules themselves. Those so-called top-down lithographic processes are well established, but to achieve possibly still smaller nanoscopic scales one needs increasing technical effort with respect to clean rooms, optical systems, chemical processes, defect handling, and process, and quality control. In addition, the interconnects between chips are also getting smaller and more complex. So one is looking for novel solutions, where bottom-up concepts based on single or self-assembled nanoobjects are discussed.

Also for organic and polymer solar cells, organic light emitting diodes (oleds), organic field effect transistors (ofets) and other devices nanoscopically thin films are required with special properties [35]. So for instance for polymer solar cells one needs thin films of p- and n-doped semiconducting polymers with preferentially a phase segregated co-continuous morphology at 10 nm scale because of limited exciton lifetimes and the need for electron conduction to upper and lower current collector. Absorption of light over the full spectral range is required which might be achieved for instance by the addition of a dye or multilayer techniques. So phase segregation at nanoscale and preferential orientation of molecules is advantageous. This again needs design of special semiconducting molecules and well-controlled processing conditions for morphology control [20, 35, 36]. Efficiencies of polymer solar cells still have not reached the 10 % level, but solution processing over large areas would provide interesting application areas. Similar conclusions hold for oleds and ofets, where however developments with organic molecules have progressed further and oled displays for instance are already used in mobile smartphones, where film thickness is at nanolevel, while lateral spatial resolution only needs to be at micrometer scale.

Quite versatile nanofilms are produced from self-assembled monolayers (SAMs), where formation of functional nanofilms is a well-established technique [4–7]. By using differently functionalized building blocks, one can deposit a nanoscopic thin film on various smooth substrates like silicon wafers or metal surfaces by self-assembly of the functional building blocks, where one end is binding to the substrate and the other end bears the functional group for further processing. A large variety of end-functional groups (amino, carboxylic, hydroxyl, epoxy, etc.) for surface functionalization are commercially available. So for instance polymer chains can be grafted onto SAMs or onto other grafted functional polymer chains, which forms a nanoscopic thin polymer monolayer and have interesting properties. If chains are grafted at high grafting density, they repel each other and stretch away from the substrate, thus forming a so-called polymer brush layer. Those brushes provide again special properties and can for instance change hydrophobicity, control adhesion of other molecules including proteins, act as nanosensors or are catalytically active [13]. If thermally or pH sensitive polymers or two different polymer chains are attached, responsive polymer brush layers are formed, which change their properties in response to an external trigger. In this way, hydrophobicity can be switched even from ultra-hydrophobic to ultra-hydrophilic state (change of contact angle by 150°), or proteins adsorb and desorb in response to the external trigger [37]. So nanoscopic thin films can change significantly surface properties and are used in many applications.

Summary and Conclusions

This short review illustrates some of the possibilities and applications of polymer nanomaterials and nano-sized polymer materials in particular. It is a fascinating area of research, where already many applications are in use in very different fields, often without being noticed. So surface materials properties can be entirely changed from original bulk materials properties by a nanoscopic thin coating, various nanoobjects can be produced by different procedures to be used as individual nano polymer devices possibly in future electronics, sensorics, or functional nanomembranes, but also as nanoadditives to polymer composites for mechanical reinforcement or other multiple functions. This is a wide field for further development, and it is quite certain that nano polymer materials will be increasingly present in our life in the future in many areas. They will widen the range of materials applications, where certain novel functionalities are introduced, which are otherwise not achieved, where rare or unfavorable materials or components can be replaced, where energy or other resources can be saved, and they can be considered to be produced at relatively reasonable price, since in many cases standard polymer matrix materials are used. This book provides a survey in this field with emphasis on nano-size polymer materials and might stimulate some of the developments which are needed to solve some of our problems.

References

1. Pal A (2015) Rebuilding the World. Progressive 79(12):62–64
2. Monitoring Report Energy of the Future (2014) BMWi 148 p. www.bmwi.de
3. International Energy Outlook, US Energy Information Administration (2013)
4. Vollrath D (2013) Nanomaterials: an introduction to synthesis, properties and applications. Wiley-VCH, 386 p. ISBN 978-3527333790
5. Cao G, Wang Y (2011) Nanostructures and nanomaterials: synthesis, properties, and applications. World Scientific, Singapore, 596 p. ISBN 978-9814322508
6. Kane DM, Micolich A (2016) Nanomaterials: science and applications. Pan Stanford, 418 p. ISBN 978-9814669726
7. Vajtai R (2013) Springer handbook of nanomaterials. Springer, New York, 1221 p. ISBN 978-3642205941
8. Thomas S, Shanks R (2015) Design and applications of nanostructured polymer blends and nanocomposite systems. William Andrew, 442 p. ISBN 978-0323394086

9. Bhattacharya SN (2007) Polymeric nanocomposites: theory and practice. Hanser-Gardner Publications, 390 p. ISBN 978-1569903742
10. Srivastava SK, Mittal V (2016) Hybrid nanostructured materials: developments in energy, environment and polymer nanocomposites. Wiley-Scrivener, New York, 400 p. ISBN 978-1119160342
11. Sattler KD (2016) Carbon nanomaterials sourcebook: nanoparticles, nanocapsules, nanofibers, nanoporous structures, and nanocomposites. CRC Press, Boca Raton, 753 p. ISBN 978-1482252705
12. Charleux B, Coperet C (2015) Chemistry of organo-hybrids: synthesis and characterization of functional nano-objects. Wiley, New York, 552 p. ISBN 978-1118379028
13. Cohen Stuart MA, Huck WTS, Genzer J, Müller M, Ober C, Stamm M, Sukhorukov GB, Szleifer I, Tsukruk VV, Urban M, Winnik F, Zauscher S, Luzinov I, Minko S (2010) Emerging applications of stimuli-responsive polymer materials. Nat Mater 9:101–113
14. Ramachandran G (2016) Assessing nanoparticle risks to human health (micro and nano technologies). William Andrew Publications, ASIN B01C44BWKG, 286 p
15. Ellenbecker MJ, Tsai CS-J (2015) Exposure assessment and safety considerations for working with engineered nanoparticles. Wiley, New York, 312 p. ISBN 0470467061
16. Stamm M (ed) (2008) Polymer surfaces and interfaces: characterisation, modification and applications. Springer, Berlin, 324 p. ISBN 978-3-540-73846-0
17. He G-L, Messina R, Löwen H, Kiriy A, Bocharova V, Stamm M (2009) Shear-induced stretching of adsorbed polymer chains. Soft Matter 5:3014–3017
18. Kiriy A, Stamm M (2012) Chain conformation and manipulation. In: Matyjaszewski K, Moelle M (eds) Polymer science: a comprehensive reference, vol 1. Elsevier, Amsterdam, pp 367–386
19. Kuila BK, Stamm M (2011) Block copolymer-small molecule supramolecular assembly in thin film: a novel tool for surface patterning of different functional nanomaterials. J Mater Chem 21:14127–14134
20. Karpov Y, Zhao W, Raguzin I, Beryozkina T, Bakulev V, Al-Hussein M, Häußler L, Stamm M, Voit B, Facchetti A, Tkachov R, Kiriy A (2015) Influence of semiconductor thickness and molecular weight on the charge transport of a naphthalenediimide-based copolymer in thin-film transistors. ACS Appl Mater Interf 7:12478–12487
21. Martínez-Criado G (2015) Application of micro- and nanobeams for materials science. In: Jaeschke E, Khan S, Schneider JR, Hastings JB (eds) Synchrotron light sources and free-electron lasers. Springer, Berlin. ISBN 978-3-319-04507-8
22. Grumezescu A (2016) Nanobiomaterials in hard tissue engineering: applications of nanobiomaterials. William Andrew, 510 p. ISBN-13: 978-0323428620
23. Grumezescu A (2016) Engineering of nanobiomaterials: applications of nanobiomaterials. William Andrew Publications, 564 p. ISBN 0323415326
24. Fratzl P, Dunlop JWC, Weinkamer R (eds) (2013) Materials design inspired by nature, function through inner architecture. The Royal Society of Chemistry, 402 p
25. Moon RJ, Martini A, Simonsen J, Youngblood J (2011) Cellulose nanomaterials review: structure, properties and nanocomposites. Chem Soc Rev 40:3941–3994
26. Kiriy A, Gorodyska G, Minko S, Jaeger W, Stepanek P, Stamm M (2002) Cascade of coil-globule conformational transitions of single flexible polyelectrolyte molecules in poor solvent. J Am Chem Soc 45:13454–13462
27. Zaman I, Manshoor B, Khalid A, Araby S (2014) From clay to graphene for polymer nanocomposites—A survey. J Polym Res 21:1–11
28. Mitchell GR, Tang BZ (2015) Electrospinning: principles, practice and possibilities. Royal Society of Chemistry, 288 p. ISBN 1849735565
29. Nandan B, Stamm M (2012) Self-assembled polymer supramolecules as templates for nanomaterials. In: Steed JW, Gale PA (eds) Supramolecular chemistry: from molecules to nanomaterials, vol 7. Wiley, Chichester, pp 3563–3586. ISBN: 978-0-470-74640-0

30. Sanwaria S, Singh S, Horechyy A, Formanek P, Stamm M, Srivastava R, Nandan B (2015) Multifunctional core-shell polymer-inorganic hybrid nanofibers prepared via block copolymer self-assembly. RSC Adv 5:89861–89868
31. Choudhury S, Agrawal M, Formanek P, Jehnichen D, Fischer D, Krause B, Albrecht V, Stamm M, Ionov L (2015) Nanoporous cathodes for high-energy Li-S batteris from gyroid block copolymer templates. ACS Nano 9:6147–6157
32. Idris NB, Norizan MN, Mohamad IS (2015) Organic solar cells: an overview on performance and fabrication techniques. Appl Mech Mater 754–755:540–545
33. Acikgoz C, Hempenius MA, Huskens J, Vancso GJ (2011) Polymers in conventional and alternative lithography for the fabrication of nanostructures. Eur Polym J 47:2033–2052
34. See e.g. at http://globalfoundries.com/docs/default-source/PDF/gf-worldwidefootprint-20150630-v02.pdf (version Jan 2016)
35. Shi Y, Peng L, Ding Y, Zhao Y, Yu G (2015) Nanostructured conductive polymers for advanced energy storage. Chem Soc Rev 44:6684–6696
36. Tkachov R, Karpov Y, Senkovskyy V, Raguzin I, Zessin J, Lederer A, Stamm M, Voit B, Beryozkina T, Bakulev V, Zhao W, Facchetti A, Kiriy A (2014) Efficient tin-free route to a donor-acceptor semiconducting copolymer with variable molecular weights. Macromolecules 47:3845–3851
37. Psarra E, Foster U, König J, You Y, Ueda M, Eichhorn K-J, Müller M, Stamm M, Revzin A, Uhlmann P (2015) Growth factor-bearing polymer brushes-versatile bioactive substrates influencing cell response. Biomacromolecules 11:3530–3542
38. Fahmi AW, Braun H-G, Stamm M (2003) Fabrication of metallized nanowires from self-assembled diblock copolymer templates. Adv Mater 15:1201–1204

Part II
Chemical Approaches for Preparation of Nano-size Polymers

Chapter 2
Nano-sized Polymer Structures via Self-assembly and Co-assembly Approaches

Grigoris Mountrichas, Petar Petrov, Stergios Pispas and Stanislav Rangelov

Introduction

The amphiphilic copolymers in selective solvents self-assemble in a much similar fashion as the conventional low-molecular weight surfactants and naturally occurring lipids. They are composed of at least one solvophobic moiety and at least one solvophilic moiety (*hydrophilic* and *hydrophobic*, respectively, if the solvent is water). The difference in the solubility of the constituent moieties as well as the constraint imposed by the chemical linkage between them govern the geometry of the self-assembled structures, whereas the sequence of the latter is dictated by the proportions of the constituent moieties. At low concentrations, the self-assembled structures are discrete and well-separated, whereas upon increasing concentration, different liquid-crystalline phases are formed. The shape of the self-assembled structures varies from spherical to cylindrical to lamellar depending on the conditions; however, the main shape-determining factor is the ratio between the constituent moieties. For polymers, it is convenient to characterize the preferred aggregate morphology by the hydrophilic fraction, *f*. The relations between the preferred geometry of the self-assembled structures and *f* are presented in Fig. 2.1 [1].

The *critical aggregation concentration*, *CAC*, is a fundamental parameter defined as the concentration at which aggregates are formed. In other words, below the CAC only unimers, that is unassociated macromolecules, exist, whereas above the CAC multimolecular aggregates are in dynamic equilibrium with the unimers. The CACs of amphiphilic polymers are typically located in the low micromolar

G. Mountrichas · S. Pispas
Theoretical and Physical Chemistry Institute, National Hellenic Research Foundation, 48 Vass. Constantinou Ave., 116 35 Athens, Greece

P. Petrov · S. Rangelov (✉)
Institute of Polymers, Bulgarian Academy of Sciences, Akad. G. Bonchev St. 103-A, 1113 Sofia, Bulgaria
e-mail: rangelov@polymer.bas.bg

S. Fakirov (ed.), *Nano-size Polymers*,
DOI 10.1007/978-3-319-39715-3_2

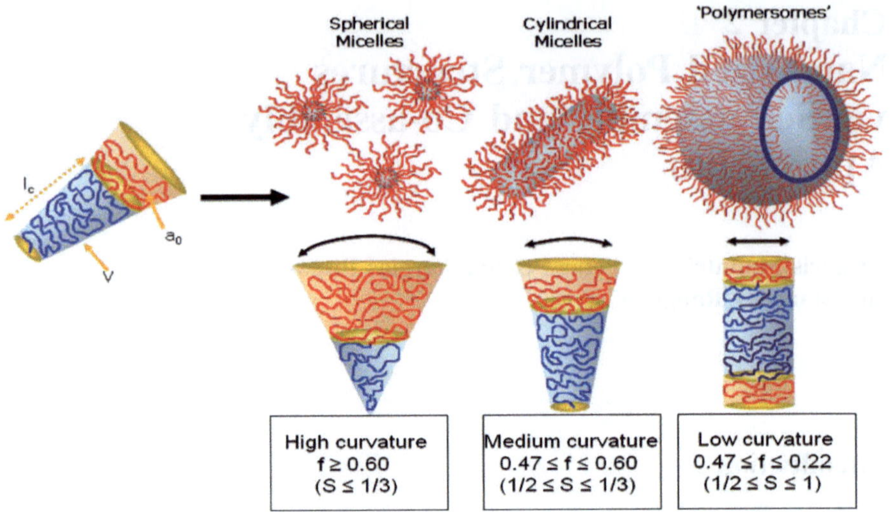

Fig. 2.1 Interrelations between the self-assembled structures formed by amphiphilic copolymers in aqueous solution and their hydrophilic fractions, *f*. The values of the corresponding surfactant packing parameter, *S*, are given in parentheses. Reproduced from [1] with permission from John Wiley and Sons

region and even lower, which implies greater stability and is advantageous for many applications since high dilution might be problematic for the conventional low-molecular-weight surfactants. The self-association can also be induced at a fixed concentration by manipulation of the solvent quality and environmental parameters. Thus, for copolymers that are composed of moieties exhibiting enhancement of hydrophobicity at certain temperature or pH, one can accordingly speak of critical temperature or pH of association.

The spherical micelles (Fig. 2.1) are the most extensively studied polymeric self-assembled structures. They represent the simplest and most widespread aggregate morphology. Cylindrical micelles as well as polymer vesicles (polymersomes) are less frequently observed (Fig. 2.1), whereas complex structures such as toroids, helices, multicore and multicompartment micelles, disks, tubules are considered as more or less exotic structures. Worth mentioning is the phenomenon of coexisting morphologies, which is believed to derive from a number of inherent for the synthetic copolymers characteristics, e.g., dispersity in molar mass.

Besides the critical aggregation concentration, temperature or pH and specific morphology, the polymeric self-assembled structures are characterized by particle molar mass and aggregation number, dimensional parameters such as hydrodynamic radius, radius of the core or membrane thickness for the polymersomes and disk-like micelles, radius of the cylindrical micelles and thickness of the corona, radius of gyration. A distinctive feature of the polymeric self-assembled particles is the *core-corona* structure. Both, the core and corona, can be considered as separate entities which are able to accommodate active substances of appropriate nature and to serve as carriers. The core is composed of strongly entangled solvophobic chains, whereas the corona is built up of solvophilic chains, which can be nonionic or charged. Depending on the macromolecular characteristics, the corona can be thick and thus largely contributing to the overall particle dimensions or relatively thin. The core–corona interface is sharp if the core- and corona-building moieties are incompatible. Accordingly, for more compatible moieties a partial mixing of chains in the boundary region is expected. Last but not least, by carefully designing the copolymer composition, intelligent properties can be conferred to the whole aggregate.

Polymeric Nanostructures Prepared via Self-assembly and Co-assembly of Preformed Copolymers

General Remarks

Block copolymers tend to form nano-sized structures due to the differences in the physicochemical characteristics of each block. As a typical example, the case of an amphiphilic block copolymer can be discussed, where only one of the blocks is water soluble. In the aforementioned example, the hydrophobic blocks tend to aggregate in aqueous media, due to hydrophobic interactions, leading to the formation of a nano-sized aggregate where the hydrophilic blocks are extended to the solvent, stabilizing the nanostructure. There is a plethora of such self-assembled polymeric nanostructures in solutions that has been described in the literature, the ones observed in aqueous solutions being significantly more interesting for biomedical applications. Among them, there are some structures that can be characterized nowadays as common since they have been obtained in several systems. These include spherical core–shell micelles, worm-like or rod-like micelles, and vesicles or polymersomes. Some other nanostructures possess less common morphologies, namely toroidal, multicompartment or bicontinuous micelles. Some of these morphologies are shown in Fig. 2.2. In addition, the formed nanostructures can be (i) dynamic, i.e., exchange of block copolymer chain is possible, (ii) stimuli responsive, i.e., the nanostructures respond by changing their structural characteristics or fully disintegrate under external stimuli, such as temperature, pH, ionic strength, light, etc., (iii) kinetically frozen, i.e., exchange of chains is not possible by changing all experimentally relevant conditions in the normally accessible range of physicochemical parameters of the system, and (iv) chemically stabilized, i.e., they may contain covalently cross-linked cores or shells and their morphology/structure is "locked."

Since the literature on the common block copolymer micellar structures is vast, only a limited number of examples will be given, in order to outline the fundamental principles behind the self-assembly of macromolecular chains. Moreover, the so-called common polymeric nanostructures have been described in detail by a number of review articles and books, which are strongly recommended for further reading [1–4]. The story of self-assembly is tightly connected with the story of

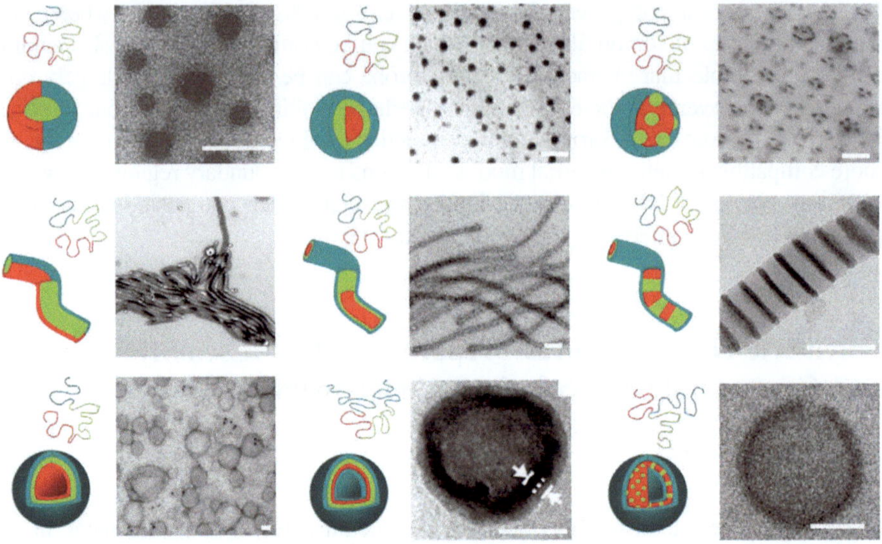

Fig. 2.2 Assemblies formed in selective solvent conditions by multiblock copolymers (from upper left to right): Janus spheres, core-shell spheres, raspberry-like spheres, Janus cylinders, core-shell cylinders, segmented cylinders, asymmetric (Janus) membrane vesicles, double-layer membrane vesicles, vesicles with hexagonally packed cylinders. Reproduced from [3] with permission from Elsevier

well-defined block copolymer synthesis, which started more than 50 years ago [5]. Meanwhile, a large number of research reports describe the effect of solvent quality in self-assembly, the connection between macromolecular characteristics and solution behavior and the effect of the environmental parameters, like temperature, salinity, and pH in the aggregation of block copolymers and the major role of polymer concentration to the self-assembly procedure. The effect of environmental parameters is one of the most appealing research subjects during the last years, because "responsiveness" is very important from a technological point of view, since it gives the opportunity to fine-tune the structures and their physicochemical properties by external stimulation of the system, a characteristic which is important for drug delivery and "switching" related applications.

Nanostructures Prepared from Stimuli-responsive Copolymers

The responsiveness of a block polymer is the key parameter for an application. Yang et al. [6] has described the potential application of a block copolymer as nano-sized drug carrier which offers fast release rate in the intestinal track. The block copolymer, namely poly(methyl methacrylate-co-methacrylic acid)-b-poly

(poly(ethylene glycol) methyl ether monomethacrylate) [P(MMA-co-MAA)-b-PPEGMA] tends to form spherical micelles upon dissolution in aqueous media with PPEGMA being the shell. Moreover, it has been found that its drug entrapment efficiency towards hydrophobic drugs reached to 90 %. The most important characteristic of the copolymer, however, is its selective responsiveness to the environmental pH. Thus, the copolymer was observed to keep its spherical micellar structure for pH lower than 5, but the micellar core can swell or even dissociate at pH values higher than 5. In an in vitro experiment, concerning the release of the drug ibuprofen, it was found that less than 20 % of the initial drug content was released in simulated gastric fluid (pH 1.2) over 12 h, but 90 % was released in simulated intestinal fluid (pH 7.4) within 6 h. The above behavior indicates an appealing candidate as drug nanocarrier, with selective release in the intestinal track. The above example clearly indicates that the responsiveness of a copolymer, and the better understanding of the relation between external stimuli and micellar response/structure is essential for the development of systems with appropriate efficacy.

Toward this end, the solution behavior of poly(2-(methacryloyloxy)ethyl phosphorylcholine)-b-poly(2-(diisopropylamino)ethyl methacrylate) (PMPC-b-PDPA) have been studied by Pearson et al. [7]. The dissociation constant (pK_a) for the conjugate acid form of the PDPA block was determined for a number of PMPC-b-PDPA copolymers with varying volume fractions of DPA over a wide range of temperatures. The polymer tends to form aggregates at pH values between 5 and 7.5. However, the size and the structure of the aggregates vary with the volume fraction of PDPA and temperature. The obtained data indicate a wide gamut of structures which can be formed from a single block copolymer, structures that are extended from simple spherical micelles to vesicles and ill-defined supramolecular structures (Fig. 2.3).

In the same context, the effect of composition, temperature, and pH has been also well-described by McCormick and coworkers in another case [8]. The authors studied the solution behavior of poly(N,N-diethylaminoethyl methacrylate)-b-poly(N-isopropyl acrylamide) (PDEAEMA-b-PNIPAM). This copolymer has a PDEAEMA pH-responsive block and a PNIPAM temperature responsive block. Therefore, macromolecular chains self-assemble into PDEAEMA-core/PNIPAM-shell spherical micelles at temperatures below the lower critical solution temperature of PNIPAM and at solution pH values greater than the pK_a of PDEAEMA. At the same time, by decreasing the pH to values lower than 7.5 and by increasing the temperature to values higher than 42 °C, the reverse structure is observed, i.e., spherical micelles with hydrophobic PNIPAM cores stabilized by a hydrophilic PDEAEMA shell. Interestingly, in the case of polymers with increased PNIPAM volume fraction, the above-mentioned reversal of structure leads to vesicles for temperatures higher than 38 °C. The aforementioned behavior, where a block copolymer can form micelles with either the first or the second block located in the core, depending on the external stimuli, is generally termed as "schizophrenic."

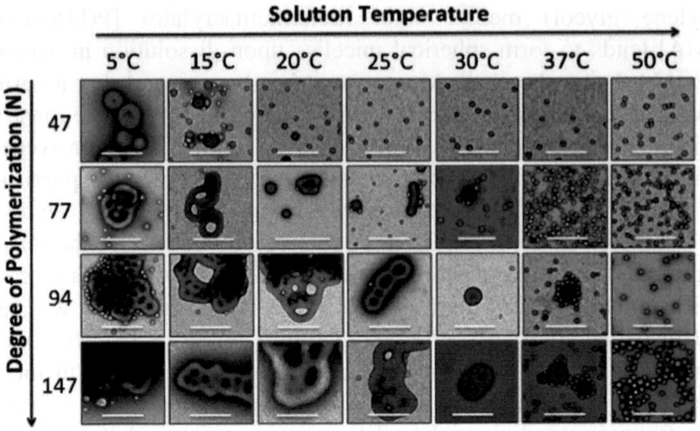

Fig. 2.3 The effect of temperature and composition of PMPC-b-PDPA on the formed structure at certain pH. Reproduced from [7] with permission from American Chemical Society

The formation of vesicles by block copolymers (also called polymersomes) is an interesting field of research due to their selectivity and bimodality in loading capacity (i.e., substances can be loaded within the polymersomes' outer layers or within the solvent pool in the center of the structures). An interesting example of a vesicle, formed by the triblock poly(N,N-diethylaminoethyl methacrylate)-b-poly(ε-caprolactone)-b-poly(N-isopropyl acrylamide) (PNIPAM-b-PCL-b-PDMAEMA) has been demonstrated by Liu et al. [9]. The polymer forms vesicles because of the presence of the PCL hydrophobic block. However, by increasing the temperature to 40 °C, the vesicle transforms to a spherical micelle with PCL and PNIPAM core and PDMAEMA corona. The transformation process seems to be reversible, since decreasing of temperature leads again to vesicles. Interestingly, the physicochemical characteristics of the vesicles respond to the presence of carbon dioxide. Therefore, under the stimulation of CO_2, the vesicular assemblies swelled while they adopted their initial characteristics upon purging with nitrogen.

The effect of CO_2 on the structural characteristics of polymer aggregates has been also demonstrated by Wang et al. [10]. In this case, a diblock copolymer containing one random copolymer block, namely the poly(ethylene oxide)-b-poly (4-vinyl pyridine-r-dimethyamino ethyl methacrylate) copolymer [PEO-b-P (4VP-r-DEAEMA)] was studied in aqueous media. It was found that the copolymer can self-assemble into vesicles in aqueous media and physiological pH, while the vesicles can be fused hierarchically into giant worm-like micelles of several micrometers in length. After bubbling CO_2 into the copolymer solution up to saturation, i.e., pH was changed to 5.43, the giant worms transformed again into large polymersomes. Notably, direct dissolution of the copolymer in acidic environment leads to small spherical micelles. Additionally, the vesicles obtained, after CO_2 saturation, could revert to worm-like aggregates after depleting CO_2. The above-described transition phenomena are mainly attributed to protonation/deprotonation of the PDEAEMA units, to the strong steric hindrance effect from the adjacent 4VP groups and to hydrogen bonding between different 4VP units and free H_2O in the interior of vesicles.

Nanostructures of Non-common Morphologies

The characteristic examples, which are given above, outline some micellar systems with the so-called common structure. However, the focus in this section will be on the formation of non-common morphologies. Initially, an example which is located at the border line between common and non-common nanostructures is given below. In particular, the hierarchical self-assembly of an amphiphilic block copolymer, poly(N,N-dimethylacrylamide)-block-polystyrene with a very short hydrophilic block (PDMA-b-PS), into large granular nanoparticles has been described by Bianchi et al. [11]. The block copolymer forms water-soluble spherical micelles, however, the corona of these micelles has a rather unusual granular shape. The reason for this shape is the partial hydrolysis of the PDMA block.

Similarly, Lodge and coworkers have presented the formation of vesicles. However, these vesicles are not really common, since they incorporate hexagonally packed cylinders [12]. The aggregates were formed by the self-assembly of a miktoarm star terpolymer, where one of the arms is fluorinated. Interestingly, it was found that the assembly of these vesicles proceeds via the formation of metastable polygonal, faceted bilayer sheets.

Located at the common/non-common morphologies border, is also the nanostructure that has been presented by Hu and Liu [13]. Even though the formation of polymeric nanocapsules has already been intensively studied, the formation of polymeric nanocapsules bearing regularly sized nanochannels is a new observation. The use of a pseudo miktoarm triblock copolymer, namely μ-poly(tert-butyl acrylate)-poly(2-cinnamoyloxyethyl methacrylate)-poly(ethylene oxide) (μ-PtBA-PCEMA-PEO) can give polymeric nanocapsules in a mixture of water/THF, where the PEO is extended to the solvent, stabilizing the capsule, PCEMA forms the capsules and PtBA forms cylinders that permeated the PCEMA wall. Following photo-cross-linking of the PCEMA wall and hydrolysis of the PtBA blocks in the cylindrical domains yield unprecedented capsules bearing regularly packed uniform poly(acrylic acid)-gated nanochannels (Fig. 2.4).

The same principle, i.e., the formation of a nanostructure, using a terpolymer, followed by selective chemical modification of one of the blocks was also followed by Zhang et al. [14]. In this case, a triblock terpolymer, consisting of one hydrophilic block and two mutually incompatible hydrophobic blocks covalently connected by a redox-responsive disulfide linkage, self-assembled into multi compartment micelles, a type of micelles with subdivided hydrophobic cores, in aqueous solution. In particular, the micellar core is composed of two polymeric phases—one that is continued and a discontinued one. The formed nanostructure is

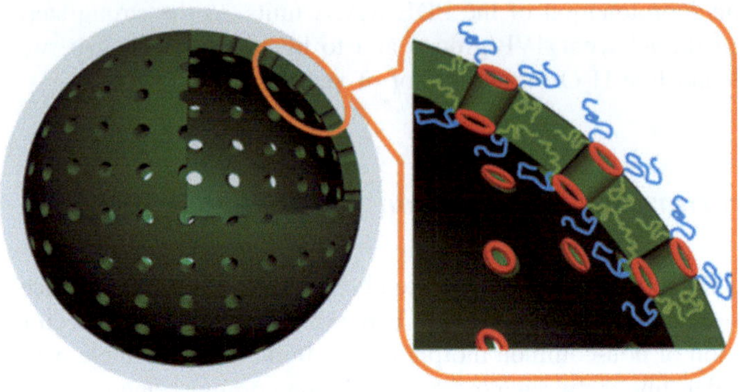

Fig. 2.4 Polymeric nanocapsules bearing regularly sized nanochannels. Reproduced from [13] with permission from American Chemical Society

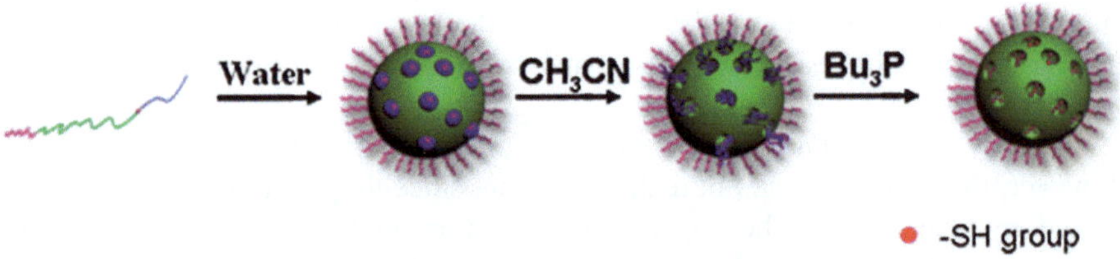

Fig. 2.5 Polymeric micelle with a mesoporous core using a block terpolymer. Reproduced from [14] with permission from American Chemical Society

subjected to cleavage of the disulfide linkage of the discontinued phase, leading to a polymeric micelle with a mesoporous core (Fig. 2.5).

The triblock copolymer poly(ethylene glycol)-b-poly(2-methyl-2-carboxyl-propylene carbonate)-b-poly(L-lactide) (mPEG-b-PMCC-b-PLA) has also been studied upon modification of the middle block with dopamine [15]. In this case, the polymer forms onion like micelles, i.e., micelles composed of a two layer core and a stabilizing shell, instead of the subdivided core that was described above. The onion like structure was subsequently stabilized by oxidative self-polymerization of dopamine at the middle block. The formed nanostructure was found to be superior, in comparison to the uncross-linked precursor, in drug loading. Generally, the cross-linking approach is frequently used in order to stabilize complex architectures or to improve the micellar functionality.

The concept of cross-linking has been used by Cohn and coworkers for the formation of temperature-responsive nanoshells [16]. The nanoshells are formed using the triblock poly(ethylene oxide)-poly(propylene oxide)-poly(ethylene oxide) (PEO-PPO-PEO) decorated with methacrylic groups at the two ends of the macromolecular chain. In particular, the triblocks form micellar structures at elevated temperature, where the methacrylate units are located at the outer phase of the shell. By performing a cross-linking reaction at the methacrylic units, the ends of the PEO block of each micelle are connected together. Following, the temperature is decreased and the PPO core is dissolved, leading to a flexible nanoshell. Various parameters, like the reaction temperature, the polymer concentration, and the presence of triblocks without methacrylic terminal groups have been found to strongly affect the structural characteristics of the nanoshells. Interestingly, even tubular nanoshells were observed by performing the reaction at elevated temperature, where the transformation of the spherical micelles to those of rod-like shape is favored.

The cross-linking approach was used for the stabilization of micellar nanostructures which are formed by the block copolymer poly[oligo(ethyleneglycol) methacrylate]-block-[poly(styrene)-co-poly(vinyl benzaldehyde)] [POEGMA-b-P(ST-co-VBA)] [17]. The interesting feature of this polymer is that it changes the micellar structure that adopts by changing the degree of polymerization (DP) of the second block. Therefore, spherical micelles are formed from the block copolymer when DP of the P(ST-co-VBA) block reaches up to 175 units. A morphology transition from micelles to worm and rod-like structures was observed for P(ST-co-VBA) blocks greater than 340 repeating units. Finally, vesicles were formed when DP of the second block reached over 500. All the above nanostructures can be stabilized using a diamine in order to cross-link the aldehyde groups.

Multicompartment Micelles

Even though cross-linking is a great tool toward the synthesis and stabilization of novel polymeric nanostructures, the advances in polymer chemistry have made possible the synthesis of macromolecules that tend to adopt some very sophisticated nanostructures without cross-linking. Multicompartment micelles are a very important class of polymeric non-common nanostructures. Generally, they are composed by a terpolymer, linear or with more complex architecture, where one of the blocks is soluble in the selective solvent, while the other two are insoluble to the solvent and, at the same time, they are strongly incompatible. Janus micelles, named after the roman god Janus who had two faces, is a typical example of multicompartment micelles. Wang et al. [18] have recently described a well-studied polymer that gives Janus micelles in aqueous solutions. The aforementioned polymer is of a A-block-B-graft-C architecture composed of biologically compatible polymers, methoxy poly(ethylene glycol) (PEG), poly(ε-caprolactone) (PCL) and poly(2-(perfluorobutyl)ethyl methacrylate) (PPFEMA). The PEG block is extended to the solution, stabilizing the nanostructure, while the PCL and PPFEMA blocks are located in the micellar core. However, the PCL segments are in crystalline state, while fluorocarbon segments are amorphous, leading to the formation of a Janus-core with adjustable compartment balance, as was revealed by complimentary techniques.

Another example of multicompartment micelles is the case where the solvophobic part of the micelle is composed by a core and some distinct nodules at the surface of the core, while the micelle is stabilized by a solvophilic corona. In the case of the above structure, both terpolymers, like in the examples described above, and block copolymer mixtures have been used. The terpolymer poly ((sulfamate-carboxylate)isoprene)-block-polystyrene-block-poly(ethylene oxide), PISC-PS-PEO, can be considered as case study for the formation of multicompartment micelles with nodules [19]. In acidic solutions, the pH-responsive terpolymers self-assemble into kinetically trapped multicompartment micelles, where the micellar solvophobic part is consisting of discrete PS nodules and PISC core, while the PEO chains are located in the shell. Because of the kinetically frozen nature of the formation, the copolymer composition of low PS weight percent can be considered as a critical parameter. Notably, the aforementioned terpolymer forms regular spherical micelles with PS core and mixed PISC/PEO corona, in alkaline solution.

Finally, the in situ polymerization of poly(4-vinylpyridine) using a capping group located at the free corona chain ends of a polystyrene-b-poly(N, N-dimethylacrylamide) (PS-b-PDMA) spherical micelles with PS core, can also lead to multicompartment micelles [20]. The formed P4VP tend to aggregate on the PS surface, creating nodules. Depending on the polymerization degree of the P4VP the nodules can grow larger, forming, finally, concentric core–shell–corona nanoparticles. This is a typical example of how the composition of the terpolymer can lead to various shaped multicompartment nanostructures.

An extreme case of multicompartment nanostructure has been recently described by Shi et al. [21]. They have studied the case of a linear triblock copolymer where the two end blocks were poly(N-(2-methacryloyloxyethyl)pyrrolidone) (PNMEP), while the middle macromolecular chain was high density grafted poly(t-butyl acrylate)-b-polystyrene (PBA-b-PS). The aforementioned copolymer was found to form soft disk-like micelles by hexagonally packing of the middle block (Fig. 2.6). It has to be noted that the hexagonal pattern of the molecular brushes aligned perpendicularly to the disk plane, while this pattern demonstrates a periodic

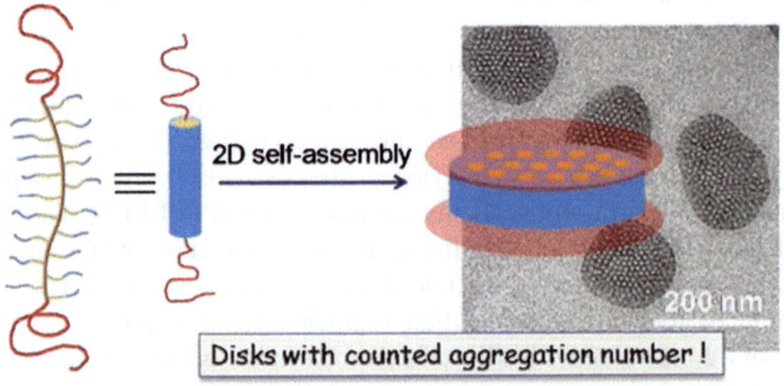

Fig. 2.6 Multicompartment nanostructure of soft disk-like micelles with perpendicular areas of hexagonally packing blocks. Reproduced from [21] with permission from American Chemical Society

spacing. The above example is indicative of the high structural complexity level that can be reached, as outcome of inspired macromolecular architectures.

Hybrid Nanostructures

Staying in the framework of multicompartment micelles, one should note that the formation of this nanostructure can be also achieved by a mixture of two block copolymers. Zhang and coworkers have described the case of a thermoresponsive multicompartment micelle composed of two block copolymers, namely poly[N-(4-vinylbenzyl)-N,N-diethylamine)]-b-polystyrene (PVEA-b-PS) and poly[2-(dimethylamino) ethyl methacrylate]-b-polystyrene (PDMAEMA-b-PS) [22]. A mixture of these copolymers, in a polar solvent, forms micelles where the PS core is decorated with PVEA nodules (Fig. 2.7). Depending on the solvent, the multi-compartment nanostructure can be either precipitated at elevated temperature, when water is used as solvent, or to be reformed into a regular micelle with PS core and mixed PVEA/PDMAEMA corona, when the solvent is methanol, just as it was described in the case of the pH-responsive terpolymer [19].

Going beyond the incompatibility of polymer chains with solvents or/and other polymers, that has been described above, the crystallization of a macromolecule can be potentially the driving force toward the formation of nanostructures. An interesting example is the case of a mixed system composed of PEO-b-PCL and PCL homopolymer [23]. In this system, the formation of nanorafts is recorded through a novel mechanism of lamella formation, involving the 2D alignment of block copolymer rod micelles. In particular, PCL chains tend to crystallize in spherical shape. The sphere-type aggregation provides the building blocks for growth along an orthogonal direction into rods. Finally, the rods align to the final 2D nanoraft product. Interestingly, the size and the aspect ratio of the formed nanostructures are highly controllable by small changes of the homopolymer content in the mixture.

Continuing in the context of mixed systems, i.e., systems composed of a block copolymer and one (or more) other entity, Betthausen et al. [24] has described a

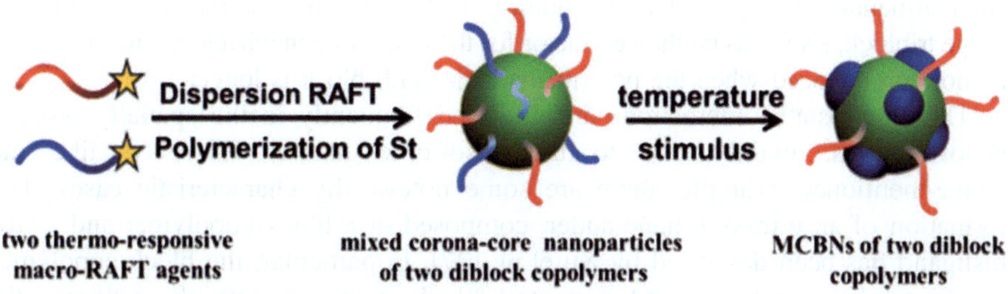

Fig. 2.7 Multicompartment micelles of PS core, decorated with PVEA nodules. Reproduced from [22] with permission from Royal Society of Chemistry

plethora of nanostructures that can be adopted by a linear ABC triblock terpolymer comprising a polyamine block: polybutadiene-b-poly(tert-butyl methacrylate)-b-poly(2-(dimethylamino)ethyl methacrylate) (PB-PtBMA-PDMAEMA) in the presence of organic di- or triacids, in mixtures of THF and water. The interaction of the PDMAEMA block with the organic acids is the driving force for the formation of nanostructures. A number of parameters, like chain architecture, amount, and functionality of added acid, the solvent quality, and the PDMAEMA block length, seem to influence the shape of the aggregates finally formed. Depending on the balance among the above parameters, a variety of structures is observed, such as spherical, disk-shaped, toroidal, ribbons featuring enlarged end-caps structures, and alongside undulated ribbons.

An even higher level of structural complexity can be achieved using binary mixtures of block copolymers. The simple solution construction of multigeometry nanoparticles, disk-sphere and disk-cylinder, through a straightforward, molecular-level blending of block copolymers has been described in an inspired work of Zhu et al. [25]. The multigeometry nanoparticles contain disk geometry in the core with either spherical patches along the disk periphery, in the case of disk-sphere particles, or cylindrical edges and handles in the case of the disk-cylinder particles. It has to be noted that the formation of the above extraordinary nanostructures is dictated not only by thermodynamic parameters, i.e., interactions between the polymers, but also by kinetic of aggregation.

Nanostructures via Electrostatic Interactions

Another possibility for obtaining multicompartment micelles, in the framework of mixed systems, is the case of electrostatic complexation. Synatschke et al. [26] have presented a system where the formation of micellar interpolyelectrolyte complexes (IPEC) leads to multicompartment micelles. In particular, the triblock polybutadiene-b-poly(1-methyl-2-vinylpyridinium methylsulfate)-b-poly(methacrylic acid) complexes with a high charge density cationic polyelectrolyte, provides a compartmentalized IPEC shell (Fig. 2.8). It was found that the complexes are not kinetically frozen and that the nature of the cationic moiety plays a minor role to the compartmentalization procedure. In contrast, the length of the anionic block, belonging to the triblock, seems to be the key factor for the compartmentalization, since the effect is more pronounced when the poly(methacrylic acid) block is longer.

The electrostatic interactions are not used frequently in the spatially ordered nanostructures, mainly due to their isotropic nature. However, like the above-mentioned example, there are some noteworthy characteristic cases. The formation of an ultralong nanoladder, composed of a block copolymer and a stiff bisligand has been described by Xu et al. [27]. In particular, the block copolymer composed by a cationic and a neutral block, interacts with the anionic stiff

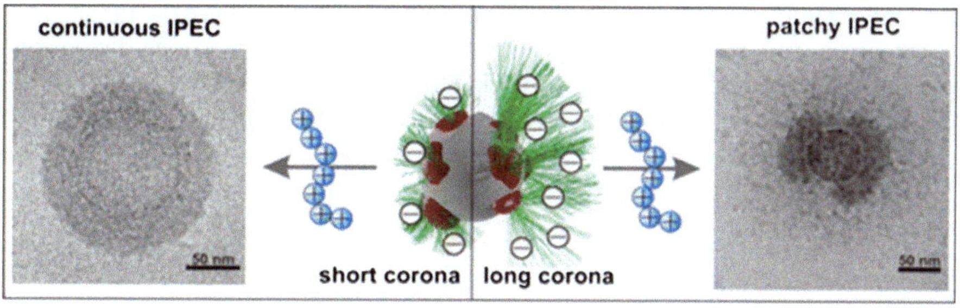

Fig. 2.8 Multicompartment micelles, assisted by electrostatic complexation. Reproduced from [26] with permission from American Chemical Society

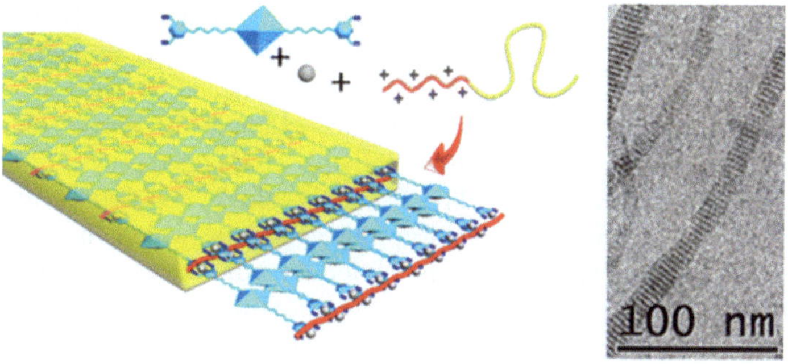

Fig. 2.9 Formation of an ultralong nanolander, composed of a block copolymer and a stiff bisligand. Reproduced from [27] with permission from American Chemical Society

bisligand, where the anionic sites are located at the two ends of the ligand. The result is nanoladders, enclosed in a matrix of the neutral block (Fig. 2.9).

The electrostatic interactions are important also in the class of polyelectrolyte aggregates. An interesting example is the case of the star block copolymer $PS_n(P2VP\text{-}b\text{-}PAA)_n$ [28]. In this case, the PS arms are always collapsed in aqueous environment, while the P2VP and PAA are positively and negatively charged, respectively, depending on the solution pH. Therefore, the electrostatic interactions between the charged moieties, in addition to the complex architecture of the star block copolymer, can lead to interesting transformations of the aggregate structures upon changes of the solution pH. Hence, at pH 1.4 unimolecular micelle are observed, which are transformed to multicore micelle (pH 1.6) and to worm-like micelles (pH 2). It is noteworthy that these transformations take place in a very narrow range of solution pH. On the other hand, in basic conditions, unimolecular micelle and network-like assembly are recorded at pH 8.5, while increasing pH to 11.8, the formation of multicompartment micelles is observed.

A whole new approach, toward the formation of complex structures by simple block copolymers, using electrostatic interactions with divalent counter ions, has been presented by Cui et al. [29]. They have proposed a technique that relies on divalent organic counter ions and solvent mixtures in order to drive the organization of the block copolymers down specific pathways into complex one-dimensional structures. In particular, using a triblock terpolymer, PAA-b-PMA-b-PS, and an organic diamine, in a proper mixture of solvents, they managed to formulate aggregates with a segmented cylinders morphology.

Hierarchical Self-assembly

Beyond the concept of multicompartment micelles, an important class of non-common polymeric nanostructures is that of nanoparticles obtained by hierarchical self-assembly. In this case the block copolymers form some initial assemblies, which, in most of the cases, are simple structures. Then, the aforementioned assemblies aggregate toward the formation of larger structures. In a third step, the aggregates can "collaborate together" in order to create a more complex nanostructure and so on. A characteristic example of these hierarchically self-assembly nanoparticles is the case of the fully conjugated poly(3-(2-ethylhexyl) thiophene)-b-polythiophene (P3EHT-PT). This block copolymer forms nanospheres which are coming together forming nanorods, which, in their turn, are aggregated in order to create nanostars, which are connected together for the formation of nanonetworks as schematically presented in Fig. 2.10 [30].

One of the most characteristic cases of hierarchically self-assembled nanoparticles is that presented by Muller and coworkers on the solution behavior of a linear

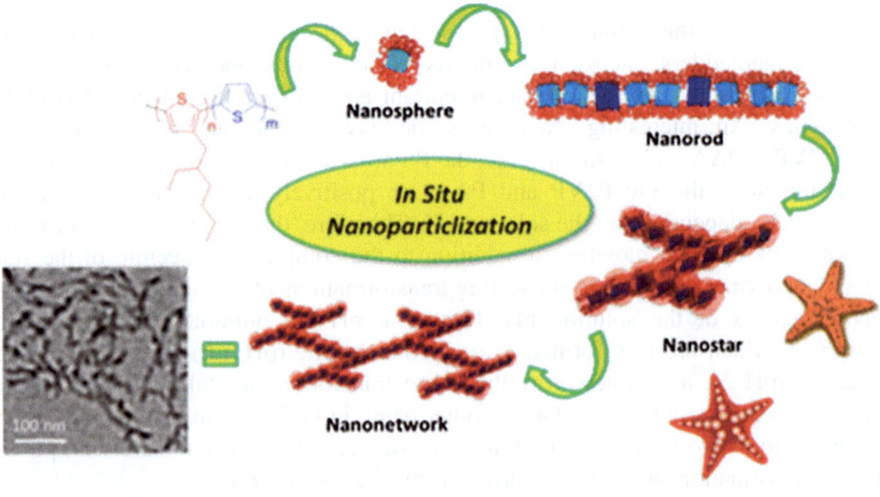

Fig. 2.10 Hierarchical self-assembly nanoparticles of a block copolymer that forms nanospheres which are coming together forming nanorods, which are aggregated to nanostars, which are connected together for the formation of a nanonetwork. Reproduced from [30] with permission from American Chemical Society

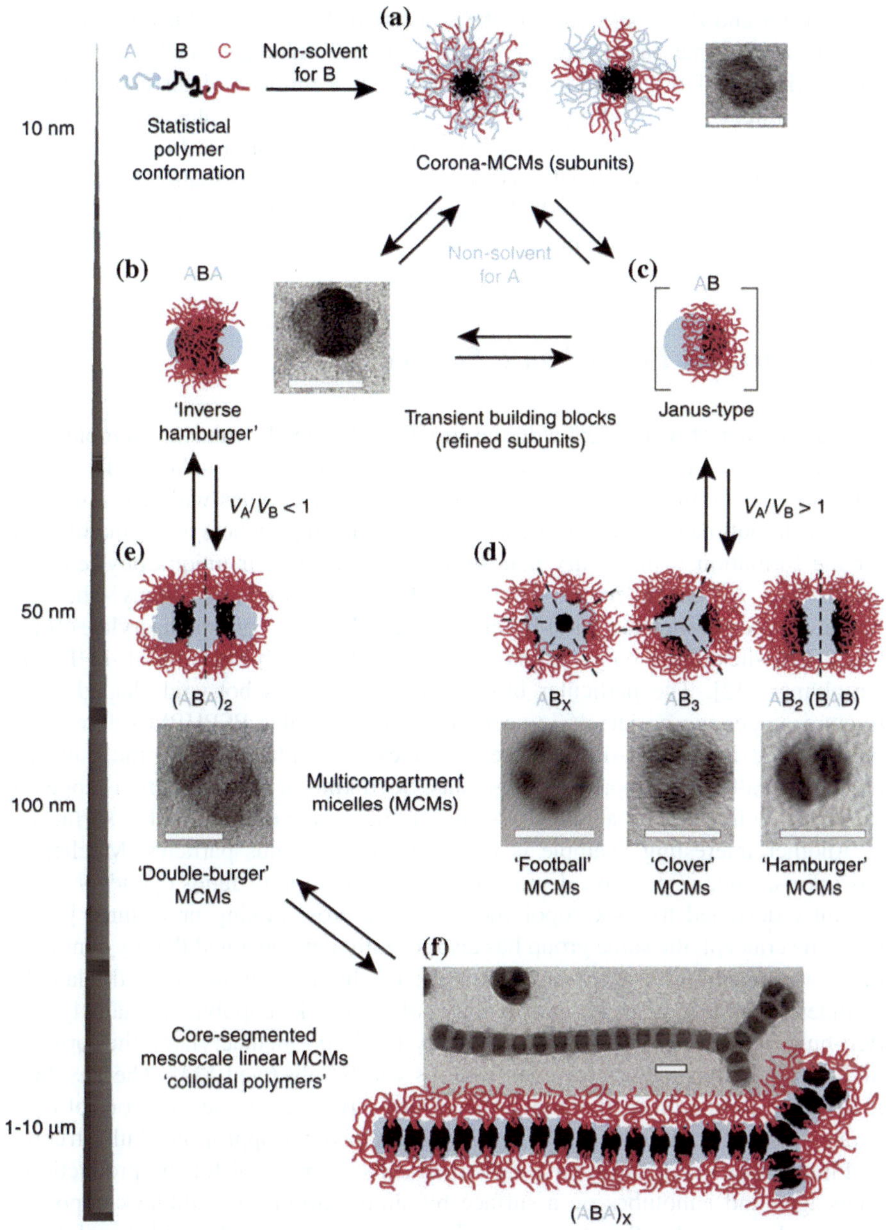

Fig. 2.11 Hierarchical self-assembly of a linear triblock copolymer. Reproduced from [31] with permission from Nature Publishing Group

triblock copolymer [31]. The used copolymer, namely PS-PB-PMMA, initially forms well-defined spherical micelles with mixed PS/PMMA corona, in a thermodynamically bad solvent only for the PB block (Fig. 2.11). The hierarchical self-assembly takes place when the micelles are dialyzed against a non-solvent for both the PS and PB blocks causing the collapse of PS blocks. This collapse can be done in a way that Janus-type or "inverse-burger" micelles can be formed. In a second degree of organization, the formation of a number of distinct multicompartment micelles, depending on the PS/PB ratio, like aggregates that resemble to double burger, soccer ball, clover or hamburger can be formed. Finally, in the last hierarchically step, the micelles link together for the formation of elongated structures of PS, covered by PMMA and doped with PB spheres. All stages of organizations are presented in Fig. 2.11.

Nanostructures Formed on a Surface

A large number of works deal with the structures formed by micelles of molecularly dissolved block copolymers when they are drying on a solid surface under appropriate conditions. Herein, only few interesting examples will be discussed, in an effort to outline this area of research. Perfluorinated polymers are frequently used for the formation of these dry nanostructures, due to their unique physical characteristics. A characteristic case is that of poly(1,3-(4-formyl-phenoxy)-2-hydroxypropyl methacrylate)-b-poly(poly(ethylene glycol) methacrylate)-b-poly (2,2,3,4,4,4-hexafluorobutyl methacrylate) (PFPHPMA-b-PPEGMA-b-PHFBMA) terpolymer [32]. The particular block terpolymer forms botryoid shaped nanostructures under appropriate drying conditions. In particular, PFPHPMA domains are miniaturized into small-sized discrete "grapes" and attached onto the outwardly branched scaffolds of fluorinated segments. Such kind of nanostructure is appealing as sensor or in catalytic applications, due to their increased accessible surface.

Another interesting example is the formation of Janus particles. Mueller and coworkers, since 2003, have demonstrated the formation of Janus cylinders using a carefully designed triblock copolymer and bulk cross-linking procedures [33]. In the same concept, the same group has also recently demonstrated that by controlling the phase transitions via pretreatment and cross-linking conditions of the lamella–cylinder equilibrium bulk morphology of a block copolymer, namely poly (tert-butoxystyrene)-b-polybutadiene-b-poly(tert-butyl methacrylate), the formation of Janus cylinders, sheets, or even ribbons can be observed [34]. The key factor toward the formation of the above mentioned structures is the fine control of the parameters, before and during drying, in order to take the appropriate bulk structure.

Finally, Petzetakis et al. have described a facile protocol for the production of nanocages and nanotubes on a surface by simple drying of a diblock copolymer solution. In particular, they have studied the case of poly(acrylic acid)-b-polylactide (PAA-PLA). This copolymer forms spherical and cylindrical micelles. However, these micelles tend to spontaneously form hollow nanocages and nanotubes upon evaporation of the solvent [35]. Additionally, the internal topology of the PAA-PLA particles could be tuned by manipulating the drying conditions to give solid or compartmentalized structures. Interestingly, upon resuspension, these reorganized nanoparticles retain their hollow structure, making the system, and the whole procedure, appealing for a number of applications.

Polymerization-Induced Self-assembly

As already described in the previous sections, the classical approach for preparation of polymeric nano-size aggregates from amphiphilic block copolymers by self-assembly in a selective solvent involves several steps. The first step is synthesis of an amphiphilic copolymer with a narrow molar mass distribution, followed by purification and characterization. Then, the self-assembly is performed by adding a non-solvent of one block into dilute copolymer solution or by direct dissolution of the copolymer in a selective solvent. Usually, the copolymer concentration varies between 1 and 10 g L^{-1} (0.1–1 wt%).

Recently, another feasible method, called *polymerization-induced self-assembly* (PISA), has been introduced for preparation of block copolymer nanoobjects with relatively high copolymer concentration (10–30 wt%). This is "one-pot" approach where the formation of block copolymers, self-assembly, and morphology transition are accomplished in the same polymerization system [36]. In general, the polymerization media is a good solvent for the first block and a non-solvent of the second block. Thus, during the chain growth of the second block, at certain degree of polymerization (critical micelle degree of polymerization, CMDP) it becomes insoluble and triggers microphase separation of the system. In other words, the block copolymers form aggregates in situ during the chain growth of the second block (Fig. 2.12).

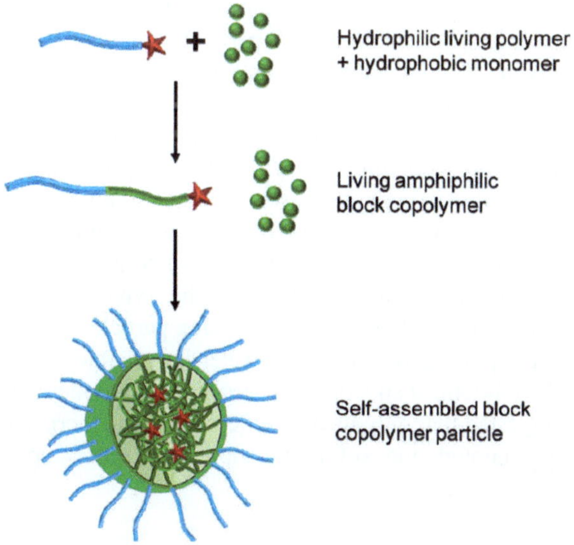

Fig. 2.12 General principle of the polymerization-induced self-assembly method. Reproduced from [36] with permission from Royal Society of Chemistry

According to the solubility of the core-forming monomer in the reaction media, two different methods—emulsion polymerization and dispersion polymerization—have been exploited to obtain self-assembled nanoparticles by PISA [37, 38]. The dispersion polymerization can be carried out either in water or in organic solvents. The emulsion polymerization starts from a monomer-in-water emulsion, where a water-soluble polymer precursor is chain-extended by polymerizing a water-immiscible monomer, resulting in self-assembled block copolymers. In contrast to the emulsion polymerization, dispersion polymerization is conceptually much simpler and the initial reaction solution is homogeneous.

Since the self-assembly phenomena requires well-defined block copolymers with narrow molecular weight distribution, only living/controlled polymerization techniques have been exploited for the preparation of polymeric nanoparticles via PISA. In particular, in aqueous media, the controlled radical polymerizations have been the methods of choice, taking the advantage of the compatibility of the reactions with water and the ability to create a wide variety of amphiphilic polymers. Atom transfer radical polymerization (ATRP), nitroxide-mediated polymerization (NMP), and reversible addition-fragmentation chain transfer (RAFT) polymerization have been the most studied techniques. The three methods possess certain advantages; however, RAFT remains particularly the most attractive due to the wide variety of polymers that can be produced in a controlled manner at low polymerization temperatures.

Ferguson, Hawkett and coworkers have pioneered the field of polymerization-induced self-assembly by employing RAFT emulsion polymerization of acrylic acid and n-butyl acrylate in aqueous media [39]. This approach led to in situ self-assembled nanoparticles formed by an amphiphilic block copolymer. First, acrylic acid was polymerized in the water phase to yield a water-soluble macro-RAFT agent, followed by the polymerization of hydrophobic n-butyl acrylate. At the completion of the polymerization, self-stabilized micelle-like PAA-b-PnBA particles of number-average diameter of 60.3 nm were obtained. When polymerization was continued by feeding of another hydrophobic monomer, styrene, spherical core-shell particles composed of poly(acrylic acid)-b-poly(butyl acrylate)-b-polystyrene triblock copolymer were formed [40].

Charleux and coworkers have made considerable progress toward the understanding of self-assembly phenomena of nanoparticles obtained via aqueous emulsion polymerization. Thus, poly(ethylene oxide) and poly(N, N-dimethylacrylamide)-based water-soluble macro-RAFT agents were chain-extended by polymerizing a water-immiscible monomer such as styrene, methyl methacrylate, or n-butyl acrylate in a batch emulsion polymerization [41–43]. Starting with the idea that the synthesized amphiphilic block copolymers would simply stabilize classical latex particles, it was realized that the good control of the polymerization of the water-immiscible monomer allowed in situ formation of well-defined spherical self-assembled polymeric nanoparticles (Fig. 2.13). It was suggested that only when the degree of polymerization of the hydrophobic block is large enough, the so-formed amphiphilic block copolymer chains can self-assemble into micelles.

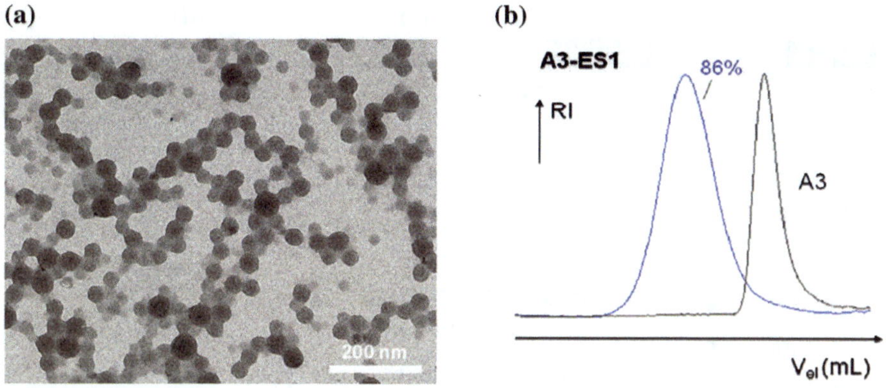

Fig. 2.13 **a** TEM image for a sample prepared by the emulsion polymerization of styrene with PDMAAm macro-RAFT agent; **b** SEC chromatograms in THF of the polymer from the same experiment. Reproduced from [43] with permission from American Chemical Society

Aqueous dispersion polymerization is another approach studied for in situ preparation of self-assembled amphiphilic block copolymer nanoparticles. In contrast to the aqueous emulsion polymerization, an important prerequisite in dispersion polymerization is the selection of a water-soluble monomer which, when polymerized, forms a water-insoluble polymer. First, An and coworkers reported on RAFT aqueous dispersion polymerization of N-isopropylacrylamide (NIPAAm) initiated with poly(N,N′-dimethylacrylamide)-based (PDMAAm) macro-RAFT agent [44]. With the progress of polymerization, an amphiphilic PDMAAm-b-PNIPAAm block copolymer was formed, and it further self-assembled into block copolymer micelles when PNIPAAm blocks became sufficiently long to collapse at the reaction temperature of 70 °C. The use of cross-linking agent was necessary to avoid dissociating of the micelles at room temperature. In the same year, Charleux and coworkers published a study of similar system with the core-forming block based on poly(N,N-diethylacrylamide) instead of NIPAAm using nitroxide-mediated polymerization [45]. Later on, Li and Armes [46] reported another aqueous dispersion polymerization formulation for in situ syntheses of core-shell nanoparticles. A poly(glycerol monomethacrylate)-based macro-RAFT agent was first synthesized and then used to initiate the propagation of the water-soluble monomer 2-hydroxypropyl methacrylate (HPMA). Since the corresponding polymer, PHPMA, is water insoluble, the microphase separation occurred when the chain length of PHPMA reached CMDP. It was demonstrated that the size of the formed nanoparticles can be controlled by varying the length of PHPMA chains. Further studies of the same group revealed that not only spherical micelles could be formed by PISA in aqueous media [47]. For example, the chain extension of highly hydrated zwitterionic poly(2-(methacryloyloxy) ethylphosphorylcholine) (PMPC) block with HPMA in water at 70 °C produced a hydrophobic PHPMA block, which triggered in situ self-assembly to form well-defined diblock copolymer spheres, worms, or vesicles. The final particle

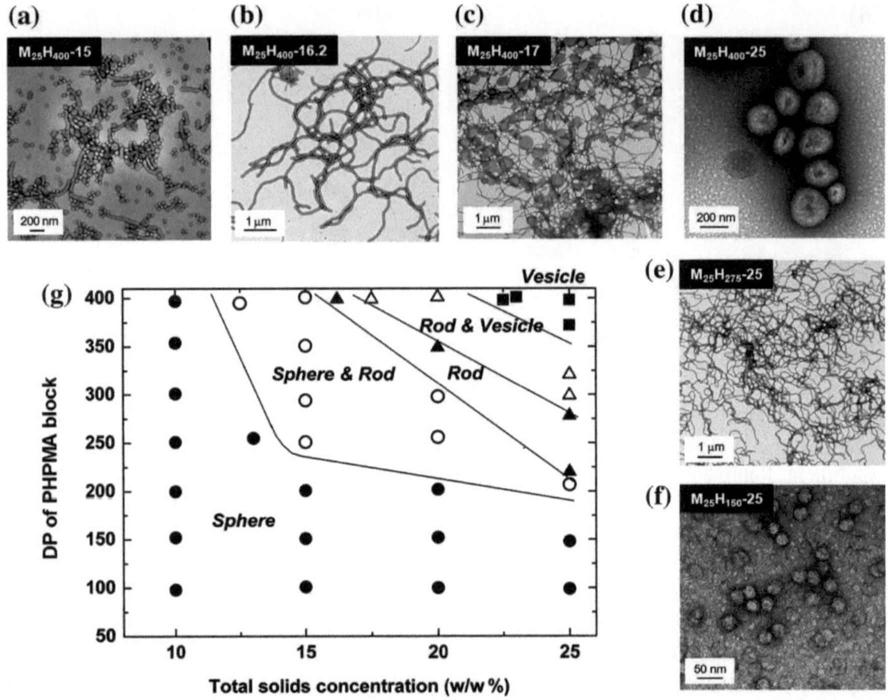

Fig. 2.14 Detailed phase diagram constructed for the $M_{25}H_x$-Y formulation (where M denotes MPC and H denotes HPMA) by systematic variation of the mean target degree of polymerization of PHPMA (x) and the total solids concentration (Y) used for each synthesis. The mean DP values of the PHPMA block shown in the phase diagram were calculated from the diblock copolymer composition determined by ^{1}H NMR spectroscopy in d_4-methanol assuming 100 % blocking efficiency for the $PMPC_{25}$ macro-CTA. TEM images for representative morphologies: **a** $M_{25}H_{400}$-15 (spheres and worms), **b** $M_{25}H_{400}$-16.2 (worms), **c** $M_{25}H_{400}$-17 (worms and vesicles), and **d** $M_{25}H_{400}$-25 (vesicles), i.e., identical diblock copolymers prepared at differing copolymer concentrations. **e** $M_{25}H_{275}$-25 and **f** $M_{25}H_{150}$-25 are two other diblock copolymers prepared at the same 25 wt% solids content used for image **d**. Reproduced from [47] with permission from American Chemical Society

morphology obtained at full monomer conversion is dictated by the degree of polymerization of the PHPMA block and the total solids concentration at which the HPMA polymerization is conducted. It is found that the onset of micellar nucleation corresponds to an enhancement in the rate of polymerization, which suggests solvation of the growing PHPMA chains by the unreacted HPMA monomer. Moreover, close monitoring of the in situ HPMA polymerization by TEM revealed a range of intermediate morphologies, which provide important information regarding the mechanism of sphere-to-worm and worm-to-vesicle transitions (Figs. 2.14 and 2.15) [47].

Pan and coworkers have published a series of papers describing the formation of spherical and worm-like micelles, vesicles, nanotubes, and some other morphologies

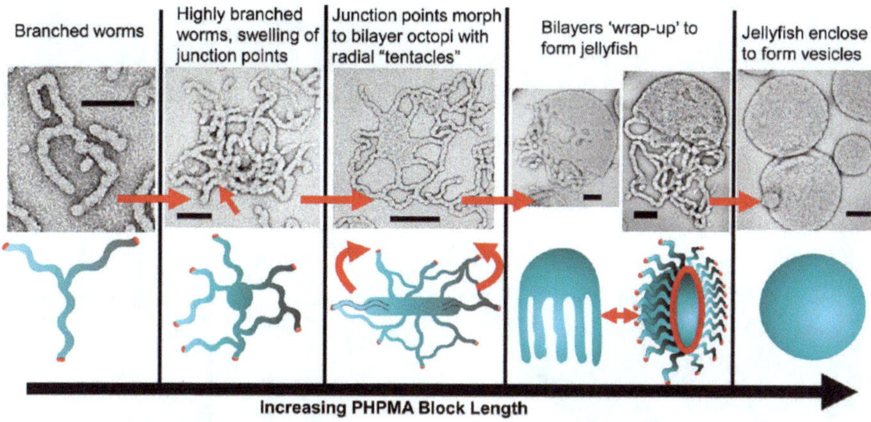

Fig. 2.15 Suggested mechanism for the polymerization-induced worm-to-vesicle transformation during the synthesis of G47-H200 by RAFT aqueous dispersion polymerization. Reproduced from [47] with permission from American Chemical Society

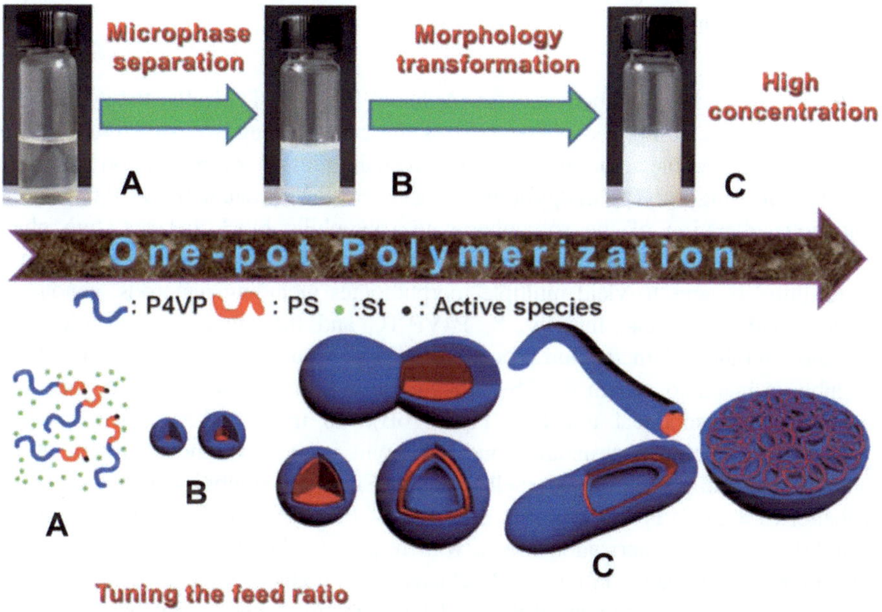

Fig. 2.16 The formation mechanism of the multiple morphologies in the RAFT polymerization of S in methanol using P4VP-TC as macro-RAFT agent, (*A*) formation of the soluble block copolymer, PS-b-P4VP; (*B*) phase separation to form spherical micelles; (*C*) re-organization of the resulting spheres to yield multiple morphologies. Reproduced from [48] with permission from Royal Society of Chemistry

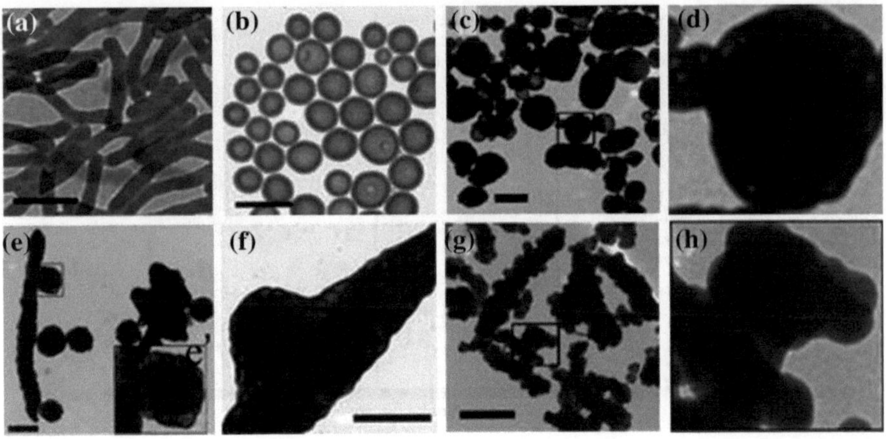

Fig. 2.17 TEM images of the aggregates prepared by RAFT polymerization of S at 80 °C for 36 h with different amounts of methanol: **a** 1.5 g; **b** 1.2 g; **c** and **d** 0.9 g; **e** and **f** 0.8 g; **g** and **h** 0.5 g. **d**, **e′** and **h** are magnified images of the parts in the square frames of **c**, **e** and **g**, respectively. Feed molar ratio: PDMAEMA/S/AIBN = 10:250,000:1, S:2.0 g. The scale bars are 500 nm (**a**, **b**) and 1 mm (**c**, **e**, **f**, **g**), respectively. Reproduced from [51] with permission from Royal Society of Chemistry

via RAFT dispersion polymerization in methanol [48, 49]. Typically, the synthesized block copolymers comprised polystyrene as the core-forming and structure-directing block. In the reversible addition-fragmentation chain transfer polymerization of styrene in methanol using poly(4-vinylpyridine) as a macro-RAFT agent (P4VP-TC), due to continuous alteration of the solvophobic to solvophilic block balance, two phase transitions occurred—phase separation to form spherical micelles and reorganization of the resulting spheres to yield multiple morphologies including nanorods, nanotubes, vesicles, and doughnuts. Initially, S, P4VP-TC and the initiator were soluble in methanol and the polymerization proceeded homogeneously (Fig. 2.16A). Since PS is insoluble in methanol, when the PS blocks grow to a critical value, phase separation occurs to form spherical micelles (Fig. 2.16B). To transfer the formed spherical micelles into other nanostructures, one key point is the continuous growth of the core-forming chains in the core–shell particles. Since the solubility parameter (d) of methanol (d = 29.7) is much higher than that of PS (d = 16.6–20.3), improving the solubility of the polymerization media with respect to the PS core is necessary for maintaining the propagation rate of PS chains in the cores of micelles. Consequently, the addition of a higher content of S (d = 19.0) in the feed acts as co-solvent before its polymerization [50]. The increase of PS chain length resulted in an increase of the packing parameter, leading to curvature decreasing of the polymeric assemblies. As a result, the spherical micelles were transferred to other morphologies along with a decrease of curvature as shown in Fig. 2.16C. Noteworthy, at high monomer conversion the morphology formed by phase separation was generally locked because the polymerization temperature was below the glass transition temperature of PS. Different morphologies can also be obtained by changing the relative content of methanol in the polymerization system [51]. TEM analysis confirmed that the RAFT polymerization of S, using a PDMAEMA-based macro-RAFT agent, at different weight ratios of methanol to S resulted in various morphologies. With the decrease of methanol content, the final structures vary from nanostrings (Fig. 2.17a) to spherical vesicles (Fig. 2.17b), to large compound vesicles (Fig. 2.17c) and to other ill-defined aggregates (Fig. 2.17e, g).

Biomedical Applications

Self-assembled block copolymers nanostructures could be particularly interesting toward biomedical applications, mainly due to their amphiphilic nature. They can serve as nanocarriers and/or stimuli to the formation of sophisticated nanostructures incorporating pharmaceutical compounds. A large number of publications deal with the potential use of block copolymers as drug carriers. Typically, a drug molecule interacts with a block copolymer, through either electrostatic or hydrophoblic interactions (depending on the nature of the drug). The polymer-drug nanoassembly is protected by the one of the blocks, typically PEO chains. Moreover, molecules and/or nanoparticles that permit the release of the drug in a specific area are generally used in order to reduce the drug toxicity.

The electrostatic complexation is widely used toward this end, because of the reversible nature of this complexation, due to changes in salinity or pH. Moreover, the complexation of an oppositely charged drug molecule and polymer chain leads to the formation of a water-insoluble polyion complex which stays in solution because of the second hydrophilic block. Following this concept, Li et al. [52] reported the complexation of the cationic drugs dibucaine, tetracaine, and procaine with anionic poly (methacrylic acid)-b-poly(ethylene oxide) (PMA-b-PEO) copolymers, which leads to micelles where the ionic drugs are located in the micellar cores. The study of the above micelles leads to the conclusion that both the electrostatic interactions and the hydrophobic interactions play an important role to the properties of the obtained micelles. In the same concept, triblocks can be also used as drug carriers, such as poly (ethylene oxide)-b-poly[sodium 2-(acrylamido)-2-methyl-1-propanesulfonate]-b-polystyrene (PEO-b-PAMPS-b-PS) triblock terpolymer [53]. Spherical micelles with zero surface charge and a stealth PEO corona were formed while the amount of the incorporated drug is controlled by the length of the charged block (PAMPS). It has to be noted that the charge ratio, [+]/[−], plays a key role on the physicochemical characteristics of the micellar complexes, as was studied by Soliman and Winnik [54]. In particular, the formation of micelles, their size and stability as well as drug uptake and release are characteristics connected to the above-mentioned ratio.

Besides the self-assembled spherical micelles, other polymeric nanostructures have been also used as drug carriers. Vesicles are of particular interest, since they can incorporate either hydrophobic drugs (in the hydrophobic polymeric part of the outer membrane), or hydrophilic drugs (inside the inner cavity). Xu et al. have presented the case of a linear triblock terpolymer, namely poly(ethylene oxide)-b-poly(acrylic acid)-b-poly(N-isopropyl acrylamide) (PEO-PAA-PNIPAM). This copolymer is molecularly dissolved at room temperature, while it forms nano-sized vesicles upon increasing the solution temperature. The formed nanostructures are stabilized in the vesicle form by cross-linking of the PAA block. The formed stabilized vesicles could be used as carriers of molecules with bioactivity, offering quite increased loading efficiency. Due to the stabilized nature of the carrier, the nanostructure is resistant to temperature and salinity changes, as well as to dilution (a very important parameter for a system that is supposed to be inserted in the systemic circulation). Even more, the carrier seems to dissociate under reductive conditions (similar to that observed at intracellural areas), making the system ideal for intracellural drug delivery [55]. Beyond vesicles, other cross-linked polymeric nanostructures have been also reported in the literature, based on the same concept as before. A characteristic example is the case of a triblock that forms onion like micelles, which are accessible to cross-linking through disulfide bonds, also suitable for intracellural delivery [56].

The structure of the polymeric nanostructure has been proven to play a significant role to the drug-carrier efficacy of a nanosystem. A comparative study has been presented by Tan et al. In this study, the authors compared the encapsulation and release of paclitaxel by two polymeric nanostructures of different architecture, namely AB_2 miktoarm star copolymers, of the types poly(ethylene glycol)-[poly(L-lactide)]$_2$ (PEG-(PLLA)$_2$) and poly(ethylene glycol)-[poly(D-lactide)]$_2$ (PEG-(PDLA)$_2$), which form hollow core spheres and nanofibers through stereocomplexation between poly(ethylene glycol)-b-poly(D-lactide) (PEG-b-PDLA) and poly(ethylene glycol)-b-poly(L-lactide) (PEG-b-PLLA) [57]. The observed drug loading was 12 and 40 wt%, respectively, showing the effect of nanostructure morphology on the loading capacity of a hydrophobic drug. Interestingly, the drug release profile was similar for the two nanostructures, without an initial rapid increase of the amount of released drug.

Even though, drug delivery is the most well-studied research area in the frame of polymeric nanostructures with appealing bioapplications, polymeric self-assemblies have also been used as vehicles for the transportation of many other molecules with biological interest, like DNA, RNA, and proteins. Following this, Varkouhi et al. has proposed the use of a diblock, consisting of the stabilizing and stealthing PEO block and a high charge density cationic polyelectrolyte as potential carrier of siRNA and plasmid DNA. Due to the high charge density of the polyelectrolyte block, the DNA complexes are particularly stable, in terms of dissociation and transfection activity. Moreover, siRNA complexes were found to show low cytotoxicity, improved siRNA delivery and high gene silencing activity [58], illustrating the importance of the molecular characteristics of the polymer, in this case the high charge density, for the particular bioapplication. It has to be noted that the results were evaluated under the prism of the comparison among the block copolymer, a homopolymer of high charge density and a reference cationic polymer, namely PDMAEMA, Fig. 2.18.

Triblock terpolymers have also been used for gene delivery. As before, the complexation takes place through a cationic block (in this case PDMAEMA), while stability is ensured by a poly(ethylene glycol) methyl ether methacrylate block, while the third block is a pH-responsive copolymer of PDMAEMA and PBMA.

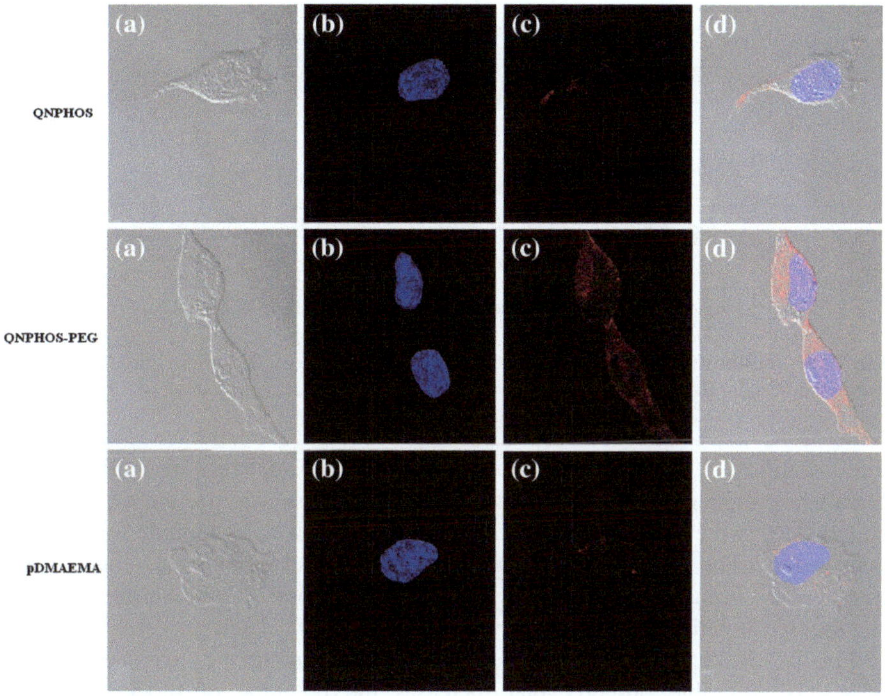

Fig. 2.18 Set of images which illustrate the successful gene transfer of RNA by cationic polymers. Confocal laser scanning microscopy images of cells incubated with complexes prepared with siRNA (*red*) and the homopolymer (QNPHOS), block copolymer (QNPHOS-PEO) or the reference polymer (PDMAEMA). **Images a** Pattern of cells with light microscopy. **Images b** stained Nuclei of cells. **Images c** Complexes with siRNA. **Images d** *Red* complexes with siRNA; *Blue* stained nuclei. Reproduced from [58] with permission from Elsevier

The above polymer forms particles of 86–216 nm upon complexation with mRNA. Depending of the relative position of each block, the polymer can be potentially used as an effective carrier due to the high transfection efficiency [59].

Besides the structure of the copolymer, its molecular characteristics seem to play a crucial role on the DNA effective transportation. Osada et al. have described this parameter very well by studying the effect of polylysine segment length, of a PEG-polylysine block copolymer, on the complexation with DNA. The obtained results indicate that packaging of plasmid DNA within both rod- and sphere-shaped polyplex micelles can be achieved. Interestingly, it was obvious that regularly folded plasmid DNA polyplex micelles are much more suitable for gene delivery than collapsed plasmid DNA polyplex micelles. The above indicates that controlling packaging of DNA, through control of the copolymer molecular characteristics, is crucial for achieving effective gene transfer [60].

Beyond the option of carrying a drug or a biomolecule in the body, the site-specific release is also an important parameter towards the formulation of an appealing nanosystem, as it was discussed above for the intracellular release of

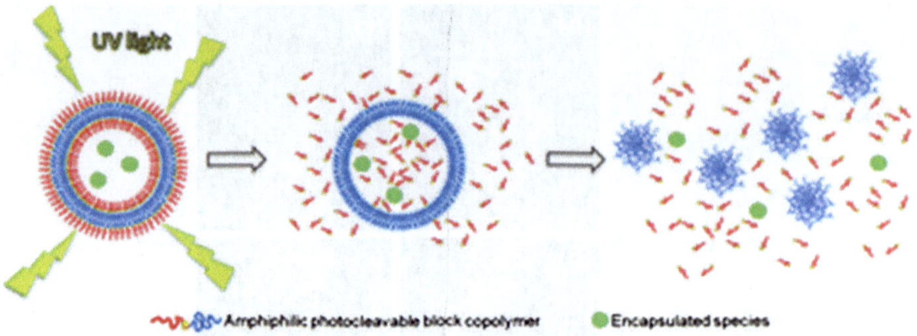

Fig. 2.19 Illustrative representation of encapsulated species release by a light-stimuli-responsive polymeric vesicle. Reproduced from [62] with permission from Royal Society of Chemistry

drugs. Pang et al. [61] have described the site-specific delivery of peptides to the brain using biodegradable polymersomes. Thiols, present in the monoclonal antibody OX26, were added to PEG-PCL polymersomes with maleimide functions on their surface. The used antibody was able to initiate endogenous receptor-mediated transcytosis of the polymersomes across the blood brain barrier, offering site-specific functionality.

Finally, the responsiveness of macromolecular self-assemblies to external stimuli is a very desirable property that increases the potential of block copolymer nanostructures for drug delivery. Toward this end, Cabane et al. [62] have presented the formation of polymeric vesicles as a light-triggered delivery system. In particular, they used a photocleavable amphiphilic block copolymer composed of PAA and poly(methyl caprolactone) as hydrophilic and hydrophobic blocks, respectively, linked by O-nitrobenzyl photocleavable segment for the formation of the vesicles. The above nanostructure is disintegrated upon UV irradiation and is rearranged to spherical aggregates, by simultaneous release of the encapsulated loaded molecules, Fig. 2.19.

Conclusions and Outlook

The amphiphilic copolymers in selective solvents self-assemble into nano-sized structures due to differences in the physicochemical characteristics of the constituent moieties. Spherical and cylindrical core–corona micelles as well as vesicles or polymersomes are the most commonly observed morphologies. Other morphologies such as toroids, bicontinuous, and multicompartment micelles, tubules, disk-like micelles, etc., are considered somewhat exotic. These structures are less frequently observed, however, they are not worth less than the commonly observed ones. The constituent moieties can be made sensitive to variations of parameters of the surrounding media such as temperature, pH, ionic strength, presence of specific additives/substances, pressure, etc., so that the resulting interactions, properties of the self-assembled structures, morphology and morphological transitions can become sufficiently complex, versatile, and tunable, which is very important from a technological point of view.

The co-assembly is an approach to influence and modify the properties of the nanoassemblies, which allows expanding the utility of the latter. Different types of forces like hydrophobic interactions, electrostatic interactions, hydrogen bonding, donor-acceptor interactions, metal-ligand coordination bonds, etc., have been found to facilitate the formation of mixed, hybrid structures and to contribute to the introduction of new functionality and properties. The incorporation of additional entities even in small amounts in the mixed/hybrid nanostructures is an excellent approach to tune the aggregate morphology and to significantly alter the aggregate characteristics [2].

The nano-sized polymer structures have proven biomedical applications. Coupled with novel strategies for targeting, biodegradability, stimuli-responsiveness, controlled release, these nanoassemblies exhibit tremendous potential for delivery of, e.g., antitumor agents, genetic material, proteins, and other biologically active substances. The required extracellular and intracellular delivery of therapeutic molecules into the disease-associated cells presents the primary roadblock for enhancement of therapeutic effects of these molecules. Therefore, the proper selection and design of specific copolymers or nanoassemblies as well as the design of adequate delivery technologies have utmost importance. The structures typically contain surface domains that shield against undesired biological interactions and enable specific host cell receptor binding as well as elements/moieties for controlled delivery functions. These multistep tasks constitute an attractive challenge for the polymer researchers.

The polymerization-induced self-assembly is a facile, efficient, and reproducible strategy for preparation of families of polymeric nanoparticles having various morphologies via a one-pot process. Direct preparation of the block copolymer aggregates in the dispersion and emulsion polymerizations at high concentration of solids provides the possibility for large-scale industrial production. Preferentially, the polymerization-induced self-assembly should be conducted in aqueous media at quantitative monomer conversion.

Acknowledgments The authors express gratitude to the EC project POLINNOVA.

References

1. Blanazs A, Armes SP, Ryan AJ (2009) Self-assembled block copolymer aggregates: from micelles to vesicles and their biological applications. Macrom Rapid Commun 30:267–277
2. Rangelov S, Pispas S (2014) Polymer and polymer-hybrid nanoparticles: from synthesis to biomedical applications. Taylor and Francis Group, Boca Raton
3. Smart T, Lomas H, Massignani M, Flores-Merino MV, Perez LM, Battaglia G (2008) Block copolymer nanostructures. Nanotoday 3:38–46
4. Paschalis A, Bjorn L (2000) Amphiphilic block copolymers: self-assembly and applications. Elsevier, Amsterdam

5. Szwarc M, Levy M, Milkovich R (1956) Polymerization initiated by electron transfer to monomer. A new method of formation of block polymers. J Am Chem Soc 78:2656–2657
6. Yang YQ, Zheng LSh, Guo XD, Qian Y, Zhang LJ (2011) pH-Sensitive micelles self-assembled from amphiphilic copolymer brush for delivery of poorly water-soluble drugs. Biomacromolecules 12:116–122
7. Pearson RT, Warren NJ, Lewis AL, Armes SP, Battaglia G (2013) Effect of pH and temperature on PMPC-PDPA copolymer self-assembly. Macromolecules 46:1400–1407
8. Smith AE, Xu X, Kirkland-York SE, Savin DA, McCormick CL (2010) "Schizophrenic" self-assembly of block copolymers synthesized via aqueous RAFT polymerization: from micelles to vesicles. Macromolecules 43:1210–1217
9. Liu B, Zhou H, Zhou S, Zhang H, Feng AC, Jian C, Hu J, Gao W, Yuan J (2014) Synthesis and self-assembly of CO_2-temperature dual stimuli-responsive triblock copolymers. Macromolecules 47:2938–2946
10. Wang W, Liu H, Mu M, Yin H, Feng Y (2015) Co_2-induced reversible morphology transition from giant worms to polymersomes assembled from a block-random segmented copolymer. Polym Chem 6:2900–2908
11. Bianchi A, Mauri M, Koynov K, Kappl M, Lieberwirth I, Butt HJ, Simonutti R (2014) Hierarchical self-assembly of PDMA-b-PS chains into granular nanoparticles: genesis and fate. Macromol Rapid Commun 35:1994–1999
12. Li Zh, Hillmyer MA, Lodge TP (2006) Laterally nanostructured vesicles, polygonal bilayer sheets, and segmented wormlike micelles. Nano Lett 6:1245–1249
13. Hu H, Liu G (2014) Miktoarm star copolymer capsules bearing pH-responsive nanochannels. Macromolecules 47:5096–5103
14. Zhang Y, Zhao C, Liu L, Zhao H (2013) Polymeric micelles with mesoporous cores. ACS Macro Lett 2:891–895
15. Wu S, Kuang H, Meng F, Wu Y, Li X, Jing X, Huang Y (2012) Facile preparation of core cross-linked micelles from catechol-containing amphiphilic triblock copolymer. J Mater Chem 22:15348–15356
16. Niu G, Djaoui AB, Cohn D (2011) Crosslinkable PEO-PPO-PEO triblocks as building blocks of thermo-responsive nanoshells. Polymer 52:2524–2530
17. Karagoz B, Esser L, Duong HT, Basuki JS, Boyer C, Davis TP (2014) Polymerization-induced self-assembly—Control over the morphology of nanoparticles for drug delivery applications. Polym Chem 5:350–355
18. Wang W, Zhang J, Li Ch, Huang P, Gao S, Han S, Dong A, Kong D (2014) Facile access to cytocompatible multicompartment micelles with adjustable Janus-cores from A-block-B-graft-C terpolymers prepared by combination ROP and ATRP. Colloids Surf B 115:302–309
19. Uchman M, Stepanek M, Prochazka K, Mountrichas G, Pispas S (2009) Multicompartment nanoparticles formed by a heparin-mimicking block terpolymer in aqueous solutions. Macromolecules 42:5605–5613
20. Huo F, Li Sh, Li Q, Qu Y, Zhang W (2014) In-situ synthesis of multicompartment nanoparticles of linear BAC triblock terpolymer by seeded RAFT polymerization. Macromolecules 47:2340–2349
21. Shi Y, Zhu W, Yao D, Long M, Peng B, Zhang K, Chen Y (2014) Disk-like micelles with a highly ordered pattern from molecular bottlebrushes. ACS Macro Lett 3:70–73
22. He X, Li Q, Shi P, Cui Y, Li S, Zhang W (2014) A new strategy to prepare thermo-responsive multicompartment nanoparticles constructed with two diblock copolymers. Polym Chem 5:7090–7099
23. Rizis G, vande Ven TGM, Eisenberg A (2014) "Raft" formation by two-dimensional self-assembly of block copolymer rod micelles in aqueous solution. Angew Chem Int Ed 53:9000–9003
24. Betthausen E, Hanske Ch, Muller M, Fery A, Schacher FH, Muller AHE, Pochan D (2014) Self-assembly of amphiphilic triblock terpolymers mediated by multifunctional organic acids: vesicles, toroids, and (undulated) ribbons. Macromolecules 47:1672–1683

25. Zhu J, Zhang S, Zhang K, Wang X, Mays JW, Wooley KL (2013) Disk-cylinder and disk-sphere nanoparticles via a block copolymer blend solution construction. Nat Commun 4:2297
26. Synatschke CV, Lobling TL, Fortsch M, Hanisch A, Schacher FH, Muller AHE (2013) Micellar interpolyelectrolyte complexes with a compartmentalized shell. Macromolecules 46:6466–6474
27. Xu L, Jiang L, Drechsler M, Sun Y, Liu Z, Huang J, Tang BZ, Li Z, Cohen Stuart MA, Yan Y (2014) Self-assembly of ultralong polyion nanoladders facilitated by ionic recognition and molecular stiffness. J Am Chem Soc 136:1942–1947
28. Iatridi Z, Tsitsilianis C (2011) pH-responsive self-assemblies from A_n-core-(B-b-C)$_n$ heteroarm star block terpolymer bearing oppositely charged segments. Chem Commun 47:5560–5562
29. Cui H, Chen Z, Zhong S, Wooley KL, Pochan DJ (2007) Block copolymer assembly via kinetic control. Science 317(5838):647–650
30. Lee I-H, Amaladass P, Yoon K-Y, Shin S, Kim Y-J, Kim I, Lee E, Choi T-L (2013) Nanostar and nanonetwork crystals fabricated by in situ nanoparticlization of fully conjugated polythiophene diblock copolymers. J Am Chem Soc 135:17695–17698
31. Gröschel AH, Schacher FH, Schmalz H, Borisov OV, Zhulina EB, Walther A, Müller AHE (2012) Precise hierarchical self-assembly of multicompartment micelles. Nat Commun 3:710
32. Deng J, Cai Y (2013) Botryoid-shaped reactive nanoparticles through spontaneous structural reorganization of terpolymer micelles. Macromol Rapid Commun 34:1459–1463
33. Liu Y, Abetz V, Mueller AHE (2003) Janus cylinders. Macromolecules 36:7894–7898
34. Wolf A, Walther A, Mueller AHE (2011) Janus triad: three types of nospherical, nanoscale Janus particles from one single triblock terpolymer. Macromolecules 44:9221–9229
35. Petzetakis N, Robin MP, Patterson JP, Kelley EG, Cotanda P, Bomans PHH, Sommerdijk NAJM, Dove A, Epps TH III, O'Reilly RK (2013) Hollow block copolymer nanoparticles through spontaneous one-step structural reorganization. ACS Nano 7:1120–1128
36. Sun J-T, Hong C-Y, Pan C-Y (2012) Formation of the block copolymer aggregates via polymerization-induced self-assembly and reorganization. Soft Matter 8:7753–7767
37. Charleux B, Delaittre G, Rieger J, D'Agosto F (2012) Polymerization-induced self-assembly: from soluble macromolecules to block copolymer nano-objects in one step. Macromolecules 45:6753–6765
38. Warren NJ, Armes SP (2014) Polymerization-induced self-assembly of block copolymers nano-objects via RAFT aqueous dispersion polymerization. J Am Chem Soc 136:10174–10185
39. Ferguson CJ, Hughes RJ, Pham BTT, Hawkett BS, Gilbert RG, Serelis AK, Such CH (2002) Effective ab ignition emulsion polymerization under RAFT control. Macromolecules 25:9243–9245
40. Ferguson CJ, Hughes RJ, Nguyen D, Pham BTT, Gilbert RG, Serelis AK, Such CH, Hawkett BS (2005) Ab ignition emulsion polymerization by RAFT-controlled self-assembly. Macromolecules 38:2191–2204
41. Rieger J, Stoffelbach F, Bui C, Alaimo D, Jérôme C, Charleux B (2008) Amphiphilic poly (ethylene oxide) macromolecular RAFT agent as a stabilizer and control agent in ab initio batch emulsion polymerization. Macromolecules 41:4065–4068
42. Rieger J, Osterwinter G, Bui C, Stoffelbach F, Charleux B (2009) Surfactant-free controlled/living radical emulsion (co)polymerization of n-butyl acrylate and methyl methacrylate via RAFT using amphiphilic poly(ethylene oxide)-based trithiocarbonate chain transfer agents. Macromolecules 42:5518–5525
43. Rieger J, Zhang W, Stoffelbach F, Charleux B (2010) Surfactant-free RAFT emulsion polymerization using poly(N, N-dimethylacrylamide trithiocarbonate macromolecular chain transfer agents. Macromolecules 43:6302–6310

44. An Z, Shi Q, Tang W, Tsung C-K, Hawker CJ, Stucky GD (2007) Facile RAFT precipitation polymerization for the microwave-assisted synthesis of well-defined, double hydrophilic block copolymers and nanostructured hydrogels. J Am Chem Soc 129:14493–14499
45. Delaittre G, Save M, Charleux B (2007) Nitroxide-mediated aqueous dispersion polymerization: from water-soluble macroalkoxyamine to thermosensitive nanogels. Macromol Rapid Commun 28:1528–1533
46. Li Y, Armes SP (2010) RAFT synthesis of sterically stabilized methacrylic nanolatexes and vesicles by aqueous dispersion polymerization. Angew Chem Int Ed 49:4042–4046
47. Blanazs A, Madsen J, Battaglia G, Ryan AJ, Armes SP (2011) Mechanistic insights for block copolymer morphologies: how do worms from vesicles? J Am Chem Soc 133:16581–16587
48. Wan W-M, Pan C-Y (2010) One-pot synthesis of polymeric nanomaterials via RAFT dispersion polymerization induced self-assembly and re-organization. Polym Chem 1:1475–1484
49. Wan W-M, Sun X-L, Pan C-Y (2010) Formation of vesicular morphologies via polymerization induced self-asembly and reorhanization. Macromol Rapid Commun 31:399–404
50. Wan W-M, Hong C-Y, Pan C-Y (2009) One-pot synthesis of nanomaterials via RAFT polymerization induced self-assembly and morphology transition. Chem Commun 5883–5885
51. Cai W, Wan W, Hong C, Huang C, Pan C (2010) Morphology transitions in RAFT polymerization. Soft Matter 6:5554–5561
52. Li Y, Ikeda S, Nakashima K, Nakamura H (2003) Nanoaggregate formation of poly(ethylene oxide)-b-polymethacrylate copolymer induced by cationic anesthetics binding. Colloid Polym Sci 281:562–568
53. Bastakoti BP, Guragain S, Yoneda A, Yokoyama Y, Yusab S, Nakashima K (2010) Micelle formation of poly(ethylene oxide-b-sodium 2-(acrylamido)-2-methyl-1-propane sulfonate-b-styrene) and its interaction with dodecyl trimethyl ammonium chloride and dibucaine. Polym Chem 1:347–353
54. Soliman GM, Winnik FM (2008) Enhancement of hydrophilic drug loading and release characteristics through micellization with new carboxymethyldextran-PEG block copolymers of tunable charge density. Int J Pharm 356:248–258
55. Xu H, Meng F, Zhong Z (2009) Reversibly crosslinked temperature-responsive nano-sized polymersomes: synthesis and triggered drug release. J Mater Chem 19:4183–4190
56. Wang YC, Li Y, Sun TM, Xiong MH, Wu J, Yang YY, Wang J (2010) Core-shell-corona micelle stabilized by reversible cross-linkage for intracellular drug delivery. Macromol Rapid Commun 31:1201–1206
57. Tan JPK, Kim SH, Nederberg F, Appel EA, Waymouth RM, Zhang Y, Hedrick JL, Yang YY (2009) Hierarchical supermolecular structures for sustained drug release. Small 5:1504–1507
58. Varkouhi AK, Mountrichas G, Schiffelers RM, Lammers T, Storm G, Pispas S, Hennink WE (2012) Polyplexes based on cationic polymers with strong nucleic acid binding properties. Eur J Pharm Sci 45:459–466
59. Cheng C, Convertine AJ, Stayton PS, Bryers JD (2012) Multifunctional triblock copolymers for intracellular messenger RNA delivery. Biomaterials 33(28):6868–6876
60. Osada K, Shiotani T, Tockary TA, Kobayashi D, Oshima H, Ikeda S, Christie RJ, Itaka K, Kataoka K (2012) Enhanced gene expression promoted by the quantized folding of pDNA within polyplex micelles. Biomaterials 33(1):325–332
61. Pang ZQ, Lu W, Gao HL, Hu KL, Chen J, Zhang CL, Gao XL, Jiang XG, Zhu CQ (2008) Preparation and brain delivery property of biodegradable polymersomes conjugated with OX26. J Controlled Release 128:120–128
62. Cabane E, Malinova V, Menon S, Palivan CG, Meier W (2011) Photoresponsive polymersomes as smart, triggerable nanocarriers. Soft Matter 7:9167–9176

Chapter 3
Direct Synthesis of Nano-size Polymers by Microemulsion Polymerization

Manas Chanda

Introduction

According to IUPAC definition, microemulsions are dispersions, which are made of water, oil, and surfactants, are optically isotropic and thermodynamically stable, and have dispersed droplet diameters varying approximately from 1 to 100 nm (usually 10–50 nm, i.e., 100–500 Å). The small particle size leads to a translucent or even to a transparent system if the particle size is a few hundred angstroms. In comparison, the average diameter of droplets in miniemulsions (obtained by shearing, usually via exposure to high power ultrasound, a mixture comprising two immiscible liquid phases, one or more surfactants and, possibly, one or more cosurfactants) is typically between 50 and 500 nm, while that in a macro-emulsion (usually referred to as an "emulsion") is in the micron range. Microemulsions, like micellar dispersions, are liquid dispersions containing surfactant aggregates. However, in micellar dispersions the aggregates are made of surfactant only and are usually dispersed in water, whereas in microemulsions the aggregates are much larger and have large liquid cores (oil or water) surrounded by a surfactant monolayer that stabilizes the dispersion. In many cases, the micellar aggregates are spherical, but they can also be tubular and, in a few cases, they can also grow very long and entangled like polymers [1].

While the aqueous component of microemulsions may contain salt(s) and/or other ingredients, the "oil" component may actually be a mixture of different hydrocarbons and olefins. Microemulsions form upon simple mixing of the components and do not require the high shear conditions generally used to make ordinary emulsions. Besides optical clarity (or translucency) of a microemulsion and small (10–50 nm) droplet size of the dispersed phase, an additional feature that

M. Chanda (✉)
Department of Chemical Engineering, Indian Institute of Science,
Bangalore 560012, India
e-mail: profmanas@gmail.com

S. Fakirov (ed.), *Nano-size Polymers*,
DOI 10.1007/978-3-319-39715-3_3

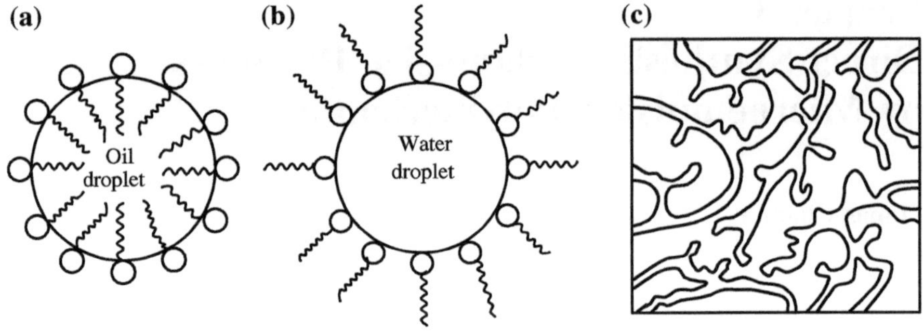

Fig. 3.1 Idealized microemulsion structures: **a** oil-in-water (o/w) microemulsion; **b** water-in-oil (w/o) microemulsion; **c** bicontinuous microemulsion

distinguishes it from ordinary emulsions is that the average drop size does not grow with time due to thermodynamic stability.

There are three basic types of microemulsions, viz., *direct* (oil dispersed in water, denoted by 'o/w'), *reversed* or *inverse* (water dispersed in oil, denoted by 'w/o'), and *bicontinuous* (i.e., regions of water and oil). The domains of the dispersed phase are either globular or interconnected (giving a bicontinuous microemulsion), as shown schematically in Fig. 3.1. These are stabilized by an interfacial film of surfactant (usually in combination with a cosurfactant), its molecules being oriented at the interface such that the hydrophilic ends are in the aqueous phase and the hydrophobic ends are in the oil phase. It may be mentioned that the term 'microemulsion' was first used by Hoar and Shulman [2], professors of chemistry at Cambridge University in 1943. However, other names are often used, such as *transparent emulsion* (implying optical clarity), *micellar solution*, and *solubilized oil*.

The stability of microemulsions is highly influenced by the nature of the surfactant and that of the oil phase in the system. The cohesive energy concept (CER) developed by Berbower and Hill [3] can also be used to predict the most suitable surfactants and oils for the production of clear and stable microemulsion systems. The concept is based on matching three properties, namely, chemical structure, hydrophilic–lipophilic balance (HLB), and molar volume. Surfactants may be classified according to their hydrophilic–lipophilic nature in terms of their HLB values, a more hydrophilic surfactant having a higher HLB value. Studies carried out to evaluate the relation between HLB and microemulsion stability have revealed [4, 5] that low HLB values of 4–6 are suitable for stable w/o and higher values for o/w microemulsion systems, an optimum HLB value being needed for the preparation of a monophasic, stable, and clear microemulsion with minimum amount of emulsifier [6]. Outside that limit, the formation of microemulsion is either difficult or requires a higher amount of surfactant. Both ionic and nonionic surfactants may be used for the preparation of a stable microemulsion, sodium dodecyl sulfate (SDS) being the most commonly used ionic surfactant. Sometimes a mixture of surfactants is used to prepare microemulsions, the surfactant with lower HLB value then performing the function of a surfactant.

Normal emulsions are "kinetically stable", whereas microemulsions are "thermodynamically stable". The thermodynamic stability of microemulsions was proposed by Ruckenstein and Chi from consideration of the free energy of formation as comprising the interfacial free energy, interaction energy between droplets, and entropy of dispersion. The interaction energy between droplets has been shown to be negligible. Therefore, the free energy of formation can be zero or even negative if the interfacial tension is of the order of 10^{-2}–10^{-3} mN/m. They also introduced the concept of hydrophilic–lipophilic balance temperature (HLBT) or phase inversion temperature (PIT) at which maximum solubilization of oil in water and ultralow interfacial tension is achieved [7–9].

To prepare microemulsion, in a simple procedure, milky emulsions can be first prepared using water, oil, and surfactant and then lower alkanols (butanol, pentanol, and hexanol) can be added in controlled amounts so as to obtain transparent or translucent solutions comprising dispersions of either water-in-oil (w/o) or oil-in-water (o/w) in nanometer or colloidal dispersions. The lower alcohols added are called cosurfactants. They lower the interfacial tension between oil and water sufficiently for almost spontaneous formation of the aforesaid microheterogeneous systems. Various surfactant to cosurfactant ratios can be used in the preparation.

Phase Behavior

The miscibility of oil, water, and amphiphile (surfactant plus cosurfactant) depends on the overall composition which, in turn, depends on the system. Ternary (water/surfactant/oil), pseudo-ternary (water/amphiphile/oil) or explicitly quaternary (water/surfactant/cosurfactant/oil) phase diagrams are usually employed to describe the phase manifestation which is essential in the study of microemulsions. These phase diagrams help define the microemulsion areas. Samples from the best combinations, i.e., those that produce the largest volume of microemulsion can be subjected to further characterization by different methods, such as polarized light microscopy, differential scanning calorimetry, zeta sizer, rheometer, etc.

While the majority of microemulsions use oil and water as immiscible liquid pairs, if a cosurfactant is used it may sometimes be represented at a fixed ratio to the surfactant as a single component and treated as a single "pseudo-component", so that the relative amounts of these three components can then be represented in a pseudo-ternary phase diagram. These diagrams can be used to depict the phase behavior of the system as a function of the volume fractions of different components.

The knowledge of phase manifestations of the pseudo-ternary (water/amphiphile/oil) or explicitly quaternary (water/surfactant/cosurfactant/oil) mixtures has been systematized. According to Winsor [10], four types of microemulsion phases exist in equilibrium. These phases are commonly referred to as Winsor phases; they are Winsor I: with two phases, the lower o/w microemulsion phase in equilibrium with

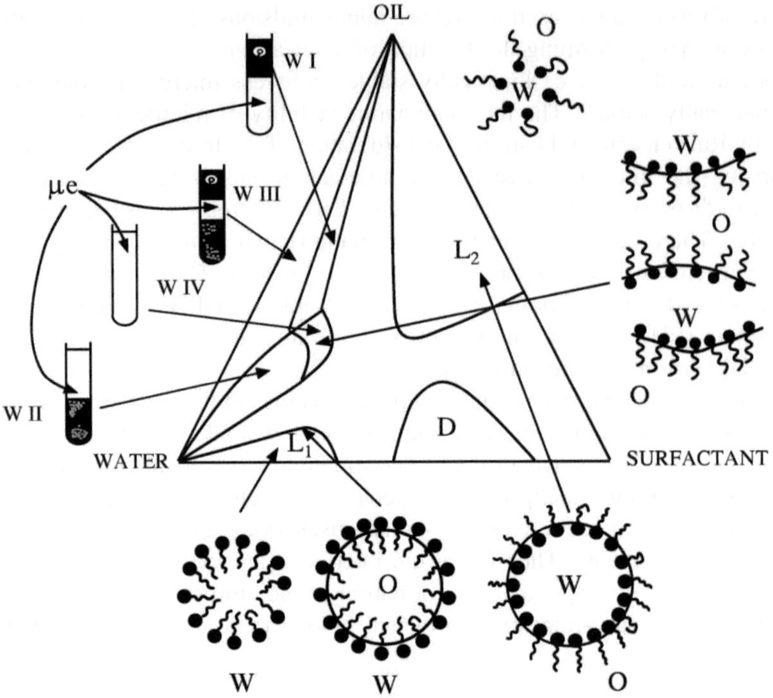

Fig. 3.2 Ternary phase diagram (schematic) of water-oil-surfactant mixtures showing Winsor classification and probable internal structures: L_1, one-phase region of normal micelles or oil-in-water (o/w) microemulsion; L_2, reverse micelles or water-in-oil (w/o) microemulsions; *D*, anisotropic lamellar liquid crystalline phase. Other symbols: μe microemulsion; *O* oil; *W* water (after [20])

the upper excess oil; Winsor II: with two phases, the upper w/o microemulsion phase in equilibrium with excess water; Winsor III: with three phases, middle microemulsion phase (o/w plus w/o, called bicontinuous) in equilibrium with upper excess oil and lower excess water; Winsor IV: in single phase, with oil, water, and surfactant homogeneously mixed. Interconversion among these phases can be achieved by varying the proportions of the components. An extension of Winsor's classification forming the fifth category is also possible where two microemulsions exist simultaneously, one in contact with water and the other in contact with oil. Figure 3.2 gives a composite representation of the aforesaid features of microemulsion forming systems.

The three components forming the system are each found at an apex of the triangle, where the corresponding volume fraction is 100 %. As one moves away from that corner, the volume fraction of that specific component decreases, while that of one or both of the other two components increases. Each point within the triangle represents the composition of a mixture of the three components, while the mixture may consist of one, two, or three phases. These points define regions with boundaries between them and represent the phase behavior of the system at constant temperature and pressure in terms of the Gibbs phase diagram. The diagram is, however, an empirical visual observation of the state of the system and may or may not, indicate the true number of phases for the given composition. For example, a clear single region may still consist of more than one isotropic phases.

The extents of formation of w/o, o/w, and bicontinuous microemulsions on mixing the components can be determined from phase equilibrium studies. However, such studies may often become complex due to the appearance of additional zones of viscous gel and liquid crystalline phases [11]. Establishment of boundary demarcations in such cases can become highly time consuming and laborious.

In contrast to the ease of preparation, the characterization of microemulsions is more complicated and requires a combination of many techniques. Thus, the characterization of microemulsions involve acquiring information on the type of dispersion (o/w, w/o, bicontinuous), dimension, shape, diffusion coefficient, polydispersity, aggregation, and dynamics of coalescence of particles, state of water pool, thermodynamics of formation, etc., of the compartmentalized systems of microemulsion. These are related to a wide variety of properties and, hence, many methods including conductance, viscosity, ultrasound attenuation [12], static and dynamic light scattering [13], nuclear magnetic resonance [14], dielectric relaxation [15], time resolved fluorescence quenching [16], freeze fracture electron microscopy [17], calorimetry [18], etc., are used. Though micrographs of microemulsions have been obtained showing reliable structures, the technique that is considered to offer the most insight to the structure of microemulsions is by measuring the self-diffusion constants of the various components of the system [19]. This technique easily distinguishes between droplet and bicontinuous structures.

The characteristic properties of microemulsions which render these organized solutions unique and include features such as spontaneous formation, optically clear appearance, large interfacial area, low interfacial tension, low viscosity, and large solubilization capability, have stimulated utilization of microemulsion systems in a plethora of chemical and industrial processes ranging from enhanced oil recovery to agricultural sprays, drug delivery systems, and reaction media [20]. An important example of the media application is polymerization in microemulsions. Millions of tons of polymer microlatexes are produced annually, notable examples being butadiene–styrene copolymers, poly(vinyl acetate), poly(vinyl chloride), acrylate ester copolymers, polyacrylamide, and derivatives. These latexes are used in coatings, adhesives, binders (papers and textile products), paints, flocculants, rheological modifiers, and biomedical applications.

Being thermodynamically stable "nano-dispersions" of water-in-oil or oil-in-water, microemulsions can be considered as microreactors to carry out chemical reactions and, in particular, to synthesize nanomaterials. Microemulsions thus have received much recent attention as media for synthesis of nanoparticles Pt, Pd, Rh, and Ir by reducing corresponding salts in the water micropools of w/o microemulsions with hydrazine or hydrogen gas and as media for polymerization to produce thermodynamically stable latexes in the nano-size range (<50 nm) not attainable with classical emulsion polymerization process. Though polymers with large molecular weights (>10^6) can be produced from monomers at fast reaction rates in both emulsion and microemulsion processes, it is only in the latter that stable latexes with polymer particles smaller than 50 nm can be easily obtained, since in this case polymerization occurs in the monomer reservoir encapsulated in a nano-size space. While microemulsions exhibit a wide variety of microstructures, it is the spherical oil-in-water (o/w) or water-in-oil (w/o) microstructures that have provoked the greatest interest to carry out polymerization in practice.

Phase Diagrams for Microemulsion Systems

The phase diagram concept provides a reliable basis for the use of microemulsion in polymerization systems. The phase diagram maps the thermodynamically stable regions of o/w and w/o microemulsions. Experimental determination of single-phase region thus precedes microemulsion polymerization. In fact, a thorough study involving composition and characterization of phase diagrams of various systems, including those that contain monomer, must be done before performing any microemulsion polymerization.

To give an example, for polymerization of methyl acrylate with ammonium persualfate, benzoyl peroxide or azobisisobutyronitrile (AIBN) as initiator at 50 °C in microemulsions using sodium dodecyl sulfate ($C_{12}H_{25}SO_4Na$) as surfactant and pentanol ($C_5H_{11}OH$) as cosurfactant, Stoffer and Bone [21] determined single-phase boundaries by titrating with water various surfactant/cosurfactant/monomer mixtures, first to the point of dissolution of surfactant (which marked one limit of the solubility region) and then to the appearance of turbidity (which defined another limit). Complementary information was obtained by titration of water/surfactant mixtures with a cosurfactant/monomer solution. Figure 3.3 thus shows a microemulsion region that contains 25 % methyl acrylate.

The phase regions comprising three components, with the content of the fourth component kept constant, can be described by a triangular phase diagram, while the complete diagram of the four components may be illustrated by a tetrahedron (Fig. 3.4). In the latter representation, the fraction of each component is equal to the ratio between the perpendicular distance from the plane that contains the other three components and the corresponding distance to the apex that represents 100 % of the component. The surface planes of the tetrahedron represent the behavior of the associated ternary system. If study is selected to a plane in which the ratio of (say) monomer (A) to surfactant-plus-cosurfactant (B + C) is constant, this plane can be obtained by addition of water (D) to compositions along line BC (in Fig. 3.5), which contains a constant fraction of monomer (A).

Since the work of Stoffer and Bone in 1980, most work was done in four or five component (including cosurfactant) microemulsion. However, the presence of a fourth component, such as alcohol cosurfactant, substantially limits the utility of microemulsion, mainly for two reasons. First, the cosurfactant complicates the

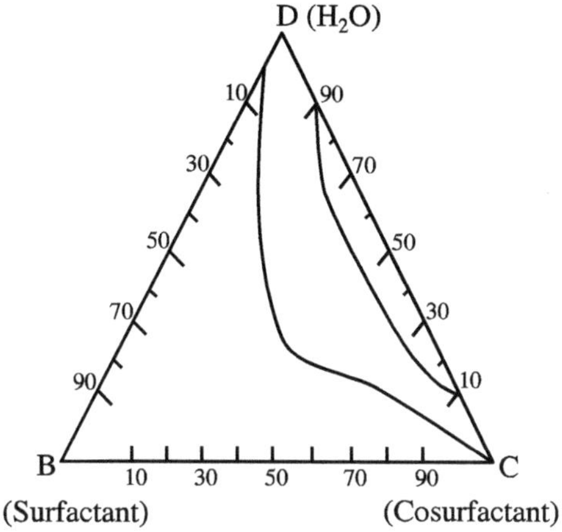

Fig. 3.3 A microemulsion region (schematic) containing water, surfactant (sodium dodecyl sulfate), cosurfactant (pentanol), and 25 % monomer (methyl methacrylate) (after [21])

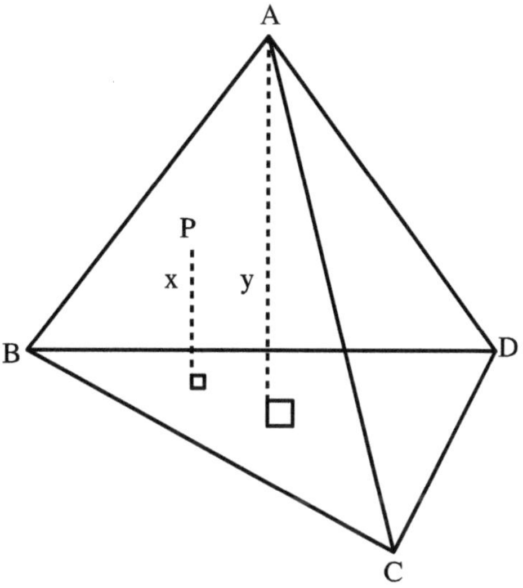

Fig. 3.4 The fraction of one component (A) in a total composition (P) is given by the ratio of the distances *x* and *y*, both perpendicular to the plane BCD

phase behavior of the microemulsion system, whereas the ternary microemulsion formulations are considerably easier to deal with. Second, alcohol present as a cosurfactant can act as a chain transfer agent, interfering with the desired polymerization and reducing the polymer molecular weight. In this context, the first report on polymerization in ternary microemulsions (without cosurfactant alcohol) was made by Pérez-Luna et al. [22] using cationic surfactant dodecyltrimethylammonium bromide (DTAB) and styrene monomer. The one-phase microemulsion region at 25 °C was determined visually by styrene titration of aqueous micellar solution of DTAB to obtain phase boundaries which were checked by preparing samples by weight with compositions below and above the titration-determined phase boundaries in sealed glass ampoules. Phase diagrams at 60 °C were made

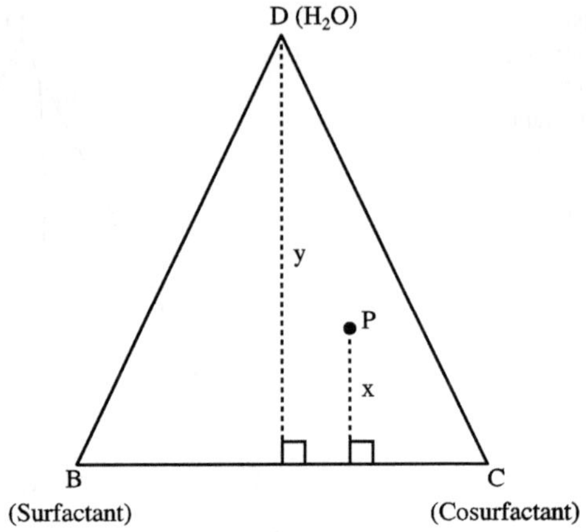

Fig. 3.5 For limited investigations, planes of surfactant (B)-cosurfactant (C)-H_2O (D) can be obtained by adding H_2O to monomer/surfactant/cosurfactant mixtures with a constant fraction of monomer, which are found along line BC

Fig. 3.6 A partial phase diagram (schematic) of DTAB/styrene/water showing extents of single-phase regions at 25 and 60 °C (after [22])

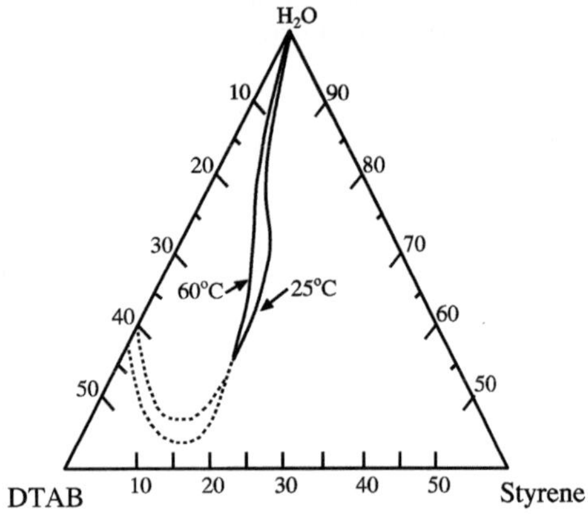

with styrene containing a few ppm of hydroquinone to inhibit thermal polymerization. The extent of single-phase region thus determined is shown in Fig. 3.6.

The fact that microemulsions can be formed with only surfactant and no addition of cosurfactant spurred new interest and systematic studies were made on ternary microemulsions based on cationic surfactants with different alkyl chain lengths. Antonietti et al. [23] reported formation of stable microemulsions of styrene in water using either of the surfactants cetyl trimethyl ammonium chloride (CTAC) and DTAB without the aid of a cosurfactant.

The polymerization of methyl methacrylate (MMA) was studied by Rodriguez-Guadarrama et al. [24] in three-component o/w microemulsion made with DTAB. One-phase regions were detected at 25 and 60 °C in this sytem near the

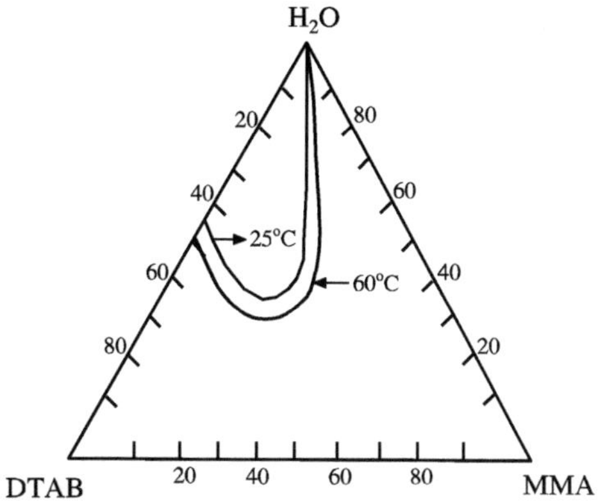

Fig. 3.7 A partial phase diagram (schematic) of DTAB/MMA/water showing extents of single-phase regions at 25 and 60 °C (after [24])

water-rich corner (Fig. 3.7). The single-phase microemulsion region was determined visibly by titrating aqueous micellar solutions of DTAB with MMA.

Unlike ionic surfactants for which the phase behavior is almost insensitive to temperature, the phase behavior of nonionic surfactants in water strongly depends on temperature and, consequently, with nonionic surfactants it is possible to observe rich phase behavior in water/oil (monomer)/surfactant mixtures over the entire range of surfactant concentrations as a function of temperature. The phase behavior is then most easily described when it is represented as an upright phase prism with temperature as the ordinate and the water-oil-surfactant Gibbs triangle as the base. However, on replacing the nonionic surfactant with an ionic surfactant such phase behavior becomes practically limited to only two variables, namely, weight percentage of oil and that of the (temperature-insensitive) surfactant. To increase the usefulness of ionic surfactants, they may be mixed with other amphiphiles or surfactants to give an additional degree of freedom in phase space. Therefore, to study the microemulsion polymerization of alkyl methacrylates, Lusvardi et al. [25] used mixtures of two cationic surfactants, such as DTAB and its double-tailed analog didodecyl dimethyl ammonium bromide (DDAB), in aqueous solutions at 60 °C. Microemulsion regions in ternary mixtures of water, polymerizable oil (alkyl methacrylate), and single-tailed surfactant DTAB were observed on the water-rich side of the phase diagram [22]. However, when DTAB was replaced with DDAB, a microemulsion region appeared on the oil-rich side of the phase diagram for a variety of alkanes [26]. Lusvardi et al. [25] mapped out the phase behavior of several alkyl methacrylates (B) in mixtures with water (A) and two surfactants, DTAB (C) and DDAB (D) using an upright phase prism (in analogy with phase prism known for systems containing nonionic surfactants [27]) with ordinate now being $\epsilon = \mathrm{D}/(\mathrm{C}+\mathrm{D})$ in weight ratio and Gibbs triangles (A–B–C or A–B–D) as the base and top (Fig. 3.8). The phase behavior may be studied conveniently either with increasing overall surfactant concentration, $\gamma = \mathrm{C}/(\mathrm{A}+\mathrm{B}+\mathrm{C})$, as function of ϵ at constant oil-to-water ratio, $\alpha = \mathrm{B}/(\mathrm{A}+\mathrm{B})$, or with increasing α as function of ϵ at constant γ. In systems with DTAB as the surfactant ($\epsilon = 0$, bottom plane), a one-phase region exists on the water-rich side

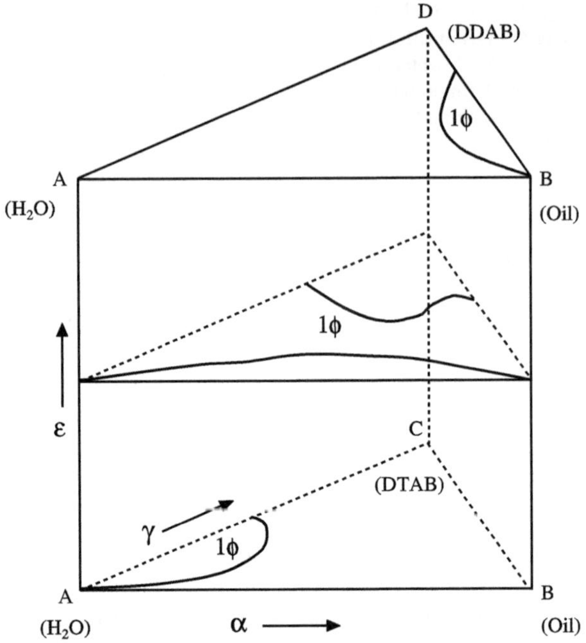

Fig. 3.8 A schematic phase prism showing the progression of one-phase (1Φ) microemulsion region as a function of mixed-surfactant composition ϵ (see text) (after [25])

and extends to fairly low oil (monomer) concentration. The top triangle is a schematic of the same system where DDAB replaces DTAB (ϵ = 100). Here, the one-phase region occurs on the oil-rich side of the triangle at relatively low surfactant and water concentrations, indicating that a one-phase channel extends from the water-rich side of the phase diagram to the oil-rich side. To determine the location of the one-phase region in ϵ and γ, a vertical section through the phase prism at $\alpha = 50$ % (i.e., equal amounts of water and oil) can be made. This two-dimensional phase map can serve as a useful guide for making a microemulsion in a particular system of water, oil (monomer), and surfactant [25].

Polymerization in Microemulsion

Polymerization reactions have been carried out in microemulsions of all types of structures. As we have noted earlier, microemulsions can be of the droplet type, either with isolated water droplets dispersed in a continuous oil phase (w/o microemulsion) that usually occur in systems with high oil content or with isolated oil droplets dispersed in a continuous water phase (o/w microemulsion), typically occurring in water-rich region. Nondroplet-type microemulsions, on the other hand, feature continuous oil and water phases intertwined in dynamic extended networks and are called bicontinuous microemulsions. A monomer can be incorporated in any of the water and oil phases of microemulsions and polymerized by normal methods. Such polymerizations can be used to obtain very small polymer particles on the order of the primary micelle size. When a hydrophobic monomer is the dispersed oil phase, the o/w microemulsion can be polymerized, typically forming a spherical stable latex composed of polymer particles as small as 15 nm in diameter. If, however, the monomer occupies the continuous phase, polymerization may produce a solid material with the dispersed phase entrapped in its matrix. On the other hand, by replacing both the oil and water phases in a microemulsion with hydrophobic and hydrophilic monomers, respectively, copolymerization may also be achieved, opening a new way to polymer synthesis with interesting possibilities. Polymerization may be carried out by any of the three methods—thermal, photochemical, and high-energy radiation methods.

Thermal Polymerization

The polymerization of hydrophobic monomers as o/w microemulsions provides an attractive way for producing polymers as latex particles of much smaller size than can be obtained by conventional emulsion polymerization. This is because polymerization occurs only in the monomer reservoir encapsulated in o/w droplets and the latter, as mentioned earlier, are usually 19–50 nm in size, compared to conventional emulsion droplets which are usually greater than 1000 nm. The first papers describing polymerization in globular microemulsions were published by Stoffer and Bone [21, 28] and by Atik and Thomas [29–31]. The former authors incorporated either methyl acrylate or methyl methacrylate in the continuous phase of w/o microemulsions stabilized by sodium dodecyl sulfate and pentanol, carried out polymerization at 50 °C using oil soluble initiator benzoyl peroxide or AIBN, and found the kinetics of polymerization to be similar to that of solution polymerization. The latter authors, however, carried out polymerization of a hydrophobic monomer in o/w microemulsion and provided the first account of a miroemulsion polymerization that produces nano-size spherical latex particles. Thus, an oil-in-water microemulsion was prepared consisting of 1.0 g of cetyl trimethyl ammonium bromide (CTAB), 1.0 g of styrene, and 0.5 g of hexanol in 50 ml of water, and polymerized by two conventional methods: (a) thermally (50 ° C) by use of 1.0×10^{-6} M AIBN and (b) radiolytically by use of a Cs γ-ray source. Both the methods yielded monodisperse latex particles, though of different sizes, the particle diameters being 350 ± 30 and 200 ± 20 Å, respectively. Figure 3.9 shows an electron microscopic picture of a styrene microemulsion polymerized by method (b). Atik and Thomas also found that the stability of microemulsions used by them was limited by the solubility of the polymer formed (see later). On the other hand, using photon correlation spectroscopy, Johnson and Guleri [32] found a good correlation between the size of latex particles formed and the droplet size in a microemulsion polymerized with the oil-soluble initiator, AIBN. However, in polymerization with a water-soluble initiator such dependence was very weak.

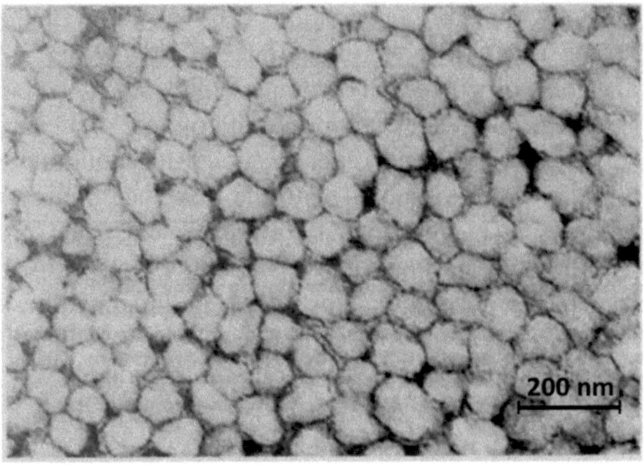

Fig. 3.9 Electron micrograph of radiation-polymerized styrene microemulsion (from [29]. With permission from American Chemical Society.)

In microemulsion polymerization, for both o/w and w/o systems, while the amount of monomer is usually restricted to 5–10 wt% with respect to the overall mass, the amount of surfactant(s) also lies within the same range or even above. The necessity of using such relatively large amounts of surfactants for microemulsion stability is a drawback of microemulsion polymerization. However, there are a few reports in which the formulation deviates from this condition and surfactant concentrations as small as less than 2 wt% are needed [33].

The main difficulty of the aforesaid polymerizations lies in retaining optical transparency and stability of microemulsions upon polymer formation. In addition to the entropic factors that contribute to the destabilization of microemulsions during polymerization, this problem arises, especially, in cases where styrene is polymerized in o/w microemulsions containing an alcohol that is a nonsolvent for the polymer. There are also other drawbacks that make alcohols an undesirable cosurfactant in microemulsion polymerization. Thus, while alcohols partition between the interfacial film and other phases, they may also modify monomer partitioning, and act as chain transfer agents and so interfere with or prevent the desired polymerization. These difficulties notwithstanding, most of the earliest studies since the pioneering work of Stoffer and Bone in 1980 [21] used alcohols as cosurfactants in the formulation of four-component o/w microemulsions and it was only in 1990 that the polymerization of a hydrophobic monomer like styrene in three-component cationic microemulsions without alcohol cosurfactant was reported when Pérez-Luna et al. [22] carried out polymerization of styrene in the one-phase microemulsion region of styrene/water/DTAB mixtures at 60 °C (see Fig. 3.6). In unpolymerized microemulsions, two apparent particle sizes were always observed: a small size of *ca.* 0.6-0.8 nm in hydrodynamic radius and a larger composition-dependent size (6–15 nm), believed to be DTAB micelles and styrene-swollen droplets, respectively. The reaction vessel was loaded with microemulsion and heated to 60 °C before addition of $K_2S_2O_8$ (1 wt% with respect to the monomer) initiator. The polymerization kinetics was followed by quasielastic light scattering (QLS), gravimetry, and dilatometry. The microlatices produced by

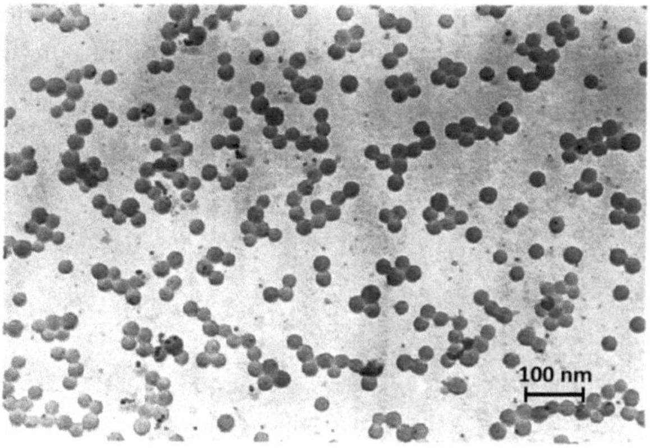

Fig. 3.10 Transmission electron micrograph of a microemulsion containing 13.8$ DTAB, 8 wt% styrene, and 78.2 wt% water after polymerization at 60 °C with $K_2S_2O_8$ initiator (1 wt% with respect to monomer) (from [22]. With permission from American Chemical Society.)

polymerization were bluish in color, remained stable with respect to coagulation for months and the particles were spherical with radii in the range 20–30 nm and apparently monodisperse when observed with TEM (Fig. 3.10).

Antonietti et al. [23] reported controlled synthesis of very fine polystyrene latices with 10 nm < R < 60 nm via polymerization in ternary o/w microemulsions formed by dispersion of styrene in water, using CTAC or DTAB as surfactants. The size of the resulting particles was controlled only by the ratio of styrene to surfactant. The limited pliability of the surfactant interface resulted in a minimum droplet size which was obtained at the limit of high surfactant concentrations.

Microgels were synthesized within the micelles of microemulsions by a cross-linking copolymerization of styrene and *m*-diisopropenylbenzene. The synthesis of microgels rather than linear polymers is advantageous in that the oil core polymerizes to *one* polymer molecule (the "microgel" particle) These microgels are easy to handle and since they retain their properties during precipitation and redissolution [34], the isolation and characterization of pure microgels allow for direct conclusions about the structure and size of the inner oil core.

Antonietti et al. [35] also compared the behavior of styrene, a nonpolar monomer, with that of polar monomers such as methyl methacrylate (MMA) in microemulsion polymerization. The comparison revealed that microemulsions of MMA are composed of micelles which are much smaller than the corresponding micelles of nonpolar monomers. No size control of the latex can be obtained for a polar monomer by altering the surfactant concentration. The behavior is explained by self-surfacting effect in which MMA itself acts as a cosurfactant.

The anionic surfactant, sodium bis (2-ethylhexyl sulfosuccinate) (AOT, represented as I) has been found to produce microemulsion without adding any cosurfactant. Thus, stable systems with cyclohexyl methacrylate monomer as the oil phase in a w/o microemulsion were obtained with AOT as surfactant [36]. Thermal polymerization of the microemulsion was carried out using AIBN as the initiator at 70 °C in a composition consisting of AOT (0.3 mol/L), water (4.5 mol/L), and AIBN (2 wt%). The microemulsions turned opaque during the course of polymerization, though no apparent phase separation occurred, and the polymerization produced solid materials which were porous in nature, with average pore size of 250 nm.

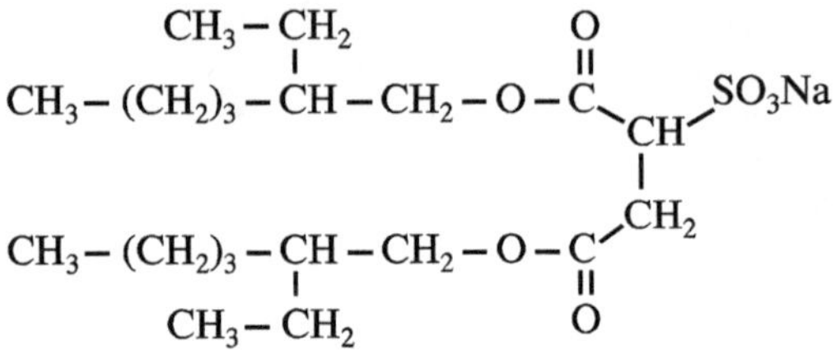

(I)

Since the physical properties of a polymer depend largely on the molecular weight and molecular weight distribution, controlling these parameters during polymerization is of utmost importance. Various polymerization methods have been used to produce rather monodisperse polymer particles. These include microemulsion polymerization with a polymerizable surfactant. Polymerizable surfactants have an advantage over nonpolymerizable surfactants since the surfactant itself becoming a part of the polymerized matrix may exert better compatibility and a strong influence over the stability and hence on the polymerization process [36]. Thus, an amphiphilic poly(ethylene oxide) (PEO) macromonomer, ω-methoxypoly(ethylene oxide)$_{40}$-undecyl-α-methacrylate (PEO-R–MA–40) having the surfactant structure as shown in **II** was synthesized.

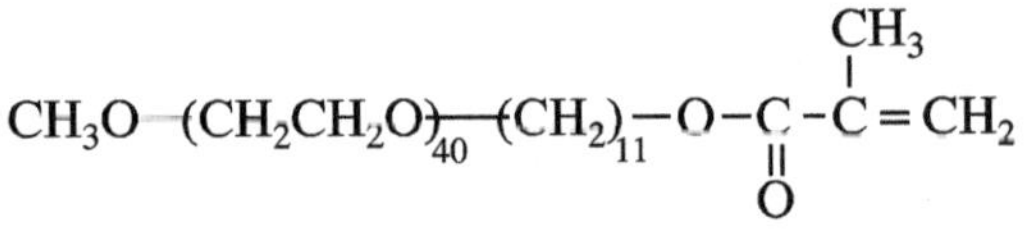

(II)

It was successfully used as a polymerizable stabilizer in the dispersion polymerization of styrene. Very stable, monodisperse polystyrene microlatexes with particle size over 200 nm were obtained. The weight ratio of polymer to polymerizable stabilizer was about 6:1.

While radical polymerization of a vinyl monomer, such as styrene, can be achieved, in principle, either in oil-in-water or water-in-oil microemulsions, the instability (turbidity) of either type of microemulsion is encountered at varying degrees of polymerization. This is related to the fact that microemulsions containing polystyrene show considerably smaller regions of stability than the ones with styrene monomer alone. While it may be viewed that the microemulsion particles cause a reduction in the number of possible conformations for the polymer (entropy effect), molecular interactions between the aromatic nucleus and polar groups of the alcohol cosurfactant also come into play. Thus, the use of butyl cellosolve (2-butoxyethanol) as a cosurfactant (instead of pentanol) has been found to improve the microemulsion stability due to its better solubility of polystyrene as compared to the pentanol system.

To demonstrate the effect of polymer solubility on the stability of microemulsion, Gan and Chew [37] studied a system containing water, MMA, poly(methyl methacrylate) (PMMA), sodium dodecyl sulfate (SDS), and acrylic acid (AA). Acrylic acid proved to be a better cosurfactant than pentanol for microemulsion stability, the difference emanating from the variation of solubility of PMMA in AA (32.4 %) and in pentanol (extremely low). Thus, a microemulsion consisting of 29.7 % AA, 19.8 % MMA, 17.5 % H_2O, and 3.3 % SDS was found to be compatible with 29.7 % PMMA of $\overline{M}_w = 81{,}800$. This clearly shows that destabilization of a microemulsion is not necessarily dictated by the entropy effect and that the solubility of the polymer in oil and cosurfactant plays an important role.

The above example highlights the important role of a polymerizable surfactant or cosurfactant in microemulsion polymerization. Gan and Chew [38] were thus able to produce transparent solid polymers by fully polymerizing 54 % MMA, 34 % AA, 10 % H_2O, and 2 % SDS or with other lower water concentrations. On further replacement of SDS by a polymerizable surfactant, sodium acrylamidoundecanoate [37], the resulting transparent solid polymers were still compatible with the similar amount of water. This prompted further studies on copolymerization in microemulsions of MMA, AA, and sodium acrylamidostearate (NaAAS), all the three components being readily polymerizable and amenable to terpolymerization in microemulsions.

NaAAS prepared by reacting acrylonitrile and oleic acid, is a saturated fatty soap with a side group of acrylamide attached to the carbon atom of stearate either at ninth or tenth position, as shown in **(III)**, and hence is polymerized readily.

$$CH_3(CH_2)_{8(7)}\text{—}\underset{\displaystyle NHCOCH{=}CH_2}{\underset{|}{CH}}(CH_2)_{7(8)}\text{—}COONa$$

(III)

The regions of solubility for the three components of the system NaAAS/H_2O/AA emanate broadly from the AA apex and converge narrowly to the water corner in a triangular phase diagram [39], while also extending quite deeply into the NaAAS region to exhibit a maximum 57 % solubility of NaAAS (Fig. 3.11a). Upon addition of MMA as the fourth component, w/o microemulsions are formed with different concentrations of MMA, showing marked reduction in water solubility (e.g., <20 % in 60 % MMA microemulsion) with increasing MMA concentration (Fig. 3.11b). In these microemulsion systems, AA performs three functions: that is, it acts as a cosurfactant, is a good solvent for the polymerizable surfactant, and also serves as a comonomer in the polymerization.

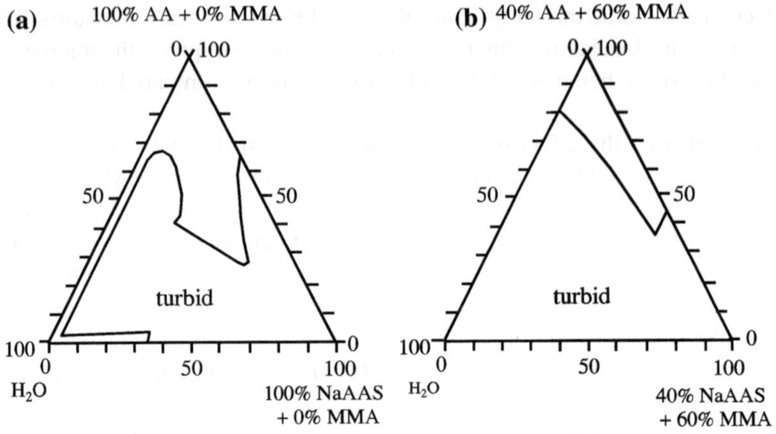

Fig. 3.11 Water-in-oil microemulsion regions (schematic) of **a** acrylic acid/sodium acrylamidostearate/H_2O and **b** methyl methacrylate/acrylic acid/sodium acrylamidostearate/H_2O at 60 °C (after [39])

The cosurfactant role of monomer in the microemulsion polymerization was also demonstrated by Leong and Candau [40], who presented the first account of inverse microemulsion polymerization of acrylamide in 1982. The inverse (w/o) microemulsion consisted of water–acrylamide mixture dispersed in toluene and stabilized by AOT surfactant without requiring the presence of any cosurfactant. It is likely that the monomer acrylamide itself served as a cosurfactant. The microemulsion, however, needed the presence of a large amount of surfactant, as can be seen from a typical composition of the microemulsion before polymerization: AOT 17.2 %, acrylamide 3.5 %, water 10.4 %, toluene 68.9 %, and AIBN/acrylamide 0.14 %. (The hydrophobic initiator AIBN was dissolved in toluene prior to the formation of microemulsion.) Photopolymerization of the inverse microemulsion was carried out leading to low-polydispersity polyacrylamide of high molecular weight ($\sim 3 \times 10^6$) confined in small (<50 nm in diameter) micellar particles. The microemulsions remained perfectly transparent and stable and no phase separation took place during the polymerization process. The polymerization was rapid and total conversion to polymer was obtained within less than 30 min. Crosslinked polyacrylamide latices, or microgels, were also prepared by using a 100:1 mixture of acrylamide-methylenebis(acrylamide).

Candau et al. [41] later studied the polymerization of AOT-stabilized acrylamide inverse microemulsion by a thermal process using either oil soluble AIBN or water soluble $K_2S_2O_8$ as the initiator and found the rate of polymerization to be first order with respect to initial monomer concentration in the presence of AIBN but 1.5 order in the presence of $K_2S_2O_8$. An inverse relationship was found between polymer molecular weight and the surfactant concentration which suggested participation of the surfactant in the initiation reaction. This was further confirmed by the observed independence of the polymer molecular weight on the concentration of the initiator.

The inverse microemulsion polymerization process appeared to combine high rate with high molecular weights (up to 10^7) and displayed a novel feature that each final latex particle consisted of one single polymer molecule in a collapsed state, suggesting that the kinetics of the reaction did not follow the Smith-Ewart theory but were characterized by continuous particle nucleation.

The polymerization of water-soluble monomers and more particularly acrylamide (AM) in w/o microemulsions has been investigated in detail [42–50]. This is because of the numerous applications of polyacrylamide. The water soluble monomers like acrylic acid and AM remain within the aqueous core of w/o microemulsions and, as we have noted earlier, these monomers can act as cosurfactants by locating partially at the w/o interface between the surfactant molecules, which leads to a considerable extension of microemulsion domains in the phase diagram. In the case of AOT/water(AM)/toluene microemulsion systems which have been studied most widely, experimental observations based on light scattering, small angle neutron scattering, and viscometry [40, 51, 52] have indicated that the interfacial localization of acrylamide molecules induces attractive interactions between the droplets. Upon further addition of acrylamide, i.e., upon increasing AM/H_2O ratio, the interaction potential becomes so attractive that transient clusters form. This has an effect on the formation of polymer latex particles and the polymerization mechanism.

To understand the process of particle nucleation in w/o microemulsion polymerization initiated in various ways (thermal, photochemical, or γ-irradiation), a thorough experimental investigation was made of the structures prior to and after polymerization of acrylamide inside water-swollen micelles stabilized by AOT. Elastic and quasielastic light scattering (QELS), viscometry, and ultracentrifugation methods were used [40, 51, 52], which yielded two very significant results. One was that the particle size of the final microlatex ($d \approx$ 20–40 nm) was much larger than that of the initial monomer-swollen droplets, which led to the conclusion that the final number of polymer latex particles was about 2 or 3 orders of magnitude smaller than the initial number of monomer droplets. The second notable result was that each latex particle contained, on average, only one polymer chain.

On the basis of the above results, it was postulated that particle nucleation took place continuously throughout the polymerization process, in sharp contrast with conventional emulsion polymerization where particle nucleation takes place only in the initial period (interval I). The results of TEM experiments performed on polyacrylamide samples taken at different extents of conversion [53] also supported the theory of continuous particle nucleation since the number of polymer particles was found to increase proportionally with conversion while the size remained approximately constant.

To account for the experimental finding, as noted above, it was proposed that only a small fraction of monomer-swollen droplets initially present in w/o microemulsion are nucleated. The nucleated particles grow as monomer is supplied from the non-nucleated monomer droplets either by diffusion through the continuous phase or by "sticky" collisions between droplets, as shown in Fig. 3.12. Because of the growth process and the presence of large amount of surfactant in the initial

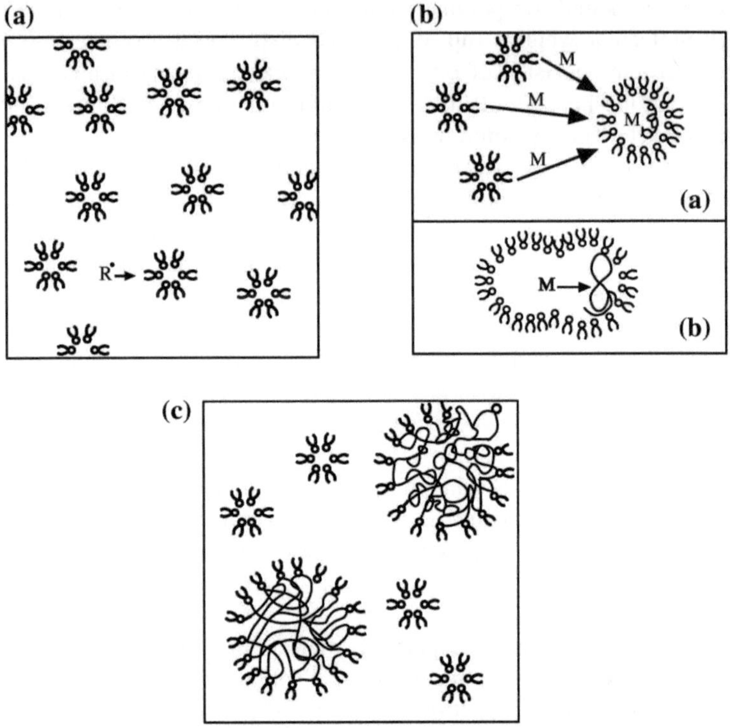

Fig. 3.12 Polymerization of acrylamide monomer (M) in AOT-stabilized w/o microemulsion: **a** micelles and water-swollen AOT droplets ($d \approx 6$ nm) before polymerization; **b** growth of polymer particle—(*a*) by monomer (M) diffusion through toluene phase and (*b*) by collision between particles; **c** polymer particles ($d \approx 40$ nm) and small micelles ($d \approx 3$ nm) (after [68])

formulation of microemulsions, small micelles are always present in the reaction mixture providing greater overall interfacial area compared to that of nucleated polymer particles. These micelles therefore capture the primary radicals preferentially and the polymer latex particles receive radicals only one each, on the average.

In the above studies, it was the monomer acrylamide in dispersed water pools that was polymerized producing latex particles, while the continuous oil phase remained intact. In contrast to these studies, Menger and Tsuno [54] used a hydrophobic monomer, styrene, as the continuous oil phase, while retaining AOT as the surfactant to stabilize the water microdoplets ("pools") in the monomer. Polymerization was carried out by irradiation for 10 h using AIBN as the initiator. This resulted in the formation of solid polymer materials containing pores that were larger (by one order of magnitude) than the radii of the original water droplets in the microemulsions. This was attributed to the collision of water droplets during the process of polymerization, leading to increase in their size as they became entrapped in the polymer matrix. Menger et al. [55] also reported on (a) the preparation of w/o systems where styrene and crosslinker divinyl benzene comprised the apolar continuous phase and

(b) the subsequent conversion of these microemulsions (stabilized by AOT) into solids by photopolymerization of optically clear mixtures by irradiation for 10 h in the presence of benzoyl peroxide initiator. This resulted in the formation of a sponge-like polymer, which was opaque, though the reacting systems were optically clear mixtures and there was no apparent phase separation.

Guo et al. [56, 57] carried out a detailed study on the polymerization of styrene in the dispersed oil phase of o/w microemulsion prepared from water/SDS/1-pentanol/styrene with water forming the continuous phase. Water soluble potassium persulfate or oil-soluble 2,2′ azobis-(2-methyl butyronitrile) (AMBN) were used as initiator at 70 °C. The polymerization gave stable latexes which were bluish and less translucent than the original microemulsions. The effects of initiator concentration, polymerization temperature, and monomer concentration on the kinetics and distributions of particle size and molecular weights were investigated. In all cases, the polymerization rate showed only two intervals: increase to a maximum and then decrease. It was proposed that the particles were nucleated by the capture of radicals from the aqueous phase for both water-soluble and oil-soluble initiators (due to solubility of AMBN in 1-pentanol and slight solubility of 1-pentanol in water), while the microemulsion droplets which did not capture radicals served as reservoirs to supply monomer to the polymer particles. The maximum polymerization rate was reached at relatively high conversion (20–25 %) as compared to conventional emulsion polymerization (2–15 %). There was no apparent constant rate period and no gel effect. This is in contrast to a conventional emulsion polymerization process, which broadly exhibits three stages of the polymerization process [58], namely, Stage I (continuous rate increase accompanied by simultaneous particle nucleation and growth), Stage II (constant overall rate of polymerization with no new particle nucleation and with constant rate of growth), and Stage III (nonlinear growth with rate decrease due to dwindling monomer concentration or rate increase due to gel effect).

Figure 3.13 presents a simplified comparison between a conventional o/w emulsion polymerization and the o/w microemulsion polymerization of a hydrophobic monomer with a water-soluble initiator, such as the one reported by Guo et al. [56, 57]. In the former case, during Stage I, the monomer is located in the following four sites: (1) in monomer droplets, (2) in inactive micelles containing monomer, (3) in active micelles that become monomer/polymer (M/P) particles where polymerization occurs, and (4) in aqueous phase as solute monomer; in Stage II: (1) no inactive micelles containing monomer, (2) free radicals enter only the M/P particles where polymerization takes place; (3) M/P particles are supplied with monomer from the emulsified monomer droplets; in Stage III: (1) no monomer droplets, (2) polymerization takes place in M/P particles with no supply of monomer via aqueous phase.

In comparison, the o/w microemulsion polymerization reported by Guo et al. [56, 57] can be represented by a simplified scheme with only two stages, as shown in Fig. 3.13b. In Stage I: particle nucleation occurs in microemulsion droplets, the fraction of droplets initiated and converted into polymer particles being determined

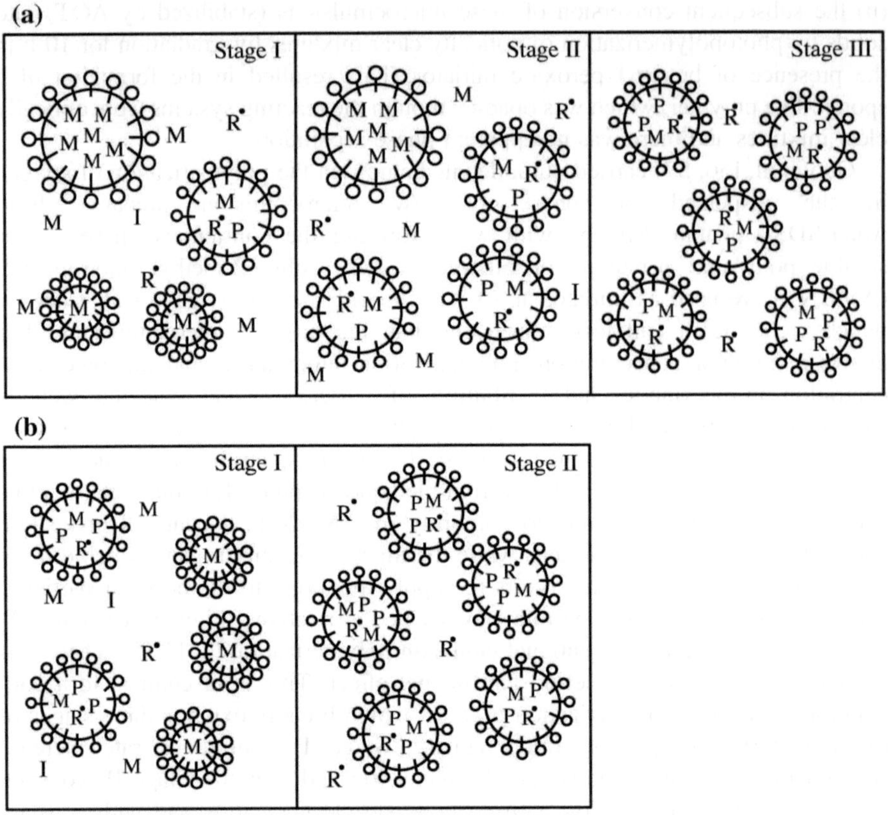

Fig. 3.13 A simplified scheme showing comparison between **a** conventional emulsion polymerization and **b** microemulsion polymerization

by the level of the initiator. The polymerization rate increases due to the increasing number of polymerizing loci with time and monomer being present in three locations, viz., (1) monomer-swollen inactive micelles, (2) active micelles, and (3) aqueous phase (solute monomer). In Stage II: (1) No monomer-swollen inactive micelles exist, having been converted into polymer particles or depleted of monomer due to diffusional migration into polymer particles; (2) polymerization occurs only in the monomer reservoir encapsulated in the particles and the rate decreases as the monomer gets depleted. No gel effect is observed, the probable reasons for which are: (i) very small size of latex particles which lead to immediate termination when the second radical enters; (ii) approximate similarity of solubility parameter for 1-pentanol [10.9 $(cal/cm^3)^{1/2}$] and polystyrene (9.0) [56, 57].

The microemulsion polymerization of styrene in o/w system with SDS surfactant and 1-pentanol cosurfactant using water soluble $K_2S_2O_8$ or oil-soluble AMBN initiator at 70 °C yielded stable latex of small size (20–30 nm) and high molecular weight (1.2×10^5) which implied that each latex particle consisted of 2 or 3 polystyrene molecules [56, 57]. The maximum polymerization rate and number of particles varied with the 0.47 and 0.40 powers of $K_2S_2O_8$ concentration and 0.39 and 0.38 powers of AMBN, respectively, in agreement with the 0.4 power predicted by Smith-Ewart Theory, Case II. This consistency was attributed to the comparable size of microemulsion droplets and micelles.

Photochemical Polymerization

Microemulsions, unlike emulsions, are transparent and hence amenable to photochemical method of polymerization. The process has an advantage over classical chemical initiation in that a very quick rate of polymerization can be achieved and, moreover, the molecular weight of polymer could be controlled easily by adjusting the intensity of light and the time of exposure, often in the presence of a photosensitizer. Graetzel et al. [59] studied the photoinduced photopolymerization of a number of monomers in cetyl trimethyl ammonium persulfate $(CTA)_2S_2O_8$ containing o/w microemulsions. The monomers employed comprised styrene, divinylbenzene (DVB), acrylamide, methyl methacrylate, and acrolein. Translucent o/w microemulsion systems capable of being polymerized were first produced, employing the monomers and initiators used in the polymerization as integral components of the microemulsions themselves, a typical wt% composition for styrene (50 %)/DVB (50 %) monomer mixture thus being: H_2O (78.3 %), 1-pentanol (9.7 %), $(CTA)_2S_2O_8$ (0.2 %), cetyl trimethyl ammonium chloride (5.5 %), hexadecane (2.5 %), styrene (1.7 %), and DVB (1.7 %). Highly efficient polymerization was found to occur under visible light by using $Ru(bpy)_3^{2+}$ or eosin Y as a sensitizer to form either polymer powders which precipitated or a gelatinous solid, as in the case of (solubilized) acrylamide. The authors found that the rate of polymerization was similar to that in the radical initiated emulsion polymerization.

Kuo et al. [60] reported photoinitiated polymerization of styrene in o/w microemulsion with SDS as surfactant, pentanol as cosurfactant, and dibenzyl ketone (DBK) as initiator. A typical o/w microemulsion recipe used in the study consisted of SDS 0.67 g, toluene 0.5 mL, styrene 0.5 mL, pentanol 0.8 mL, and water 10 mL. The DBK concentration was 0.05–2 mM based on water content and the irradiation was carried out by UV light at 313 $\pm$ 10 nm. About 80 % monomer conversion was attained with 2.0 mM DBK concentration and 100 % light transmission. The molecular weights of the polymer produced were in the order of 10^5 with the polydispersity indexes (PDIs) in the range 1.6–2.2, while the latex particle sizes were in the range 30–60 nm with narrower PDIs of 1.5–1.06. The authors found that the polymerization proceeds with the following dependence of the degree of polymerization (D_p) and the rate of polymerization (R_p) on oil-soluble DBK concentration and light intensity (L): $D_p \propto [DBK]^{-0.4} L^{-0.2}$ and $R_p \propto [DBK]^{0.2} L^{0.2}$. The molecular weight of polystyrene varies inversely with the photoinitiator concentration. As the DBK concentration increases, the number of polymer chains initiated also increases, enhancing the probability for radicals to terminate the propagating polymer chains and thereby decreasing the polymer molecular weight.

The most interesting feature of the above polymerization is that the microemulsion remained transparent during the whole polymerization process. The existence of toluene in the oil phase played a significant role in maintaining the transparency of the microemulsion. Without toluene the microemulsion became cloudy after irradiation of only about 5 min and after 2 h of irradiation the conversion was below 10 %.

The photochemical UV radiation method was first employed by Leong and Candau [41] for the radical polymerization of acrylamide in inverse microemulsions stabilized by Aerosol OT. The polymerization was carried out using AIBN initiator and induced by UV irradiation. It was shown that the use of a microemulsion rather than an emulsion led to stable and clear microlatices ($d \approx 50$ nm) of uniform size, thus providing a way to overcome some of the problems of conventional inverse emulsion polymerization, such as instability of the latexes resulting in rapid flocculation and a broad particle size distribution.

Unlike in conventional emulsion polymerization, where, as noted earlier, particles are nucleated only in the initial stages of polymerization, the polymerization of acrylamide in microemulsion involves continuous particle nucleation during the whole course of the polymerization because of the much higher concentration of the surfactant. This feature of continuous particle nucleation, as well as the finding that each latex particle contains, on the average, only one polymer molecule signify that the kinetics of the inverse microemulsion polymerization are markedly different from those of emulsion polymerization and are rather similar to those of solution polymerization, in so far as the nucleation in solution polymerization is continuous throughout the course of polymerization and the polymer particles generated are single molecules [44]. On the basis of this simplification, homogeneous kinetics may be applied to write the rate of polymerization as [58]

$$R_p = k_p[\mathrm{M}]\left(\frac{\Phi I_a}{k_t}\right)^{1/2} \tag{3.1}$$

for biradical termination, and

$$R_p = k_p[\mathrm{M}]\left(\frac{2\Phi I_a}{k_t'[\mathrm{T}]}\right) \tag{3.2}$$

for monoradical termination, where k_p is the propagation rate constant, k_t and k_t' are rate constants for biradical and monoradical termination, respectively. [T] is the terminating agent concentration, and $2\Phi I_a$ is the rate of photoinitiation where Φ is the quantum yield and I_a is the absorbed light intensity. The monoradical termination may be attributed to a degradative transfer to some terminating agent present in the solution, which could be one of the solution components, including the monomer, the surfactant, or the solvent.

A drawback of the polymerization in AOT-stabilized microemulsions is that the concentration of polyacrylamide in the obtained latex particles is much lower than that resulting from classical emulsion polymerization. Since high solid contents are usually desirable in most industrial applications, microemulsions made with other surfactants were investigated. Thus, Candau et al. [61] prepared water-soluble copolymers of acrylamide and sodium acrylate of variable compositions by radical copolymerization in microemulsions stabilized by a nonionic emulsifier blend and dispersed in an isoparaffinic oil—a narrow-cut isoparaffinic mixture, Isopar M, having boiling range 207–257 °C. The emulsifier was a blend of sesquioleate solution (Arlacal 83, HLB = 3.7) and polyoxyethylene sorbitol hexaoleate with 40 ethylene oxide residues (G-1086, HLB = 10.2). With both the emulsifiers, Arlacal 83 and G-1086, containing identical and large hydrocarbon tails (oleate chains), the molecules of the two emulsifiers associate at the water/oil interface to form a more condensed interfacial film than each surfactant taken separately. On the other hand, since the chemical structure and the solubility parameter of the isoparaffinic oil Isopar M are close to those of the aliphatic tails of the surfactants, it allows maximization of the steric stabilization of the systems, by increasing the cohesive forces (London forces) between the lipophilic parts of the emulsifiers and the organic solvent [6]. Interestingly, the polymerization carried out by UV irradiation at 20 °C proceeded with the appearance of turbidity and a notable increase in the viscosity of the medium. At the end (with high conversions to polymer in a few minutes), the system became clear and fluid again. The authors proposed that microemulsions with high monomer contents (~25 %) exhibit a bicontinuous character and polymerization in these systems produces a transformation of the random disordered structure toward a concentrated dispersion of spherical latex particles. The inverse latexes were highly stable and showed no settling over months. The dimensions of the particles were rather low (d ~ 60 nm) with a narrow size distribution.

Fouassier et al. [62], on the other hand, studied photopolymerization of acrylamide in AOT-stabilized w/o inverse microemulsion using decane as the continuous oil phase. The polymerization was carried out in the presence of benzophenone (BP) and its water-soluble derivatives BP^- and BP^+ (**IV**) as photoinitiators

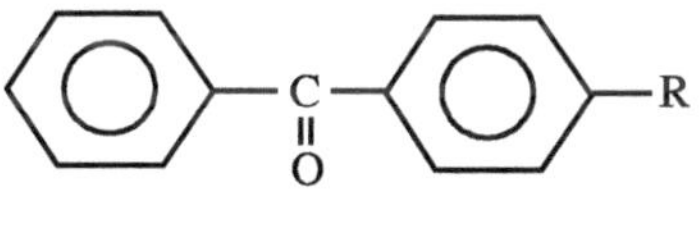

R = H	BP
R = CH_2SO_3Na	BP^-
R = $CH_2N(CH_3)_3Cl$	BP^+

(IV)

After dissolving appropriate amounts of AOT in decane and adding a precise volume of 50 % (w/w) acrylamide/water solution, more water necessary to dissolve BP^+ and BP^- was added and microemulsions were then obtained by adding cyclohexanol. All the three photoinitiators were found to be located at the o/w interface. The rate of polymerization of acrylamide in the AOT-decane-water system was found to be proportional to the square root of the incident light intensity. The rate was also found to increase with monomer concentration according to the relationship $R_p \propto [\mathrm{M}]^{\alpha}$ in which $\alpha = 1.3$ for BP^{+} and 1.4 for BP, while for BP^{-} α was very close to unity. With both BP^{+} and BP^{-}, the rates were very much higher ($\sim$100 times) than those obtained in homogeneous solution. A kinetic scheme was postulated where the initiation in the micelles takes place mainly through the quenching of the excited states of the initiators by the oil phase decane. That the rate of polymerization in inverse micelles was found to be higher at greater concentration of decane for BP and BP^{-} is in agreement with this postulate. Furthermore, when decane was replaced by toluene, the rate of polymerization decreased which substantiated the theory of initiation by decane-derived radicals.

The kinetics of polymerization of acrylamide in w/o inverse microemulsion has been found to be greatly influenced by the nature of the solvent (oil phase). Thus, for inverse microemulsion polymerization in three different systems (e.g., with solvents toluene, heptane, and benzene), the dependence of the rate of polymerization (R_p) on incident light intensity ($R_p \propto I^x$, where x is the light intensity exponent) was found to be x = 1.06, 0.73, and 0.55, respectively. The increase in x in the order benzene < heptane < toluene corresponds to the increasing susceptibility of the hydrogen atoms in the solvent molecules to abstraction by the free radicals. In toluene, hydrogen abstraction occurs readily and results in the formation of fairly stable benzyl radical, which is unlikely to be involved in further polymerization. The monoradical termination that was found previously for acrylamide polymerization in toluene systems is therefore likely due to the degradative transfer to toluene and no degradative chain transfer to monomer possibly occurs. Benzene, on the other hand, has no labile atom and hence no abstraction can take place, thus allowing biradical chain termination to occur. Heptane then represents an intermediate case. Comparison of R_P for acrylamide in benzene and toluene inverse microemulsions have shown that for a given initiator concentration and similar compositions, the polymerization in benzene is faster than that in toluene. This can also be attributed to the degradative scavenging of primary and polymer radicals by toluene.

In the aforesaid studies of inverse microemulsion polymerization, it is the water-soluble monomer in the dispersed water phase that is polymerized, while the oil phase remains intact resulting in latex particles. In contrast to these systems, Menger et al. [55] prepared solid polymers containing embedded water pools by replacing the oil phase by a polymerizable hydrophobic monomer and using AOT as the surfactant. AOT is so effective that one can readily dissolve 10–20 % water in hydrophobic octane to give optically clear solutions, the size of the resulting water pools depending on the $[H_2O]/[AOT]$ ratio (R). For example, hydrodynamic radii of 3.6 and 14.5 nm in octane are obtained with R values of 11 and 56, respectively. Using styrene/divinylbenzene (6:4 v/v) in place of octane as the apolar continuous phase, polymerization was carried out by irradiating the optically clear mixture for 10 h with 2 % benzoyl peroxide initiator. The reverse micellar systems were thus rigidified with no apparent phase separation. The solid polymers thus produced can be used as such or ground into porous powder. The pores in such polymers were found to be 1 order of magnitude larger than the radii of the water pools in the monomer phase. This was a clear indication of the propensity for pools to assemble as they become encased in polystyrene.

High-Energy Radiation Polymerization

High-energy radiation can be used to carry out polymerization in microemulsion systems. It has the advantages of simplicity of handling and requiring no initiators for initiation, unlike UV initiation, thereby avoiding the problem of destabilization of microemulsions which sometimes arises from the presence of organic initiators. Atik and Thomas [29] were the first to report high-energy radiation induced polymerization of microemulsion to spherical latex particles. An o/w microemulsion consisting of cetyl trimethyl ammonium bromide (CTAB) surfactant, styrene monomer (oil), and hexanol (cosurfactant) in water (1.0 g CTAB, 1.0 g styrene, 0.5 g hexanol, 50 mL water) was polymerized radiolytically by the use of a Cs γ-ray source, besides performing conventional thermal polymerization with AIBN initiator at 60 °C. These two different polymerization techniques afforded monodisperse latex particles of diameters 20 and 35 nm, respectively, as measured by electron photomicrograph. Lianos [62] also used γ-radiation for the polymerization of o/w microemulsion containing water, sodium dodecyl sulfate, 1-pentanol, and styrene in various proportions. The interaction between the polymerized microemulsion and pyrene was studied by fluorescence probe to find the solubilization site for pyrene.

During polymerization of microemulsion, transformation occurs with the passage of time from water/surfactant/monomer into water/surfactant/monomer/polymer and finally to water/surfactant/polymer at the end. Thus, there arises a situation of destabilization due to conflict between the microemulsion droplets and the conformational arrangement of polymer chains. Phase separation would therefore be expected during the course of polymerization of microemulsion. The system may become turbid or milky as the polymerization proceeds and may even result in the appearance of two or three different phases of liquids and solid polymer [61]. Several studies have been reported on the problem of solubility of polymers in microemulsions. The presence of nonpolar polymers has been found to have a drastic effect on the stability of microemulsions, even in miniscule concentrations [63]. Thus, there occurs a pronounced reduction of the w/o microemulsion solubility area when polystyrene of average molecular weight 22,000 is added in a concentration of only 0.1 % based on the styrene content, the solubility area being reduced by 50 % [63]. While this drastic influence of nonpolar polymers is intuitively referred to as a space restriction conformational problem since the presence of solid particles is viewed as causing a reduction in the number of conformation possibilities for the polymer and the corresponding reduced entropy should give a higher free energy and destabilization, Gan et al. [63] have shown that this is not the only mechanism of destabilization. They have proved this fact by using oligomers of styrene to determine the influence of molecular weight and the results showing an influence on the solubility even of the smallest oligomer (e.g., the styrene dimer), which can have no space restriction effect. Apart from the conformational entropy component of the polymer free energy, molecular interactions between the aromatic nucleus and polar groups also plays an important role in determining the stability of microemulsion, as it has been shown that this interaction lowers the water concentration in w/o microemulsion [34, 64].

With regard to the behavior of polar polymers in microemulsions, it has been reported that polymerization in methyl methacrylate microemulsions with non-polymerizable surfactants and cosurfactants [28] leads to phase separation of the system in spite of the fact that the poly(methyl methacrylate) is soluble in its monomer. On the other hand, polar polymers such as amphiphilic block copolymers, e.g., poly(2-vinyl pyridine–*b*–ethylene oxide), have been used as cosurfactants to stabilize microemulsions [65, 66] and a specific mechanism has been proposed for polar polymers as stabilizers for emulsions and microemulsions [67].

To determine the influence of polymer solubility on the stability of microemulsions for polar polymers, Gan and Chew [38] carried out studies on a w/o microemulsion system containing water, methyl methacrylate (MMA), PMMA, SDS, and acrylic acid (AA) and found that a microemulsion consisting of 29.7 % AA, 19.8 % MMA, 17.5 % H_2O, and 3.3 % SDS is compatible with 29.7 % PMMA ($\bar{M}_w$ = 81,800). They were also able to produce transparent solid polymers by fully polymerizing a microemulsion containing 54 % MMA, 34 % AA, 10 % H_2O, and 2 % SDS [38].

Microemulsion-Based Polymers: Characteristics and Applications

The characteristics of the final products of microemulsion polymerization depend critically on the composition of the microemulsion system prior to polymerization. Both latex and solid material may be produced depending upon the composition. Polymerization of a microemulsion containing a monomer (or monomers) in the dispersed phase yields microlatex containing the polymer, whereas a solid material is obtained by the polymerization of a microemulsion in which a monomer or monomers constitute the continuous phase. Microlatexes obtained from o/w microemulsions are often used in the same form to utilize their properties, while w/o systems are commonly used to produce water-soluble polymers of high molecular weight. As a general rule, the higher the surfactant/monomer ratio in the initial formulation, the smaller the particle size of the final latex product. Therefore, both high solid contents and small size particles can hardly be achieved simultaneously. Moreover, the size of microlatex particles (usually determined by quasielastic light scattering, QELS, and transmission electron microscopy techniques) significantly exceeds that of the precursor microemulsion droplets. The latex particles are typically 20–60 nm compared to 4–5 nm diameter globular (o/w or w/o) microemulsions. However, the final latex particles are still bigger (around 50–150 nm) if the polymerizing microemulsions are bicontinuous, simply due to a larger monomer content of the latter. As a general rule, the particle size increases on increasing the monomer content or decreasing the surfactant content and/or the initiator concentration [68].

The latexes formed by microemulsion polymerization of monomers in the dispersed state can be precipitated in a large excess of non-solvent and dried under vacuum. The molecular weights of the polymers thus obtained are high, usually ranging 10^6–10^7 and, in some cases, exceptionally high molecular weights (M_w 2.5–3.3 $\times$ 10^7) have been reported [69, 70]. However, when alcohols are used as cosurfactants in the formulation, chain transfer reactions can occur which reduce the molecular weight. As a few polymer chains of high molecular weight are confined in the microlatex particle, these must be highly collapsed in order to fill the nano-size space of the particle ($d \sim 40$ nm).

One of the main objectives of using microemulsions as reaction media for polymerization is to utilize their microstructures as templates to produce polymers with similar characteristics. For example, polymerization of a large amount of hydrophobic monomers in the continuous phase of w/o microemulsions could lead to solid polymers containing the preexisting aqueous disperse phase in a swiss-cheese like formation. This would permit inclusion of materials in the disperse phase that would otherwise be insoluble in the polymer, e.g., colloidal particles of metals as catalysts. In the case of bicontinuous microemulsions, however, both hydrophobic and hydrophilic monomers can be incorporated. The morphology of the final product would then depend on the microemulsion composition and on the nature of the incorporated monomers.

Solid porous materials can be prepared by polymerization of all three types (o/w, w/o, and bicontinuous middle phase) of microemulsions [71]. Gupta and Singh [34] obtained porous polymers by polymerization of styrene/divinylbenzene as the continuous phase in an oil-continuous (w/o) microemulsion. Which was prepared using AOT as surfactant. The polymerization was carried out thermally at 70 °C using benzoyl peroxide as initiator. The porous materials may eventually be transformed into porous membranes. Such membranes have many applications in the field of separation science, composites, medicines, and biotechnology. Depending on the pore size, the porous polymers could be used for the separation of dust as also microparticles such as virus, bacteria, pigments, colloidal particles, etc.

The difference in relative permeability of gases through membranes offers an attractive method of separation and enrichment of gases. The above types of porous membranes can be used for this purpose. For example, N_2 shows much higher permeability than O_2 through porous polystyrene membrane which therefore has potential for application in the production of oxygen-enriched air. Porous polymers can also be used for the extraction of rare earth metals from weak solutions such as processing streams. For example, a ligand-bearing hydrophilic monomer can be polymerized in the porous polymer matrix to produce a structure where the porous polymer acts as a scaffold and the inner wall of pores is coated with a polymer gel capable of binding metal ions. By allowing the industrial effluent to pass through the porous structure, the ligands pick up metal ions from solution, thereby providing an attractive alternative to the presently used tedious separation processes [34].

The porosity of solid polystyrene produced by polymerization in a middle-phase (bicontinuous) microemulsion is greater than that obtained by polymerization in either water-continuous or oil-continuous microemulsion. The first account of a middle-phase microemulsion-based porous polymer was reported by Haque and Qutubuddin in 1988 [71]. The microemulsions were formulated with styrene, water, sodium dodecyl sulfate (SDS), and 2-pentanol or butyl cellosolve as the cosolvent. (Since butyl cellosolve has greater solubility than 2-pentanol in polystyrene, it increases the stability of SDS microemulsion.) Figure 3.14 shows the structure of polystyrene when obtained from middle-phase microemulsion polymerization at 60 °C for 36 h, the composition (wt%) before polymerization being: SDS 10 %, 2-pentanol 25 %, styrene 40 %, and water 25 %. The polymerized structure shows pores in both micron and submicron ranges. The observed greater porosity of this solid compared to the solids obtained from polymerization of oil-continuous microemulsion (SDS 10 %, 2-pentanol 25 %, styrene 55 %, water 10 %) and water-continuous microemulsion (SDS 10 %, 2-pentanol 25 %, styrene 5 %, water 60 %) is apparently related to the fact that middle-phase microemulsions contain interconnected domains of both water-continuous and oil-continuous regions.

The pore morphology of microemulsion-based solid polymer depends on the initial microstructure of the microemulsion as determined by the type of surfactant and cosurfactant in addition to composition. Some examples of nonionic surfactants used for microemulsions are: Emsorb 6916 (sorbitan monolaurate), Neodol 91-5 (primary alcohol ethoxylate), and Trycol (polyoxyethylene alkylphenol), while sodium dodecyl sulfonate is commonly used as the anionic surfactant. Figure 3.15

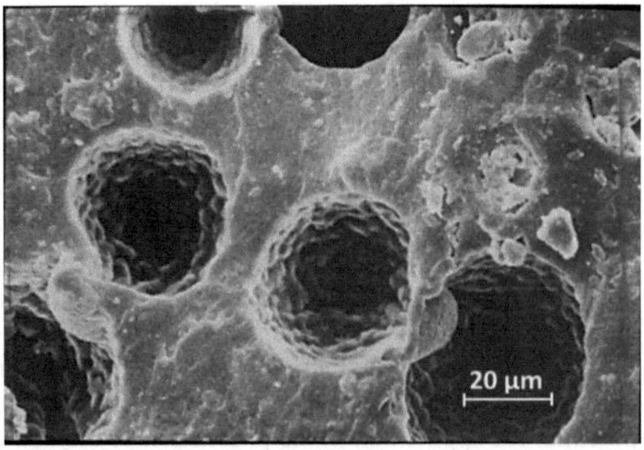

Fig. 3.14 Electron micrograph of styrene polymerized in middle-phase microemulsion containing anionic surfactant SDS (from [71]. With permission from John Wiley & Sons, Inc.)

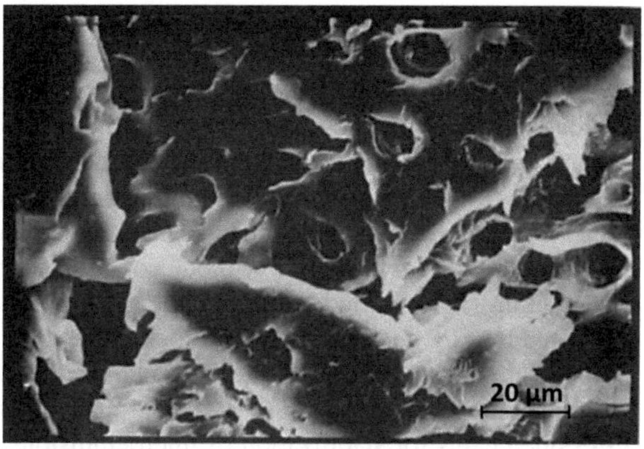

Fig. 3.15 Electron micrograph of styrene polymerized in middle-phase microemulsion containing nonionic surfactant Neodol 91-5 (from [71]. With permission from John Wiley & Sons, Inc.)

shows the structure of polystyrene obtained from nonionic surfactant (Neodol 91-5)-based middle-phase microemulsion [71]. The structure has different morphology as compared to the anionic system (Fig. 3.14). Polymers obtained from microemulsions based on anionic surfactant are found to be ionically conductive, whereas nonionic surfactant-based polymers are nonconductive. Anionic surfactant-based polymers also exhibit a higher glass transition temperature.

A marked dependence of pore size of the solid polymer on the composition of the microemulsion also emerged from the studies of Menger et al. [55], who showed that the pore size in the material is highly dependent on the water/surfactant ratio in the microemulsion and further that the size of pores in the solid material is always larger than that of the original water droplets in microemulsion. This latter aspect has already been dealt with earlier.

The nature of surfactant used in microemulsion has a pronounced effect on the above mentioned properties of polymer products. Thus, if the surfactant has any polar or electrostatic interaction with the polymer, this may lead to an improvement in its tensile properties. However, if the surfactant is not compatible with the polymerized matrix, it may simply act as a low-molecular additive (plasticizer) and reduce the mechanical strength of the material. For polystyrene, it has been observed that the polymer produced from a nonionic surfactant-based microemulsion is ductile while that obtained from anionic surfactant (e.g., SDS)-based microemulsion is brittle in nature and has much higher glass transition temperature than the bulk-polymerized polymer.

Microemulsion polymerization could provide materials with many interesting properties since various possibilities exist for choosing the composition of polymerizable microemulsions and also the method of polymerization. A significant development in microemulsion polymerization has thus been the use of block copolymers to stabilize the system. If an AB type diblock copolymer contains water soluble A block (vvvvvvAAAAvvvvvv) and oil-soluble B block (vvvvvvBBBBvvvvvv), it may have emulsifying properties similar to low-molecular weight surfactants. Such block polymers may thus be referred to as "macrosurfactants." Their surfactant properties are controlled by the size of the block as it determines the overall hydrophilic/hydrophobic balance. Riess et al. [72] were the first to show that poly(styrene–*b*–ethylene oxide) block copolymers acting as surfactants can lead in the presence of cosurfactants (such as isopropanol and butylamine) to the formation of microemulsions. The research group of Riess [73, 74] further developed a series of block copolymers containing polystyrene (PS), polyoxyethylene (POE), and poly(dimethylsiloxane) (PDMS) blocks and showed that microemulsions could be stabilized by using these block copolymers in the presence of cosurfactants. Leong et al. [75] reported the polymerization of acrylamide in w/o microemulsion stabilized by a PS and POE based triblock copolymer in the presence of isopropanol. Polyacrylamide microlatexes with molecular weights in the range of 50,000–1,30,000 were obtained. Candau et al. [76] studied graft copolymers having a PS backbone and PEO side chains for stabilization of microemulsions. In addition, Maria and Gallot [77] reported microemulsions formed with polystyrene–*b*–poly(vinyl-2-pyridinium chloride) AB block copolymers, i.e., where the B part of the chain has a polyelectrolyte nature.

Percolation is an important phenomenon in microemulsion polymerization and can influence the structure and properties of the final product. It has been studied in detail in different systems. A w/o globular microemulsion having electrolytic surfactants can display electrical conductivity due to the charge exchange between nano-size droplets. When the interaction potential between the droplets is sufficiently attractive, a significant number of dimers and higher order clusters could form. At a percolating threshold where an infinite cluster of these droplets is formed, the electrical conductivity can increase by several orders of magnitude. For multicomponent microemulsions, conduction is dependent upon the nature of the interfacial zone as well as the total volume fraction of particles. In a w/o microemulsion, initially the droplets are isolated from each other by the insulating oil phase, but above the threshold value the interfaces are in contact with each other to form water channels between the droplets, allowing the current to flow. This phenomenon was observed in toluene/AOT/(acrylamide + H_2O) water-in-oil microemulsion [78]. It was found that the addition of acrylamide monomer to toluene/AOT/H_2O microemulsion produces a sharp rise in conductivity with increasing acrylamide concentration centered around an inflection point. It was attributed to a percolation process. As mentioned earlier, acrylamide functions as a cosurfactant and is located at the interface. It also performs two other functions: (i) it enhances the attraction between the droplets and (ii) it increases the flexibility of the interface, thereby allowing the micelles to open and form transient conducting water channels from particle to particle. Therefore, polymerization of a percolating microemulsion may give rise to a channelized structure in the polymerized product.

Porous polymeric membranes derived from microemulsion polymerization have found significant use in the field of bio- and medical technology. Thus, in a number of biochemical operations, these porous membranes have been found to perform the function of a porous microcarrier for culturing living cells and enzymes. The membranes have the advantage that they are able to support a higher cell density per unit bed volume.

Polymerization in microemulsions containing "water pools" (water micro-droplets) is the basis of a simple strategy to synthesize hydrophobic polymers (such as polystyrene) bearing chemically active, electrophilic functionalities on the surface. Though functional groups can be easily introduced into polymer chains via polymerization of a monomer already bearing the desired group, this "tailored" monomer approach has the problem that only a small percentage of the groups end up on the polymer surface (the remainder lying effectively buried in the solid polymer), thereby seriously curtailing the chemical reactivity. This is especially relevant for insoluble polymers lacking porosity. In contrast, the aforesaid microemulsion polymerization strategy enables synthesis of high surface area insoluble polymers in which most of the chemically active appendages reside on the polymer surface thereby providing high accessibility. Menger and Tsuno [54] thus prepared porous polystyrene with chemically active moiety on the surface. A w/o microemulsion containing water pools was prepared using styrene/divinylbenzene (6:4 w/w) as the oil phase, AOT or DDAB (dimethyl dioctadecyl ammonium bromide) as the surfactant (which solubilizes water microdroplets in the oil), and an entity **(V)** with the structure $XArCH{=}CH_2$, where X is a chemically active moiety, serving as a comonomer. This comonomer (in the presence of equimolar HCl added to water) would likely orient at the water–hydrocarbon interface with XH^+Cl^- immersed in or near the pools. If this occurs and if the pools ultimately transform into surface irregularities during polymeriztion, then the polar groups will become fixed to the polymer exterior [54], as shown schematically in Fig. 3.16.

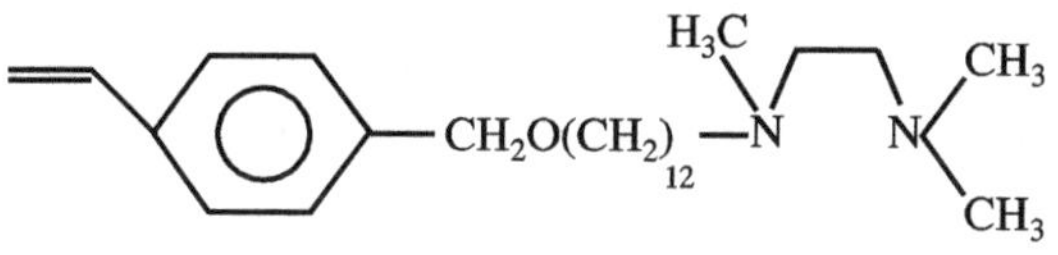

(V)

The polymerization of the microemulsion, in the above case, was carried out, with no apparent phase separation, by exposure to light in the presence of AIBN initiator. The solid polymer thus obtained was ground into powder, washed with methanol to remove surfactant and then dried. The final product was a porous polymer with active surfaces which had the ability to form complexes with Cu^+ displaying heterogeneous catalysis in the hydrolysis of *p*-nitrophenol diphenyl phosphate.

The water pool strategy could also be used for entrapment of drugs. For example, the drug to be entrapped could be dispersed into the water micropools of a w/o microemulsion which could then be polymerized using any of the conventional techniques. The drug is thus confined to the droplet microcapsule and will be released by diffusion through the polymerized matrix. This process is particularly useful where a slow release of drug is desired. Menger and Tsuno [55] demonstrated

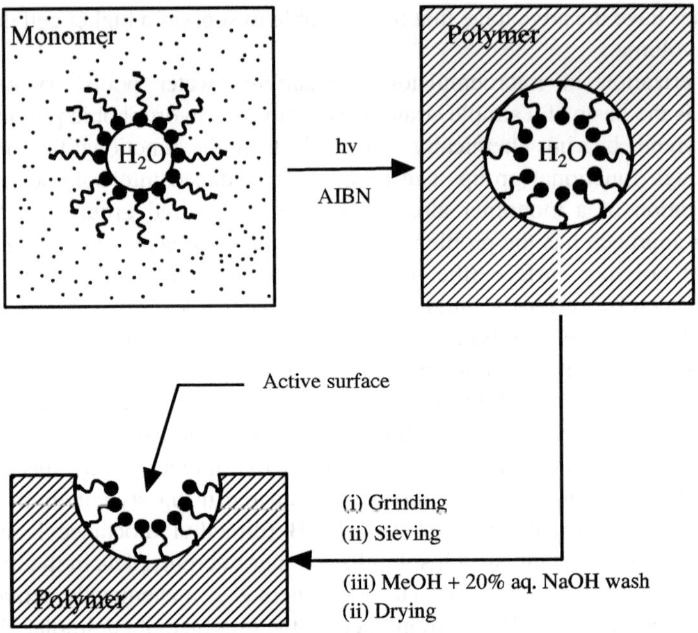

Fig. 3.16 Formation of porous polymers with chemically active surfaces. The *shaded circles* represent diamine groups of polar side chains of polymerizable surfactant which absorbs/adsorbs into or onto water pools solubilized (dispersed) in styrene/divinyl benzene mixture (after [54])

the controlled release property of the aforesaid porous polystyrene using $CuCl_2$. Gupta and Singh [34] also studied the in vitro controlled release of drugs from porous polystyrene and observed that timed release could be achieved for a period of 8 days.

Besides homopolymers, a variety of copolymers have been successfully synthesized by microemulsion polymerization of monomer mixtures, thereby opening a new way to manufacture copolymer latexes with new properties. The polymerization has been achieved by using either a water-soluble monomer mixture in the water phase or an oil-soluble monomer mixture in the oil phase of a microemulsion. Thus, Atik and Thomas [30] copolymerized styrene and divinylbenzene (DVB) by photochemical initiation in a microemulsion consisting of cetyl trimethyl ammonium bromide, hexanol, styrene, and DVB. This yielded stable latex particles in the range 20–40 nm.

On the other hand, Candau et al. [6] copolymerized acrylamide and sodium acrylate of variable compositions by radical polymerization in microemulsions, stabilized by a nonionic emulsifier blend and dispersed in isoparaffinic oil. Microemulsions with high monomer contents ($\sim$25 %) exhibited a bicontinuous character. However, polymerization produced a transformation of the random disordered structure toward a concentrated dispersion of spherical latex particles. The inverse latexes obtained were highly stable and had low particle size ($d \sim 60$ nm) with a narrow size distribution.

In all the above cases, both the copolymerizing monomers are either water-soluble or oil-soluble and they are both in the dispersed water or oil phase. An interesting variation of such polymerization is where polymerizing monomers are incorporated in both the phases. Such a polymerization would yield polymers with interesting properties. Thus, Puig et al. [79] reported the copolymerization of oil-soluble styrene and water-soluble acrylic acid located at different microdomains of a cationic microemulsion made with dodecyl trimethyl ammonium bromide (DTAB). The acrylic acid remains mostly in the aqueous domains and is adsorbed at the micelle surface, whereas styrene is solubilized in the interior of swollen DTAB micelles. Polymerization was carried out at 60 °C using a water-soluble initiator $K_2S_2O_8$. The product was washed with hot water and then with toluene in order to remove DTAB and any homopolymers of acrylic acid and styrene, while the copolymer formed did not dissolve in these solvents. The overall conversion was 60 % in 2 h and the resulting latex was a stable monodisperse dispersion of spherical particles of 21 nm in radius. The copolymer backbone consisted of isolated acrylic acid units randomly distributed among polystyrene blocks.

For the reaction mechanism it was hypothesized that the potassium persulfate free radicals react with acrylic acid molecules at or close to the micelle surface. The acrylic acid radicals in turn initiate reaction in the micelles where styrene begins to react. To explain the formation of monodisperse latexes it was assumed that a fixed number of micelles are initiated without further nucleation during the growth period and the initiated micelles grow by recruiting styrene, acrylic acid, and surfactant from non-initiated micelles. However, since styrene monomer is more reactive than acrylic acid, the acrylic acid radicals entering the micelle and reacting with styrene will react with more styrene, producing the structure.

While latexes prepared from conventional emulsion polymerization and having latex particle size in the range 0.1–10 μm are already used for such purposes as in immune-assays, protein adsorbents, enzyme immobilization, and controlled release in drug delivery, stable microlatexes from microemulsion polymerization in nano-size range (20–50 nm) may be preferred for many of these applications. Thus, nano-size polymers have been extensively studied as particulate carriers in the pharmaceutical and medical fields because of their subcellular size, sustained release properties, and biocompatibility with tissue and cells [80]. The polymeric nanoparticles have been classified according to whether the particle is formed by a polymerization reaction or is derived from a preformed polymer. Procedures have been proposed based on inverse microemulsion polymerization for the preparation of nanoparticles and nanocapsules [81]. De and Hoffman [82] prepared stable polyacrylic acid nanoparticles, averaging approximately 50 nm and having a narrow size range, by a reverse microemulsion polymerization process. The particles were isolated and lyophilized in dry powder form and were redispersible as individual particles in buffer. The drug timolol maleate was loaded into the nanoparticles from aqueous drug solutions and when the drug-loaded particles were dispersed in a phosphate buffer solution, the drug was found to release slowly from the nanoparticles over several hours.

Microemulsion-based polymers play an important role in the application of enhanced oil recovery (EOR) technology. In fact, it is likely that the concept of polymerization in microemulsions appeared as a consequence of studies performed on microemulsion systems after the 1974 oil crisis. Chemical injection, usually known as chemical flooding, is divided into two different methods—polymer flooding and surfactant polymer flooding. Surfactants are added to injection water to wash out more oil in the reservoir by reducing the interfacial tension between oil and water, while polymers are added to raise the viscosity of the injection water. Water soluble polymers are generally used, the most common being polyacrylamide. The latex obtained from w/o microemulsion polymerization of water-soluble monomers is used in EOR as drive fluids to achieve better displacement and volumetric sweep efficiency in pushing micellar flood through underground reservoirs.

Conclusions

While microemulsions have found many uses and applications due to their unique properties, that include large interfacial area, ultralow interfacial tension, thermodynamic stability, and the ability to solubilize otherwise immiscible liquids, particularly significant is the use of microemulsions as polymerization media as it offers new prospects for yielding nano-size polymeric materials with novel and interesting properties. The main limitation, however, arises from the necessity of using rather high levels of surfactant (about 10 % of the total mass) in order to stabilize a large internal surface area. This has promoted extensive investigations on new surfactants and there are reports of microemulsion formulations with surfactant concentrations less than 2 wt%. Industrial polymers made from microemulsions are now commercially available.

Acknowledgements I wish to express my gratitude to Professor Ganapathy Ayappa, Chairman, Department of Chemical Engineering, Indian Institute of Science, Bangalore, for his constant support and encouragement. The trust deposed in me by Professor S. Fakirov, the editor of this book, for writing this chapter is greatly appreciated.

References

1. Langevin D (1992) Micelles and microemulsions. Annu Rev Phys Chem 43:341–369
2. Hoar TP, Schulman JH (1943) Transparent water-in-oil dispersions: the oleopathic hydro-micelle. Nature 152:102
3. Holtzscherer C, Candau F (1988) Application of the cohesive energy ratio concept (CER) to the formation of polymerizable microemulsions. Colloids Surf 29:411–423

4. Baade W, Reichert KH (1984) Kinetics of dispersion polymerization of acrylamide. Eur Polym J 20:505–510
5. Candau F, Leong YS, Pouyet G, Candau S (1984) Inverse microemulsion polymerization of acrylamide—characterization of the water-in-oil microemulsions and the final microlatexes. J Coll Interface Sci 101:167–183
6. Candau F, Zekhnimi Z, Durand JP (1986) Copolymerization of water-soluble monomers in nonionic bicontinuous microemulsions. J Colloid Interface Sci 114:398–408
7. Ruckenstein E (1975) The origin of thermodynamic stability of microemulsions. Chem Phys Lett 57:517–521
8. Shinoda K, Saito H (1969) Stability of o/w type emulsions as function of temperature and HLB of emulsifier—emulsification by PIT method. J Colloid Interface Sci 30:258–265
9. Shinoda K, Friberg S (1975) Microemulsions—colloidal aspects. Adv Colloid Interface Sci 4:281–300
10. Winsor PA (1954) Solvent properties of amphiphilic compounds. Butterworth, London
11. Mukherjee K, Mukherjee DC, Moulik SP (1997) Reaction kinetics in microemulsion medium. 4. Hexacyanoferrate (III)-iodide reaction in water/aerosol-OT/heptane microemulsion and mixed solvents. Bull Chem Soc Jpn 70:1245–1253
12. Dukhin AS, Goetz PJ (2006) Ultrasound for characterizing emulsions and microemulsions (Chap 9). In: Sjöblom J (ed) Emulsions and stability (2nd edn). CRC Press, Boca Raton
13. Rouch J, Safouane A, Tartaglia P, Chen SH (1989) Static and dynamic light scattering studies of water-in-oil microemulsions in the critical region—evidence of a crossover effect. J Chem Phys 90:3756–3764
14. Pena AA, Hirasaki, GJ (2006) NMR characterization of emulsions (Chap. 8). In: Sjöblom J (ed) Emulsions and stability (2nd edn). CRC Press, Boca Raton
15. Bansal VK, Chinnaswamy K, Ramachandran C, Shah DO (1979) Structural aspects of microemulsions using dielectric relaxation and spin-label techniques. J Colloid Interface Sci 72(3):524–537
16. Almgren M, Mays H (1999) Time-resolved luminescence quenching in microemulsions. In Kumar P, Mittal, KL (eds) Handbook of microemulsion science and technology. CRC Press, Boca Raton
17. Strey WJR (1988) Microstructure of microemulsions by freeze fracture electron microscopy. J Phys Chem 92(8):2294–2301
18. Senatra D (2006) Dielectric analysis and differential scanning calorimetry of water-in-oil microemulsions. Adv Colloid Interface Sci 123:415–424
19. Guering P, Lindman B (1985) Droplet and bicontinuous structures in microemulsions from multicomponent self-diffusion measurements. Langmuir 1(4):464–468
20. Paul BK, Moulik SP (2001) Uses and applications of microemulsions. Curr Sci 80(8):990–1001
21. Stoffer JO, Bone T (1980) Polymerization in water-in-oil microemulsion systems. J Polym Sci Polym Chem Ed 18:2641–2648
22. Perez-Luna VH, Puig JE, Castano VM, Rodriguez BE, Murthy AK, Kaler EW (1990) Styrene polymerization in 3-component cationic microemulsions. Langmuir 6:1040–1044
23. Antonietti M, Bremser W, Müschenborn D, Rosenauer C, Schupp B (1991) Synthesis and size control of polystyrene latices via polymerization in microemulsion. Macromolecules 24:6636–6643
24. Rodriguez-Guadarrama LA, Mendizabal E, Puig JE, Kaler EW (1993) Polymerization of methyl methacrylate in 3-component cationic microemulsion. J Appl Polym Sci 48:775–786
25. Lusvardi KM, Schubert KV, Kaler EW (1995) Phase behavior and microstructure of polymerizable microemulsion. Langmuir 11:4728–4734
26. Strom P, Anderson DM (1992) The cubic phase region in the system didodecyldimethylammonium bromide water styrene. Langmuir 8:691–709
27. Kahlwait M, Strey R, Busse G (1993) Weakly to strongly structured mixtures. Phys Rev 47:4197–4209

28. Stoffer JO, Bone T (1980) Polymerization in water-in-oil microemulsion systems containing methyl methacrylate. J Disp Sci Technol 1:37–54
29. Atik SS, Thomas KJ (1981) Polymerized microemulsions. J Am Chem Soc 103:4279–4280
30. Atik SS, Thomas KJ (1982) Photochemistry in polymerized microemulsion systems. J Am Chem Soc 104:5868–5874
31. Atik SS, Thomas KJ (1983) Photoinduced reactions in polymerized microemulsion. J Am Chem Soc 105:4515–4519
32. Johnson PL, Gulari E (1984) Characteristics of microemulsion polymerized styrene with water-soluble versus oil-soluble initiators. J Polym Sci Polym Chem Ed 22:3967–3982
33. Ferrick MR, Murtagh J, Thomas JK (1989) Synthesis and characterization of polystyrene latex particles. Macromolecules 22:1515–1517
34. Gupta B, Singh H (1992) Polymerization in microemulsion systems. Polym Plast Technol Eng 31:635–658
35. Antonietti M, Bremser W, Schmidt M (1990) Microgels—model polymers for the cross-linked state. Macromolecules 23:3796–3805
36. Xu XJ, Siow KS, Wong MK, Gan LM (2001) Microemulsion polymerization of styrene using a polymerizable nonionic surfactant and a cationic surfactant. Colloid Polym Sci 279:879–886
37. Gan LM, Chew CH (1984) W/o microemulsion of polymerizable components. J Disp Sci Technol 5:179–191
38. Gan LM, Chew CH (1983) Polymerization in the transparent water-in-oil solutions. 1. Methyl methacrylate and the copolymerizable cosurfactant. J Disp Sci Technol 4:291–312
39. Chew CH, Gan LM (1985) Polymerization of water-in-oil microemulsions—methyl methacrylate, acrylic acid, and sodium acrylamidostearate. J Polym Sci Polym Chem Ed 23:2225–2232
40. Leong YS, Candau F (1982) Inverse microemulsion polymerization. J Phys Chem 86:2269–2273
41. Candau F, Leong YS, Fitch RM (1985) Kinetic study of the polymerization of acrylamide in inverse microemulsion. J Polym Sci Polym Chem Ed 23:193–214
42. Carver MT, Dreyer U, Knoesel R, Candau F, Fitch RM (1989) Kinetics of photopolymerization of acrylamide in AOT reverse micelles. J Polym Sci Polym Chem Ed 27:2161–2177
43. Carver MT, Candau P, Fitch RM (1989) Effect of solution components on the termination mechanism in acrylamide microemulsion polymerizations. J Polym Sci Polym Chem Ed 27:2179–2188
44. Fouassier JP, Longnot DJ, Zuchowicz I (1986) Reverse micelle radical photopolymerization of acrylamide: excited states of ionic initiators. Eur Polym J 11:933–938
45. Vaskova V, Juranicova V, Barton J (1990) Polymerization in inverse microemulsions 1. Homopolymerization of water- and oil-soluble monomers in inverse microemulsions. Makromol Chem 191:717–723
46. Vaskova V, Juranicova V, Barton J (1991) Polymerization in inverse microemulsion 2. Copolymerization of water- and oil-soluble monomers initiated by 2,2′-azobisisobutyronitrile. Makromol Chem 192:989–997
47. Barton J, Tino J, Hlouskova Z, Stillhammerova M (1994) Effect of percolation on free radical polymerization of acrylamide in inverse microemulsion. Polym Int 34:89–96
48. Lasik I, Barton J, Warr GG (1995) Use of fluorescence to study inverse microemulsion polymerization of acrylamide. Macromol Chem Phys 196:2223–2236
49. Leong YS, Riess G, Candau F (1981) Polymerization of acrylamide inside macromolecular micelles. J Chim Phys 78:264–279
50. Daubresse C, Grandfils C, Jerome R, Teyssie PJ (1994) Enzyme immobilization in nanoparticles produced by inverse microemulsion polymerization. Colloid Interface Sci 168:222–229
51. Candau F, Leong YS, Pouyet G, Candau SJ (1984) Inverse microemulsion polymerization of acrylamide—characterization of the water-in-oil microemulsions and the final microlatexes. J Colloid Interface Sci 101:167–183

52. Holtzcherer C, Candau F, Ottewill RH (1990) A small-angle neutron scattering study on AOT toluene (water + acrylamide) micellar solutions. Prog Colloid Polym Sci 81:81–86
53. Craver MT, Hirsch E, Wittmann JC, Fitch RM, Candau F (1989) Percolation and particle nucleation in inverse microemulsion polymerization. J Phys Chem 93:4867–4873
54. Menger FM, Tsuno T (1990) Synthesis of porous polystyrene with chemically active surfaces. J Am Chem Soc 112:6723–6724
55. Menger FM, Tsuno T, Hammond GS (1990) Cross-linked polystyrene incorporating water pools. J Am Chem Soc 112:1263–1264
56. Guo JS, El-Aasser MS, Vanderhoff JW (1989) Microemulsion polymerization of styrene. J Polym Sci Polym Chem Ed 27:691–710
57. Guo JS, El-Aasser MS, Vanderhoff JW (1989) Microemulsion polymerization of styrene. ACS Symp Ser 384:86–99
58. Chanda M (2013) Introduction to polymer science and chemistry (2nd edn, Chap. 6). CRC Press, Boca Raton
59. Graetzel CK, Jirousek M, Graetzel M (1986) Photoredox-induced polymerization of microemulsion droplets. Langmuir 2:292–296
60. Kuo PL, Turo NJ, Tseng CM, El-AAsser MS, Vanderhoff JW (1987) Photoinitiated polymerization of styrene in microemulsions. Macromolecules 20:1216–1221
61. Qutubuddin S, Haque E, Benton WJ, Fendler EJ (1989) Preparation and characterization of porous polymers from microemulsions. ACS Symp Ser 384 (Polym Assoc Struct):64–83
62. Lianos P (1982) Fluorescence probe study of the interaction between pyrene and microemulsion-polymerized styrene. J Phys Chem 86:1935–1937
63. Gan LM, Friberg SE, Higashimura T (1981) Solubility of non-polar polymers in microemulsions. 1. Solubility of oligomers. J Polym Sci Polym Chem Ed 19:1585–1587
64. Friberg SE, Romg G, Yang CC, Yang Y (1989) Polymerization in microemulsions. ACS Symp Ser 384:34–46
65. Marie P, Gallot Y (1979) Micellar solutions obtained with amphiphilic copolymers in the presence of water, oil and 2-propanol. Makromol Chem 180:1611–1615
66. Marie P, Duplessix R, Gallot Y, Picot C (1979) Micellar solutions obtained with amphiphilic block copolymers in the presence of water, oil and alcohol. 1. Small angle neutron scattering structure investigation in the case of poly(1-vinyl pyridine-b-ethylene oxide) block copolymers. Macromolecules 12:1180–1186
67. Taubman AB, Peregudova LJ (1979) Microemulsions and the stabilization of epoxy-resin emulsions with surfactants. Colloid J USSR 41:522–526
68. Pavel FM (2004) Microemulsion polymerization. J Disp Sci Technol 25:1–16
69. Gan LM, Chew CH, Lye I (1992) Styrene polymerization in oil-in-water microemulsions—kinetics of polymerization. Makromol Chem Macromol chem Phys 193:1249–1260
70. Gan LM, Chew CH, Lim JH, Lee KC, Gan LH (1994) Styrene polymerization in ternary microemulsions—effect of water-soluble and oil-soluble initiators. Colloid Polym Sci 272:1082–1089
71. Haque E, Qutubuddin S (1988) Novel polymeric materials from microemulsions. J Polym Sci Part C: Polym Lett 25:429–432
72. Riess G, Nervo J, Rogez D (1977) Emulsifying properties of block copolymers—oil-water emulsions and microemulsions. Polym Eng Sci 17:634–638
73. Barker MC, Vincent B (1984) The preparation and characterization of polystyrene-poly (ethylene oxide) AB-block copolymers. Colloid Surf 8:289–296
74. Haesslin HW, Eicke HF, Riess G (1984) Dimethyl siloxane-ethylene oxide block copolymers. 1. Microphase separation of low-segment mass copolymers and their compatibility with water and oil. Macromol Chem 185:2625–2645
75. Leong YS, Riess G, Candau F (1981) Polymerization of acrylamide inside macromolecular micelles. J Chim Phys 78:279–284
76. Candau F, Boutillier J, Tripler F, Wiltmann JC (1979) Structure of colloidal particles in water oil mixtures stabilized by polymer emulsifiers. 1. Phase diagrams and electron microscope studies. Polymer 20:1221–1226

77. Marie P, Gallot Y (1977) Microemulsions containing block copolymer surfactants. CR Acad Sci (Paris) Ser C 284:327–330
78. Carver MT, Hirsch E, Wittmann JC, Firch RM, Candau F (1989) Percolation and particle nucleation in inverse microemulsion polymerization. J Phys Chem 93:4867–4873
79. Puig JE, Corona-Galvan E, Maldonado A, Schulz PC, Rodriguez BE, Kaler EW (1990) Microemulsion copolmerization of styrene and acrylic acid. J Colloid Interface Sci 137:308–310
80. Pinto Reis C, Neufeld RJ, Ribeiro AJ, Veiga F (2006) Nanoencapsulation 1. Methods for preparation of drug-loaded polymeric nanoparticles. Nanomedicine 2(1):6–21
81. Gin H, Dupuy B, Bourignon D (1990) Biocompatibility of polyacrylamide microcapsules implanted in peritoneal-cavity or spleen of the rat—effect on various inflammatory reactions in vitro. Biomat Artif Cells and Artif Organs 18:25–42
82. De TK, Hoffman AS (2001) A reverse microemulsion polymerization method for preparation of bioadhesive polyacrylic acid nanoparticles for mucosal drug delivery: loading and release of timolol maleate. Artif Cells Blood Substit Immobil Biotechnol 29(1):31–46

Part III
Physical Approaches for Preparation of Nano-size Polymers

Chapter 4
Electrospinning: Current Status and Future Trends

Soheila Mohammadzadehmoghadam, Yu Dong, Salim Barbhuiya, Linjun Guo, Dongyan Liu, Rehan Umer, Xiaowen Qi and Youhong Tang

Introduction

Fibres with diameters in the range of 1–100 nm are generally referred to as nanofibres in scientific literature. Presently, nanofibre manufacturing is one of the key advancements in nanotechnology. In recent years, various approaches for producing polymeric nanofibres have been employed such as melt blowing [1], template synthesis [2], phase separation [3], self-assembly [4] and electrospinning [5]. Among these methods, electrospinning or electrostatic spinning is recognised as the cheapest and the most straightforward benchtop technique for fabricating continuous nanofibres with diameters ranging from several micrometres down to tens of nanometers [6].

Various outstanding characteristics of electrospun nanofibres are featured including high surface area per unit mass (about 1–100 m^2/g), high porosity (about

S. Mohammadzadehmoghadam · Y. Dong (✉) · L. Guo
Department of Mechanical Engineering, Curtin University, Perth, WA 6845, Australia
e-mail: Y.Dong@curtin.edu.au

S. Barbhuiya
Department of Civil Engineering, Curtin University, Perth, WA 6845, Australia

D. Liu
Institute of Metal Research, Chinese Academy of Sciences, Shenyang 110016, China

R. Umer
Department of Aerospace Engineering, Khalifa University of Science, Technology & Research, Abu Dhabi, United Arab Emirates

X. Qi
School of Mechanical Engineering, Yanshan University, Qinhuangdao 066004, China

Y. Tang
Centre for NanoScale Science and Technology, School of Computer Science, Engineering and Mathematics, Flinders University, Adelaide, SA 5001, Australia

S. Fakirov (ed.), *Nano-size Polymers*,
DOI 10.1007/978-3-319-39715-3_4

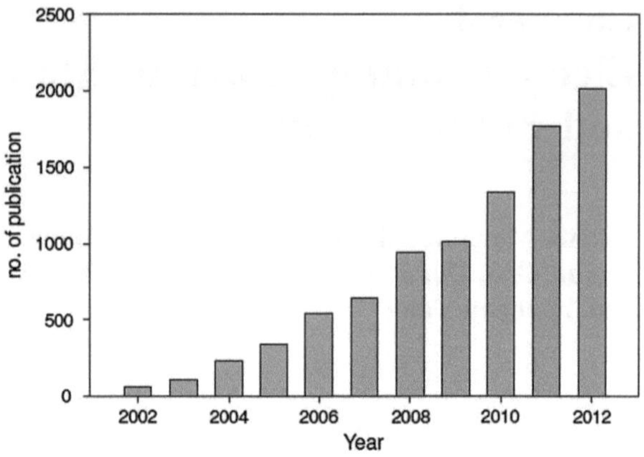

Fig. 4.1 Number of electrospinning publications from 2002 to 2012. Reproduced from Ref. [20]

90 %), light weight, tuneable pore size, flexibility in surface functionalities, relatively good mechanical strength, high permeability, high aspect ratio up to 1000 [7, 8].

Moreover, these properties boost a key role of electrospun nanofibres as an ideal material candidate for a wide range of applications such as biomedicine (e.g. tissue engineering, drug delivery, wound dressing and release control) [9–12], protective clothing [13], filtration [14], reinforcement of composite materials [15], microelectronics (e.g. batteries, supercapacitors, transistors, sensors and display devices) [16, 17], space applications [18] and microwave absorption [19]. Due to ongoing research commitments in this field (Fig. 4.1), significant innovations in set-up, characterisation methods and their applications have been introduced over a short period of time. Hence, there are high demand and great interest in understanding the recent progress in this open-ended field of electrospinning.

Processing and Fabrication

Electrospinning History and Principle

The technology to produce synthetic filaments with the aid of electrostatic forces has been utilised for over a hundred years. This process is called "electrospinning" that was initially derived from "electrostatic spinning" and developed from the electrospraying method. This fundamental technique was observed by Rayleigh [6] in 1897, which was followed by Morton [21] and Cooley [22] that patented methods to disperse fluids using electrostatic forces. In 1914, Zeleny [23] reported the behaviour of conductive liquid droplets at the end of metallic tubes in the presence of an electrostatic force. Further developments were made by Formhals [6, 24] for the fabrication of textile yarns and described in a sequence of patents from 1934 to 1944. After that, the focus shifted to developing a better

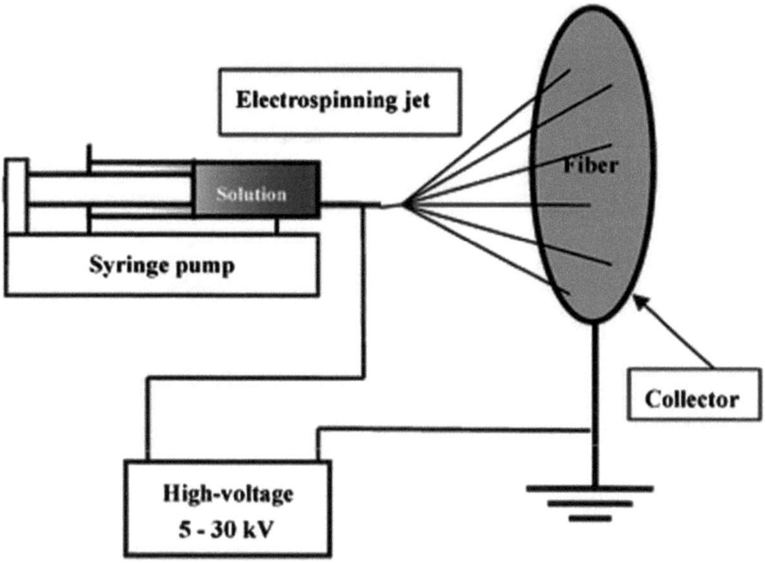

Fig. 4.2 Schematic diagram of a basic electrospinning apparatus. Reproduced from Ref. [29]

understanding of electrospinning process. In 1969, Taylor [25] further studied electrospinning by means of mathematically modelling the conical shape formed by fluid droplets under the effect of an electric field. This cone has since been known as the "Taylor cone". Several years later, the first investigation of electrospinning melting polymers was carried out by Larrondo and Manley [26]. The method was not well recognised until early 1990s when researchers started to realise the huge potential of electrospinning process in nanofibre production [27].

The basic principle for a polymer solution or melt to be electrospinnable is its ability to carry an electric charge and have sufficient viscosity to be stretched without breaking up into droplets. The basic framework of electrospinning is shown in Fig. 4.2. In a typical electrospinning process, three main components are predominantly utilised, namely a syringe filled with a polymer solution, a high-voltage supplier to provide the required electric force for stretching the liquid jet, and a grounded collection plate to hold nanofibre mats. At the initial stage, polymer melt or solution is introduced into a capillary tube. A high electrical potential (e.g. 10–50 kV) is applied between droplets of polymer solution or melt at the end of a needle and a grounded collector. By increasing the intensity of electric field, the pendant drop of polymer solution becomes highly electrified, thus inducing the electric charge on the liquid surface, which results in the deformation of liquid drops into a conical shape, known as "Taylor cone". When the electric voltage reaches a critical value, the electrostatic repulsion forces prevail over the solution surface tension so that a charged jet of solution is ejected. A polymer jet is then ejected from the tip of "Taylor cone" and travels rapidly to the metal collector.

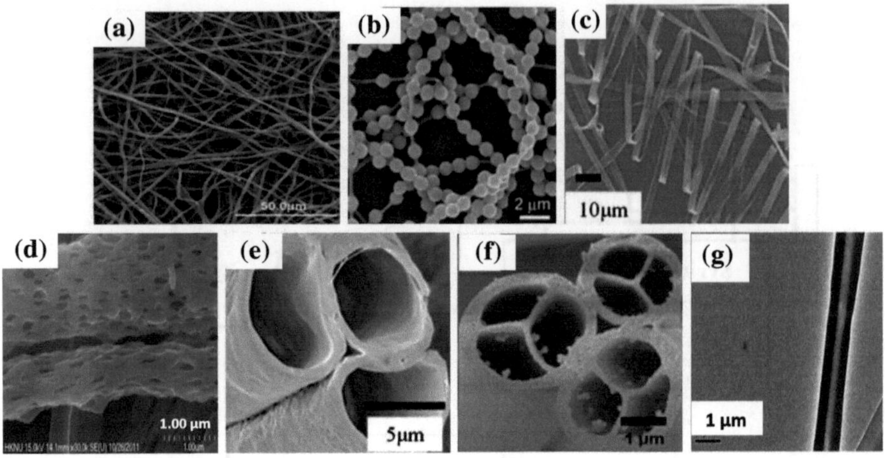

Fig. 4.3 Different morphologies of electrospun polymeric fibres: **a** random fibres [30], **b** necklace-like fibres [31], **c** ribbon fibres [32], **d** porous fibres [33], **e** hollow fibres [34], **f** multichannel tubular fibres [35] and **g** core/shell structure [33]

Before reaching the collector, the jet undergoes a series of electrically driven bending instabilities and gradually becomes thinner in air due to fibre elongation and solvent evaporation. The charged jet is finally collected on a ground collector in the formation of stretched fibrous structures [6, 28].

A typical electrospinning process creates long and continuous fibres that can vary in diameters along the lengths and generally exhibit a solid interior and smooth surface. However, after the recent developments, different nanofibres with specific structures like core/shell, porous, hollow, necklace-like ribbon and multichannel tubular structures can also be prepared (Fig. 4.3) [30–35].

Electrospinnable Polymers

A wide range of polymers from solution, sol-gel suspension, or melt can be electrospun into nanofibres. To date, over 200 types of various materials including natural polymers, synthetic polymers and hybrid blends have been used to obtain electrospun fibres [36]. Due to the practicality, mouldability, flexibility, lightness, durability and chemical and physicochemical stability, natural polymers are more preferable to synthetic polymers. Therefore, four major classes of biopolymers including proteins, polysaccharides, deoxyribonucleic acids (DNAs) and lipids as well as their derivatives have been fabricated into electrospun scaffolds [37, 38]. The most popular natural polymers include chitosan, collagen, gelatin, casein, hyaluronic acids, silk protein, chitin and fibrinogen [37–43].

Synthetic polymers often provide many advantages over natural polymers in that they can be tailored according to desired properties for specific applications and

Table 4.1 Different polymers and solvents used for electrospinning

Type of polymer	Polymer	Solvent	Reference
	Silk fibroin	Formic acid	[42]
Natural polymers	Chitosan	Trifluoroacetic acid(TFA)	[39]
	Gelatin	Acetic acid	[41]
	Collagen	Hexafluoroisopropanol (HFIP)	[40]
	Fibrinogen	HFIP/10 × minimal essential medium	[43]
Synthetic polymers	PCL	Chloroform/dimethylformamide (DMF)	[44]
	PLA	Dichloromethane (DCM)	[45]
	PVA	Water	[46]
	PVP	Methanol	[47]
	PAN	DMF	[48]
	Nylon-6	Formic acid	[49]
	PET	TFA/DCM	[50]
	PU	DMF	[51]
	PI	N,N-dimethylace tamide (DMAc)	[52]
	EVOH	80 % propan-2-ol/Water	[53]
	CA	Acetic acid/Water	[54]
	PGA	Water	[55]
	PEO	Water	[56]

also present good uniformity. In addition, they are much cheaper and serve as a more reliable source of raw materials than that of natural polymers. These polymers include polycaprolactone (PCL) [44], polylactic acid (PLA) [45], poly(vinyl alcohol) (PVA) [46], polyvinylpyrrolidone (PVP) [47], polyacrylonitrile (PAN) [48], nylon-6 [49], polyethylene terephthalate (PET) [50], polyurethane (PU) [51], polyimide (PI) [52], poly(ethylene-co-vinyl alcohol) (EVOH) [53], cellulose acetate (CA) [54], polyglycolic acid (PGA) [55], polyethylene oxide (PEO) [56] and so on. Several commonly used natural and synthetic polymers are listed in Table 4.1.

Types of Electrospinning

A majority of electrospun nanofibres are typically fabricated from polymer solutions and organic solvents have to be used in most cases to form homogeneous polymer solutions [57]. Despite many attempts, electrospinning process suffers from many drawbacks including low productivity (up to 300 mg/hr) [58], the requirement of an extraction process for additional solvents and environmental concerns due to the use of toxic solvents [59]. To tackle these issues, several designs for the types of electrospinning have been employed as highlighted in the following sections.

Melt Electrospinning

Electrospinning of molten polymers, also known as melt electrospinning, is an effective approach to overcome some disadvantages of conventional electrospinning process such as environmental concerns as well as electrospinning of non-soluble polymers like polypropylene (PP) and polyethylene (PE) [23]. Larrondo and Manley [26] applied an electrostatic force to molten polymers for the first time, leading to the production of microfibres. Generally, the set-up of melt electrospinning is similar to that of conventional electrospinning in addition to the provision of polymer melting [59]. More impressively, melt spinning can be used without the requirement for a ventilation system. In some cases, melt electrospinning can be performed without the use of syringe pump where a polymer melt is pushed out of the spinneret using an air pressure system [60] or based on a solid polymer filament that is pushed out of the device [61, 62]. Several heating systems with benefits and disadvantages regarding safety and efficiency can be employed for heating polymers such as circulating fluids, electrical systems, hot air or laser melting devices (Fig. 4.4) [59, 62, 63]. Among these, electrical heating systems are most commonly used [63–65]. However, if high voltage is applied, the heater should be isolated from the high-voltage source that requires a complicated configuration [63]. An efficient hybrid system combining electrical heating and hot air was developed in Joo's laboratory [66]. In such a combined system, a stream of air was used to assist in maintaining the electrified molten jet and the process was named gas-assisted melt electrospinning (GAME). Various polymers such as PLA [66], polylactide [67], PP [68], PET [69] and PCL [70] were fabricated by melt electrospinning.

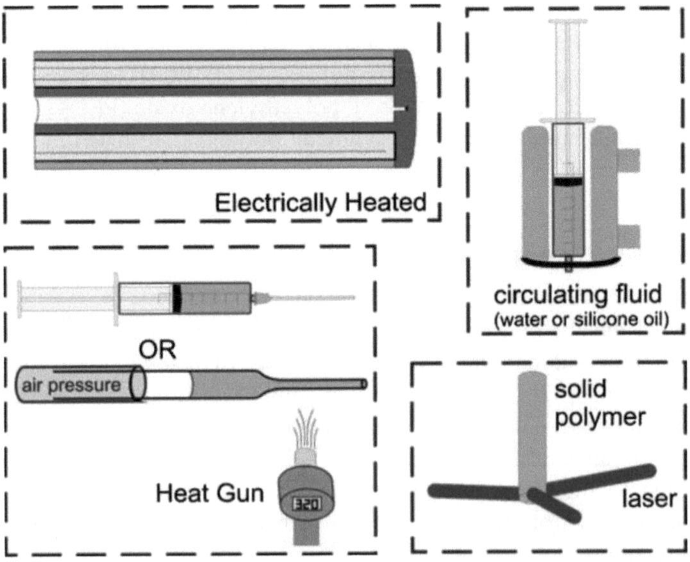

Fig. 4.4 Schematic diagram of different heating approaches used for melt electrospinning. Reproduced from Ref. [62]

Despite numerous advantages of melt electrospinning such as no recycling/removal of toxic solvents, high throughput rate without a mass loss by solvent evaporation, ease of fabricating polymeric fibre blends, this technique is still accompanied by some limitations such as the requirement of a high-temperature melting system, an electric discharge problem associated with the melt and low conductivity of the melt. In addition, it has been noted that diameters of melt electrospun nanofibres are typically over 10 microns owing to the low charge density, high viscosity of the melt as well as rapid solidification of polymers between the needle and collector [59, 71]. Many studies have been carried out to control and reduce the diameter of melt electrospun nanofibres. For example, the use of special additives for viscosity reduction like Irgatec CR 76 to PP can decrease fibre diameter to <1 μm [72]. In another study, poly(ethylene glycol) (PEG) was found as an effective plasticiser for reducing the diameter of electrospun PLA melt [73]. Another approach utilised to decrease melt electrospun fibre diameter is by incorporating polar additives such as stearic acid and sodium stearate [74], though at higher concentrations of additives, agglomeration may result in less uniform distribution and lower stretching force across the electrospinning jet length leading to larger fibre diameters. Doping the melt with salt additive to enhance its conductivity has also been shown to be very effective in reducing the diameter of melt electrospun fibres to less than a micron [57]. Recently, Hochleitner et al. [75] was able to produce PCL fibres with diameters of about 800 nm without the use of additives in polymers. This is achieved by applying a voltage of 2.9 kV, collector distance of 1.5 mm and spinneret heating temperature of 84 °C at the PCL molecular weight of 83×10^3 g/mol. To control fibre diameters, the effect of processing parameters was studied as well, but the results were not consistent. For example, in a comprehensive study by Brown et al. [76] across various parameters, there was no consistent trend observed for the effect of increasing collector distance on fibre diameter. Whereas, Fang et al. [57] demonstrated that at the same applied voltage, with increasing spinning distance, the fibre diameter increased. This is probably due to reduced electric field intensity at a larger distance. Zhou et al. [77] has reported that temperatures at the spinneret and in the spinning region are critical to produce submicron-sized fibres. An important parameter that has been found to considerably affect melt electrospinning is the molecular weight of polymers. Polymer melt with lower molecular weight was found to result in a corresponding reduction in fibre diameter [62]. A 20-fold decrease in fibre diameter from over 60 μm to below 3 μm was demonstrated by utilising heated gas to facilitate the elongation and thinning of electrospinning jet [65]. Ko et al. [78] reported an increase in fibre diameter at a larger tip-to-collector distance, which was ascribed to the reduction in electric field intensity.

Needleless Electrospinning

Recently, a new electrospinning technique has been developed with the capability to produce mass production of nanofibres, which is known as needleless

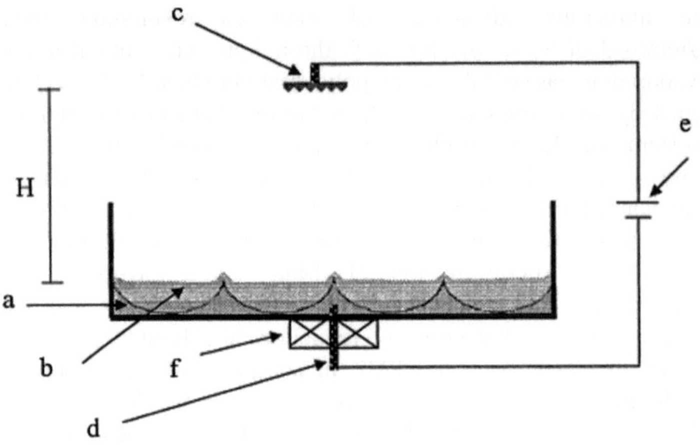

Fig. 4.5 Schematic diagram of experimental set-up for needleless electrospinning. *a* A layer of magnetic liquid, *b* a layer of polymer solution, *c* counter electrode located at a distance H from the free surface of polymer, *d* electrode submerged into magnetic fluid, *e* high-voltage source and *f* strong permanent magnet or electromagnet. Reproduced from Ref. [82]

electrospinning or free surface electrospinning (Fig. 4.5). The set-up strategy is based on concurrent generation of numerous jets from an open liquid surface without the influence of capillary effect that is usually associated with needle-like nozzles [58, 79]. In 1979, needleless electrospinning appeared when Simm et al. [80] filed a patent on using rings to electrostatically spin fibres for filtration applications. Later, Jirsak [81] patented their needleless electrospinning design using a rotating horizontal cylinder as the fibre generator for the mass production of nanofibres, which was rapidly commercialised by Elmarco s.r.o., Czech Republic, under the brand name of Nanospider™ [58, 79]. In needleless electrospinning, the productivity and fibre morphology are determined by a fibre generator. Depending on applied methods to initiate the spinning process, this technique can be divided into rotating or stationary electrospinning [79]. Rotating needleless electrospinning is set up based on loading a thin layer of polymer solution by the rotation of spinneret onto the spinneret surface. The rotation and perturbance create conical spikes on the surface of solution layer. When a high voltage is applied to the spinneret, these spikes tend to concentrate charges and amplify the perturbance. Thus the fluid around the spikes is drawn to these spikes under high electric forces, resulting in "Taylor cones". Fine solution jets are then ejected from the tips of those "Taylor cones" with sufficient electrical forces [79].

Since 2009, various rotating spinnerets for needleless electrospinning have been built in different shapes such as bowel edge [83], conical wire coil [58], metal plate [84], splashing spinneret [85], rotary cone [86] and moving bead chain [88] (Fig. 4.6).

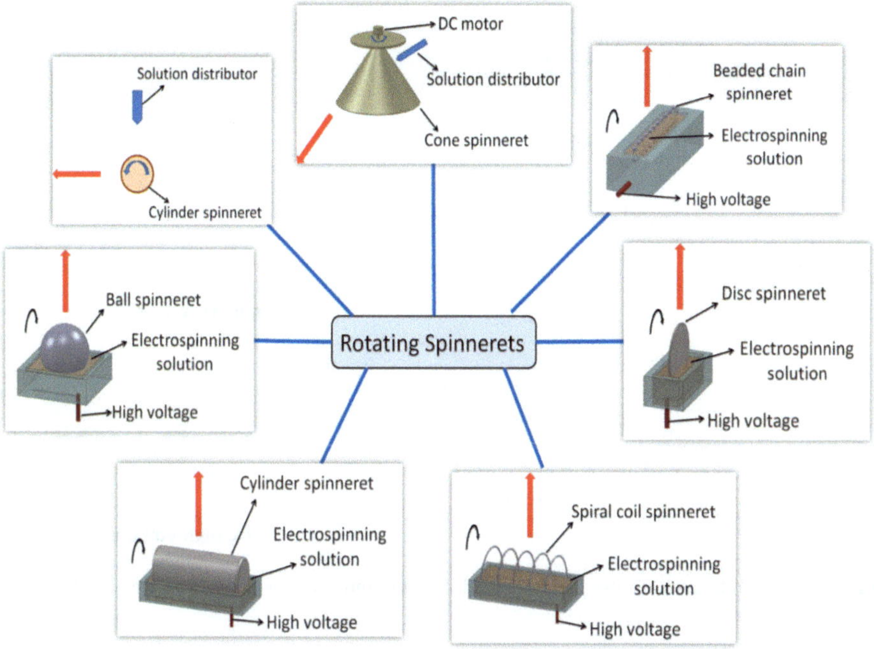

Fig. 4.6 Schematic diagram of needleless rotating spinnerets (electrospinning direction along the *red arrow*). Reproduced from Ref. [87]

Rotation spinnerets can automatically introduce instability from spinneret movement and vibration of transmission components. Whereas, stationary spinnerets need the help of external mechanism such as magnetic field, gravity and gas bubble [79]. In 2004, Yarin and Zussman [82] used a magnetic-field-assisted needleless electrospinning for instability. In 2007, air bubbles were introduced to initiate a non-steady state (Fig. 4.7). Such an electrospinning technique is also called "bubble electrospinning" [89]. Other work investigated the effect of applied voltage on fibre diameter and morphology in bubble electrospinning [90–92].

Wu et al. [93] employed a stationary cylinder spinneret that improved the mass production of nanofibres by 260 times with respect to conventional electrospinning system. Later, fibre uniformity was improved using a strip-distributed solution delivery method. This type of electrospinning is named tipless electrospinning [94]. Figure 4.8 demonstrates different stationary spinnerets.

Today the research work on needleless electrospinning generally focuses on four areas: (1) investigation on the physical principle of needleless electrospinning process to understand the mechanism of controlling fibre quality, (2) experimental study on production parameters, (3) development of electrospinning machine at laboratory, pilot and production levels and (4) development of final fibre products for various purposes [87].

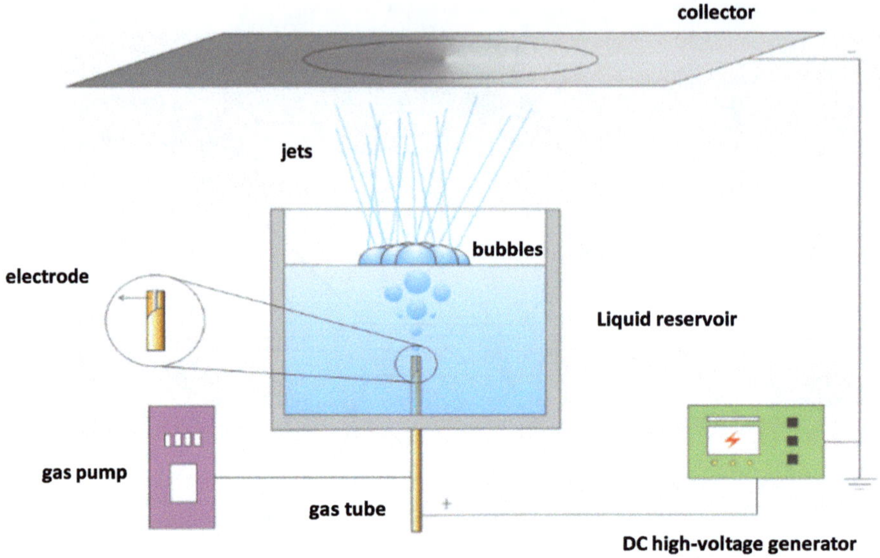

Fig. 4.7 Illustrative set-up used in bubble electrospinning. Reproduced from Ref. [92]

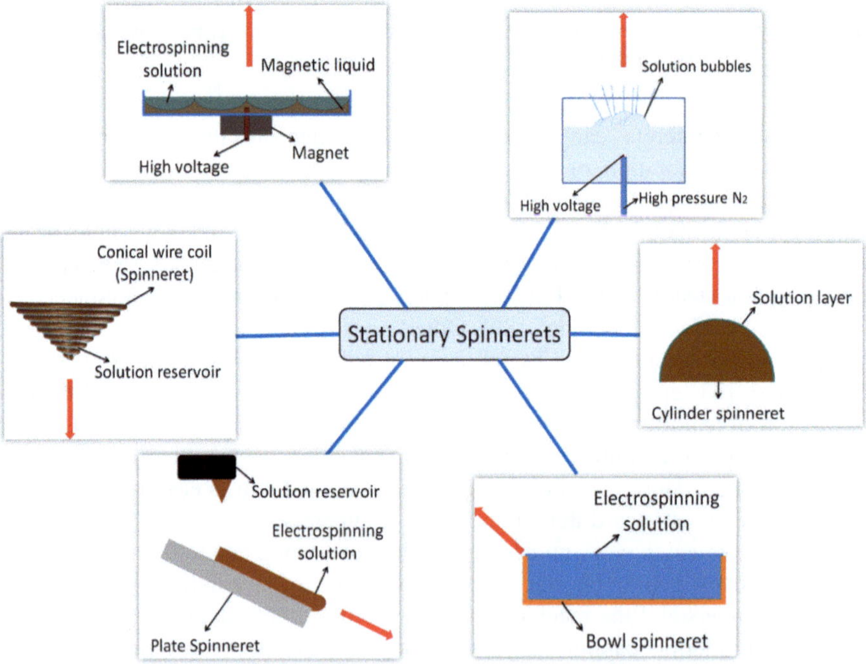

Fig. 4.8 Schematic diagram of different stationary needleless spinnerets (electrospinning direction along the *red arrow*). Reproduced from Ref. [87]

Multijet Electrospinning

Multijet electrospinning is used for upscaling nanofibre production by means of a single needle or multiple needles. The former approach is based on splitting the polymer into separate subjets during its trajectory to the collector. Yamashita et al. [95] observed a multijet formation for the first time during the electrospinning of polybutadiene by a single home-made stainless steel needle with a grooved tip mounted on a glass syringe. Vaseashta [96] used a curved collector to generate multijet from multiple "Taylor cones" in a single-needle electrospinning system. The mechanism of jet splitting has not been fully analysed yet and more experimental studies are required [59].

The electrospinning of nanofibres using a multiple needle approach has been carried out for various materials and configurations (Fig. 4.9). In this system, the needle configuration, number of needles and needle gauge should be considered. The needle arrangement can be made in a linear or two dimensional manner with elliptical, circular, triangular, square and hexagonal arrays [59]. Ding et al. [97] adopted a few syringes concurrently used for electrospinning from different polymer solutions. Theron et al. [98] showed a model that, under comparable conditions, nine jets could be electrospun steadily from separate nozzles located with a pitch of 10 mm on a square of 400 mm^2. To correlate the configuration with the efficiency of the process, Tomaszewski and Szadkowski [99] used three types of multijet electrospinning heads, namely series, elliptic and concentric types. Considering the efficiency and quality of the process, the concentric electrospinning head was reported as the best type. In case of the linear configuration, a failure was observed for central needles, resulting from a low internozzle distance at 3 mm.

The main issue in the multijet electrospinning process is jet deviation, which causes instability problems such as dripping of polymer solution and difficulties in

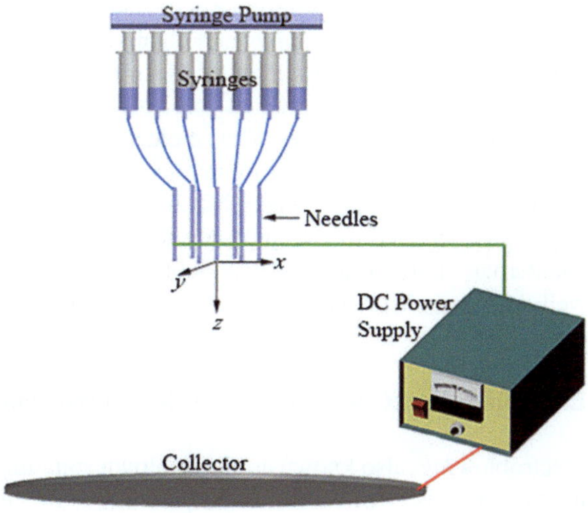

Fig. 4.9 Schematic diagram of multi-needle electrospinning. Reproduced from Ref. [100]

fibre collection [100]. Many configurations, such as extra electrode [101] and auxiliary plate [102], have been employed into multijet electrospinning to modify the electric field and control the process. Angammana and Jayaram [103] investigated strength variations in the electric field at the tip of the needles in multi-needle arrangements. It was revealed that the deterioration of local field at the needle tips degraded the electrospinning processing considerably and produced significant variations in fibre morphology. Yamashita et al. [104] presented several multijet electrospinning devices including an industrial plant with 1000 nozzles at a coated width of 0.6 m. The study also addressed some issues related to the industrialisation of electrospinning. The multi-needle set-up generally requires a large operating space, and optimisation of relative locations of needles to avoid the strong charge repulsion between adjacent solution jets, solvent recovery from nanofibres and atmosphere, and a regular nozzle clean-up, which makes the whole set-up inapplicable when thousands of needles are used for nanofibre production [105].

Multihole Electrospinning

Varabhas et al. [106] used a cylindrical porous tube to form multiple jets from air-pressurised polymer solutions inside the tube through its inherent pores. The tube containing pores with an average diameter of 20–40 μm was oriented horizontally. Due to the size limit of holes, Dosunmu et al. [107] improved the process by drilling multiple holes with larger diameters on PE tube surface to achieve a higher production rate. It has been reported that consequent mass production rate is 250 times greater than that of conventional single-needle electrospinning. Recently, a seven-hole spinneret with a metal flat electrode was used to scale up the production rate of electrospun nanofibres [108]. Compared with conventional electrospinning, a more uniform electric field was generated by this multihole system, in turn leading to the formation of finer fibres. In another work [109], 19-hole multistep electrospinning set-up was developed to strengthen the electric field at the central position of the spinneret and obtain a more uniform electric field. It has been found that the line distribution of electric field, as determined by the spinneret configuration, plays a dominant role to influence jet repulsion. Recently, Zheng et al. [100] investigated the effect of electric field distribution on jet motion and jet deviation in two multijet electrospinning systems (e.g. multihole and multi-needle electrospinning configurations). Much larger envelop cones and outer jet deviation angles were created by the multi-needle system when compared with the multihole counterpart. Furthermore, it was noted that jet behaviour could be controlled by well-designed shape and tailored strength of the electric field.

Electroblowing (Gas-Assisted/Gas Jet Electrospinning)

Electroblowing, also known as electroblown spinning, is an electrospinning process in the presence of a controlled airflow. In this process, two forces are

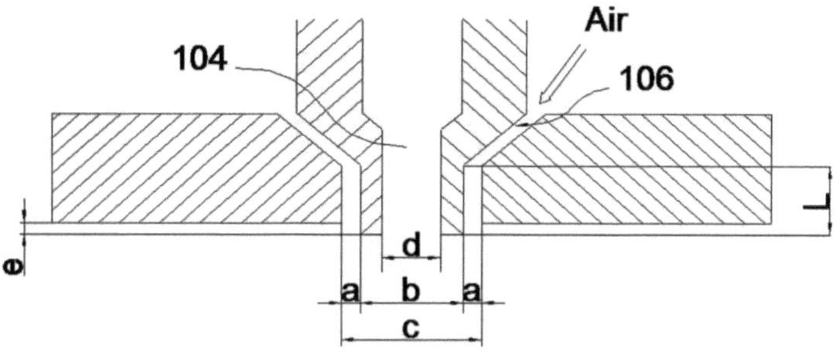

Fig. 4.10 Schematic diagram of electroblowing. Reproduced from Ref. [111]

simultaneously applied for producing nanofibres, namely electrical force and air-blowing shear force [59]. The initial work was traced back to 2004 when Wang et al. [110] implemented an air-blowing system to conventional electrospinning set-up. This method is especially useful for highly viscous solutions in which applying high voltage alone is not sufficient to overcome surface tension. Kim et al. [111] designed an electrospinning spinneret with an air nozzle to produce nanofibre webs from thermoplastic or thermosetting polymers. Due to the injection of compressed air, a higher productivity over conventional electrospinning set-up was reported (Fig. 4.10).

Another set-up was patented by Arora et al. [112] where compressed and heated air was supplied from air nozzles positioned around the sides of spinning nozzle. The air forwarded the newly ejected polymeric solution from the nozzle and attenuated to nanofibres that were collected on a grounded porous collection belt. In an extension of electroblowing, heated gas may be used instead to reduce the solution viscosity and achieve greater fibre elongation [113]. Using airflow to increase the production rate of electrospinning has also been examined on the set-up of needleless electrospinning. Wang et al. [114] investigated the effect of airflow on the production of electrospun fibres from a ring, which rotated and dipped into a solution reservoir. The fibre production rate appeared to increase only at a higher airflow rate. Bubakir et al. [115] developed the mass production of melt electrospinning using hot air, which was facilitated using fibre drawing. With the aid of PP, the hot air-assisted electrospinning was able to reduce the fibre diameter from a range of 1–2 μm to 0.2–0.8 μm and obtained a melt flow rate of 12 kg/h, which is much higher than any other reported mass production of electrospinning.

Centrifugal Electrospinning

For the first time, Liao et al. [116] reported electrospinning process that was combined with the use of an electrical field and a centrifugal field (Fig. 4.11). Compared with pure centrifugal spinning that is required to rotate at thousands of revolutions per minutes (rpm), the rotational speed in centrifugal electrospinning

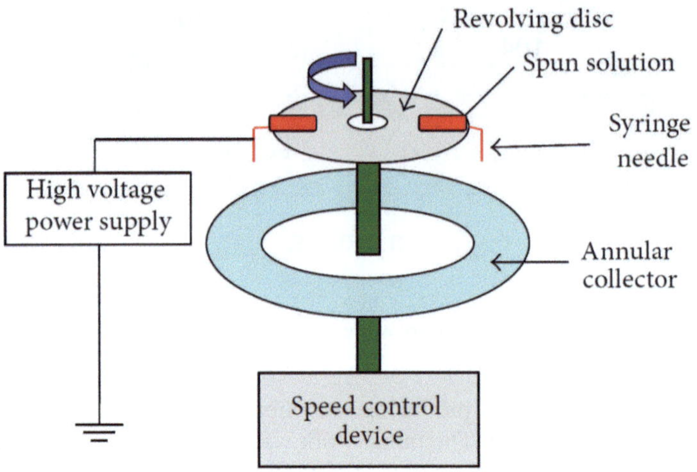

Fig. 4.11 Schematic illustration of a centrifugal electrospinning apparatus. Reproduced from Ref. [117]

can decrease to 300–600 rpm with more aligned fibres [117–119]. Centrifugal electrospinning combines centrifugal and electrical forces into the spinning process. Hence it leads to the further orientation of polymer chains in nanofibres and a higher production rate can be achieved at a lower working voltage or slower rotating velocity [117].

As opposed to conventional electrospinning where the nozzle is stationary, centrifugal electrospinning applies a centrifugal force on the solution when it leaves the nozzle. Consequently, the force that pushes the solution towards the collector increases, which can reduce the flight duration of electrospinning jet, and hence the amount of fibre stretching. It was shown that by increasing tip-to-collector distance from 13 to 17 cm, the fibre diameter was reduced from 700 to 550 nm [120]. Although there are several advantages such as easy maintenance and high volume output, this method still requires relatively expensive equipment and resulting fibres may be non-continuous.

Near-Field Electrospinning (NFES)

Near-field electrospinning (NFES) is a potential approach for easier and more predictable location control for the deposition of nanofibres, which is impossible to be achieved by conventional electrospinning due to the whipping action of fibres. In this method, by reducing the distance between the nozzle and the substrate to less than a few millimetres, fibres land on the substrate before the onset of whipping [121–123]. In particular, NFES, using the stable region of the electrospun jet close to the metallic needle, has been studied to fabricate PEO [124], PVP [125], PCL [126] (with diameters down to 16 nm) as well as light-emitting conjugated polymers, and assembled even in complex three-dimensional patterns [127]. Hellmann et al. [128] reported the impact of NFES processing parameters on the deposition morphology of nanofibres when nonwoven nanofibres were deposited precisely along a predetermined pattern. Chang et al. [122] showed a piezoelectric polymeric nanogenerator with high energy conversion efficiency using PEO based on NFES where mechanical energy was converted into electrical energy. The use of patterned silicon structures was found to increase the deposition precision of nanofibres [129].

The applied voltage, required to achieve an electrical field strength to initiate spinning, is much lower (~200 V) due to the significantly reduced distance between the nozzle and the substrate (from 500 to 3 mm) as well as the fine nozzle size [121, 130]. NFES can largely outperform conventional electrospinning for depositing conjugated polymeric fibres with high spatial control. In addition, the control of fibre morphology is improved by this method, which is crucial for the exploitation of single active fibres for photonic or sensing applications [127].

An alternative approach for controlling the deposition of oriented nanofibres is based on using a microfabricated scanning tip as an electrospinning source. The technique is called scanning tip electrospinning. In this process, the tip is dipped into a polymer solution to gather a droplet as a source material. A voltage applied to the tip causes the formation of a "Taylor cone", and at sufficiently high voltages, a polymer jet is extracted from droplets. By moving the source relative to a surface, acting as a counter electrode, oriented nanofibres can be deposited and integrated with microfabricated surface structures [131].

Coaxial Electrospinning

Coaxial electrospinning is an innovatively extended form of electrospinning, which uses two concentrically aligned capillaries to enforce the fibre formation with a core-shell structure (Fig. 4.12). Coaxial electrospinning is of particular interest for those core materials that cannot form fibres via electrospinning by themselves such as conductive polymers, metals, or some natural polymers. Its set-up can be made in both horizontal and vertical directions and its process is conceptually similar to that of conventional electrospinning [132]. This technique was first demonstrated by Loscertales et al. [133] in 2002, where two dissimilar materials were delivered independently through the coaxial capillary and drawn to generate nanofibres in a core-shell configuration. The shell and the core may or may not be miscible due to the short process duration at which the jet becomes solidified in the fibres to prevent fluids from considerable mixing [134].

Due to the similar processes between coaxial electrospinning and conventional electrospinning, all parameters to determine fibre morphology and quality in conventional processing can also affect the performance of coaxial electrospinning. Although the shell and core solutions are in contact and undergo the same bending instability and whipping motion, the degree of dissimilarity between them, in terms of composition, and physical and rheological properties, has an important role in the

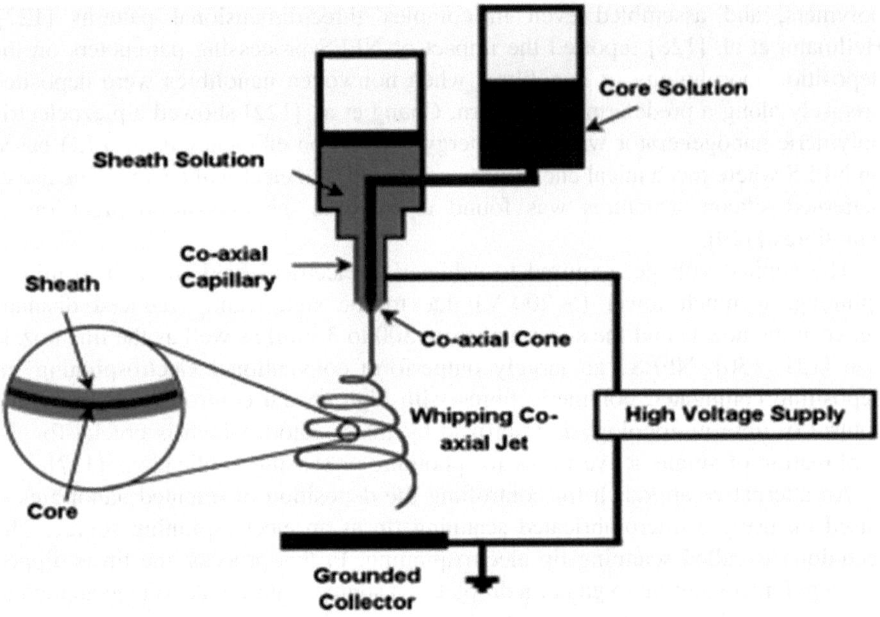

Fig. 4.12 Schematic diagram of coaxial electrospinning set-up for core-shell fibres. Reproduced from Ref. [132]

formation of nanofibre structures [133]. The concentrations of both phases should exceed a threshold value to maintain uniform morphology. Moreover, flow rates of both core and shell should be matched [135]. Hollow nanofibres can be electrospun using this method. McCann et al. [136] fabricated several nanofibres with walls made of polymer/inorganic composites or ceramics by electrospinning two immiscible liquids through a coaxial, two-capillary spinneret, followed by the selective removal of cores. To obtain hollow nanofibres, an extraction process or a thermal treatment is required. With the removal of the core phase, the specific surface area has been enlarged significantly to nearly 200 %, which enhances their performance for different applications [137, 138]. Jiang et al. [139] reported the enhancement of fabrication productivity over 100 times when compared with conventional manufacturing of core-shell nanofibres. A large number of coaxial jets can be generated simultaneously on the edges of stepped pyramid-shaped spinneret, resulting in manufacturing core-shell nanofibres with a high throughput (Fig. 4.13a).

Coaxial spinning could also be conducted using polymer melts, in which a heating system is used to surround the reservoir [140]. Wang et al. [141] designed a slightly different and much simpler set-up, which eliminates the requirement to insert one capillary into the other. The design comprises two separate syringes with different sized capillaries where the smaller capillary is inserted from outside into the "Taylor cone" formed at the exit of the bigger capillary (Fig. 4.13b). The extension of coaxial spinneret to more complex configuration can be further

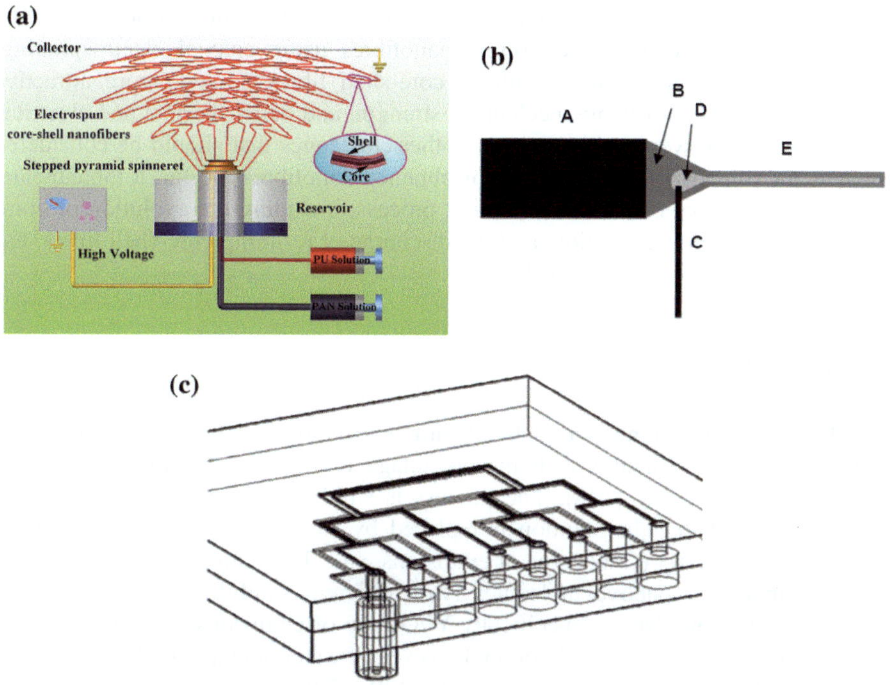

Fig. 4.13 **a** Schematic diagram of free surface coaxial electrospinning apparatus using a stepped pyramid spinneret, **b** schematic diagram of alternative set-up for coaxial electrospinning, **c** illustration of polydimethylsiloxane (PDMS) microfluidic device source using branching microchannel architecture and coaxial spinnerets for parallel electrospinning hollow and core/shell nanofibres. Reproduced from Refs. [139, 132, 145], respectively

favoured in the fibre formation with various structures and functionalities. For example, triple-layered core-shell-corona fibres could be produced by triaxial electrospinning. Compared with coaxial electrospinning, more coupled parameters impact triaxial electrospinning process and the properties of resulting nonwoven mats. In addition, the alignment of three concentric capillaries is also complex work [142–144].

Srivastava et al. [145] designed and constructed a prototype coaxial microfluidic device through soft lithography technique to fabricate multicomponent nanofibres and to scale up the production of electrospinning. Two layers of non-intersecting, stacked microchannels arranged in a branching tree pattern were created within the device to provide constant flow of core/shell fluids to each of the eight outlet spinnerets (Fig. 4.13c). Another study by the same group [146] employed a hydrodynamic focusing method to generate a coaxial stream of two immiscible fluids within a microfluidic device, in turn eliminating the requirement for concentric spinneret.

Nearly all polymer solutions and their polymer composite matrices could be fabricated into core-shell and hollow nanofibres using coaxial electrospinning. Compared with single-structure fibres, core-shell fibres possess more attractive merits including controllable mechanical strength, and better thermal and electrical conductivities. However, there are also other challenges that should be considered such as the dimension and the order of obtained nanofibres. So far, a fundamental understanding of coaxial electrospinning process, including the evolution of compound jets, solvent evaporation, and the discharge of resulting core-shell fibres, has not yet been achieved [135, 147].

Emulsion Electrospinning

Emulsion electrospinning is a relatively simple technique for fabricating nanofibres with a core-shell structure with the advantage of having a simpler set-up than coaxial electrospinning [148]. The process is similar to normal solution electrospinning, except that the solution is replaced by an emulsion with a water-in-oil (W/O) or oil-in-water (O/W) type of emulsions, and allows for the encapsulation of a wide range of bioactive molecules (with different solubilities) into polymeric nanofibres. The emulsions (specifically W/O type of emulsions) comprise a water phase, which is mostly a drug/protein dissolved in water, and an oil phase, which is a polymer dissolved in an organic solvent [149, 150].

Xu et al. [151] presented electrospun W/O emulsions where the water drops contained an anticancer drug. Whereas, the continuous organic phase was a poly (ethylene glycol)-poly(L-lactic acid) diblock copolymer dissolved in chloroform. This study demonstrated the potential use of emulsion electrospinning for the fabrication of continuous core-shell fibres. During electrospinning, aqueous drops were stretched into elliptical shapes in the axial direction of fibres and gave rise to a continuous core. The organic solvent evaporates relatively fast when compared to the water phase and its viscosity increases more rapidly, resulting in the solidification and creation of a core-shell structure [152]. Surfactants are capable of controlling fibre morphology via electrostatic and hydrogen bonding, depending on their chemical structures [150, 153].

On the other hand, Angeles et al. [154] showed a reverse case of O/W emulsion electrospinning using polystyrene (PS)/mineral oil as drop/core phase and PEO as continuous/shell phase. Emulsion electrospinning assists in fabricating nanofibres from less concentrated polymer solutions compared to conventional electrospinning. In addition, this method does not require a specific type of spinneret/set-up for manufacturing core-shell nanofibres. It is considered as a user-friendly and economically viable technique. The proper viscosity of drop phase and stable emulsion play an important role in emulsion electrospinning [132, 149].

Material and Processing Parameters

Although the electrospinning principle is simple, the process itself can be quite complicated because many parameters influence fibre diameters and final structures of nanofibres. These parameters are generally classified into three groups [6, 155–157]:

- Solution parameters: viscosity, concentration, molecular weight, surface tension and electrical conductivity
- Processing parameters: feed (flow) rate, electric field strength, tip-to-collector distance, needle (tip) shape as well as collector composition and geometry
- Ambient parameters: temperature, humidity and airflow.

Nanofibre diameter, surface morphology, mechanical properties, porosity and pore size distribution highly depend on these parameters.

Polymer concentration determines solution spinnability, thus a threshold concentration of polymer solution is required. Higher concentration leads to the nanofibre formation with less irregularities but larger diameters. The latter is due to the higher number of entanglements between polymer chains, which opposes jet stretching in the applied electric field. Whereas, too dilute solution results in the combination of beads and fibres. Hence it is important to find an optimum concentration to obtain uniform bead-free fibres [6, 156]. To achieve this, a power law empirical relationship was discovered, which states that fibre diameter can increase with increasing the solution concentration [158]. Megelski et al. [159] found that increasing the PS concentration in tetrahydrofuran (THF) not only increased the fibre diameter, but also narrowed the pore size distribution. The polymer concentration influences both solution viscosity and its surface tension, which are two important processing parameters in electrospinning [6, 157]. The solutions with high viscosity cannot be ejected from the spinneret. Whereas, the solutions with low viscosity do not produce fibres [6]. It has been proven that the formation of pores, micropores and beaded structures are less likely to form during electrospinning with high solution viscosity [157, 160]. Polymer solutions have a different viscosity range for spinnability. Maximum spinning viscosities have been reported to be in the range of 1–215 poise [6]. The molecular weight of polymers determines fibre morphology as well as many electrical and physical properties of the solution including electrical conductivity, solution viscosity, dielectric strength, surface tension and the number of entanglements for polymer chains in a solution [161]. A high molecular weight often leads to large fibre diameters, and a low molecular weight results in pores and beaded structures, which are generally considered as defects in the fibre formation [155, 157]. Surface tension is the primary force to counteract the electrostatic repulsive force applied through high power voltage during the electrospinning process to determine the electrospinnability [162]. The previous results [163] indicate that the lower surface tension yields the lower tendency for the bead formation. Additionally, the minimum voltage for fabricating nanofibres increases with the surface tension of solution but not always in a linear manner [162, 164]. Another solution parameter that should be considered for better fibre structure and morphology is the electrical conductivity of polymer solution. This processing parameter is governed by the type of polymer, solvent and the availability of ionisable salts [155]. The jet radius becomes varied inversely with the cube root of solution electrical conductivity [6]. Generally, higher conductivities lead to the formation of thinner nanofibres, and thus supressing the bead generation [165]. Polymer solutions with a very low electrical conductivity cannot be electrospun because there is no electric charge at the surface of fluid droplets to form a "Taylor cone". Similarly, very high conductivity leads to a depleted tangential electric field along the surface of fluid droplets, preventing the formation of "Taylor cone" as well [6]. Limited studies have reported the effect of dielectric constants of solutions on nanofibre morphology. For successful electrospinning, the dielectric constant of solutions should be high enough in order to obtain required thin nanofibres [163, 166].

Among processing parameters, applied voltage is the critical factor because it provides surface charge on electrospinning jet and affects nanofibre diameters. Higher voltage mostly causes the reduction in fibre diameter due to the generation of greater repulsive electric field with more possibility of beads [6]. The following correlation was given to demonstrate the relationship between fibre diameter d and electrical potential V [157]:

$$\mathrm{d} \sim \mathrm{V}^{-1/2} \tag{4.1}$$

Some studies [167–169] demonstrate the formation of larger diameter fibres with increasing the applied voltage or even reported no considerable effect on fibre diameter. The jet velocity and the rate of material transfer are affected by the flow rate of polymer solution, thus contributing to fibre diameter and morphology [155, 156]. A minimum flow rate is necessary for nanofibre fabrication in which lower feeding rates can hinder electrospinning [6, 164]. Beyond this value, it leads to the formation of thicker nanofibres because the amount of polymer in the electrospun jet increases. Moreover, due to the unavailability of proper solvent evaporation time prior to reaching the collector, the generation of beads or deposition of non-dried nanofibres on the collector can be observed. Ideally, feed rate should match the solution-removing rate from the tip [162, 166]. The effect of collector distance on fibre diameters has been well investigated. A minimum distance at approximately 8–15 cm is generally required to provide sufficient time to dry fibres before reaching the collector, and avoid the bead formation when separated by smaller or greater distances [155, 170]. Furthermore, the design of needle tip and its position as well as collector geometry also influences fibre morphology and diameter. It has been reported that the reduction of orifice size leads to the decrease in fibre diameter [6, 171].

The effects of temperature and humidity as two major ambient parameters are worth noting even though electrospinning is often conducted at room temperature. Increasing solution temperature can significantly reduce mean fibre diameters, which can be attributed to its indirect effect on solvent evaporation rate and solution

Table 4.2 Effects of electrospinning parameters on fibre morphology

Parameter	Effect on fibre morphology
Applied voltage ↑	Fibre diameter ↓ initially, then ↑ (not monotonic)
Flow rate ↑	Fibre diameter ↑ (beaded morphologies occur if the flow rate is too high)
Distance between capillary and collector ↑	Fibre diameter ↓ (beaded morphologies occur if the distance between the capillary and collector is too short)
Polymer concentration (viscosity) ↑	Fibre diameter ↑ (within optimal range)
Solution conductivity ↑	Fibre diameter ↓ (broad diameter distribution)
Solvent volatility ↑	Fibres exhibit microtexture (pores on their surfaces, which increase surface area)

Reproduced from Ref. [156]

viscosity [172, 173]. The effect of humidity on fibre diameter depends on polymer composition. Regardless of moisture absorption of polymers, high humidity can either decrease or increase fibre diameter. Rapid solvent evaporation condenses the moisture present in air, leaving imprints in the form of micro- and nanopores on the fibre surface [157, 163, 174]. Table 4.2 summarises the effect of different processing parameters on nanofibre morphology.

Fibre Alignment

Electrospun fibres are usually collected as a nonwoven mat in which the fibres have random orientation. However, to expand electrospun nanofibre applications, recent interest has been focused on well-aligned and highly ordered structures. For example, in the field of photonics, uniaxially aligned nanofibres can be used as optical polarisers as well as for decoupling sample emission and excitation signals in lab-on-chip devices [175, 176]. Whereas, for bio-related applications it has been known that scaffolds with aligned fibres favour cell growth along a preferential direction [177].

The most common method for collecting aligned nanofibres is by using rotating mandrel (Fig. 4.14a) [178–181]. The degree of alignment and fibre diameter can be tailored and adjusted based on the rotational speed of mandrel or drum [182, 183]. Depending on the systems used, the threshold speed for alignment can be varied. By increasing the mandrel speed, fibre breakage can occur due to exerted tensile forces. Whereas, at a slower speed poor alignment has been reported [7, 184]. Although rotating cylinder allows for the coverage of comparatively large surface area, it is difficult to obtain perfect nanofibre alignments using only a simple drum collector. However, with the help of airflow, the alignment can be improved along the airflow direction. To achieve the further alignment, Kiselev et al. [185] eliminated the whipping motion by means of a back electrode, which leads to nearly

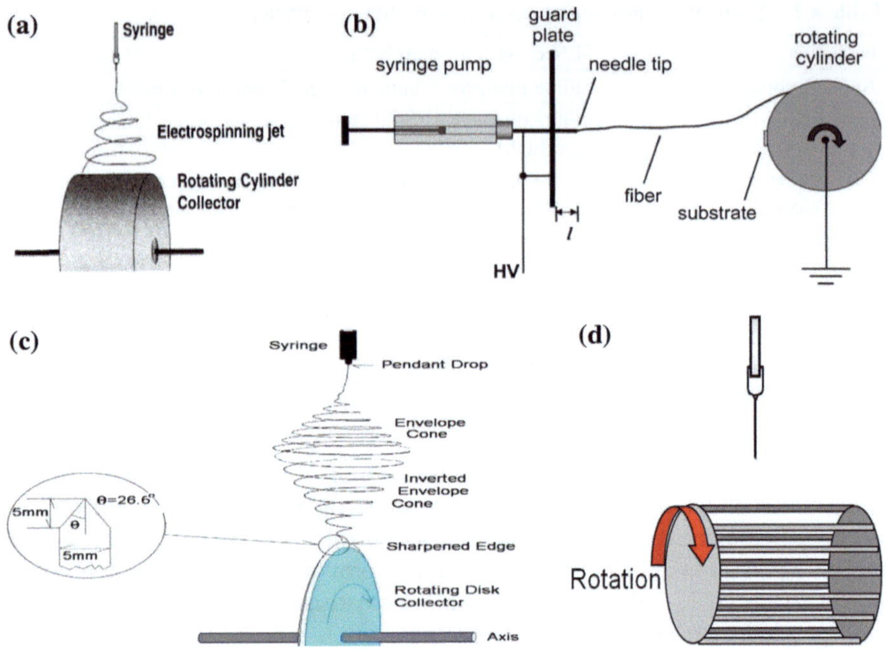

Fig. 4.14 Schematic diagrams of different collectors: **a** rotating cylinder collector, **b** rotating cylinder collector with a back electrode, **c** rotating disk collector and **d** wire drum collector. Reproduced from Refs. [188, 185, 186, 189], respectively

perfect alignment of fibres that are collected onto fast-rotating cylindrical collectors (Fig. 4.14b).

The rotating disk collector is different from rotating drum collector, which helps to improve the alignment of nanofibres. Theron et al. [186] used a wheel-like disk collector in which the strength of electrostatic field significantly increased near the disk edge (Fig. 4.14c). Compared with drum collector where jet travels in a conical path, the jet trajectory of a disk collector is both conical and inversely conical [7]. The main drawback of disk collector is the reduced area onto which fibres can be deposited due to its geometrical limitation [187]. Katta et al. [182] used copper wires with even space in form of a circular drum as a collector for manufacturing electrospun nanofibres (Fig. 4.14d). Aligned nanofibre sheets can be collected easily without affecting the aligned structure.

Another approach for fabricating aligned fibres is by the manipulation of electric field in the collecting area [184]. The most common method is based on gap or gap-like method that employs two electrodes spaced apart to create highly aligned fibres [190–192]. Li et al. [190] showed that using a pair of split collectors (composed of two conductive strips and separated by a void gap with the width varying from micrometres to several centimetres), aligned nanofibres can be formed (Fig. 4.15). The fibre alignment and array density are influenced by the gap size,

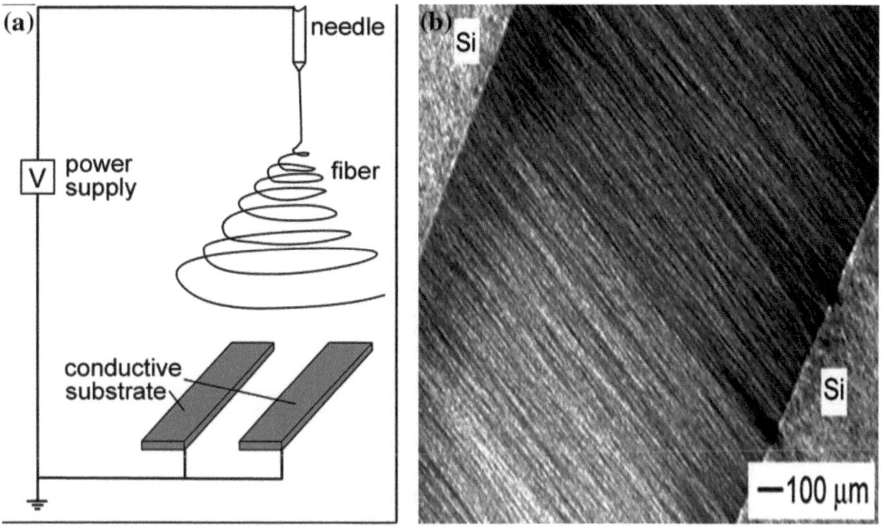

Fig. 4.15 **a** Set-up for electrospinning aligned nanofibres using the parallel plate technique, **b** Dark field optical micrograph of PVP nanofibres collected across a void gap formed between two silicon strips. Reproduced from Ref. [190]

but the results are not consistent according to previous findings [193, 194]. Moreover, the fibre alignment was affected by the thickness of deposited layer that decreased as more fibres landed. The fibres were reported to break due to the inability to sustain their own weight and repulsive charges from other fibres [182, 184]. Later to increase the size of electrospun membrane, an array of parallel strip electrodes was developed [195]. In another work [196], the collector was also modified by introducing two collecting rings in order to obtain the alignment across the ring gap with limited fibre lengths.

Park et al. [187, 197] improved the alignment quality of electrospun nanofibres by a simple reconfiguration using an inclined gap (Fig. 4.16a). This modification allows for easy transfer onto other solid substrates or devices. Compared with planar gap, fibre distribution with better alignment and homogeneity were achieved by the inclined gap (Fig. 4.16b, c).

Target electrode configurations based on electric field concentrations to achieve fibre alignment were introduced with noticeable alignment improvements [198–200]. Teo et al. [198] demonstrated that the electric field could be focused on by positioning a knife edge blade at a 45° angle to the collector and adding a steel blade to the syringe needle to create a uniformly electrostatic field. This set-up generally produces excellent fibre alignment, but a good control over the bending instability of electrospinning jet still presents a challenge. In another work [200], a single counter electrode was used, which was positioned behind a rotating collection mandrel and set at an equal and opposite voltage to the spinneret.

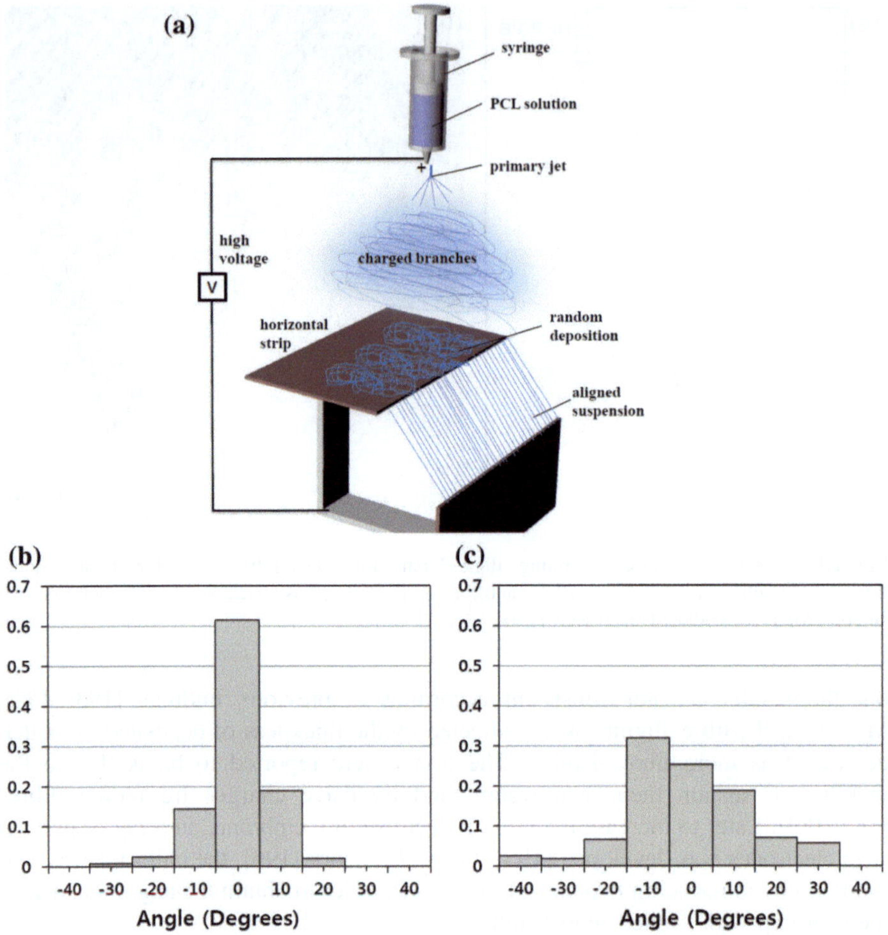

Fig. 4.16 **a** Schematic diagram of the electrospinning set-up using an inclined gap method; histograms of angle distribution of fibre alignment in the arrays transferred from **b** inclined gap collector and **c** planar gap collector. Reproduced from Refs. [187, 197], respectively

Since the electrospinning jet is charged, it can also be controlled by introducing auxiliary electrodes with various shapes to manipulate the external electric field for controlling the path and deposition area of electrospun fibres [201–204]. For instance, Nurfaizey et al. [202] used two pairs of rod electrodes positioned adjacent and parallel to the line of flight of spun fibres to generate an auxiliary field. This method could be used to control the flight path of an electrospun fibre. Arras et al. [205] upgraded a conventional electrospinning set-up by two turnable plate-like auxiliary high-voltage electrodes to allow for aligned fibre deposition in adjustable directions (Fig. 4.17a). At the target speed of 0.9 ms^{-1}, 90 % of the fibres had

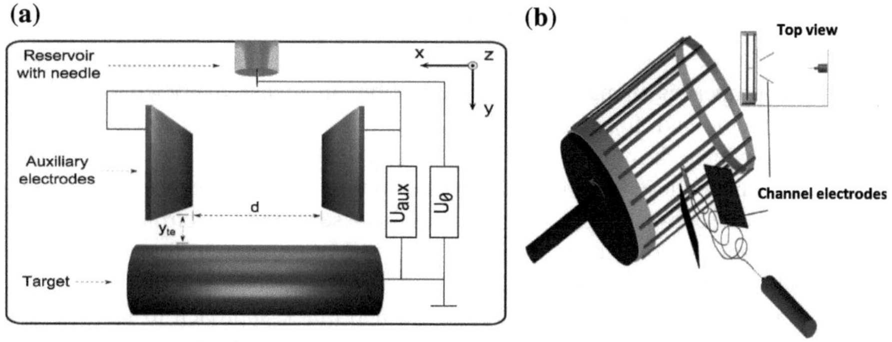

Fig. 4.17 **a** Top view of the electrospinning set-up with two turnable plate-like auxiliary high-voltage electrodes and **b** pole electrospinning with channel electrodes. Reproduced from Refs. [205, 203], respectively

aligned within 2°. Whereas, the angular spread was 70° without the use of auxiliary electrodes. Later such a method was further improved, which relies on controlling both bending and buckling instabilities by electrical field modifications regardless of target design and its movement [206]. Jafari et al. [203] proposed the three-pole configuration with a channel electrode. The channel shape pole was made of two thin rectangular aluminium plates to form an open angle of about 60° (Fig. 4.17b). Using this method not only yields aligned nanofibres with regular diameters but also helps to scale up for industrial mass production.

An alternative approach is represented by conveyor belts, made of insulating or metallic materials sliding on grounded metallic guides, which offers significant possibilities of scaling up the material production with good fibre alignment [207]. Another powerful strategy is "near-field electrospinning" or "precision electrospinning" in which collecting fibres before bending instability occurs by reducing the distance between syringe orifice and collector [121, 128].

Material Characterisation and Properties

Electrospun nanofibres still represent a relatively new class of advanced nanomaterials. It is desirable to consider not only the possibility of their preparation and applications but also their detailed characterisation and properties. Despite the difficulty in obtaining single nanofibres, their characterisation is also a tough task to achieve. Various features of electropsun nanofibres have been characterised, such as their morphological structure, mechanical properties, thermal behaviour, as well as chemical and electrical properties.

Fibre Geometry and Morphology

Geometric properties of nanofibres include fibre diameter, diameter distribution, fibre orientation and fibre morphology (e.g. cross-sectional shape and surface roughness), which show critical effects on physical properties and their applications. To measure these features, different instrumental techniques such as scanning electron microscopy (SEM), field emission scanning electron microscopy (FESEM), transmission electron microscopy (TEM) and atomic force microscopy (AFM) have been widely used [6, 208].

Among these methods, the most common and quick approach to identify fibre diameters and morphology is SEM. In case of sample preparation, the sample should be very small in size and electrically conductive so that the prepared sample can be coated with gold or platinum in order to precisely measure fibre diameters at high magnifications with good contrasts. Due to small fibre dimensions, high-quality images with appropriate magnifications are required. To obtain statistically reliable data, it is necessary to take a large number of pictures on various places of mats for statistically measuring fibre diameters. Although it is a painstaking and time-consuming process, the data are still unreliable due to small sampling areas compared with the total mat size [209]. TEM is another useful technique for detecting fibre morphology with very small diameters (<300 nm). Since TEM does not require samples in a dry state as opposed to SEM, it is possible to observe nanofibres directly electrospun from a polymer solution [6]. The only limitation is that nanofibres need to be collected onto carbon-coated Cu grids [208]. On the other hand, AFM is an alternative approach to characterise the morphology of nanofibres, enabling to provide the information of nanofibre morphology, topography, and surface roughness [208]. By employing the phase imaging mode on the AFM, surface topography can be imaged. For example, the relative roughness of poly (L-lactic acid) (PLLA) nanofibres before and after surface modifications were comparatively measured by Chen and Su [210] with AFM. Figure 4.18 shows nanofibre structures observed through SEM, TEM and AFM, respectively.

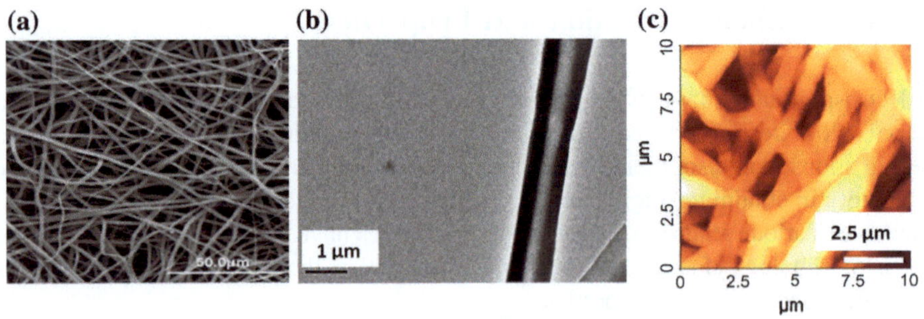

Fig. 4.18 **a** SEM image of PCL nanofibres, **b** TEM image of co-electrospun PEG/PLA and **c** AFM image for the surface morphology of PLLA nanofibres. Reproduced from Refs. [30, 33, 210]

The fibre diameter and its distribution can be obtained by conventional manual collecting method in a very time-consuming and tedious process [211]. The obvious drawback is its sensitivity to human errors and inapplicability for quality control. To address this issue, image analysis has been employed using two methods, namely distance transform [209] and direct tracking [212]. Typically, these methods first isolate each fibre in the images and then measure the diameter of each isolated fibre. Fibre isolation is an error-prone process. Öznergiz et al. [213] reported an automated calculation of nanofibre diameter without fibre isolation based on image processing and analysis algorithms.

Membrane thickness is another feature that plays an important role in some nanofibre applications such as air filtration. Due to the high porosity of electrospun nanofibres, they can be easily compressed. Hence, conventional equipment for thickness measurement such as a micrometer screw gauge may introduce a systematic error in the readings. Two common methods for thickness measurement of electrospun mats are SEM and measurement using a digital micrometer [214, 215]. Although SEM can provide good accuracy, the membrane may be distorted in the sample preparation step, and thus the process becomes quite time-consuming. Using a micrometer is a fast and easy method to measure thicknesses of any materials, but due to the direct contact with the sample and exerting force, the results become uncertain. Affandi et al. [216] measured the thickness of electrospun nanofibrous membrane (PAN and nylon) using a white light profilometer. The results were compared with micrometer measurements as well. As expected, the thickness measured using the former is significantly larger than that measured from the latter. The thickness values [216] of electrospun PAN and nylon 6 as measured by white light profilometry and a digital micrometer are shown in Table 4.3 for comparison.

Another geometric parameter is porosity, which is one of important parameters in filter design and filter performance. There are a few methods for measuring the porosity such as conventional methods using apparent density and bulk density, image analysis and mercury porometry [208, 211, 217].

Table 4.3 Thickness of electrospun PAN and nylon 6 as measured by white light profilometry and a digital micrometer

Electrospun membrane	Collection time (min)	Thickness (μm)	
		White light profilometry (step height)	Digital micrometer
PAN-(30 min)	30	21 ± 1.1	10 ± 0.7
PAN-(60 min)	60	46 ± 0.8	30 ± 0.1
Nylon 6-(30 min)	30	10 ± 0.3	7 ± 0.6
Nylon 6-(60 min)	60	16 ± 0.1	6 ± 0.1
Glass slide[a]	–	1100 ± 0.1	1100 ± 0.1

[a]control Reproduced from Ref. [216]

Based on apparent density, the porosity is determined by the given equation:

$$\text{Porosity} = \left(1 - \left[\frac{\text{Apparent density of the mat}}{\text{Bulk density of polymer}}\right]\right) \times 100\,\% \tag{4.2}$$

The apparent density of nanofibre mats can be determined by measuring the weight per unit volume, and the bulk density is defined for each type of polymer.

A capillary flow porometer can also be used to analyse the porosity and pore size of nanofibrous mats. This method is based on the difference in flow rates of a gas through the dry membrane and through the membrane wetted with a low surface energy fluid [208, 211, 218]. The pore diameter (D) can be calculated from the differential pressure by the following equation:

$$D = \frac{4\gamma \cos\theta}{p} \tag{4.3}$$

where γ is the surface tension of a wetting liquid, θ is the contact angle of a wetting liquid and p is the differential pressure. While it is useful for transport applications, this method requires an extreme care to assure that air flow does not disrupt the pore structure of the sample. Furthermore, finding an ideal liquid without interactions with sample or swelling is not easy at all. This technique also does not quantify the distribution of pore volume within the membrane/mat [211, 217].

Mercury intrusion porosimetry (or liquid extrusion porosimetry) is another method to study the pore structure of nanofibres. In addition to pore size and its distribution, total pore volume and total pore area can also be determined [217]. Mercury intrusion depends on the volumetric measurement of a non-wetting liquid. In this case, mercury is intruded into membrane pores by increasing the pressure. Whereas, liquid extrusion relies on the volumetric measurement of a wetting liquid that is extruded from the membrane pores by increasing the pressure. Whether a liquid is wetting or not depends on its contact angle θ with the used material for the membrane. The pressure p_i at which the wetting liquid is extruded, or the non-wetting liquid intruded into the pores, is calculated by the Washburn equation [217]:

$$p_i = \alpha / D \tag{4.4}$$

where D is the pore diameter and $\alpha = \pm 4\ \gamma \cos\theta$. γ is the surface energy of the liquid. '+' corresponds to liquid extrusion where $\cos\theta > 0$, and '−' refers to liquid intrusion where $\cos\theta < 0$. Applying this method is accompanied by some inaccuracies and concerns such as sample collapse and compression due to the high pressure, as well as not considering closed pores due to the inability of mercury to intrude into them. The other drawbacks of using mercury are its cost and toxicity. In general, pore size in the range of 1.8 nm–400 μm can be studied by this technique [211, 217].

With the aid of image analysis, it is possible to explore porosity measurement of electrospun nanofibre mats as well [219, 220]. In relation to this, micrographs of nanofibre mats, usually obtained by SEM, TEM or AFM with high-quality and appropriate magnifications, are required to produce binary images. Ziabari et al. [219] applied this method and evaluated the effect of mat density, fibre diameter and its variation on pore characteristics of mats. The results demonstrated that mat density and fibre diameter influenced the pore characteristics considerably despite insignificant effect of fibre diameter variations. Image analysis is a simple, comprehensive and fast method that can be used to directly measure parameters of pore structures. Recently, Sreedhara et al. [221] determined the porosity of nanofibre mats using a pycnometer. The volume of nanofibres was measured by a pycnometer, and thus porosity was determined as a void fraction in the total fibre volume. This total volume was calculated from FESEM image thickness. This method provides some advantages like minimum damage to the fibres, chemically inert, measurement of total porosity including micro and mesopores due to the ease of penetration of helium gas, and independence of the medium direction as the static pressure is applied.

Mechanical Properties

Mechanical testing of an individual nanofibrous sample appears to be the most straightforward method for investigating its mechanical behaviour. Although electrospun nanofibres are ultrathin, they are usually collected in form of nonwoven bundles, causing the difficulty in mechanical characterisation. A wide spectrum of methods have been proposed mostly based on AFM for mechanical measurement of individual polymeric nanofibres. These techniques are mainly based on tensile, stretching, bending, nanoindentation, resonance frequency and shear modulation frequency of fibres [222–225].

Tensile Tests

Two main approaches, based on AFM and nano/microtensile apparatus, can be conducted, which involves measuring the applied load and fibre elongation for a given cross-sectional area [225]. In case of AFM based method, typically one end of the nanofibre is fastened onto a substrate (e.g. a silicon wafer) by the adhesive, which acts as a pulling element. The other end is tethered to the AFM tip that works as a force sensing element. The microscopic tensile force is applied through the motion of the AFM tip with the stress–strain behaviour determined from the length of the nanofibre that is simultaneously imaged by SEM [224] or optical microscopy [226]. The applied tensile force is further determined from cantilever deflection and its spring constant (Fig. 4.19).

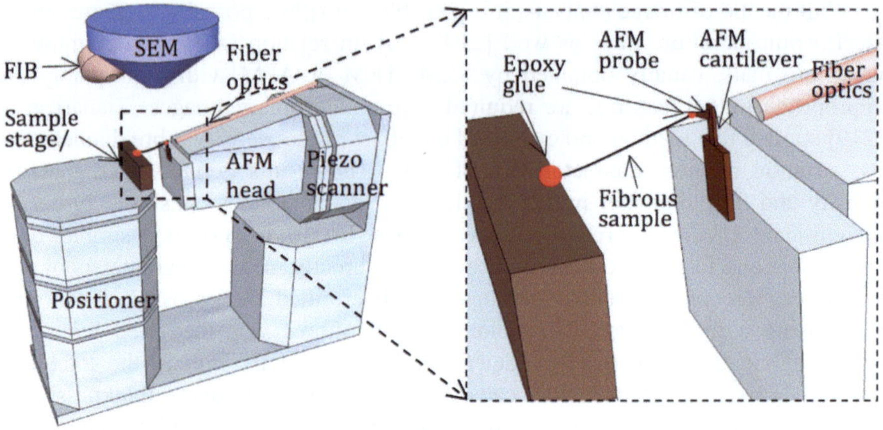

Fig. 4.19 In situ configuration of tensile tester using a nanofibre sample by the combined AFM and SEM mode. Reproduced from Ref. [224]

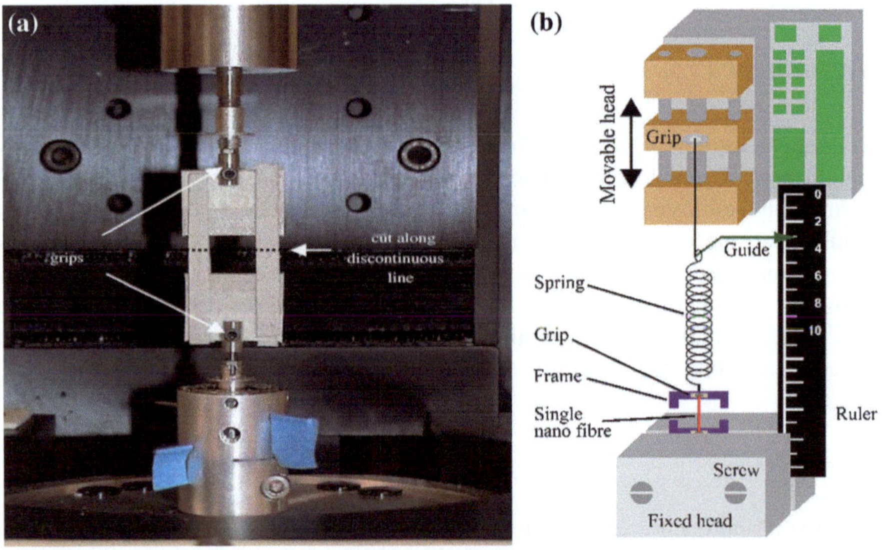

Fig. 4.20 **a** Nanotensile tester mounted with a single PCL nanofibre (NanoBionix, MTS, USA) and **b** schematic diagram of laboratory designed tensile tester. Reproduced from Refs. [228, 223], respectively

Alternative approaches based on tensile testing are using commercialised universal micro/nanotensile testers such as Nano Bionix® and Nano UTM® by Agilent® (Fig. 4.20a) [227–229] and microelectromechanical systems (MEMS)-based microtensile testers [230–232]. Bazbouz et al. [223] designed a laboratory set-up for tensile test using the configuration of two linear springs in series based on the fundamental vibration principle (Fig. 4.20b).

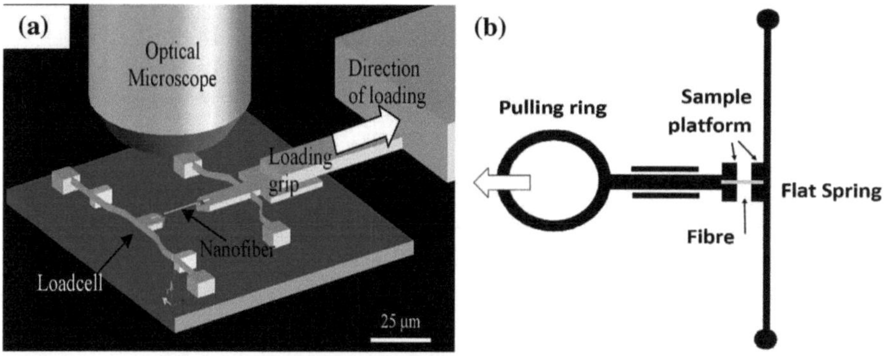

Fig. 4.21 **a** Microtensile testing platform for mechanical characterisation of single electrospun nanofibre and **b** set-up of MEMS-based tensile tester. Reproduced from Refs. [231, 232] respectively

Along the same lines, MEMS devices with on-chip force sensors and actuators have been developed to measure the load–displacement relationship of single nanofibres (Fig. 4.21a) [231]. A MEMS system which is composed of two movable components, namely pulling and bending cantilevers was used by Jaeger et al. (Fig. 4.21b) [232]. Fee et al. [230] utilised a precise screw-type linear actuator and a 1.1 N load cell that were both controlled by a customised LabView® program.

Stretching

In the stretching method, fibres are suspended over grooves in a striated and transparent substrate. The AFM, situated above the sample, is used to laterally stretch the fibres. To measure the applied force, the fluorescence microscope/optical microscope, situated below the sample, is usually employed to visualise the stretching process (Fig. 4.22a) [233–235].

Recently, Gestos et al. [235] introduced a method for fixing individual nanofibres across a gap that does not require additional fibre handling. The fibres were electrospun onto TEM grids with previously glued parallel bars (Fig. 4.22b).

Bending Tests

Most bending techniques reported in the literature are three-point bending tests that have been conducted on suspended nanofibres by spanning a gap, with their two ends fixed on a substrate. An AFM tip exerted a force at the centre point of suspended fibre, and subsequently the elastic modulus of fibres was determined by measuring the fibre deflection (Fig. 4.23a) [236–238].

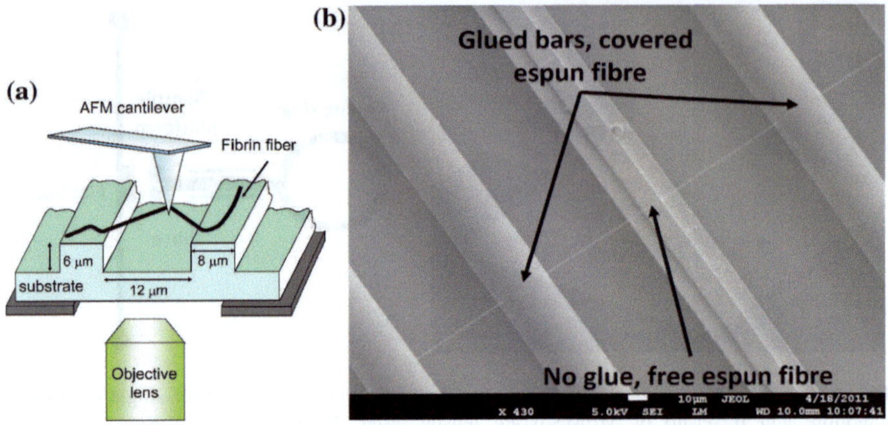

Fig. 4.22 **a** Schematic diagram of AFM/optical microscopic set-up. The AFM tip was used to stretch fibres that were suspended over the grooves of a striated substrate. **b** SEM image of an electrospun poly(acrylic acid) (PAA) fibre spanning the bars of a TEM grid with and without glue holding the nanofibre in place. Reproduced from Refs. [233, 235], respectively

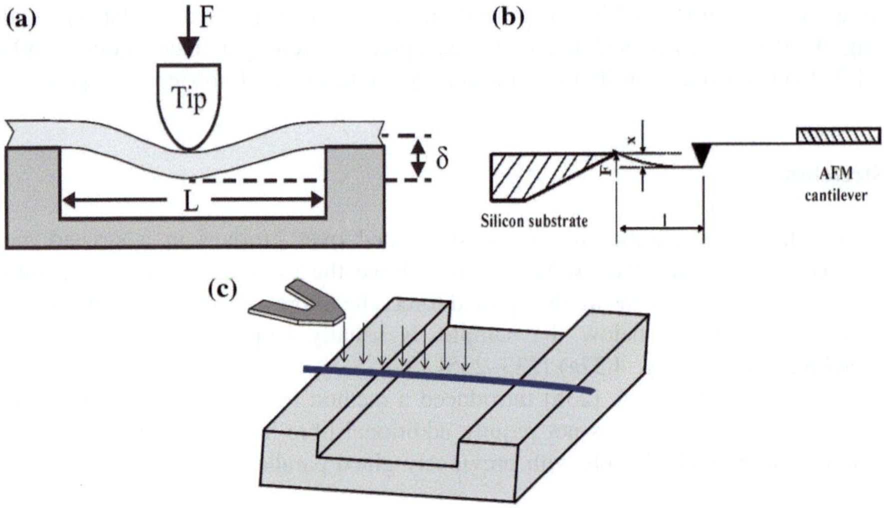

Fig. 4.23 **a** Schematic diagram of three-point bending test, **b** two- point bending test and **c** tipless multi-point bending test. Reproduced from Refs. [236, 239, 241], respectively

Gu et al. [239] performed a two-point bending test on single electrospun PAN nanofibre. Single electrospun nanofibre was attached to the end of an AFM cantilever using epoxy glue and the other end was freely loaded by a substrate edge. To eliminate the requirement to find the exact centre point of the fibre, Guhados et al. [240] introduced a multi-point bending test in which a known force was applied by the cantilever at multiple points along a suspended fibre while the fibre deflection was measured and recorded for each point (Fig. 4.23b). Later this method was further developed using a tipless cantilever to avoid damage to the fibre surface (Fig. 4.23c) [241].

Resonance Frequency Method

The mechanical resonance technique is a non-destructive method that applies an electrically or mechanically induced periodic force (approaching the nanofibre's resonance frequency) for the excitation of a nanofibre clamped to the cantilever tip inside a SEM. The resulting frequency responses of the nanofibre are then recorded and the resonance peaks are obtained in order to calculate Young's modulus (Fig. 4.24a) [242–244].

Yuya et al. [243] extracted the elastic modulus of single electrospun PAN nanofibre dynamically through the natural frequencies of a pair of AFM micro-cantilevers linked by a nanofibre segment (Fig. 4.24b). The theory of this technique is based on the dynamic relationship between the fibre stiffness (i.e. spring constant) and the resonance frequencies of cantilever vibration mode. On the other hand, Liu et al. [244] used atomic force acoustic microscopy (AFAM) based on ultrasonic frequency oscillations to excite an AFM cantilever when the tip was in contact with a sample. A different approach based on a model of the resonant frequency that is dependent on the bob's free flight was employed to measure the elastic modulus of as-spun nylon 6, 6. A ball was glued to a nanofibre and suspended from a cantilever beam that was attached to a piezoelectric-actuated base [245].

Another approach is nanoindentation tests where the Young's modulus can then be obtained through probing the localised curvature created on the fibre surface after indentation [246, 247]. Although this method is easy to perform, many factors should be considered with some uncertainties [222]. In shear modulation force

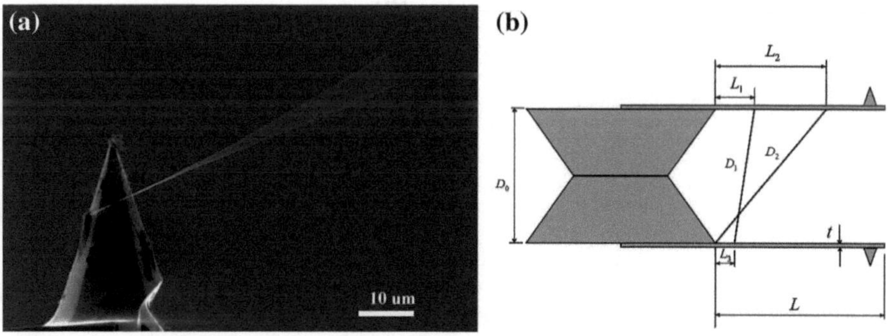

Fig. 4.24 **a** Typical example of a nanofibre driven at its fundamental resonance by the piezoelectric actuator and **b** schematic diagram that shows how the nanofibres are attached and the distance is obtained from the fixed end of the cantilever. Reproduced from Refs. [242, 243], respectively

microscopy (SMFM) method, an additional small oscillation of the tip is introduced parallel to the sample axis [244, 248]. Based on the Hertz model and assumption of a totally elastic deformation, the amplitude of the lateral deflection is then used to calculate the apparent modulus. Lin et al. [249] suggested a different method to measure mechanical properties of nanofibres through their interaction with air streams. The nanofibre was deflected by a laminar airflow along a direction perpendicular to the fibre axis at a measured velocity.

In general, during electrospinning process, nanofibres are subjected to a great elongation and drawing, which may yield in highly aligned molecular chains along the fibre axis. Hence electrospun nanofibres demonstrate much higher mechanical properties in comparison with their bulk counterparts [225, 250].

Measuring mechanical properties of single electrospun nanofibres with above-mentioned techniques reveals a feature in electrospun nanofibres that are different from their conventional counterparts. It has been found that axial modulus and ultimate tensile strength of electrospun polymeric nanofibres increase abruptly when the fibre diameter is below a certain value (Fig. 4.25) [229, 251–253].

This behaviour is ascribed to various reasons such as surface tension, chain alignment and the formation of crystalline structures as well as core/shell structure of nanofibres [251, 252, 254, 255]. For instance, Arinsten et al. [253, 256] suggested a model based on the concept of a supramolecular structure of the amorphous phase, consisting of oriented fragments of polymer chains. A manifested mechanism of confinement effect was shown to provide the size-dependent elastic modulus of electrospun nanofibres.

Another phenomenon that has been observed was surface rippling when nanofibres undergo axial stretching. Such behaviour was detected by Naraghi et al. [231, 257] when studying the deformation process of PAN nanofibres by MEMS-based microtension tests. A cascade of periodical ripples (necks) was found on the surface of these nanofibres when the axial stretch reached a certain value.

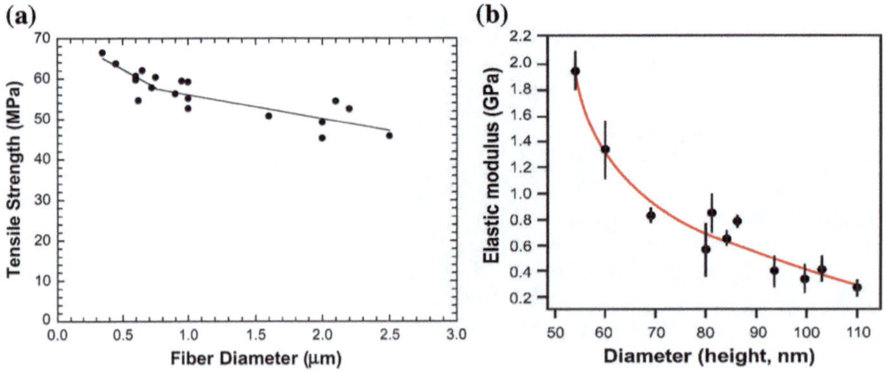

Fig. 4.25 **a** Tensile strength and tensile modulus versus diameter and **b** variations of measured elastic modulus as a function of diameter for poly (2-acrylamido-2-methyl-1-propanesulfonic acid) (PAMPS) nanofibre. Reproduced from Refs. [251, 252], respectively

The axial strain to failure (in sense of engineering strain) measured in single-fibre tension tests was up to 60–130 % for as-electrospun PAN nanofibres, which is several folds of that measured from PAN microfibres produced via drawing or dry-jet-wet spinning. Such large axial strain to failure may be ascribed to the concurrent surface rippling of nanofibres and relevant global softening. Therefore, the plastic strain became much larger than that due to single necking failure in conventional polymer microfibres [258].

Chemical Properties

For chemical characterisation of nanofibres, Fourier transform infrared (FTIR), nuclear magnetic resonance (NMR), circular dichroism (CD), differential scanning calorimetry (DSC), X-ray diffraction (XRD) and X-ray scattering can be used. For molecular structural analysis, vibrational spectroscopic technique such as FTIR is employed. This technique allows to find the chemical reaction between constituent polymers in case of polymer blends [6, 259]. FTIR results demonstrated that in as-spun fibres, silk sericin was present in a random coil conformation, but after heat treatment, the molecular structure was transformed into a β-sheet based structure [260].

Raman spectroscopy is another well recognised approach for studying structural properties of carbonaceous materials. Raman spectra provide information on the perfect crystalline of graphite-based materials [261]. Sadrjahani et al. [262] used this method to measure the molecular orientation of electrospun PAN nanofibres. A maximum chain orientation parameter of 0.25 was determined for nanofibres collected at a take up velocity of 59.5 m/min. NMR is considered as a complimentary tool to vibrational spectroscopic technique for characterising molecular structure of a nanofibre [259]. Solid-state ^{13}C NMR has been deemed to be more effective analytical tool for demonstrating the conformational transition of silk fibroin than Infrared spectroscopy [263]. Surface chemical properties of nanofibres can be determined by its hydrophilicity, which can be measured by the water contact angle analysis for the surface of nanofibre membrane and can also be determined via X-ray photoelectron spectroscopy (XPS), and attenuated total reflectance FTIR (ATR-FTIR) spectroscopy [2, 256]. XPS is capable of element detecting up to the depth of 100 Å [264]. The use of this technique confirms that the shell of electrospun nanofibres within core-shell structures does not form blend or chemically react with the core [265].

Optical birefringence, wide angle X-ray diffraction (WAXD), small angle X-ray scattering (SAXC) and DSC are employed to study the supermolecular structures that are described as the macromolecular configuration in a nanofibre [2]. WAXD is used for determining the crystal type and crystalline phase. Whereas, SAXC is employed for investigating lamellar structure of semicrystalline polymers. The latter is not used extensively due to the complexity in the interpretation of its patterns [259]. Applying small-area electron diffraction (SAED) technique is another advanced evaluation of fibre structures. Using this technique, crystallite size and misorientation angle of molecular chains within nanofibres can be obtained quantitatively as well [259, 266].

Thermal Properties

DSC can be used to determine thermal properties of electrospun nanofibres such as their crystallisation and melting processes. To prepare samples for this method, first electrospun mats about 10 mg are weighed in sealed aluminium pans. There are holes in pan covers to maintain atmospheric pressure and to allow for the evaporation of residual solvents. Subsequently, samples are heated from 30 to 300 °C, maintaining a heating rate of typically 10°C/min with constant flow of dry nitrogen. The percent crystallinity (χ_c) is determined by the following equation [208]:

$$\chi_c(\%) = \frac{\Delta H_f - \Delta H_c}{\Delta H_f^\circ} \times 100 \quad (4.5)$$

where the melting enthalpy and enthalpy of crystallisation (ΔH_f and ΔH_c) are values obtained from DSC traces. The heat of fusion ΔH_f° is a thermal property of completely crystalline substance [2, 208]. DSC analysis of electrospun nylon-6,6 nanofibres demonstrated an increase in the crystallinity of electrospun samples compared to that of neat polymer, which could be due to the high shear stress applied to a polymeric jet during electrospinning [263]. Similar observation was reported by Peresin et al. [267] that after electrospinning, the degree of crystallinity of PVA increased considerably together with a slight increase in the corresponding melting temperature (about 2 °C). These effects are explained as the result of alignment and enhanced crystallisation of polymer chains within individual PVA fibres subjected to very high shear stresses during electrospinning. In another study [268], DSC thermal characterisation reveals that by optimising the solution and processing parameters during the electrospinning process, it may be possible to tailor the degree of PLLA crystallinity [268]. The glass transition temperature (T_g) was reported to be lower compared to that of as-received polymer, in good accordance with previous results for electrospun fibres [269, 270].

Electrical Properties

Electrodes pre-patterned onto the substrate or evaporated on the top of electrospun materials have been employed for the electrical characterisation of single fibres and nonwoven mats [271]. Two-probe [272, 273] or four-probe measurements

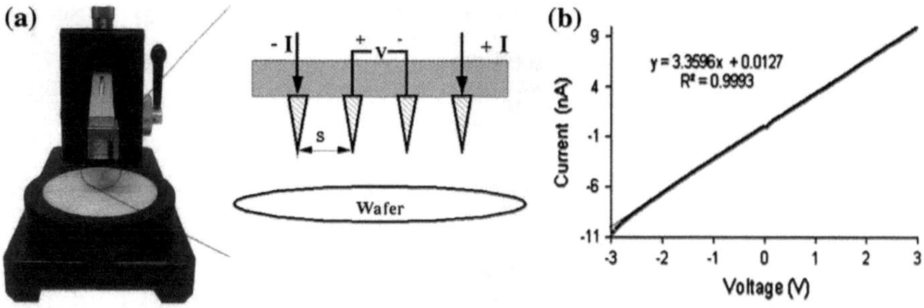

Fig. 4.26 **a** Four-point probe set-up and **b** I–V curve at a voltage in the range from −3 to +3 V for electrospun PLLA/polyaniline (PANi) nanofibres. Reproduced from Refs. [276, 272], respectively

[274, 275] are used to measure the electrical conductivity values from I-V curves [271]. Four-point probe involves having four equally spaced probes in contact with the material of unknown resistance (Fig. 4.26).

Calculations based on this method assume a thin film rather than a highly porous fibre network. Therefore, reported conductivity values could be much lower than actual value for bulk films. The penetration depth (height) of the pins into fibre mats is another source of uncertainty for electrical conductivity [276].

Agend et al. [274] used SEM images of carbonised PAN nanofibres for the evaluation of their electrical properties. One sample was coated with gold, and the other was not exposed to the sputtering process. It has been clearly shown that the carbonised PAN nanofibres possessed electrical properties (Fig. 4.27a, b). Under the evaluation of the conductivity behaviour of nanofibres with four-point probe, a sharp increase in conductivity was reported with increasing the pyrolysis temperature (Fig. 4.27c).

Alternative technique for determining the electrical conductivity of electrospun nanofibres is based on using interdigitated electrodes [277, 278]. Zhang and Rutledge [277] have found that using hot pressure of fibres onto the electrodes, a fine electrical contact can be obtained. Then the contact resistance was estimated by measuring the total fibre resistance on interdigitated electrodes with variable finger spacing and by extrapolating the resistance value at zero spacing (Fig. 4.28a, b). The electrical conductivities of fibres were found to increase exponentially with the weight percentage of doped PANi in the fibres, as well as with values as high as 50 ± 30 S/cm for as-electrospun fibres of 100 % doped PANi and as high as 130 ± 40 S/cm upon further solid-state drawing (Fig. 4.28c) [277].

The electrospinning of conductive polymers mainly focuses on PANi and its blends. Highly conductive sulfuric acid-doped electrospun PANi fibres can be prepared using a mixture of PANi and different conventional polymers such as PEO, PS, PAN, etc. [273, 279].

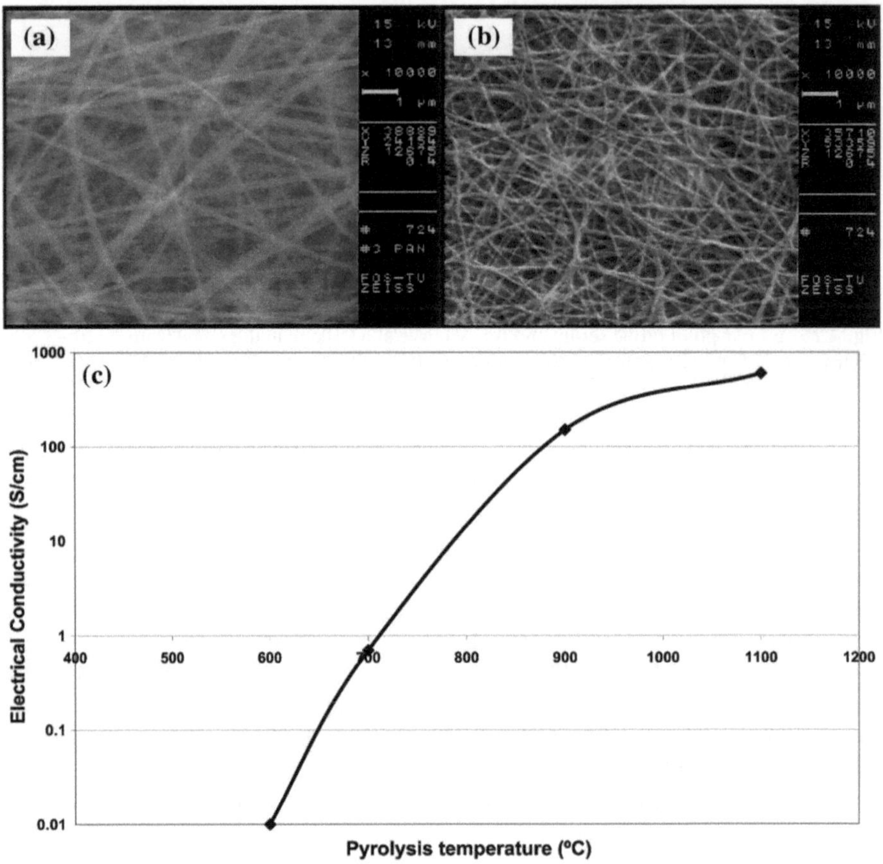

Fig. 4.27 SEM images of **a** PAN, **b** carbon nanofibres without a gold coating and **c** pyrolysis temperature versus the conductivity. Reproduced from Ref. [274]

Optical Properties

Nagata et al. [280] performed a series of optical characterisation methods using UV-Vis spectrophotometer and luminescence spectrofluorometer to investigate optical properties of poly[2-methoxy- 5-(2-ethylhexyloxy)-1,4-phenylenevinylene] (MEH-PPV) electrospun nanofibres. A significant red shift was observed for all concentrations of MEH-PPV nanofibres compared to thin film (Fig. 4.29). Photoluminescence (PL) measurements also confirmed red shift with increasing the polymer concentration. This behaviour was reported in the other study as well by Babel et al. [281] to detect the optical properties of electrospun nanofibres based on conjugated polymer blends using UV/vis/near-IR spectrophotometry and photoluminescence spectroscopy. Their results demonstrate that binary blends of conjugated polymers have tuneable, composition-dependent optical properties to be

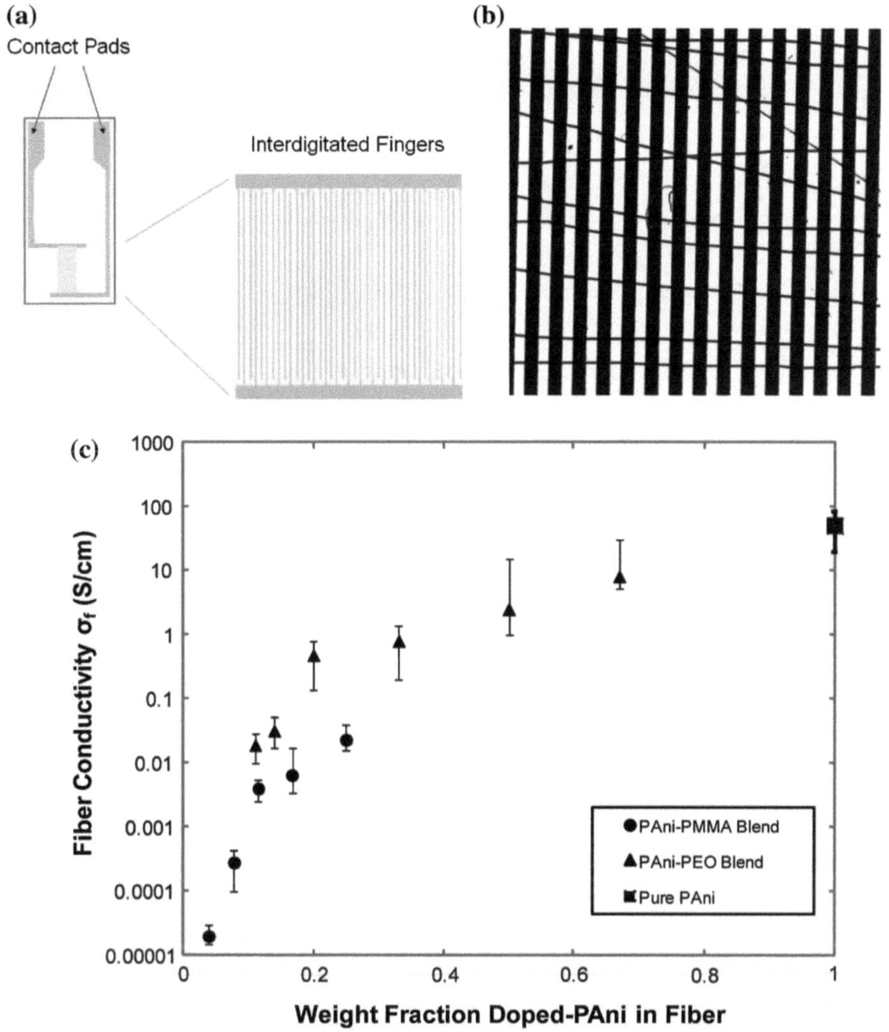

Fig. 4.28 **a** Illustration of interdigitated electrodes (IDE) and a magnified view of the fingers, **b** optical microscope image of electrospun PAni–PEO blend fibres deposited on IDE and hot pressed and **c** electrical conductivity of as-spun polyaniline fibres. Reproduced from Ref. [277]

potentially exploited in field-effect transistors. Balderas et al. [282] produced similar absorption red shift in fibres electrospun from a blend of MEH-PPV and poly(9-vinylcarbazole) (PVK).

Nanofibres obtained by colloidal electrospinning have special optical properties that have drawn much attention recently. For example, Kumar et al. [283] fabricated anatase TiO_2 nanofibres with average diameters of 60, 100 and 150 nm by controlled electrospinning of a polymeric solution and subsequent sintering of as-spun

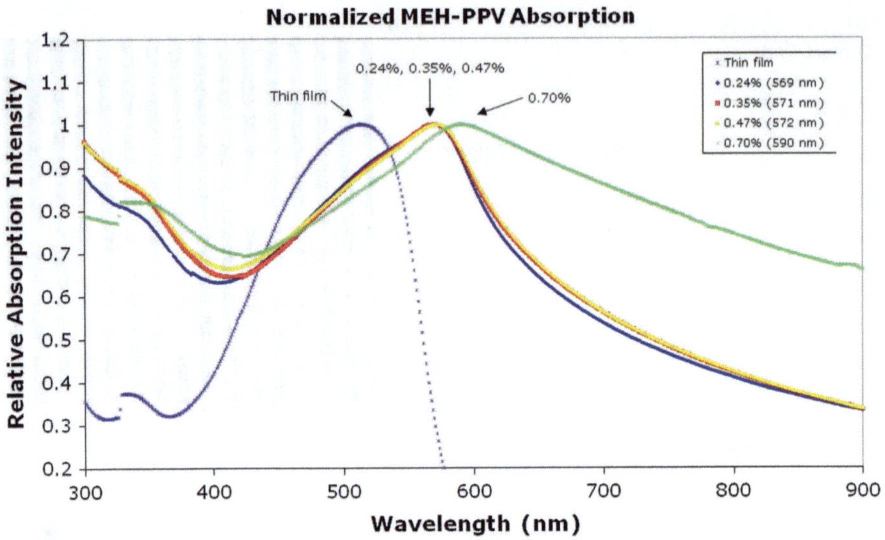

Fig. 4.29 MEH-PPV absorptions of 0.24, 0.35, 0.45, and 0.7 % concentrations in chloroform. All *curves* were normalised to their maximum value. Reproduced from Ref. [280]

fibres. It was found that the absorption spectra of fibres had a red shift with an increase in the fibre diameter, which is ascribed to an increase in the surface stress with a decrease in fibre diameter.

Applications

In recent years, research interest on electrospinning has shifted from fabrication at the laboratory level to applications. The extremely small diameters of electrospun nanofibres, coupled with the high surface area and interconnected fibrous networks, make them desirable for a wide range of applications [6].

Fibre Reinforcement

One of the useful techniques for improving the mechanical strength and other performance of composite materials is fibre-based reinforcement. Due to the unique properties of electrospun nanofibres including high aspect ratios, high specific surface area, electrospun nanofibres can be regarded as a promising material candidate for the fabrication of unique nanocomposites [225]. Kim and Reneker [284] investigated the reinforcing ability of electrospun nanofibres for the first time in 1999. The reinforcing effect of electrospun polybenzimidazole (PBI) nanofibres

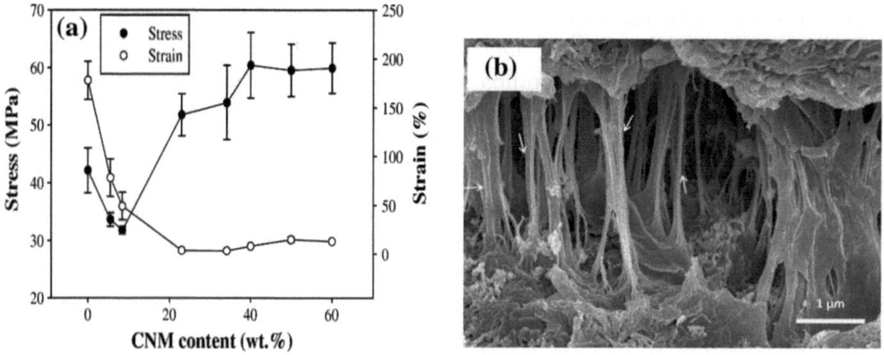

Fig. 4.30 **a** *Stress and strain curves* of PVA/CNM composites as a function of CNM content and **b** SEM micrograph of the cryofractured surface of composites reinforced with 0.26 wt% nanofibres. Reproduced from Refs. [285, 286], respectively

both in an epoxy matrix and in a styrene-butadiene rubber (SBR) matrix was investigated. With increasing the fibre content, the fracture energy increased significantly. The reinforcement effect of PBI nanofibres was proven as evidenced by a tenfold increase in Young's modulus and the tear strength was twice as large as that of unfilled rubber. It was demonstrated that by the impregnation of 40 wt% non-transparent cellulose nanofibrous mats (CNM) into PVA film, both mechanical strength and Young's modulus of composites were improved by 50 and 600 %, respectively, compared to those of neat PVA (Fig. 4.30a) [285]. Moreover, due to the intimate contact and strong interfacial adhesion, fabricated nanocomposites exhibited a high visible light transmittance of 75 %.

Lu et al. [286] prepared high density polyethylene (HDPE) based composites containing electrospun nylon 66 nanofibres by a hot-compaction method. Simultaneous improvements in tensile strength, modulus and toughness were reported. The increase in the crystallinity and interlocked morphology of matrices and nanofibres due to the impregnation of HDPE melt in interfibre voids during hot compaction caused an enhancement in mechanical properties (Fig. 4.30b). The improvement of toughness is mainly ascribed to the increase in amorphous thickness since toughness generally depends on the chain mobility in the amorphous phase. There is a threshold value for fibre content to obtain improvement in mechanical properties. Whereas, beyond this threshold further improvement would be difficult to achieve owing to more interfacial defects [225, 287]. Interestingly, dynamic mechanical analysis (DMA) proved that the presence of nanofibres enhanced the storage modulus of nanocomposites as well. For example, for composite films containing randomly distributed CNMs at 18 and 22 wt%, the E′ values were 92.3 and 205 MPa above T_g, respectively, resulting in 57- and 127-fold increases over those of epoxy resin [288]. Tang et al. [289] demonstrated the effect of fibre diameter on properties of nanocomposites. The addition of 4 wt%

nanofibres with diameters of 520 and 250 nm increased the Young's moduli of composites by 11 and 14 times, and improved the stiffness at the melting zone of 160 °C by 136 and 170 times compared to those of neat PVA, respectively. Jiang et al. [290] incorporated long and short nanofibres for composite reinforcement. It was found that to achieve similar mechanical properties, much higher fibre content (38 wt%) is required for long fibres with respect to their short counterparts (2 wt%). Another factor is nanofibre orientation that has impact on mechanical performance of resulting nanocomposites. The tensile strength and Young's modulus of poly (methyl methacrylate) (PMMA) composite films reinforced with 16 wt% PAN aligned fibres achieved 40 and 30 % increases, as opposed to those of randomly oriented fibres [291]. Meng et al. [292] designed a new 3D architecture with growing thorns on an electrospun nanofibre surface. The thorns could tie molecules and interlock with the surrounding epoxy, leading to the formation of a stronger fibre–matrix interface and enhancing the flexural strength and modulus by 36 and 38 %, respectively. Liao et al. [293] demonstrated the effect of surface chemistry and roughness of electrospun fibres on fibre/matrix interfacial structures and the corresponding macroscopic properties of composite films. It was reported that the difference in interfacial interactions not only influenced the mechanical performance of composites but also their visible light transmittance.

Additionally, the incorporation of an entangled nanofibrous layer within laminated composites may contribute to the improvement of interlaminar fracture resistance as well [294]. Magniez et al. [295] fabricated poly(hydroxyether of bisphenol A) (phenoxy) nanofibres electrospun directly onto a pre-impregnated carbon fibre material (Toray G83C) at various concentrations of 0.5–2 wt%. The fracture toughness in mode I and mode II was improved by up to 150 and 30 %, respectively. These improvements were ascribed to the formation of inverse phase domain morphology through complex micromechanisms of failure. The critical effect of polymerisation-induced phase separation for toughening interlayers by thermoplastic nanofibres was highlighted by Zhang et al. [296]. It was shown that both PCL [296] and polyetherketone cardo (PEK-C) [297] nanofibres improved fracture toughness of epoxy/carbon composites due to the phase separation of nanofibres with epoxy matrices during curing. The increases of strain energy release rate for crack initiation ($G_{IC\text{-}INI}$) for PEK-C nanofibres with diameters of 450, 750 and 950 nm were 65, 51 and 60 %, respectively. In comparison, PCL nanofibre modified composites achieved 55, 92 and 87 % increases in $G_{IC\text{-}INI}$, when nanofibres with diameters of 103, 125 and 210 nm were used. The influence of nanofibre configuration on the interlaminar toughening of epoxy/glass fibre composite laminates reinforced by PCL nanofibre nonwovens was demonstrated by van der Heijden et al. [298] using resin transfer moulding. A configuration in which the nanofibres were directly electrospun on both sides of unidirectional glass fibre mats was superior to other tested configurations due to a crack path deflection mechanism during the initiation of delamination.

Tissue Scaffolding

In recent years, tissue engineering has been recognised as a new approach for repairing the tissue or organ using bioresorbable synthetic scaffolds that help and promote the development of new tissues in vitro, and subsequently implanting them in vivo. In order to assist the formation of new tissue, the fabricated scaffold should follow the extracellular matrix (ECM) structure, which is a 3D nanoscaled fibrous network with an open porosity that provides structural support for developing tissues [299–301]. A wide range of electrospun nanofibre mats from different polymers such as synthetic and natural polymers, biodegradable and non-biodegradable polymers have been used as tissue scaffolds and related to cell growth performance. Tissue engineering techniques have been used to various types of tissue and organ including skin, bone, liver, intestine, heart valve muscle and tongue. Current research focus in this field is to find appropriate polymeric scaffolds that serve as three-dimensional templates for cell attachment, proliferation, differentiation, and tissue formation [155, 302]. Pelipenko et al. [303] investigated the effect of electrospun nanofibre diameter on the proliferation and mobility of keratinocytes and skin fibroblasts and found that nanofibres could affect the cell response. Therefore, this parameter should be considered for biomedical application, such as wound dressing or tissue scaffolds. Doustgani et al. [304] demonstrated that the alignment of nanofibres enhanced the osteogenic differentiation of stem cells. Gomes et al. [305] compared the performance of electrospun nanofibre mats from three different polymers regarding cell–scaffold interactions and wound healing promotion. According to in vitro tests, cells adhered and proliferated in all scaffolds. Whereas, cells deep into the scaffold were only found for PCL and chitosan scaffolds. The investigation made by Gluck et al. [306] showed the advantages of combining both natural and synthetic polymers to create a coaxial scaffold that is capable of withstanding dynamic culture conditions and enhancing cellular migration to the scaffold interior. Kim et al. [307] reported the successful bridging of a 17 mm nerve gap in rats using electrospun fibres. Shalumon et al. [308] fabricated a scaffold of PCL and natural polymer chitosan (CS) by electrospinning. The scaffold was shown to support the growth of various cell lines such as murine aneuploidy fibro sarcoma (L929), human osteosarcoma cells (MG63), and mouse embryo fibroblasts (NIH3T3), demonstrating prospective applications in skin and bone tissue engineering.

Drug Delivery

One of the main research areas in electrospinning for biomedical applications is drug delivery, where the fibres are used for the encapsulation of therapeutic agents. Electrospun fibres have also been used for applications in transdermal drug delivery systems or as wound dressings [28, 309]. Many drugs have been incorporated into electrospun polymeric fibres such as antibiotics [310], anticancer agents [311], anti-inflammatory agents [312], proteins [313], DNA [314] and RNA [315] and delivered to desired targets in the body. It is possible to change the drug release kinetics by the selection of polymers and through the control over nanofibre diameter, porosity, geometry, and morphology by adjusting various processing variables during electrospinning [309]. Nanofibres are also capable of site-specific delivery of more than one drug into the body. Xu et al. [316] and Ranganath et al. [317] fabricated drug delivery implants that provide site-specific sustained release of an anticancer drug at a tumour. Kenawy et al. [318] compared the release profile of ketoprofen from electrospun mats with cast films of different formulations. It was reported that electrospinning technique had superior benefits for the delivery of such drugs. Lin et al. [319] used electrospun PCL-based PU nanofibres as a carrier for water-insoluble drug nifedipine. It was found that compared with a spin-coated film, electrospun nanofibres are an ideal drug carrier with more effective controlled drug release. Zamani et al. [320] incorporated metronidazole benzoate (MET) in electrospun PCL nanofibres. In vitro drug release studies have demonstrated that solvent ratio and drug concentration have an influence on the drug release rate. The solubility and compatibility of drugs in the drug/polymer/solvent system were shown to be critical factors for the preparation of electrospun fibre formulation with a constant drug release [309]. Wang et al. [321] fabricated both poly(D,L-lactic acid) PDLLA/poly(3-hydroxybutyrate) (PHB) and PHB/PDLLA core-shell nanofibres by reversing polymer solutions in the inner and outer capillaries. The nanofibres used as a carrier for the drug delivery of dimethyloxalylglycine (DMOG) from the core section. Two-stage drug release kinetics was observed for core-shell structure, in contrast to the burst release from electrospun nanofibres of either PDLLA or PHB alone. Another effective approach to achieve sustained release is blending hydrophilic–hydrophobic polymers using different polymer combinations, thus increasing the drug-loading efficiency considerably and subsequently decreasing the burst release of drugs [322].

Nanofiltration

Electrospun nanofibrous structure possess several attractive features, such as high porosity and interconnected open pore structure, submicron pore sizes, and a large surface area-to-volume ratio. These characteristics make them an ideal material candidate for removing solid substances from air or liquid [6]. To demonstrate the filtration performance of electrospun nanofibre membrane, a thin coat of electrospun PAN nanofibres over a metal mesh has been shown to collect more than 95 % of PM2.5 (i.e. particulate matter in size of 2.5 μm) in a polluted city environment with 90 % transparency [323]. Filtration efficiency, which is closely associated with fibre fineness, is one of the most important concerns for the filter performance [6]. Molaeipour et al. [324] used ultrafine cellulose acetate nanofibres with different diameters to evaluate their filtration efficiencies for tar removal. It was found that nanofibre with diameters of 280 nm demonstrated significantly higher filtration efficiency (10 % and greater) than fibres with diameters of 370 and 620 nm. In the study investigated by Li et al. [325] where PVA nanofibres with different thickness values were used for aerosol filter media, it was shown that nanofibre layer thicker than 6.3 μm did not possess the significant improvement in filtration efficiency. For air filtration media, it has been recommended that an approximately 0.02–0.07 g/m^2 nanofibre layer with a diameter of 100–400 nm on a substrate is optimum [326]. In case of liquid filtration, electrospun polyvinyl chloride (PVC) nanofibre membranes were employed for the removal of divalent metal ions from aqueous solutions [327]. Not only the rejection rates up to 91 % was achieved for the removal of cadmium ions, but also copper and lead ions were removed with efficiencies of 73 and 82 %, respectively. Daels et al. [328] evaluated the use of a functionalised nanofibrous microfiltration membrane as an antibacterial water filter, which was effective for killing the bacteria during the intimate contact. Wang et al. [329] fabricated a high-flux ultrafiltration medium for oil/water emulsion separation, consisting of a top layer of nonporous hydrophilic nanocomposites, a middle layer of electrospun PVA nanofibrous substrate, and a conventional nonwoven microfibrous support. Combined with the top barrier layer, oil/water emulsion tests showed a high flux rate (up to 330 L/m^2h at the feed pressure of 100 psi) and an excellent total organic solute rejection ratio of 99.8 %. Many modifications and functionalisations including adding other materials to the spinning solution, or thermal or chemical post-treatment with the aim of modifying important features such as pore size, hydrophobicity, electrical conductivity and/or mechanical integrity have also been applied to improve the filtration performance of electrospun nanofibres [330].

Nanosensors

By virtue of the large surface area, tailored pore structures, large stacking and easy surface modification of electrospun nanofibrous structures, their applications for sensors achieve high sensitivity, fast response and recovery and good reversibility. Electrospun nanofibres have been used for different sensing techniques such as optical change, electrical resistivity, electrochemical sensing and acoustic waves, depending on used materials and their properties. Electrospun fibres are often more sensitive than cast film sensors of the same material [172, 331]. A sensor material may be electrospinnable on its own. For example, electrospun PVA that is sensitive to water vapour may be used as a humidity sensor [332]. Lin [333] fabricated a surface acoustic wave (SAW) humidity sensor by casting electrospun polyaniline/poly(vinyl butyral) nanofibres on the SAW resonator; as-prepared sensor could detect 0.5 % relative humidity in 1 s, demonstrating its potentials for low humidity detection. Adewuyi et al. [334] constructed a fluorescent nanofibre probe for the detection of Ni^{2+} by electrospinning a covalently functionalised pyridylazo-2-naphthol-poly(acrylic acid) polymer. The sensing system exhibited high sensitivity and good selectivity in its fluorescence “turn-off” response towards

Ni^{2+} against other metal ions. Poltue et al. [335] used the electrospun dimethylglyoxime/poly(caprolactone) (DMG/PCL) blend nanofibres as an optical sensor for the Ni(II) detection based on the formation of red Ni(DMG)2 complex. It was found that the as-spun sensors could be used to detect Ni(II) in the concentration of 1 ppm with a good linear response between 1 and 10 ppm. Li et al. [336] produced polyphenol biosensor by blending electrospun carbon naofibres with laccase and Nafion. The results showed that the sensitivity of the biosensor was 41 μA mM^{-1}, the detection limit was 0.63 μM, the linear range was 1–1310 μM and the response time was within 2 s, which excelled when compared to those of most other laccase-based biosensors reported. Additives may be incorporated into electrospun fibres to amplify the reaction of its sensing material or to give it an alternative detection property [331].

Protective Clothing

Due to the superior properties of electrospun nanofibres such as thin diameter, small pore sizes, high surface area-to-volume ratio as well as good mechanical properties, incorporating them into the design of protective clothing appears to be promising for the performance improvement and significant weight reduction. These characteristics in turns allow for high penetration resistance of particles, good breathability (evaporative cooling) of materials, low areal mass and low pressure drop [337, 338]. Another attractive feature is the direct application of electrospun mats to garment systems, which can eliminate costly manufacturing steps and solve seam-sealing problems that have been limiting factors in protective garments [339]. The large surface area of electrospun nanofibres make them capable of neutralising chemical agents without impeding their air or water vapour permeability. The U.S. Army Natick Soldier Centre investigated enhancement of barrier materials via a fine layer of electrospun fibres, focusing on preventing the penetration of chemical warfare agents in aerosol form [340]. It was reported that electrospun mats of nylon 6,6, PBI, PAN and PU presented good aerosol particle protection, without a significant change in moisture vapour transport of the system. Lee et al. [341] reported that with only 0.001 kg/m^2 inclusion of electrospun PU fibres onto nonwoven substrate, the penetration of pesticide mixtures was reduced from 85 to less than 25 %. Moreover, by introducing electrospun nanofibres, not only the barrier performance of the filter was improved considerably, but also their air permeability was much higher than that of most materials currently in use for protective clothing [341]. Faccini et al. [342] fabricated layered textiles by electrospinning polyamide 6 (PA6) nanofibres with different thickness onto a nonwoven viscose substrate (Fig. 4.31). By increasing the thickness of nanofibre mat, the filtration efficiency increased to over 99 % along the whole nanoparticle range. The results highlighted the potential of nanofibres in the development of efficient personal protective equipment against nanoparticles.

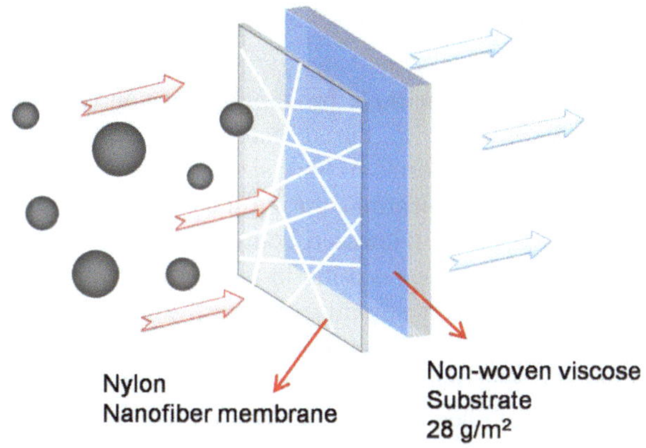

Fig. 4.31 Schematic representation of nanofibre-based protective textiles developed by Faccini et al. [342]. Reproduced from Ref. [342]

Vitchuli et al. [343] deposited electrospun nylon 6 nanofibres over nylon/cotton woven fabric and investigated the filtration efficiency by NaCl particles at the size of 300 nm for the application of protective clothing. An efficiency of more than 99.5 % without sacrificing air permeability and pressure drop was reported.

Cosmetics

Electrospun nanofibres have the potential to be used in different cosmetic applications such as perfumes, deodorants, antiperspirants and facial masks like care mask, skin healing and skin cleansing with active agents (cosmetics). It also involves employing the active agents used in cosmetics, body care supplements through a drug delivery system, which makes the cosmetics and drug delivery closely related areas [6, 37]. Taepaiboon et al. [344] loaded vitamin E (α tocoferol) and vitamin A (retinoic acid) onto both electrospun fibre mats and cast films. Unlike the burst release detected in cast film, electrospun nanofibres presented a gradual and monotonous increase in the cumulative release of vitamins over the testing periods. Emulsion electrospinning has been used to fabricate membranes containing volatile fragrance. PVA fibrous matrix has been demonstrated to release encapsulated (R)-(+)-limonene over 15 days under ambient conditions [345]. Fathi-Azarbayjani et al. [346] evaluated the skin penetration of active ingredients in an electrospun antiwrinkle nanofibre face mask. The high surface area-to-volume ratio of nanofibre mask ensured the maximum contact with the skin surface and helped to enhance the skin permeation to restore its healthy appearance. Sheng et al. [347] demonstrated electrospinning a blend of water soluble vitamin E (VE) and silk fibroin (SF) to produce nanofibrous membranes with up to 8 wt% of vitamin E. In vitro tests showed that the mouse skin fibroblasts (L929 cells) cultured on the VE-loaded SF nanofibrous mats spread and proliferated much better than those on cover slips, suggesting a promising applicative potential of this novel product on personal skin care.

Summary

Among several methods for fabricating nanofibres, electrospinning has drawn an extensive interest to the research community due to its relatively impressive characteristics in terms of fibre size, economy of the production, scale-up capability and ease of modifications during the operation. In the past few decades, enormous progress has been witnessed both in a fundamental understanding of the details for complex processes, resulting in the fibre formation and the development of a broad range of technical spinning devices. Along with monoaxial electrospinning, the occurrence of diverse set-ups for electrospinning, such as coaxial electrospinning, emulsion electrospinning, side-by-side electrospinning, facilitates the fabrication of an abundant class of polymers from a solution or melt with different structures. To extend the application of electrospun nanofibres, the modification has been employed to the collector, resulting in the production of aligned structures. In addition, needleless electrospinning and multijet electrospinning demonstrate the great potential in electrospinning for mass production. In order to obtain desired features for specific applications, fibre morphology and structure can be tailored by the manipulation of processing parameters such as solution concentration, viscosity, applied voltage, tip-to-collector distance, flow rate, etc. As far as the properties of electrospun nanofibres are concerned, many characterisation methods have been developed extensively. One of the important features revealed is the size dependency on their mechanical properties. For example, tensile properties such as increases of elastic modulus and strength with decreasing fibre diameter. Due to the extraordinary characteristics of electrospun nanofibres, they are attractive for a wide range of applications such as life science (with tissue engineering, drug delivery, wound healing), protective clothing, sensor, filtration, reinforcement of composite materials, microelectronics, space applications, and microwave absorption.

Future Trends and Recommendations

Unquestionably, in the area of electrospinning, more substantial progress has been made in the last decade than ever before, and technological advances continue to evolve. In spite of significant breakthroughs, numerous problems need to be tackled in the future.

- **Processing and fabrication**: It remains a challenge to ensure uniform nanofibres to be fabricated repeatedly and at large mass-production levels with specific morphologies, mechanical and chemical properties that are tailored for the requirements of end-user applications. Moreover, the performance of electrospun nanofibres can be further optimised by understanding how processing conditions and solutions parameters can control fibre mat properties for different polymers. It is well known that interactions of numerous processing parameters make electrospinning complicated. Hence, further investigations are required to achieve expected outcomes from the optimisation of processing parameters. Natural polymers, in some cases, exhibit poor chemical and mechanical properties, but they are less commonly used in different applications compared to synthetic polymers. Hence the improvement is required to find novel hybrid polymer systems based on synthetic and natural polymers that are electrospinnable with improved functionalities to suit a wide spectrum of applications, especially in the field of biotechnology.

Furthermore, coaxial electrospinning is demonstrated as an attractive processing method to tailor physical properties of electrospun nanofibres. Nonetheless, the process control and mechanism for core-shell structures remain to be explored in a systematic manner.

In addition, due to existing drawbacks in solution electrospinning such as environmental and health issues, as well as productivity complications, it is crucial to seek substitute methods to fabricate such kinds of polymer nanofibres. Under this circumstance, melt electrospinning has more advantages over solution electrospinning, including the absence of toxic solvents as well as high productivity. However, great attempts are required to reduce fibre diameters while concurrently creating a high degree of orientation in the structure to consequently obtain theoretically predicted strengths.

Regarding the mass production of electrospun nanofibres, needleless spinnerets show great potential in electrospinning nanofibres on large scales, but needleless electrospinning of bicomponent nanofibres is still a challenge. Additionally, in spite of the proposal of variously different spinnerets for the mass production, they may not be ideal and some of them still need further experimental verifications in terms of the ability to control the fibre quality and electrospinning process. A comprehensive understanding is required to evaluate the effect of polymer types and solution properties, particularly for those using organic solvent systems on the electrospinning process and productivity. Hence the design and construction of process equipment for controllable and reproducible electrospinning can act as a stimulus to provide novel products based on electrospinning technology. Further, it has been found that the jet path can be controlled by the manipulation of electrostatic attraction and repulsion. With enhanced approaches, it may be feasible to produce woven or braided fabrics directly from the electrospinning station.

- **Material characterisation and properties**: Recent developments in characterisation of electrospun nanofibres at a single level reveal mechanical behaviours such as the size dependency of their mechanical properties. To explain such behaviour, different models have been proposed, but still a holistic experimental validation is lacked and they are incapable to explain all the observed behaviours. Studies have particularly concentrated on fabricating desired nanofibres by controlling processing parameters. Over 200 types of polymers have been successfully electrospun into nanofibres, but limited studies can be found on the macromolecular orientation and crystalline structures of the fibres. Moreover, the role of crystalline phase and molecular orientation on mechanical properties is still not clear. Indeed, the development in characterisation tools such as atomic force microscopy–infrared (AFM-IR) spectroscopy, nanofocus XRD, confocal Raman spectroscopy and SAED to facilitate structural investigations at the single-fibre level will contribute to deeper understanding and broad exploitation of electrospun nanofibres up to their full potential.

- **Applications**: Although electrospun nanofibres are shown to be a good candidate for composite reinforcements but little work has been done in this area mainly due to their lower mechanical properties compared with their conventional counterparts such as carbon fibres or glass fibres. Therefore, it may be appropriate to develop the mechanical properties of electrospun nanofibres by the post-treatment of adding nanoparticles.

  In the field of tissue engineering, one of the main challenges that hinders the expansion of electrospinning applications is increasing scaffold thickness and pore size. Therefore fundamental research into the physics of electrospinning becomes important. Besides, it is difficult to fabricate identical scaffolds between research groups, due to many parameters that contribute to fibre properties and resulting cell responses, which in turn narrows down the use of electrospun nanofibre mats for tissue engineering. Future research in this field should focus on the convergence of several disciplines to regenerate complex human tissues and organs. Although it is well established that nanofibres can interfere with cell proliferation, differentiation and orientation, such detailed information is still not sufficient to expand fully functional tissue replacements.

  A systematic study should be conducted to investigate the relationships between drug-controlled release profiles and nanofibre structures. Mathematical models of drug release from various nanofibres would be beneficial to elucidate the drug transport mechanism, and to estimate the drug release kinetics as a function of nanofibre structure. Moreover, reports on the release of antibacterial agents or anticancer drugs from electrospun fibres have been mostly carried out in vitro, but in vivo studies have been rarely seen.

  As mentioned in this chapter, extraordinary characteristics of electrospun nanofibres have been revealed for sensor applications. To enhance the sensitivity, further investigation is suggested to improve the surface area and pore structures of nanofibrous membranes. Besides, the integration of such nanomaterials into useful devices is a challenge, which requires materials with the well-controlled orientation, size, as well as reproducibility to relocate them in specific positions and orientations.

  Finally, with respect to filtration applications, potential development in this area may include employing various polymers and/or post-treatment methods to control the formation of pores on fibres, along with a better understanding of transport mechanism through fibres with microporous rough surfaces.

Additionally, there is a demand to thoroughly study the long-term stability, particularly for modified electrospun membranes. Electrospinning does not produce dense membranes required for the separation in diffusion-based processes such as nanofiltration or reverse osmosis, though they may provide a better support layer than currently used phase inversion membranes.

References

1. Ellison CJ, Phatak A, Giles DW, Macosko CW, Bates FS (2007) Melt blown nanofibers: fiber diameter distributions and onset of fiber breakup. Polymer 48:3306–3316
2. Li H, Ke Y, Hu Y (2006) Polymer nanofibers prepared by template melt extrusion. J Appl Polym Sci 99:1018–1023
3. Ma PX, Zhang R (1999) Synthetic nano-scale fibrous extracellular matrix. J Biomed Mater Res 46:60–72
4. Yang Z, Xu B (2007) Supramolecular hydrogels based on biofunctional nanofibers of self-assembled small molecules. J Mater Chem 17:2385–2393
5. Reneker DH, Chun I (1996) Nanometre diameter fibres of polymer, produced by electrospinning. Nanotechnology 7:216–223
6. Bhardwaj N, Kundu SC (2010) Electrospinning: a fascinating fiber fabrication technique. Biotechnol Adv 28:325–347
7. Baji A, Mai Y-W, Wong S-C, Abtahi M, Chen P (2010) Electrospinning of polymer nanofibers: effects on oriented morphology, structures and tensile properties. Compos Sci Technol 70:703–718
8. Chronakis IS (2005) Novel nanocomposites and nanoceramics based on polymer nanofibers using electrospinning process—A review. J Mater Process Technol 167:283–293
9. Lannutti J, Reneker D, Ma T, Tomasko D, Farson D (2007) Electrospinning for tissue engineering scaffolds. Mat Sci Eng C Mater 27:504–509
10. Zeng J, Yang L, Liang Q, Zhang X, Guan H, Xu C, Chen X, Jing X (2005) Influence of the drug compatibility with polymer solution on the release kinetics of electrospun fiber formulation. J Controlled Release 105:43–51
11. Kenawy E-R, Bowlin GL, Mansfield K, Layman J, Simpson DG, Sanders EH, Wnek GE (2002) Release of tetracycline hydrochloride from electrospun poly (ethylene-co-vinylacetate), poly(lactic acid), and a blend. J Controlled Release 81:57–64
12. Khil M-S, Cha D-I, Kim H-Y, Kim I-S, Bhattarai N (2003) Electrospun nanofibrous polyurethane membrane as wound dressing. J Biomed Mater Res Part B 67B:675–679
13. Gorji M, Jeddi AAA, Gharehaghaji AA (2012) Fabrication and characterization of polyurethane electrospun nanofiber membranes for protective clothing applications. J Appl Polym Sci 125:4135–4141
14. Qin X-H, Wang S-Y (2006) Filtration properties of electrospinning nanofibers. J Appl Polym Sci 102:1285–1290
15. Fong H (2004) Electrospun nylon 6 nanofiber reinforced BIS-GMA/TEGDMA dental restorative composite resins. Polymer 45:2427–2432
16. Lee SW, Choi SW, Jo SM, Chin BD, Kim DY, Lee KY (2006) Electrochemical properties and cycle performance of electrospun poly(vinylidene fluoride) based fibrous membrane electrolytes for Li-ion polymer battery. J Power Sources 163:41–46
17. Pinto NJ, Johnson AT, MacDiarmid AG, Mueller CH, Theofylaktos N, Robinson DC, Miranda FA (2003) Electrospun polyaniline/polyethylene oxide nanofiber field-effect transistor. Appl Phys Lett 83(20):4244–4246

18. Zhang G, Kataphinan W, Teye-Mensah R, Katta P, Khatri L, Evans EA, Chase GG, Ramsier RD, Reneker DH (2005) Electrospun nanofibers for potential space-based applications. Mater Sci Eng B-Adv Funct Solid-State Mater 116(3):353–358
19. Xiang J, Chu Y, Zhang X, Shen X (2012) Magnetic and microwave absorption properties of electrospun $Co_{0.5}Ni_{0.5}$ Fe_2O_4 nanofibers. Appl Surf Sci 263:320–325
20. Rafiei S, Maghsoodloo S, Noroozi B, Mottaghitalab V, Haghi AK (2013) Mathematical modeling in electrospinning process of nanofibers: a detailed review. Cellulose Chem Technol 47(5–6):323–338
21. Morton WJ (1902) US patent, 705691
22. Cooley JF (1902) US patent, 692631
23. Zeleny J (1914) Phys Rev 3:69–91
24. Formhals A (1934) US patent, 1975504
25. Taylor GI (1969) Electrically driven jets. Proc R Soc Lond A Math Phys Sci (1934–1990) 313:453–475
26. Larrondo L, St John Manley R (1981) Electrostatic fiber spinning from polymer melts. I. Experimental observations on fiber formation and properties. J Polym Sci Polym Phys Ed 19:909–920
27. Li D, Xia Y (2004) Electrospinning of nanofibers: reinventing the wheel? Adv Mater 16:1151–1170
28. Hu X, Liu S, Zhou G, Huang Y, Xie Z, Jing X (2014) Electrospinning of polymeric nanofibers for drug delivery applications. J Control Release 185:12–21
29. Wang L, Liu X, Hou Z, Li C, Yang P, Cheng Z, Lian H, Lin J (2008) Electrospinning Synthesis and Luminescence Properties of One-Dimensional $Zn_2SiO_4:Mn^{2+}$ Microfibers and Microbelts. J Phys Chem C 112:18882–18888
30. Alves da Silva ML, Martins A, Costa-Pinto AR, Costa P, Faria S, Gomes M, Reis RL, Neves NM (2010) Cartilage tissue engineering using electrospun PCL nanofiber meshes and MSCs. Biomacromolecules 11:3228–3236
31. Jin Y, Yang D, Kang D, Jiang X (2009) Fabrication of necklace-like structures via electrospinning. Langmuir 26(2):1186–1190
32. Koombhongse S, Liu W, Reneker DH (2001) Flat polymer ribbons and other shapes by electrospinning. J Polym Sci Part B Polym Phys 39:2598–2606
33. Nguyen TT, Ghosh C, Hwang SG, Chanunpanich N, Park JS (2012) Porous core/sheath composite nanofibers fabricated by coaxial electrospinning as a potential mat for drug release system. Int J Pharm 439:296–306
34. Dror Y, Salalha W, Avrahami R, Zussman E, Yarin A, Dersch R, Greiner A, Wendorff J (2007) One-step production of polymeric microtubes by co-electrospinning. Small 3:1064–1073
35. Zhao Y, Cao X, Jiang L (2007) Bio-mimic multichannel microtubes by a facile method. J Am Chem Soc 129:764–765
36. Agarwal S, Wendorff JH, Greiner A (2009) Progress in the field of electrospinning for tissue engineering applications. Adv Mater 21:3343–3351
37. Zanin MHA, Cerize NNP, de Oliveira AM (2011) Production of nanofibers by electrospinning technology: overview and application in cosmetics. In: Beck R, Guterres S, Pohlmann A (eds) Nanocosmetics and nanomedicines new approaches for skin Care. Springer, Berlin, pp 311–332
38. Liang D, Hsiao BS, Chu B (2007) Functional electrospun nanofibrous scaffolds for biomedical applications. Adv Drug Deliv Rev 59(14):1392–1412
39. Haider S, Al-Zeghayer Y, Ahmed Ali FA, Haider A, Mahmood A, Al-Masry WA, Imran M, Aijaz MO (2013) Highly aligned narrow diameter chitosan electrospun nanofibers. J Polym Res 20(105):1–11
40. Rho KS, Jeong L, Lee G, Seo BM, Park YJ, Hong S-D, Roh S, Cho JJ, Park WH, Min B-M (2006) Electrospinning of collagen nanofibers: effects on the behavior of normal human keratinocytes and early-stage wound healing. Biomaterials 27:1452–1461
41. Maleknia L, Rezazadeh Majidi Z (2014) Electrospinning of gelatin nanofiber for biomedical application. Orient J Chem 30(4):2043–2048

42. Bai S, Han H, Huang X, Xu W, Kaplan DL, Zhu H, Lu Q (2015) Silk scaffolds with tunable mechanical capability for cell differentiation. Acta Biomater 20:22–31
43. Baker S, Sigley J, Helms CC, Stitzel J, Berry J, Bonin K, Guthold M (2012) The mechanical properties of dry, electrospun fibrinogen fibers. Mater Sci Eng B 32:215–221
44. Wang X, Zhao H, Turng L-S, Li Q (2013) Crystalline morphology of electrospun poly (ε-caprolactone) (PCL) nanofibers. Ind Eng Chem Res 52:4939–4949
45. Natarajan L, New J, Dasari A, Yu S, Manan MA (2014) Surface morphology of electrospun PLA fibers: mechanisms of pore formation. RSC Adv 4:44082–44088
46. Park JC, Ito T, Kim K-O, Kim KW, Kim BS, Khil MS, Kim HY, Kim IS (2010) Electrospun poly (vinyl alcohol) nanofibers: effects of degree of hydrolysis and enhanced water stability. Polym J 42:273–276
47. Chuangchote S, Sagawa T, Yoshikawa S (2009) Electrospinning of poly(vinyl pyrrolidone): effects of solvents on electrospinnability for the fabrication of poly(p-phenylene vinylene) and TiO_2 nanofibers. J Appl Polym Sci 114:2777–2791
48. Wang R, Liu Y, Li B, Hsiao BS, Chu B (2012) Electrospun nanofibrous membranes for high flux microfiltration. J Membr Sci 392–393:167–174
49. Zhang H, Zhang L, Jia Q, Shi C, Yang J (2015) Preparation of porous nylon 6 fiber via electrospinning. Polym Eng Sci 55(5):1133–1141
50. Veleirinho B, Lopes-da-Silva JA (2009) Application of electrospun poly (ethylene terephthalate) nanofiber mat to apple juice clarification. Process Biochem 44:353–356
51. Vlad S, Ciobanu C, Macocinschi D, Filip D, Butnaru M, Gradinaru LM, Nistor A (2010) Polyurethane nanofibers by electrospinning for biomedical applications. CAS Proc 2:353–356
52. Chisca S, Irina Barzic A, Sava I, Olaru N, Bruma M (2012) Morphological and rheological insights on polyimide chain entanglements for electrospinning produced fibers. J Phys Chem B 116:9082–9088
53. Xu C, Xu F, Wang B, Lu T (2011) Electrospinning of poly(ethylene-co-vinyl-alcohol) nanofibres encapsulated with Ag nanoparticles for wound healing. J Nanomater 2011, Article ID 201834
54. Han SO, Youk JH, Min KD, Kang YO, Park WH (2008) Electrospinning of cellulose acetate nanofibers using a mixed solvent of acetic acid/water: effects of solvent composition on the fiber diameter. Mater Lett 62:759–762
55. Boland ED, Wnek GE, Simpson DG, Pawlowski KJ, Bowlin GL (2001) Tailoring tissue engineering scaffolds using electrostatic processing techniques: a study of poly(glycolic acid) electrospinning. J Macromol Sci Pure 38:1231–1243
56. Gensheimer M, Becker M, Heep AB, Wendorff LH, Thauer RK, Greiner A (2007) Novel biohybrid materials by electrospinning: nanofibers of poly(ethylene oxide) and living bacteria. Adv Mater 19:2480–2482
57. Fang J, Zhang L, Sutton D, Wang X, Lin T (2012) Needleless melt-electrospinning of polypropylene nanofibres. J Nanomater 2012, Article ID 382639
58. Wang X, Niu H, Lin T (2009) Needleless electrospinning of nanofibers with a conical wire coil. Polym Eng Sci 49:1582–1586
59. Nayak R, Padhye R, Kyratzis IL, Truong YB, Arnold L (2011) Recent advances in nanofibre fabrication techniques. Text Res J 82(2):129–147
60. Hunley MT, Karikari AS, McKee MG, Mather BD, Layman JM, Fornof AR, Long TE (2008) Taking advantage of tailored electrostatics and complementary hydrogen bonding in the design of nanostructures for biomedical applications. Macromol Symp 270:1–7
61. Tian S, Ogata N, Shimada N, Nakane K, Ogihara T, Yu M (2009) Melt electrospinning from poly(L-lactide) rods coated with poly(ethylene-co-vinyl alcohol). J Appl Polym Sci 113:1282–1288
62. Hutmacher DW, Dalton PD (2011) Melt electrospinning. Chem Asian J 6(1):44–56
63. Muerza-Cascante ML, Haylock D, Hutmacher DW, Dalton PD (2015) Melt electrospinning and its technologization in tissue engineering. Tissue Eng Part B Rev 21(2):187–202

64. Farrugia BL, Brown TD, Upton Z, Hutmacher DW, Dalton PD, Dargaville TR (2013) Dermal fibroblast infiltration of poly (ε-caprolactone) scaffolds fabricated by melt electrospinning in a direct writing mode. Biofabrication 5:025001
65. Li H, Chen H, Zhong X, Wu W, Ding Y, Yang W (2014) Interjet distance in needleless melt differential electrospinning with umbellate nozzles. J Appl Polym Sci 131:40515
66. Zhmayev E, Cho D, Joo YL (2010) Nanofibers from gas assisted polymer melt electrospinning. Polymer 51:4140–4144
67. Ogata N, Shimada N, Yamaguchi S, Nakane K, Ogihara T (2007) Melt-electrospinning of poly(ethylene terephthalate) and polyalirate. J Appl Polym Sci 105:1127–1132
68. Nayak R, Kyratzis IL, Truong YB, Padhye R, Arnold L (2012) Melt-electrospinning of polypropylene with conductive additives. J Mater Sci 47:6387–6396
69. Lyons J, Li C, Ko F (2004) Melt-electrospinning part I: processing parameters and geometric properties. Polymer 45:7597–7603
70. Detta N, Brown TD, Edin FK, Albrecht K, Chiellini F, Chiellini E, Dalton PD, Hutmacher DW (2010) Melt electrospinning of polycaprolactone and its blends with poly (ethylene glycol). Polym Int 59:1558–1562
71. Góra A, Sahay R, Thavasi V, Ramakrishna S (2011) Melt-electrospun fibers for advances in biomedical engineering, clean energy, filtration, and separation. Polym Rev 51:265–287
72. Dalton P, Grafahrend D, Klinkhammer K, Klee D, Moller M (2007) Electrospinning of polymer melts: phenomenological observations. Polymer 48:6823–6833
73. Yoon YI, Park KE, Lee SJ, Park WH (2013) Fabrication of microfibrous and nano-/microfibrous scaffolds: melt and hybrid electrospinning and surface modification of poly (L-lactic acid) with plasticizer. Biomed Res Int 2013:30904
74. Chen Z, He J, Zhao F, Liu Y, Liu Y, Yuan H (2014) Effect of polar additives on melt electrospinning of non-polar polypropylene. J Serb Chem Soc 79:587–596
75. Hochleitner G, Jungst T, Brown TD, Hahn K, Moseke C, Jakob F, Dalton PD, Groll J (2015) Additive manufacturing of scaffolds with sub-micron filaments via melt electrospinning writing. Biofabrication 7:035002
76. Brown TD, Edin F, Detta N, Skelton AD, Hutmacher DW, Dalton PD (2014) Melt electrospinning of poly(ε-caprolactone) scaffolds: phenomenological observations associated with collection and direct writing. Mater Sci Eng C 45:698–708
77. Zhou H, Green TB, Joo YL (2006) The thermal effects on electrospinning of polylactic acid melts. Polymer 47:7497–7505
78. Ko J, Mohtaram NK, Lee PCD, Willerth SM, Jun MB (2013) Parametric studies of melt electrospinning poly-e-(caprolactone) fibers for tissue engineering applications. In: 8th ICOMM, 25–28 March 2013, pp 526–531
79. Lin T (2012) Needleless Electrospinning: a practical way to mass production of nanofibers. J Textile Sci Eng 2(6):1–3
80. Simm W, Gosling C, Bonart R, Falkai BV (1979) US patent, 4143196
81. Jirsak O, Sanetrnik F, Lukas D, Kotek V, Martinova L, Chaloupek J (2005) WO 2005/024101 A1
82. Yarin A, Zussman E (2004) Upward needleless electrospinning of multiple nanofibers. Polymer 45:2977–2980
83. Thoppey NM, Bochinski JR, Clarke LI, Gorga RE (2011) Edge electrospinning for high throughput production of quality nanofibers. Nanotechnology 22:345301
84. Thoppey NM, Bochinski JR, Clarke LI, Gorga RE (2010) Unconfined fluid electrospun into high quality nanofibers from a plate edge. Polymer 51:4928–4936
85. Tang S, Zeng Y, Wang X (2010) Splashing needleless electrospinning of nanofibers 50:2252–2257
86. Lu B, Wang Y, Liu Y, Duan H, Zhou J, Zhang Z, Wang Y, Li X, Wang W, Lan W, Xie E (2010) Superhigh-throughput needleless electrospinning using a rotary cone as spinneret. Small 6:1612–1616
87. Niu H, Lin T (2012) Fiber generators in needleless electrospinning. J Nanomater 2012, Article ID 725950

88. Green TB, King SL, Li L (2011) US 2011/0223330 A1
89. Liu Y, He JH (2007) Bubble electrospinning for mass production of nanofibers. Int J Nonlinear Sci Numer Simul 8:393–396
90. Yang R, He J, Xu L, Yu J (2009) Bubble-electrospinning for fabricating nanofibers. Polymer 50:5846–5850
91. Liu Y, Ren Z-F, He J-H (2010) Bubble electrospinning method for preparation of aligned nanofibre mat. Mater Sci Technol 26:1309–1312
92. Liu Y, Dong L, Fan J, Wang R, Yu JY (2011) Effect of applied voltage on diameter and morphology of ultrafine fibers in bubble electrospinning. J Appl Polym Sci 120:592–598
93. Wu D, Huang X, Lai X, Sun D, Lin L (2010) High throughput tip-less electrospinning via a circular cylindrical electrode. J Nanosci Nanotechnol 10:4221–4226
94. Wang X, Hu XW, Qiu XC, Huang X, Wu D, Sun D (2013) An improved tip-less electrospinning with strip-distributed solution delivery for massive production of uniform polymer nanofibers. Mater Lett 99:21–23
95. Yamashita Y, Ko F, Tanaka A, Miyake H (2007) Characteristics of elastomeric nanofiber membranes produced by electrospinning. J Textile Eng 53:137–142
96. Vaseashta A (2007) Controlled formation of multiple Taylor cones in electrospinning process. Appl Phys Lett 90:093115
97. Ding B, Kimura E, Sato T, Fujita S, Shiratori S (2004) Fabrication of blend biodegradable nanofibrous nonwoven mats via multi-jet electrospinning. Polymer 45:1895–1902
98. Theron SA, Yarin AL, Zussman E, Kroll E (2005) Multiple jets in electrospinning: experiment and modeling. Polymer 46:2889–2899
99. Tomaszewski W, Szadkowski M (2005) Investigation of electrospinning with the use of a multi-jet electrospinning head. Fibres Text East Eur 13:22–26
100. Zheng Y, Hugh Gong R, Zeng Y (2015) Multijet motion and deviation in electrospinning. RSC Adv 5:48533–48540
101. Yang Y, Jia Z, Hou L, Liu J, Wang L, Guan Z (2010) A shield ring enhanced equilateral hexagon distributed multi-needle electrospinning spinneret. IEEE Trans Dielectr Electr Insul 17:1592–1601
102. Xie S, Zeng Y (2012) Effects of electric field on multineedle electrospinning: experiment and simulation study. Ind Eng Chem Res 51:5336–5345
103. Angammana CJ, Jayaram SH (2011) The effects of electric field on the multijet electrospinning process and fiber morphology. IEEE Trans Dielectr Electr Insul 47:1028–1035
104. Yamashita Y, Ko F, Miyake H, Higashiyama A (2008) Establishment of nanofiber preparation technique by electrospinning. Sen'i Gakkaishi 64:24–28
105. Varesano A, Rombaldoni F, Mazzuchetti G, Tonin C, Comotto R (2010) Multi-jet nozzle electrospinning on textile substrates: observations on process and nanofibre mat deposition. Polym Int 59:1606–1615
106. Varabhas JS, Chase GG, Reneker DH (2008) Electrospun nanofibers from a porous hollow tube. Polymer 49:4226–4229
107. Dosunmu OO, Chase GG, Kataphinan W, Reneker DH (2006) Electrospinning of polymer nanofibres from multiple jets on a porous tubular surface. Nanotechnology 17:1123–1127
108. Zheng Y, Liu X, Zeng Y (2013) Electrospun nanofibers from a multihole spinneret with uniform electric field. J Appl Polym Sci 130:3221–3228
109. Zheng Y, Zeng Y (2014) Electric field analysis of spinneret design for multihole electrospinning system. J Mater Sci 49:1964–1972
110. Wang X, Um I, Fang D, Okamoto A, Hsiao B, Chu B (2005) Formation of water-resistant hyaluronic acid nanofibers by blowing-assisted electro-spinning and non-toxic post treatments. Polymer 46:4853–4867
111. Kim YM, Ahn KR, Sung YB (2003) KR Patent WO03080905
112. Arora P, Chen G, Frisk S, Graham D, Marin R, Suh H (2009) US Patent 2009/0261035
113. Ahmad B, Stride E, Stoyanov S, Pelan E, Edirisinghe M (2012) Electrospinning of ethyl cellulose fibres with a heated needle and heated air using a co-axial needle: a comparison. J Med Biol Eng 1:1–3

114. Wang X, Lin T, Wang X (2014) 3D electric field analysis of needleless electrospinning from a ring coil. J Ind Text 44:463–476
115. Bubakir MM, Li H, Wu W, Li X, Ma S, Yang W (2014) Applications of web produced by hot air assisted melt differential electrospinning method. IOP conf ser Mater Sci Eng 64:012052
116. Liao CC, Hou SS, Wang CC, Chen CY (2010) Electrospinning fabrication of partially crystalline bisphenol A polycarbonate nanofibers: the effects of molecular motion and conformation in solutions. Eur Polym J 51:2887–2896
117. Liu SL, Long YZ, Zhang ZH, Zhang HD, Sun B, Zhang JC, Han WP (2013) Assembly of oriented ultrafine polymer fibers by centrifugal electrospinning. J Nanomater 2013, Article ID 713275
118. Edmondson D, Cooper A, Jana S, Wood D, Zhang M (2012) Centrifugal electrospinning of highly aligned polymer nanofibers over a large area. J Mater Chem 12:18646–18652
119. Dabirian F, Ravandi SAH, Pishevar AR, Abuzade RA (2011) A comparative study of jet formation and nanofiber alignment in electrospinning and electrocentrifugal spinning systems. J Electrostat 69:540–546
120. Kancheva M, Toncheva A, Manolova N, Rashkov I (2014) Advanced centrifugal electrospinning setup. Mater Lett 136:150–152
121. Sun D, Chang C, Li S, Lin L (2006) Near-field electrospinning. Nano Lett 6:839–842
122. Chang C, Limkrailassiri K, Lin L (2008) Continuous near-field electrospinning for large area deposition of orderly nanofiber patterns. Appl Phys Lett 93:123111
123. Bisht G, Nesterenko S, Kulinsky L, Madou M (2012) A Computer-controlled near-field electrospinning setup and its graphic user interface for precision patterning of functional nanofibers on 2D and 3D substrates. J Lab Autom 17:302–308
124. Zheng J, Long YZ, Sun B, Zhang ZH, Shao F, Zhang HD, Zhang ZM, Huang JY (2012) Polymer nanofibers prepared by low-voltage near-field electrospinning. Chin Phys B 21(4):048102
125. Chang C, Tran VH, Wang J, Fuh YK, Lin L (2010) Direct-write piezoelectric polymeric nanogenerator with high energy conversion efficiency. Nano Lett 10:726–731
126. Zhou FL, Hubbard PL, Eichhorn SJ, Parker GJM (2011) Jet deposition in near-field electrospinning of patterned polycaprolactone and sugar-polycaprolactone core–shell fibres. Polymer 52:3603–3610
127. Camillo DD, Fasano V, Ruggieri F, Santucci S, Lozzi L, Camposeo A, Pisignano D (2013) Near-field electrospinning of light-emitting conjugated polymer nanofibers. Nanoscale 5:11637–11642
128. Hellmann C, Belardi J, Dersch R, Greiner A, Wendorff JH, Bahnmueller S (2009) High precision deposition electrospinning of nanofibers and nanofiber nonwovens. Polymer 50:1197–1205
129. Zheng G, Li W, Wang X, Wu D, Sun D, Lin L (2010) Precision deposition of a nanofibre by near-field electrospinning. J Phys D Appl Phys 43:415501
130. Luo CJ, Stoyanov SD, Stride E, Pelan E, Edirisinghe M (2012) Electrospinning versus fibre production methods: from specifics to technological convergence. Chem Soc Rev 41: 4708–4735
131. Kameoka J, Orth R, Yang Y, Czaplewski D, Mathers R, Coates GW, Craighead HG (2003) Scanning tip electrospinning source for deposition of oriented nanofibres. Nanotechnology 14:1124–1129
132. Fazley Elahi MD, Lu W, Guoping G, Khan F (2013) Core-shell fibers for biomedical applications—A review. J Bioeng Biomed Sci 3:1–14
133. Loscertales IG, Barrero A, Guerrero I, Cortijo R, Marquez M, Ganan-Calvo AM (2002) Micro/nano encapsulation via electrified coaxial liquid jets. Science 295:1695–1698
134. Amler E, Mickova A, Buzgo M (2013) Electrospun core/shell nanofibers: a promising system for cartilage and tissue engineering? Nanomedicine 8:509–512
135. Qu H, Wei S, Guo Z (2013) Coaxial electrospun nanostructures and their applications. J Mater Chem A 1:11513–11528

136. McCann JT, Li D, Xia YN (2005) Electrospinning of nanofibers with core-sheath, hollow, or porous structures. J Mater Chem 15:735–738
137. Anka FH, Balkus KJ (2013) Novel Nanofiltration hollow fiber membrane produced via electrospinning. Ind Eng Chem Res 52:3473–3480
138. Lee BS, Son SB, Park KM, Seo JH, Lee SH, Choi IS, Oh KH, Yu WR (2012) Fabrication of Si core/C shell nanofibers and their electrochemical performances as a lithium-ion battery anode. J Power Sources 206:267–273
139. Jiang G, Qin X (2014) An improved free surface electrospinning for high throughput manufacturing of core–shell nanofibers. Mater Lett 128:259–262
140. McCann JT, Marquez M, Xia Y (2006) Melt coaxial electrospinning: a versatile method for the encapsulation of solid materials and fabrication of phase change nanofibers. Nano Lett 6:2868–2872
141. Wang M, Jing N, Su CB, Kameoka J, Chou CK, Hung MC, Chang KA (2006) Electrospinning of silica nanochannels for single molecule detection. Appl Phys Lett 88:033106
142. Kalra V, Lee JH, Park JH, Marquez M, Joo YL (2009) Confined assembly of asymmetric block-copolymer nanofibers via multiaxial jet electrospinning. Small 5:2323–2332
143. Chen HY, Wang N, Di JC, Zhao Y, Song YL, Jiang L (2010) anowire-in-microtube structured Core/Shell fibers via multifluidic coaxial electrospinning. Langmuir 26:11291–11296
144. Han D, Steckl AJ (2013) Triaxial electrospun nanofiber membranes for controlled dual release of functional molecules. Interfaces 5:8241–8245
145. Srivastava Y, Loscertales I, Marquez M, Thorsen T (2008) Electrospinning of hollow and core/sheath nanofibers using a microfluidic manifold. Microfluid Nanofluid 4:245–250
146. Srivastava Y, Rhodes C, Marquez M, Thorsen T (2008) Electrospinning hollow and core/sheath nanofibers using hydrodynamic fluid focusing. Microfluid Nanofluid 5:455–458
147. Jiang H, Wang L, Zhu K (2014) Coaxial electrospinning for encapsulation and controlled release of fragile water-soluble bioactive agents. J Control Release 193:296–303
148. Wang C, Tong SN, Tse YH, Wang M (2012) Conventional electrospinning vs. emulsion electrospinning: a comparative study on the development of nanofibrous drug/biomolecule delivery vehicles. Adv Mater Res 410:118–121
149. Vasita R, Gelain F (2013) Core-sheath fibers for regenerative medicine. In: Tiwari A, Tiwari A (eds) Nanomaterials in drug delivery, imaging, and tissue engineering. Wiley, Hoboken, pp 493–534
150. Hua J, Prabhakaran MP, Ding X, Ramakrishna S (2015) Emulsion electrospinning of polycaprolactone: influence of surfactant type towards the scaffold properties. J Biomater Sci Polym Ed 26:57–75
151. Xu XL, Yang LX, Xu XY, Wang X, Chen XS, Liang QZ, Zeng J, Jing XB (2005) Ultrafine medicated fibers electrospun from W/O emulsions. J Control Release 108:33–42
152. Viry L, Moulton SE, Romeo T, Suhr C, Mawad D, Cook M, Wallace GG (2012) Emulsion-coaxial electrospinning: designing novel architectures for sustained release of highly soluble low molecular weight drugs. J Mater Chem 22:11347–11353
153. Wang C, Wang M (2014) Formation of core–shell structures in emulsion electrospun fibres: a comparative study. Aust J Chem 67:1403–1413
154. Angeles M, Cheng HL, Velankar SS (2008) Emulsion electrospinning: Composite fibers from drop breakup during electrospinning. Polym Adv Technol 19:728–733
155. Ingavle GC, Leach JK (2014) Advancements in electrospinning of polymeric nanofibrous scaffolds for tissue engineering. Tissue Eng Part B Rev 20:277–293
156. Sill TJ, von Recum HA (2008) Electrospinning: applications in drug delivery and tissue engineering. Biomaterials 29:1989–2006
157. Khan WS, Asmatulu R, Ceylan M, Jabbarnia A (2013) Recent progress on conventional and non-conventional electrospinning processes. Fiber Polym 14:1235–1247
158. Ki CS, Baek DH, Gang KD, Lee KH, Um IC, Park YH (2005) Characterization of gelatin nanofiber prepared from gelatin-formic acid solution. Polymer 46:5094–5102

159. Megelski S, Stephens JS, Chase DB, Rabolt JF (2002) Micro- and nanostructured surface morphology on electrospun polymer fibers. Macromolecules 35:8456–8466
160. Huang L, Nagapudi K, Apkarian RP, Chaikof EL (2001) Engineered collagen-PEO nanofibers and fabrics. J Biomater Sci Polym Ed 12:979–993
161. Haghi AK, Akbari M (2007) Trends in electrospinning of natural nanofibers. Phys Status Solidi 204:1830–1834
162. Andrady AL (2008) Factors affecting nanofiber quality. In: Science and technology of polymer nanofibers. Wiley, Hoboken, pp 81–110
163. Pelipenko J, Kristl J, Jankovic B, Baumgartner S, Kocbek P (2013) The impact of relative humidity during electrospinning on the morphology and mechanical properties of nanofibers. Int J Pharm 456:125–134
164. Okutan N, Terzi P, Altay F (2014) Affecting parameters on electrospinning process and characterization of electrospun gelatin nanofibers. Food Hydrocolloids 39:19–26
165. Liu Y, He JH, Yu JY, Zeng HM (2008) Controlling numbers and sizes of beads in electrospun nanofibers. Polym Int 57:632–636
166. Pham QP, Sharma U, Mikos AG (2006) Electrospinning of polymeric nanofibers for tissue engineering applications: a review. Tissue Eng 12:1197–1211
167. Zhang C, Yuan X, Wu L, Han Y, Sheng J (2005) Study on morphology of electrospun poly (vinyl alcohol) mats. Eur Polym J 41:423–432
168. Demir MM, Yilgor I, Yilgor E, Erman B (2002) Electrospinning of polyurethane fibres. Polymer 43:3303–3309
169. Reneker DH, Chun L (1996) Nanometre diameters of polymer produced by electrospinning. Nanotechnology 7:216–223
170. Geng X, Kwon K-H, Jang J (2005) Electrospinning of chitosan dissolved in concentrated acetic acid solution. Biomaterials 26:5427–5432
171. Katti DS, Robinson KW, Ko FK, Laurencin CT (2004) Bioresorbable nanofiber-based systems for wound healing and drug delivery: optimization of fabrication parameters. J Biomed Mater Res Part B Appl Biomater 70:286–296
172. Agarwal S, Greiner A, Wendorff JH (2013) Functional materials by electrospinning of polymers. Prog Polym Sci 38:963–991
173. Vrieze SD, Camp TV, Nelvig A, Hagstrom B, Westbroek P, Clerck KD (2009) The effect of temperature and humidity on electrospinning. J Mater Sci 44:1357–1362
174. Casper CL, Stephens JS, Tassi NG, Chase DB, Rabolt JF (2004) Controlling surface morphology of electrospun polystyrene fibers: effect of humidity and molecular weight in the electrospinning process. Macromolecules 37:573–578
175. Li D, Wang YL, Xia Y (2003) Electrospinning of polymeric and ceramic nanofibers as uniaxially aligned arrays. Nano Lett 3:1167–1171
176. Pagliara S, Camposeo A, Polini A, Cingolani R, Pisignano D (2009) Electrospun light-emitting nanofibers as excitation source in microfluidic devices. Lab Chip 9:2851–2856
177. Agarwal S, Greimer A, Wendorff JH (2009) Electrospinning of manmade and biopolymer nanofibers—Progress in techniques, materials, and applications. Adv Funct Mater 19: 2863–2879
178. Matthews JA, Wnek GE, Simpson DG, Bowlin GL (2002) Electrospinning of collagen nanofibers. Biomacromolecules 3:232–238
179. Vaz CM, van Tuijl S, Bouten CVC, Baaijens FPT (2005) Design of scaffolds for blood vessel tissue engineering using a multi-layering electrospinning technique. Acta Biomater 1:575–582
180. Putti M, Simonet M, Solberg R, Peters GWM (2015) Electrospinning poly(ε-caprolactone) under controlled environmental conditions: influence on fiber morphology and orientation. Polymer 63:189–195
181. Chan KHK, Wong SY, Tiju WC, Li X, Kotaki M, He CB (2010) Morphologies and electrical properties of electrospun poly (R)-3- hydroxybutyrate-co-(R)-3-hydroxyvalerate/ multiwalled carbon nanotubes fibers. J Appl Polym Sci 116(2):1030–1035

182. Katta P, Alessandro M, Ramsier RD, Chase GG (2004) Continuous electrospinning of aligned polymer nanofibers onto a wire drum collector. Nano Lett 4:2215–2218
183. Sundaray B, Subramanian V, Natarajan TS, Xiang RZ, Chang CC, Fann WS (2004) Electrospinning of continuous aligned polymer fibers. Appl Phys Lett 84:1222–1224
184. Beachley V, Katsanevakis E, Zhang N, Wen X (2012) In: Jayakumar R, Nair S (eds) Biomedical applications of polymeric nanofibers, vol 246. Springer, Berlin, pp 171–212
185. Kiselev P, Rosell-Llompart J (2012) Highly aligned electrospun nanofibers by elimination of the whipping motion. J Appl Polym Sci 125:2433–2441
186. Theron A, Zussman E, Yarin AL (2001) Electrostatic field-assisted alignment of electrospun nanofibres. Nanotechnology 12:384–390
187. Park SH, Yang DY (2011) Fabrication of aligned electrospun nanofibers by inclined gap method. J Appl Polym Sci 120:1800–1807
188. Ramakrishna S, Fujihara K, Teo WE, Lim TC, Ma Z (2005) Electrospinning process. In: An introduction to electrospinning and nanofibers. World Scientific Publishing Company, Singapore, pp 90–154
189. Teo WE, Ramakrishna S (2006) A review on electrospinning design and nanofibre assemblies. Nanotechnology 17:R89–R106
190. Li D, Wang YL, Xia Y (2004) Electrospinning Nanofibers as uniaxially aligned arrays and layer-by-layer stacked films. Adv Mater 16:361–366
191. Teo WE, Ramakrishna S (2005) Electrospun fibre bundle made of aligned nanofibres over two fixed points. Nanotechnology 16:1878–1884
192. Kakade MV, Givens S, Gardner K, Lee KH, Chase DB, Rabolt JF (2007) Electric field induced orientation of polymer chains in macroscopically aligned electrospun polymer nanofibers. J Am Chem Soc 129:2777–2782
193. Pokorny M, Niedoba K, Velebny V (2010) Transversal electrostatic strength of patterned collector affecting alignment of electrospun nanofibers. Appl Phys Lett 96:193111
194. Bazbouz MB, Stylios GK (2008) Alignment and optimization of nylon 6 nanofibers by electrospinning. J Appl Polym Sci 107:3023–3032
195. Kim G, Kim W (2006) Formation of oriented nanofibers using electrospinning. Appl Phys Lett 88:233101
196. Dalton PD, Klee D, Moller M (2005) Electrospinning with dual collection rings. Polymer 46:611–614
197. Park SH, Hong JW, Shin JH, Yang DY (2011) Quantitatively controlled fabrication of uniaxially aligned nanofibrous scaffold for cell adhesion. J Nanomater, Article ID 201969
198. Teo WE, Kotaki M, Mo XM, Ramakrishna S (2005) Porous tubular structures with controlled fibre orientation using a modified electrospinning method. Nanotechnology 16:918–924
199. Carnell LS, Siochi EJ, Holloway NM, Stephens RM, Rhim C, Niklason LE, Clark RL (2008) Aligned mats from electrospun single fibers. Macromolecules 41:5345–5349
200. Zussman E, Theron A, Yarin AL (2003) Electrostatic field-assisted building 3-D nano-structures from electrospun nanofibers. Appl Phys Lett 82:973–975
201. Wu Y, Carnell LA, Clark RL (2007) Control of electrospun mat width through the use of parallel auxiliary electrodes. Polymer 48:5653–5661
202. Nurfaizey AH, Stanger J, Tucker N, Buunk N, Wallace A, Staiger MP (2012) Manipulation of electrospun fibres in flight: the principle of superposition of electric fields as a control method. J Mater Sci 47:1156–1163
203. Jafari A, Jeon JH, Oh IK (2011) Well-aligned nano-fiberous membranes based on three-pole electrospinning with channel electrode. Macromol Rapid Commun 32:921–926
204. Acharya M, Arumugam GK, Heiden PA (2008) Dual electric field induced alignment of electrospun nanofibers. Macromol Mater Eng 293:666–674
205. Arras MML, Grasl C, Bergmeister H, Schima H (2012) Electrospinning of aligned fibers with adjustable orientation using auxiliary electrodes. Sci Technol Adv Mater 13:035008
206. Grasl C, Arras MML, Stoiber M, Bergmeister H, Schima H (2013) Electrodynamic control of the nanofiber alignment during electrospinning. Appl Phys Lett 102:053111

207. dos Santos AM, Dierck J, Troch M, Podevijn M, Schach E (2011) Production of continuous electrospun mats with improved mechanical properties. Macromol Mater Eng 296:637–644
208. Unser AM, Xie Y (2012) Electrospinning of nanofibers. In: Xie Y (ed) The nanobiotechnology handbook. CRC Press, Boca Raton, pp 293–320
209. Shin EH, Cho KS, Seo MH, Kim H (2008) Determination of electrospun fiber diameter distributions using image analysis processing. Macromol Res 16:314–319
210. Chen JP, Su CH (2011) Surface modification of electrospun PLLA nanofibers by plasma treatment and cationized gelatin immobilization for cartilage tissue engineering. Acta Biomater 7:234–243
211. Haghi AK, Zaikov GE (2010) Nanofiber research: advances in theory and practice. Nova Science Publishers. Inc, New York, pp 59–106
212. Ziabari M, Mottaghitalab V, Haghi AK (2009) Application of direct tracking method for measuring electrospun nanofiber diameter. Braz J Chem Eng 26:53–62
213. Öznergiz E, Kiyak YE, Kamasak ME, Yildirim I (2014) Automated nanofiber diameter measurement in SEM images using a robust image analysis method. J Nanomater 2014, Article ID 738490
214. Barhate RS, Loong CK, Ramakrishna S (2006) Preparation and characterization of nanofibrous filtering media. J Membr Sci 283:209–218
215. Aussawasathien D, Teerawattananon C, Vongachariya A (2008) Separation of micron to sub-micron particles from water: electrospun nylon-6 nanofibrous membranes as pre-filters. J Membr Sci 315:11–19
216. Affandi NDN, Truong YB, Kyratzis IL, Padhye R, Arnold L (2010) A non-destructive method for thickness measurement of thin electrospun membranes using white light profilometry. J Mater Sci 45:1411–1418
217. Rutledge GC, Lowery JL, Chia-Ling P (2009) Characterization by mercury porosimetry of nonwoven fiber media with deformation. J Eng Fiber Fabr 4:1–13
218. Frey MW, Li L (2007) Electrospinning and porosity measurements of nylon-6/poly(ethylene oxide) blended nonwovens. J Eng Fiber Fabr 2:31–37
219. Ziabari M, Mottaghitalab V, Khodaparast Haghi A (2008) Evaluation of electrospun nanofiber pore structure parameters. Korean J Chem Eng 25:923–932
220. Ghasemi-Mobarakeh L, Semnani D, Morshed M (2007) A novel method for porosity measurement of various surface layers of nanofibers mat using image analysis for tissue engineering applications. J Appl Polym Sci 106:2536–2542
221. Sreedhara SS, Tata NR (2013) Novel method for measurement of porosity in nanofiber mat using pycnometer in filtration. J Eng Fiber Fabr 8:132–137
222. Tan EPS, Lim CT (2006) Mechanical characterization of nanofibers—A review. Compos Sci Technol 66:1102–1111
223. Bazbouz MB, Stylios GK (2010) The tensile properties of electrospun nylon 6 single nanofibers. J Polym Sci 48:1719–1731
224. Hang F, Lu D, Bailey RJ, Jimenez-Palomar I, Stachewicz U, Cortes-Ballesteros B, Davies M, Zech M, Bodefeld C, Barber AH (2011) In situ tensile testing of nanofibers by combining atomic force microscopy and scanning electron microscopy. Nanotechnology 22:365708
225. Mohammadzadehmoghadam S, Dong Y, Daives IJ (2015) Recent progress in electrospun nanofibers: reinforcement effect and mechanical performance. J Polym Sci Pt B-Polym Phys 53:1171–1212
226. Zussman E, Burman M, Yarin AL, Khalfin R, Cohen Y (2006) Tensile deformation of electrospun nylon-6,6 nanofibers. J Polym Sci Part B Polym Phys 44:1482–1489
227. Li Y, Lim CT, Kotaki M (2015) Study on structural and mechanical properties of porous PLA nanofibers electrospun by channel-based electrospinning system. Polymer 56:572–580
228. Tan EPS, Ng SY, Lim CT (2005) Tensile testing of a single ultrafine polymeric fiber. Biomaterials 26:1453–1456
229. Papkov D, Zou Y, Andalib MN, Goponenko A, Cheng SZD, Dzenis YA (2013) Simultaneously strong and tough ultrafine continuous nanofibers. ACS Nano 7:3324–3331

230. Fee TJ, Dean DR, Eberhardt AW, Berry JL (2012) A novel device to quantify the mechanical properties of electrospun nanofibers. J Biomech Eng 134:104503
231. Naraghi M, Chasiotis I, Kahn H, Wen Y, Dzenis Y (2007) Mechanical deformation and failure of electrospun polyacrylonitrile nanofibers as a function of strain rate. Appl Phys Lett 91:151901
232. Jaeger D, Schischka J, Bagdahn J, Jaeger R (2009) Tensile testing of individual ultrathin electrospun poly(L-lactic acid) fibers. J Appl Polym Sci 114:3774–3779
233. Carlisle CR, Coulais C, Namboothiry M, Carroll DL, Hantgan RR, Guthold M (2009) The mechanical properties of individual, electrospun fibrinogen fibers. Biomaterials 30: 1205–1213
234. Baker S, Sigley J, Helms CC, Stitzel J, Berry J, Bonin K, Guthold M (2012) The mechanical properties of dry, electrospun fibrinogen fibers. Mater Sci Eng C 32:215–221
235. Gestos A, Whitten PG, Spinks GM, Wallace GG (2013) Tensile testing of individual glassy, rubbery and hydrogel electrospun polymer nanofibres to high strain using the atomic force microscope. Polym Test 32:655–664
236. Tan EPS, Lim CT (2004) Physical properties of a single polymeric nanofiber. Appl Phys Lett 84:1603–1605
237. Liao CC, Wang CC, Chen CY, Lai WJ (2011) Stretching-induced orientation of polyacrylonitrile nanofibers by an electrically rotating viscoelastic jet for improving the mechanical properties. Polymer 52:2263–2275
238. Bellan LM, Kameoka J, Craighead HG (2005) Measurement of the Young's moduli of individual polyethylene oxide and glass nanofibres. Nanotechnology 16:1095–1099
239. Gu SY, Wu QL, Ren J, Vancso GJ (2005) Mechanical properties of a single electrospun fiber and its structures. Macromol Rapid Commun 26:716–720
240. Guhados G, Wan W, Hutter JL (2005) Measurement of the elastic modulus of single bacterial cellulose fibers using atomic force microscopy. Langmuir 21:6642–6646
241. Croisier F, Duwez AS, Jérôme C, Léonard AF, van der Werf KO, Dijkstra PJ, Bennink ML (2012) Mechanical testing of electrospun PCL fibers. Acta Biomater 8:218–224
242. Zussman E, Chen X, Ding W, Calabri L, Dikin DA, Quintana JP, Ruoff RS (2005) Mechanical and structural characterization of electrospun PAN-derived carbon nanofibers. Carbon 43:2175–2185
243. Yuya PA, Wen Y, Turner JA, Dzenis YA, Li Z (2007) Determination of the Young's modulus of individual electrospun nanofibers by microcantilever vibration method. Appl Phys Lett 90:111909
244. Liu Y, Chen S, Zussman E, Korach CS, Zhao W, Rafailovich M (2011) Diameter-dependent modulus and melting behavior in electrospun semicrystalline polymer fibers. Macromolecules 44:4439–4444
245. Burman M, Arinstein A, Zussman E (2008) Free flight of an oscillated string pendulum as a tool for the mechanical characterization of an individual polymer nanofiber. Appl Phys Lett 93:193118
246. Wang W, Peijs T, Barber AH (2010) Indentation induced solid state ordering of electrospun polyethylene oxide fibres. Nanotechnology 21:035705
247. Chen YQ, Zheng XJ, Mao SX, Li W (2010) Nanoscale mechanical behavior of vanadium doped ZnO piezoelectric nanofiber by nanoindentation technique. J Appl Phys 107:094302
248. Ji Y, Li B, Ge S, Sokolov JC, Rafailovich MH (2006) Structure and nanomechanical characterization of electrospun PS/clay nanocomposite fibers. Langmuir 22:1321–1328
249. Lin Y, Clark DM, Yu X, Zhong Z, Liu K, Reneker DH (2012) Mechanical properties of polymer nanofibers revealed by interaction with streams of air. Polymer 53:782–790
250. Hwang KY, Kim SD, Kim YW, Yu WR (2010) Mechanical characterization of nanofibers using a nanomanipulator and atomic force microscope cantilever in a scanning electron microscope. Polym Test 29:375–380
251. Wong SC, Baji A, Leng S (2008) Effect of fiber diameter on tensile properties of electrospun poly (ε-caprolactone). Polymer 49:4713–4722

252. Shin MK, Kim SI, Kim SJ, Kim SK, Lee H, Spinks GM (2006) Size-dependent elastic modulus of single electroactive polymer nanofibers. Appl Phys Lett 89:231929
253. Arinstein A, Burman M, Gendelman O, Zussman E (2007) Effect of supramolecular structure on polymer nanofibre elasticity. Nat Nanotechnol 2:59–62
254. Wu XF, Dzenis YA (2007) Size effect in polymer nanofibers under tension. J Appl Phys 102:044306
255. Camposeo A, Greenfeld I, Tantussi F, Pagliara S, Moffa M, Fuso F, Allegrini M, Zussman E, Pisignano D (2013) Local mechanical properties of electrospun fibers correlate to their internal nanostructure. Nano Lett 13:5056–5062
256. Arinstein A (2013) Confinement mechanism of electrospun polymer nanofiber reinforcement. J Polym SciPart B Polym Phys 51:756–763
257. Naraghi M, Chasiotis I, Kahn H, Wen Y, Dzenis Y (2007) Novel method for mechanical characterization of polymeric nanofibers. Rev Sci Instrum 78:085108
258. Wu XF, Zhou Z, Rahman A, Bedarkar A (2013) Mechanical properties of continuous nanofibers: characterization and mechanics. In: Dong Y (ed) Nanostructures: properties, production methods and applications. Nova Science Publishers Inc, New York, pp 247–286
259. Tan EPS, Zhang YZ, Ramakrishna S, Teck LC (2007) Polymer nanofibers: fabrication, applications and characterization. In: Mohammad F (ed) Specialty polymers: materials and applications. I. K. International Pvt Ltd, New Delhi, pp 77–116
260. Rahman Khan MM, Tsukada M, Zhang XH, Morikawa H (2013) Preparation and characterization of electrospun nanofibers based on silk sericin powders. J Mater Sci 48:3731–3736
261. Lespade P, Marchand A, Couzi M, Cruege MF (1984) Caracterisation de materiaux carbones par microspectrometrie Raman. Carbon 22:375–385
262. Sadrjahani M, Hoseini SA, Mottaghitalab V, Haghi AK (2010) Development and characterization of highly oriented pan nanofiber. Braz J Chem Eng 27:583–589
263. Kim SH, Nam YS, Lee TS, Park WH (2003) Silk Fibroin Nanofiber. Electrospinning Prop Struct Polym J 35:185–190
264. Ratner BD, Chilkoti A, Castner DG (1992) Contemporary methods for characterizing complex biomaterial surfaces. Clin Mater 11:25–36
265. Zhang Y, Huang ZM, Xu X, Lim CT, Ramakrishna S (2004) Preparation of core-shell structured PCL-r-gelatin bi-component nanofibers by coaxial electrospinning. Chem Mater 16:3406–3409
266. Abbasi A, Nasef MM, Takeshi M, Faridi-Maj R (2014) Electrospinning of nylon-6,6 solutions into nanofibers: rheology and morphology relationships. Chin J Polym Sci 32:793–804
267. Peresin MS, Habibi Y, Zoppe JO, Pawlak JJ, Rojas OJ (2010) Nanofiber composites of polyvinyl alcohol and cellulose nanocrystals: manufacture and characterization. Biomacromolecules 11:674–681
268. Ero-Phillips O, Jenkins M, Stamboulis A (2012) Tailoring Crystallinity of electrospun PLLA fibres by control of electrospinning parameters. Polymers 4:1331–1348
269. Cui WG, Li XH, Zhu XL, Yu G, Zhou SB, Weng J (2006) Investigation of drug release and matrix degradation of electrospun poly(DL-lactide) fibers with paracetanol inoculation. Biomacromolecules 7:1623–1629
270. Zong XH, Ran SF, Kim KS, Fang DF, Hsiao BS, Chu B (2003) Structure and morphology changes during in vitro degradation of electrospun poly(glycolide-co-lactide) nanofiber membrane. Biomacromolecules 4:416–423
271. Luzio A, Canesi EV, Bertarelli C, Caironi M (2014) Electrospun polymer fibers for electronic applications. Materials 7:906–947
272. Prabhakaran MP, Ghasemi-Mobarakeh L, Jin G, Ramakrishna S (2011) Electrospun conducting polymer nanofibers and electrical stimulation of nerve stem cells. J Biosc Bioeng 112:501–507
273. Chronakis IS, Grapenson S, Jakob A (2006) Conductive polypyrrole nanofibers via electrospinning: electrical and morphological properties. Polymer 47:1597–1603

274. Agend F, Naderi N, Fareghi-Alamdari R (2007) Fabrication and electrical characterization of electrospun polyacrylonitrile-derived carbon nanofibers. J Appl Polym Sci 106:255–259
275. Srivastava Y, Marquez M, Thorsen T (2007) Multijet electrospinning of conducting nanofibers from microfluidic manifolds. J Appl Polym Sci 106:3171–3178
276. Zhang Y (2013) Electrospun nanofibers with tunable electrical conductivity. Ph.D dissertation, Massachusetts Institute of Technology, USA
277. Zhang Y, Rutledge GC (2012) Electrical conductivity of electrospun polyaniline and polyaniline-blend fibers and mats. Macromolecules 45:4238–4246
278. McCullen SD, Stevens DR, Roberts WA, Ojha SS, Clarke LI, Gorga RE (2007) Morphological, electrical, and mechanical characterization of electrospun nanofiber mats containing multiwalled carbon nanotubes. Macromolecules 40:997–1003
279. MacDiarmid AG, Jones WE Jr, Norris ID, Gao J, Johnson AT Jr, Pinto NJ, Hone J, Han B, Ko FK, Okuzaki H, Llaguno M (2001) Electrostatically-generated nanofibers of electronic polymers. Synth Met 119:27–30
280. Nagata S, Atkinson GM, Pestov D, Tepper GC, Mcleskey Jr JT (2013) Electrospun polymer-fiber solar cell. Adv Mater Sci Eng 2013, Article ID 975947
281. Babel A, Li D, Xia Y, Jenekhe SA (2005) Electrospun nanofibers of blends of conjugated polymers: morphology, optical properties, and field-effect transistors. Macromolecules 38:4705–4711
282. Balderas U, Falcony C, Moggio I, Ariasc E, Mondragón M (2013) A photoluminescence study of electrospun fibers of conjugated poly[2-methoxy-5-(2-ethylhexyloxy)-1,4-phenylenevinylene] blended with poly(9-vinylcarbazole). Polymer 54:2062–2066
283. Kumar A, Jose R, Fujihara K, Wang J, Ramakrishna S (2007) Structural and optical properties of electrospun TiO_2 nanofibers. Chem Mater 19:6536–6542
284. Kim JS, Reneker DH (1999) Mechanical properties of composites using ultrafine electrospun fibers. Polym Compos 20:124–131
285. Tang C, Liu H (2008) Cellulose nanofiber reinforced poly (vinyl alcohol) composite film with high visible light transmittance. Compos Part A-Appl S 39:1638–1643
286. Lu B, Zheng G, Dai K, Liu C, Chen J, Shen C (2015) Enhanced mechanical properties ofpolyethylene composites with low content of electrospun nylon-66 nanofibers. Mater Lett 140:131–134
287. Sun W, Cai Q, Li P, Deng X, Wei Y, Xu M, Yang X (2010) Post-draw PAN-PMMA nanofiber reinforced and toughened bis-GMA dental restorative composite. Dent Mater 26:873–880
288. Liao H, Wu Y, Wu M, Zhan X, Liu H (2012) Aligned electrospun cellulose fibers reinforced epoxy resin composite films with high visible light transmittance. Cellulose 19:111–119
289. Tang C, Wu M, Wu Y, Liu H (2011) Effects of fiber surface chemistry and size on the structure and properties of poly(vinyl alcohol) composite films reinforced with electrospun fibers. Compos Part A Appl S 42:1100–1109
290. Jiang S, Duan G, Schöbel J, Agarwal S, Greiner A (2013) Short electrospun polymeric nanofibers reinforced polyimide nanocomposites. Compos Sci Technol 88:57–61
291. Wu M, Wu Y, Liu Z, Liu H (2012) Optically transparent poly(methyl methacrylate) composite films reinforced with electrospun polyacrylonitrile nanofibers. J Compos Mater 46:2731–2738
292. Meng F, Zhao R, Zhan Y, X Liu X (2011) Design of thorn-like micro/nanofibers: fabrication and controlled morphology for engineered composite materials applications. J Mater Chem 21:16385–16390
293. Liao H, Wu Y, Wu M, Liu H (2011) Effects of fiber surface chemistry and roughness on interfacial structures of electrospun fiber reinforced epoxy composite films. Polym Compos 32:837–845
294. Dzenis Y (2008) Structural nanocomposites. Science 319:419–420
295. Magniez K, Chaffraix T, Fox B (2011) Toughening of a carbon-fibre composite using electrospun poly (hydroxyether of bisphenol a) nanofibrous membranes through inverse phase separation and inter-domain etherification. Materials 4:1967–1984

296. Zhang J, Yang T, Lin T, Wang CH (2012) Phase morphology of nanofibre interlayers: critical factor for toughening carbon/epoxy composites. Compos Sci Technol 72:256–262
297. Zhang J, Lin T, Wang X (2010) Electrospun nanofiber toughened carbon/epoxy composites: effects of polyetherketone cardo (PEK-C) nanofibre diameter and interlayer thickness. Compos Sci Technol 70:1660–1666
298. van der Heijden S, Daelemans L, Schoenmaker BD, Baere ID, Rahier H, Paepegem WV, Clerck KD (2014) Interlaminar toughening of resin transfer moulded glass fibre epoxy laminates by polycaprolactone electrospun nanofibres. Compos Sci Technol 104:66–73
299. Langer R, Vacanti J (1993) Tissue engineering. Science 260:920–926
300. Cen L, Liu W, Cui L, Zhang W, Cao Y (2008) Collagen tissue engineering: Development of novel biomaterials and applications. Pediatr Res 63:492–496
301. Dvir T, Timko BP, Kohane DS, Langer R (2011) Nanotechnological strategies for engineering complex tissues. Nat Nanotechnol 6:13–22
302. Ribba L, Parisi M, D'Accorso NB, Goyanes S (2014) Electrospun nanofibrous mats: from vascular repair to osteointegration. J Biomed Nanotechnol 10:1–28
303. Pelipenko J, Kocbek P, Kristl J (2015) Nanofiber diameter as a critical parameter affecting skin cell response. Eur J Pharm Sci 66:29–35
304. Doustgani A, Vheghani-Farahani E, Soleimani M (2013) Aligned and random nanofibrous nanocomposite scaffolds for bone tissue. Nanomed J 1:20–27
305. Gomes SR, Rodrigues G, Martins GG, Roberto MA, Mafra M, Henriques CMR, Silva JC (2015) In vitro and in vivo evaluation of electrospun nanofibers of PCL, chitosan and gelatin: a comparative study. Mater Sci Eng C Mater Biol Appl 46:348–358
306. Gluck JM, Rahgozar P, Ingle NP, Rofail F, Petrosian A, Cline MG, Jordan MC, Roos KP, MacLellan WR, Shemin RJ, Heydarkhan-Hagvall S (2011) Hybrid coaxial electrospun nanofibrous scaffolds with limited immunological response created for tissue engineering. J Biomed Mater Res B 99B:180–190
307. Kim YT, Haftel VK, Kumar S, Bellamkonda RV (2008) The role of aligned polymer fiber-based constructs in the bridging of long peripheral nerve gaps. Biomaterials 9: 3117–3127
308. Shalumon KT, Anulekha KH, Chennazhi KP, Tamura H, Nair SV, Jayakumar R (2011) Fabrication of chitosan/poly (caprolactone) nanofibrous scaffold for bone and skin tissue engineering. Int J Biol Macromol 48:571–576
309. Pillay V, Dott C, Choonara YE, Tyagi C, Toma L, Kumar P, du Toit LC, Ndesendo VMK (2013) A review of the effect of processing variables on the fabrication of electrospun nanofibers for drug delivery applications. J Nanomater 2013, Article ID 789289
310. Gilchrist SE, Lange D, Letchford K, Bach H, Fazli L, Burt HM (2013) Fusidic acid and rifampicin co-loaded PLGA nanofibers for the prevention of orthopedic implant associated infections. J Control Release 170:64–73
311. Lu T, Jing X, Song X, Wang X (2012) Doxorubicin-loaded ultrafine PEG-PLA fiber mats against hepatocarcinoma. J Appl Polym Sci 123:209–217
312. Tungprapa S, Jangchud I, Supaphol P (2007) Release characteristics of four model drugs from drug-loaded electrospun cellulose acetate fiber mats. Polymer 48:5030–5041
313. Li X, Su Y, Liu S, Tan L, Mo X, Ramakrishna S (2010) Encapsulation of proteins in poly(L lactide- co-caprolactone) fibers by emulsion electrospinning. Colloid Surf B 75:418–424
314. De Laporte L, Shea LD (2007) Matrices and scaffolds for DNA delivery in tissue engineering. Adv Drug Deliv Rev 59:292–307
315. Rujitanaroj P, Wang YC, Wang J, Chew SY (2011) Nanofiber-mediated controlled release of siRNA complexes for long term gene-silencing applications. Biomaterials 32:5915–5923
316. Xu X, Chen X, Ma P, Wang X, Jing X (2008) The release behavior of doxorubicin hydrochloride from medicated fibers prepared by emulsion-electrospinning. Eur J Pharm Biopharm 70:165–170
317. Ranganath SH, Wang CH (2008) Biodegradable microfiber implants delivering paclitaxel for post-surgical chemotherapy against malignant glioma. Biomaterials 29:2996–3003

318. Kenawy ER, Abdel-Hay FI, El-Newehy MH, Wnek GE (2009) Processing of polymer nanofibers through electrospinning as drug delivery systems. Mater Chem Phys 113:296–302
319. Lin X, Tang D, Du H (2013) Self-assembly and controlled release behaviour of the water-insoluble drug nifedipine from electrospun PCL-based polyurethane nanofibres. J Pharm Pharmacol 65:673–681
320. Zamani M, Morshed M, Varshosaz J, Jannesari M (2010) Controlled release of metronidazole benzoate from poly epsilon-caprolactone electrospun nanofibers for periodontal diseases. Eur J Pharm Biopharm 75:179–185
321. Wang C, Yan KW, Lin YD, Hsieh PCH (2010) Biodegradable core/shell fibers by coaxial electrospinning: processing, fiber characterization, and its application in sustained drug release. Macromolecules 43:6389–6397
322. Meng ZX, Xu XX, Zheng W, Zhou HM, Lia L, Zheng YF, Lou X (2011) Preparation and characterization of electrospun PLGA/gelatin nanofibers as a potential drug delivery system. Colloids Surf B Biointerfaces 84:97–102
323. Liu C, Hsu PC, Lee HW, Ye M, Zheng G, Liu N, Li W, Cui Y (2015) Transparent air filter for high-efficiency PM2.5 capture. Nat Commun 6, Article number: 6205
324. Molaeipour Y, Gharehaghaji AA, Bahrami H (2014) Filtration performance of cigarette filter tip containing electrospun nanofibrous filter. J Indus Text. doi:10.1177/1528083714528016
325. Li J, Gao F, Liu LQ (2013) Zhang Z. Needleless electro-spun nanofibers used for filtration of small particles, eXPRESS Polym Lett 7:683–689
326. Jaroszczyk T, Petrik S, Donahue K (2009) Recent development in heavy duty engine air filtration and the role of nanofiber filter media. J KONES 16:207–216
327. Sang Y, Li F, Gu Q, Liang C, Chen J (2008) Heavy metal-contaminated groundwater treatment by a novel nanofiber membrane. Desalination 223:349–360
328. Daels N, De Vrieze S, Sampers I, Decostere B, Westbroek P, Dumoulin A, Dejans P, De Clerck K, Van Hulle SWH (2011) Potential of a functionalised nanofibre microfiltration membrane as an antibacterial water filter. Desalination 275:285–290
329. Wang XF, Chen XM, Yoon K, Fang DF, Hsiao BS, Chu B (2005) High flux filtration medium Based on nanofibrous substrate with hydrophilic nanocomposite coating. Environ Sci Technol 39:7684–7691
330. Ahmed FE, Lalia BS, Hashaikeh R (2015) A review on electrospinning for membrane fabrication: challenges and applications. Desalination 356:15–30
331. Wang X, Li Y, Ding B (2014) Electrospun nanofiber-Based sensors. In: Ding B, Yu J (eds) Electrospun nanofibers for energy and environmental applications. Springer, Berlin, pp 267–297
332. Khoshaman AH (2011) Application of electrospun thin films for supra-molecule based gas sensing. Master of applied sciences thesis, Simon Fraser University
333. Lin Q, Li Y, Yang M (2012) Highly sensitive and ultrafast response surface acoustic wave humidity sensor based on electrospun polyaniline/poly(vinyl butyral) nanofibers. Anal Chim Acta 748:73–80
334. Adewuyi S, Ondigo DA, Zugle R, Tshentu Z, Nyokong T, Torto N (2012) A highly selective and sensitive pyridylazo-2-naphthol-poly(acrylic acid) functionalized electrospun nanofiber fluorescence “turn-off” chemosensory system for Ni^{2+}. Anal Method 4:1729–1735
335. Poltue T, Rangkupan R, Dubas ST, Dubas L (2011) Nickel (II) ions sensing properties of dimethylglyoxime/poly(caprolactone) electrospun fibers. Mater Lett 65:2231–2234
336. Li D, Pang Z, Chen X, Luo L, Cai Y, Wei Q (2014) A catechol biosensor based on electrospun carbon nanofibers. Beilstein J Nanotechnol 5:346–354
337. Nurfaizy AH, Tucker N, Stanger J, Staiger MP (2012) Functional nanofbers in clothing for protection against chemical and biological hazards. In: Wei Q (ed) Functional nanofibers and their applications. Woodhead Publishing, Cambridge, pp 236–261
338. Ko FK, Wan Y (2014) Nanofiber technology. In: Introduction to nanofiber materials. Cambridge University Press, Cambridge, pp 44–63
339. Gibson P, Schreuder-Gibson H, Rivin D (2001) Transport properties of porous membranes based on electrospun nanofibers. Colloids Surf A Physicochem Eng Asp 187:469–481

340. Schreuder-Gibson H, Gibson P, Senecal K, Sennett M, Walker J, Yeomans W, Ziegler D, Tsai PP (2002) Protective textile materials based on electrospun nanofibers. J Adv Mater 34 (3):44–55
341. Lee S, Obendorf SK (2007) Use of Electrospun nanofiber web for protective textile materials as barriers to liquid penetration. Text Res J 77:696–702
342. Faccini M, Vaquero C, Amantia D (2012) Development of protective clothing against nanoparticle based on electrospun nanofibers. J Nanomater 2012, Article ID 892894
343. Vitchuli N, Shi Q, Nowak J, McCord M, Bourham M, Zhang X (2010) Electrospun ultrathin nylon fibers for protective applications. J Appl Polym Sci 116:2181–2187
344. Taepaiboon P, Rungsardthong U, Supaphol P (2007) Vitamin-loaded electrospun cellulose acetate nanofiber mats as transdermal and dermal therapeutic agents of vitamin A acid and vitamin E. Eur J Pharm Biopharm 67:387–397
345. Camerlo A, Corinne VN, Rossi RM, Popa AM (2013) Fragrance encapsulation in polymeric matrices by emulsion electrospinning. Eur Polym J 49:3806–3813
346. Fathi-Azarbayjani A, Qun L, Chan YW, Cha SY (2010) Novel vitamin and gold-loaded nanofiber facial mask for topical delivery. AAPS Pharm Sci Tech 11:1164–1170
347. Sheng X, Fan L, He C, Zhang K, Mo X, Wang H (2013) Vitamin E-loaded silk fibroin nanofibrous mats fabricated by green process for skin care application. Int J Biol Macromol 56:49–56

Chapter 5
Isolation of Cellulose Nanowhiskers and Their Nanocomposites

Dongyan Liu, Yu Dong and Guoxin Sui

Introduction

Cellulose and Cellulose Nanowhiskers (CNWs)

Since past decade, there has been a growing interest in developing biodegradable products using natural materials from sustainable resources to reduce the dependence of plastics on fossil fuel when environment friendly products are generated. Cellulose, one of the most abundant natural polymers on Earth, yields annual biomass production about one trillion tons [1]. Cellulose is a linear homopolymer of glucose $(C_6H_{10}O_5)_n$, which is linked by 1,4-β-glycosidic bonds with cellobiose, a dimer, as its repeating unit. Cellulose is bound together by lignin and hemicelluloses [2]. Cellulose fibers are superior to other engineering fibers due to their biodegradability, abundance, low cost, low density, and acceptable mechanical properties. Cellulose fibers have been widely used as engineering materials for a long time in textile, paper and pulps, building construction, chemical filter, and fiber-reinforced composites [3, 4]. It is regarded as an inexhaustible source of raw materials for environment friendly bio-products.

Cellulose nanowhiskers (CNWs), also known as cellulose nanocrystals (CNCs) or nanocrystalline cellulose, are usually produced by the acid hydrolysis of natural cellulosic material after removing noncellulosic substances such as dewaxing, hemicelluloses, and lignin. Most CNWs show a high crystallinity index and a lower aspect ratio. Thus it is expected that CNWs extracted from tunicates exhibit a high

D. Liu · G. Sui (✉)
Institute of Metal Research (IMR), Chinese Academy of Sciences (CAS),
Shenyang 110016, China
e-mail: gxsui@imr.ac.cn

Y. Dong
Department of Mechanical Engineering, Curtin University,
Perth, WA 6845, Australia

S. Fakirov (ed.), *Nano-size Polymers*,
DOI 10.1007/978-3-319-39715-3_5

aspect ratio of 38 [5]. The degree of crystallinity, size, and morphology depend on the source of raw materials and preparation methods. CNWs are categorized as "nanocellulose," which usually refers to cellulose materials with at least one dimension smaller than 100 nm. Isolated cellulose fibers or whiskers possess lateral dimensions around several nanometers and longitudinal dimension in a wide range from tens of nanometers to several micrometers. The round-shaped cellulose nanoball with diameters of 20–40 nm is extracted from sweet potato residue [6]. The rod-like polymeric nanomaterials have a low coefficient of thermal expansion, high thermal stability, high aspect ratio, and low density of 1.5 g/cm^3.

Cellulose nanofibers (CNF) are obtained by mechanical disintegration after the raw cellulosic materials are chemically or enzymatically pretreated to obtain pure cellulose. Therefore, the CNF has a high aspect ratio with a length over 1000 nm including crystalline and amorphous celluloses [7]. Chen et al. [8] produced ultralong CNFs in form of cellulose I crystals (>1 mm in length and 30–80 nm in diameter), which were prepared from bamboo fibers using conventional chemical pretreatment combined with high intensity ultrasonication, and then followed by freeze-drying techniques [8]. CNFs are more available for commercial production than chemically or enzymatically hydrolyzed CNWs [9]. CNWs as a natural nanomaterial have recently attracted great attention from scientists and industrial engineers because they have high tensile modulus (138 GPa), which is higher than that of S-glass (86–90 GPa) and comparable to Kevlar (131 GPa). CNWs are considered as good natural polymeric nanomaterials [10] owing to their unique properties such as good renewability, excellent mechanical properties, high specific surface area, good biodegradability, and biocompatibility [11]. Moreover, CNWs with rich hydroxyl groups have good affinity with a variety of polymers and nanomaterials.

Raw Materials

Cellulose nanowhiskers (CNWs) can be extracted from numerous raw natural materials on Earth. It shows potentials in nanocomposites, paper making, coating additives, food packaging, cosmetics, and gas barrier fields. However, the production of CNWs in an economic and environmental way and exploration of its functional products are the major tasks for the researchers. It will eventually promote the development of CNWs-based hybrid nanostructures.

CNWs were initially isolated from natural fibers by Mukherjee and Woods [12] in 1953. They can be prepared from a variety of sources such as wood pulp, plant fibers (e.g., hemp, sisal, flax, ramie, jute, algae, cotton) [13, 14], tree leaves [15], microbial (acetobacter xylinum) [16], sea creatures (tunicates) [17, 18], fruit skins (banana and grape) [19], fruit husks (coconuts) [20], and even agricultural products (e.g., wheat straws and soy hulls) [21], which makes them more attractive and applicable. Three methods are available for producing nanocellulose, namely, chemical acid hydrolysis, chemical treatment in combination with mechanical refining and the enzymatic methods.

Isolation Methods of Cellulose Nanowhiskers

There are mainly three methods used for isolating CNWs from raw cellulosic materials, including acid hydrolysis, enzymatic hydrolysis and 2,2,6,6-tetramethylpiperidine-1-oxyl radical (TEMPO) mediated oxidation. In cellulosic materials, it is only pure in cotton fibers, whereas cellulose in wood and other plants always exists in combination with other materials such as lignin and hemicelluloses.

No matter what methods are used, the pretreatment to remove noncellulosic substances such as water soluble substance, hemicellulose, lignin, pectin, wax, etc., is essential to obtain highly purified cellulose fibers. The raw cellulose containing fibers is usually boiled in toluene/ethanol (volume ratio of 2:1) mixture after water-washing process to eliminate the dirt, water soluble extractives, and wax, respectively [22]. The alkali-bleaching treatment is the most preferentially conventional chemical treatments to purify cellulose fibers in that NaOH/KOH, sodium chlorite, and acetic acid are easy to access. The purpose of the alkali pulping and subsequent bleaching treatment is to eliminate lignin and hemicellulose, thus enabling the cellulose fibers broken up into smaller parts, toward freeing individual fibers with more uniform width and even cellulose microfibrils. Vázquez et al. [23] used 0.7 w/v% sodium chlorite solution and 17.5 w/v% NaOH solution to remove lignin and holocellulose. Enzyme is also used to degrade lignin and hemicellulose with the maintained cellulose portion. The combination of alkaline and acid pretreatment is able to solubilize lignin, hemicellulose, and pectin [18, 21]. Cellulose fibers from Xanthoceras sorbifolia bung husk (XSBH) were obtained by alkaline soaking and H_2O_2/acetic acid (volume ratio of 1:1) boiling twice [24]. The coconut husk fiber was delignified using an organosolv process, which was followed by alkaline bleaching (5 % (w/w) H_2O_2 + 4 % (w/w) NaOH) at 50 °C for 90 min [25].

Acid Hydrolysis Method

The most commonly used technique for isolating cellulose nanowhiskers is acid hydrolysis. Sulfuric acid is usually selected for hydrolyzing cellulose fibers because stable suspensions can be obtained. This is mainly caused by the electrostatic repulsion between negatively charged whiskers on their surfaces. The morphology is mostly dependent on the acid concentration, cellulose/acid ratio, temperature, and hydrolysis time. The temperature is usually ranging between 30 and 105 °C and the hydrolysis time is from 20 min to 4 h depending on the reaction temperature, which is affected by the acidic concentration. The concentration of sulfuric acid solution usually ranges from 30 to 65 wt% according to the source of raw materials and pretreatment conditions. Cellulose can be seriously destroyed and degraded to sugar molecules if the concentration is over 65 %, whereas the too slow concentration of sulfuric acid can produce poorly dispersed fibers and aggregates. The concentration of sulfuric acid CNWs with the sizes of 72 $\pm$ 88 nm in length and 8 $\pm$ 3 nm in width was previously obtained from coconut fibers by 30 wt% sulfuric acid hydrolysis for 6 h at 60 °C [25]. MCCs were hydrolyzed in 64 wt% H_2SO_4 and used to hydrolyze microcrystalline cellulose at 40 °C for 60 min at an acid to cellulose amount of 8.75 ml/g under the strong agitation [26]. Cellulose whiskers with sizes of 4 ± 2 and 8 ± 3 nm in diameter and 255 ± 55 nm in length (extracted from 5 g bleached sugarcane bagasse) was hydrolyzed in 100 ml of 6 M sulfuric acid at 45 °C for 30 min and 75 min, respectively [27]. Mandal and Chakrabarty [28] extracted CNWs by refluxing with 60 % (w/v) sulfuric acid (fiber to liquor ratio of 1:20) for 5 h at 50 °C under strong agitation.

In addition to sulfuric acid, HCl, HNO_3, and H_3PO_4 are also used to hydrolyze cellulose into nanowhiskers. Other acids such as hydrochloric acid, hydrobromic acid as well as mixed acetic and nitric acids are capable of hydrolyzing cellulose into CNCs without esterifying the surfaces. Transverse dimensions in the range of 7–8 nm and with lengths of 100–200 nm were obtained from Whatman #1 filter paper using 100 ml of 1.5–4.0 M HBr at 100 °C for 1–4 h prior to the ultrasonication treatment [29]. Cellulose whiskers with a diameter of 15 nm, isolated from steam exploded banana fibers, were obtained by 11 wt% oxalic acid in an autoclave under a pressure of 20 lb [30]. The CNWs hydrolyzed from these acid solutions tend to aggregate due to the lack of negative charges on their surfaces. However, they show higher initial decomposition temperatures in TGA curves [31].

TEMPO Mediated Oxidation Method

2,2,6,6-tetramethylpiperidine-1-oxyl radical (TEMPO) mediated oxidation is one of the regioselective chemical modifications of primary hydroxyl groups [32]. When native cellulose is oxidized with the TEMPO system, only C6 hydroxyl groups exposed on the microfibril surfaces are converted to sodium carboxylated groups. Similar to sulfuric acid hydrolyzed method, CNWs prepared by TEMPO oxidation are able to be completely dispersed at the individual nanofibril level in water by electrostatic repulsion and/or osmotic effects due to anionically charged sodium carboxylate groups densely present on the fiber surfaces [33]. The TEMPO-oxidized wood celluloses with carboxylate contents of more than ~1 mmol g^{-1} showed higher aspect ratios with uniform dimensions at 3–4 nm in width and over 500 nm in length by using mild mechanical disintegration in water [34, 35], which are opposed to acid hydrolyzed CNWs with the lengths less than 500 nm. CNWs with ultrathin diameters (3–10 nm) were extracted from chemically pretreated jute fibers with the combination of selective oxidization of TEMPO/NaBr/NaClO system and mechanical homogenization methods. Sun et al. [36] compared the structures and properties of CNWs prepared by TEMPO-oxidizing and sulfuric acid hydrolyzing bleached wood pulps. The average lengths of 200.7 and 163.0 nm, the diameters of 5.8 and 15.6 nm, and aspect ratios of 34.4 and 10.4 have been reported for TEMPO-oxidization and sulfuric acid hydrolysis methods, respectively [36].

Microbial Hydrolysis Method

The concentrated acid used in acid hydrolysis is toxic, hazardous, and corrosive; thus highly corrosion-resistant reactor and extreme care in material handling are required in the process. This makes acid pretreatment an expensive option. In addition, the concentrated acid must be recovered after pretreatment to make the process economically and environmentally feasible. As compared with acid hydrolysis method, enzymatic hydrolysis technique is more environmental friendly without the involvement of chemicals. However, the hydrolysis of cellulose by enzyme is a slow process because it has to diffuse from the surface into innermost of cellulose fibers [37, 38]. Enzymes in fiber processing have been mainly applied in degrading and modifying hemicellulose and lignin, but still retaining the cellulosic portion. The conventional method of acid hydrolysis resulted in sulfating on the CNWs surface while the surface chemistry of CNWs prepared by fungal hydrolysis remains unaltered. This can further improve their performance as nanofillers in composites. Moreover, it enhances the biocompatibility of CNWs and their scope in biomedical applications and pharmaceuticals. Although enzymatic route for nanocellulose synthesis offers the potential for higher yields, higher selectivity, lower energy costs, and milder operating conditions than chemical processes, such a technology was still hindered by economical (i.e., costly cellulase enzyme) and technical (rate limiting step of cellulose degradation with a long processing period) barriers. The slow rate of enzymatic hydrolysis is easily influenced by several factors that comprise structural features resulting from pretreatment and enzyme mechanism [39].

The Characterization and Properties of Cellulose Nanowhiskers (CNWs)

CNWs are nano-sized natural polymer with particular characteristics and properties including high crystallinity, chemical reactivity, high surface area, and aspect ratio, as well as high mechanical modulus. A variety of characterization methods have been used to evaluate morphologies, physical and chemical structures, thermal behavior, and mechanical properties of cellulose nanowhiskers.

Physical Structures

Cellulose has no taste, is odorless, is hydrophilic, and insoluble in water and most organic solvents. It consists of crystalline and amorphous regions. The amorphous regions can be partly broken down by concentrated acid attacking, leaving the nano-sized crystalline region. Thus CNWs are semicrystalline polymer with high degree of crystallinity. There are several different crystalline structures of cellulose, corresponding to the location of hydrogen bonds between and within strands. Natural cellulose is cellulose I, with structures Iα and Iβ. Cellulose produced by bacteria and algae is enriched in Iα while cellulose of higher plants consists mainly of Iβ. X-ray is used for characterizing physical structure of CNWs. The characteristic peaks are positioned at 2θ angles of 14.8, 16.5, and 22.6°, which correspond to reflections of (1, −1, 0), (1, 1, 0), and (2, 0, 0), respectively (Fig. 5.1). CNWs usually show a high crystallinity degree, which can be calculated using an empirical method proposed by the Segal equation [40].

$$I_c[\%] = \frac{I_{(200)} - I_{(am)}}{I_{(200)}} \times 100 \tag{5.2}$$

where $I_{(200)}$ is the counts at peak intensity at 2θ = 22.6°, representing the crystalline and amorphous phases. $I_{(am)}$ is the counts at peak intensity at 2θ = 18.6°, which is indicative of the amorphous phase in cellulose fibers. The acid hydrolysis attacked the amorphous regions and led to the increase of crystallinity degree. Consequently, CNWs show a highly crystallinity index (i.e., degree of crystallinity) between 50 and 91 % [41]. The crystalline degree depends on the source of cellulosic materials and preparation conditions as well as experimental techniques. The crystallinity degree of CNWs was found to be 82 % after isolated from flax fibers using acid hydrolysis for 4 h in 55 wt% sulfuric acid at 55 °C, a little lower than those of 83 and 85 % for non-hydrolyzed cellulose and commercial microcrystalline cellulose, respectively, as shown in Fig. 5.1 [42]. CNWs showed a crystallinity index of 84 %

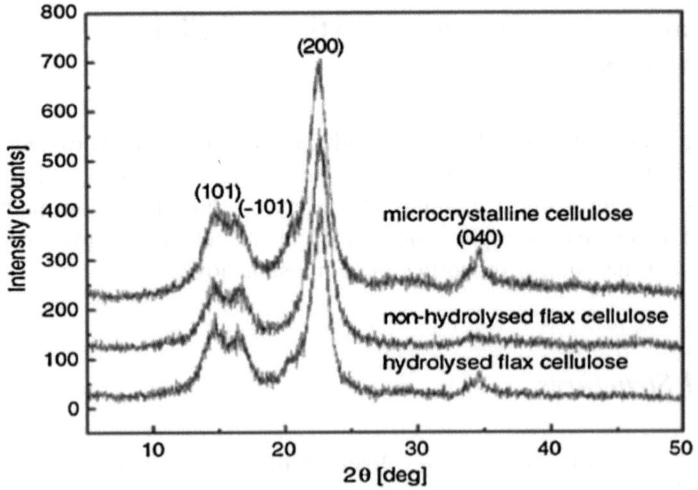

Fig. 5.1 X-ray diffraction (XRD) patterns of non-hydrolysed flax cellulose (bleached flax yarns) and freeze-dried flax cellulose naowhiskers (CNWs) and commercial microcrystalline cellulose. Reproduced from Ref. [42]

using 64 wt% acid at 40 °C for 1 h [43]. CNWs yielded a crystallinity degree of 50.79–56.88 % at different pre-acid hydrolysis treatments [44]. So far, Martins et al. [45] has reported the highest crystalline index of 91 %.

The crystallinity index can also be calculated by the following equation:

$$I_c = \frac{A_{cryst}}{A_{total}} \tag{5.3}$$

where I_c is the crystalline index, $A_{crystal}$ is the sum of crystalline band areas, and A_{total} is the total area under the diffractograms. The crystallinity degrees of CNWs can vary from 62 to 66 % at different bleaching conditions according to the above equation [46].

Chemical Structures

Cellulose is an organic compound with the formula $(C_6H_{10}O_5)n$, a polysaccharide consisting of a linear chain of several hundred to many thousands of β(1 → 4) linked D-glucose units [47]. Its chemical formula is shown in Fig. 5.2.

FTIR spectroscopy has been widely used to obtain the information about chemical structure of polymers. The advantages of this technique are small sample sizes, rapid analysis time, and nondestructivity. The spectra were usually divided into two regions, i.e., the region from 4000 to 2600 cm^{-1}, and the "fingerprint" region which is assigned to different stretching vibrations of different from 1800 to 800 cm^{-1}. The bands at around 3340 and 2900 cm^{-1} are ascribed to –OH stretching, $-CH_3$, and $-CH_2$, respectively. Several bands are seen in the "fingerprint" region. The broad and weak band is related to adsorbed water in cellulose. The bands at 1427, 1373, 1337, and 1318 cm^{-1} belong to $-CH_2$ symmetric bending, CH bending, in-plane –OH bending, $-CH_2$ rocking vibration. On the other hand, the bands at 1162, 1108, 1057, 1033, 897 cm^{-1} are assigned to asymmetric C–O–C bridge stretching, anhydroglucose ring asymmetric stretching, C–O stretching, in-plane C–H deformation, and C–H deformation of cellulose, respectively [48]. The peak at 1726 cm^{-1} associated with carboxylic acids and aliphatic ketones present in hemicellulose and 1515 cm^{-1} assigned to C=C aromatic skeletal

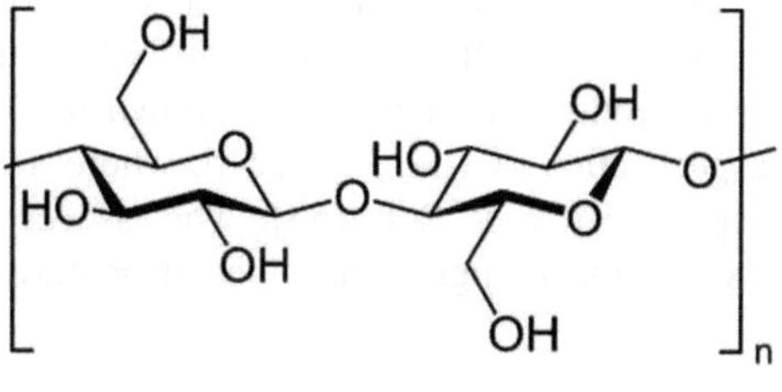

Fig. 5.2 Chemical formula of cellulose

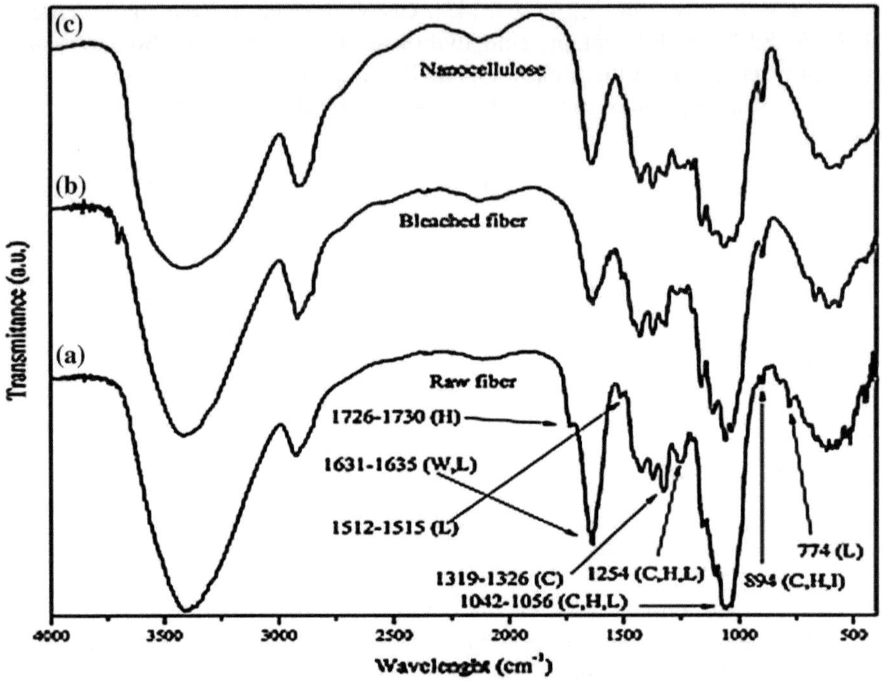

Fig. 5.3 FTIR spectra of raw (**a**) and bleached (**b**) banana pseudostem fiber (BPF) (**c**) cellulose nanowhiskers (CNWs). Reproduced from Ref. [49]

vibration in lignin are not present in the spectra of CNWs because of their removal during the pretreatment and acid hydrolysis process [49] (Fig. 5.3).

Morphological Properties

Transmission electronic microscopy (TEM) and atomic force microscopy (AFM) were usually used to examine the morphologies of CNWs. Sulfuric acid hydrolysis of cellulose gave rise to the breakdown of fibers into rod-like fragments. The amorphous phases were selectively hydrolyzed while the crystalline counterparts remained unaffected. The average diameter and length for individual cellulose fibers were usually ranging from several to several tens nm and a few hundred nm, respectively [50]. The sizes of CNWs depend on the source of raw materials and their production methods, as shown in Fig. 5.4 CNWs from cotton showed lower aspect ratios of 10–12 [51, 52], whereas CNWs from tunicin showed much higher aspect ratios of 200 [53]. The aspect ratio of CNWs prepared by TEMPO-oxidization is higher than that of prepared by sulfuric acid hydrolysis [54].

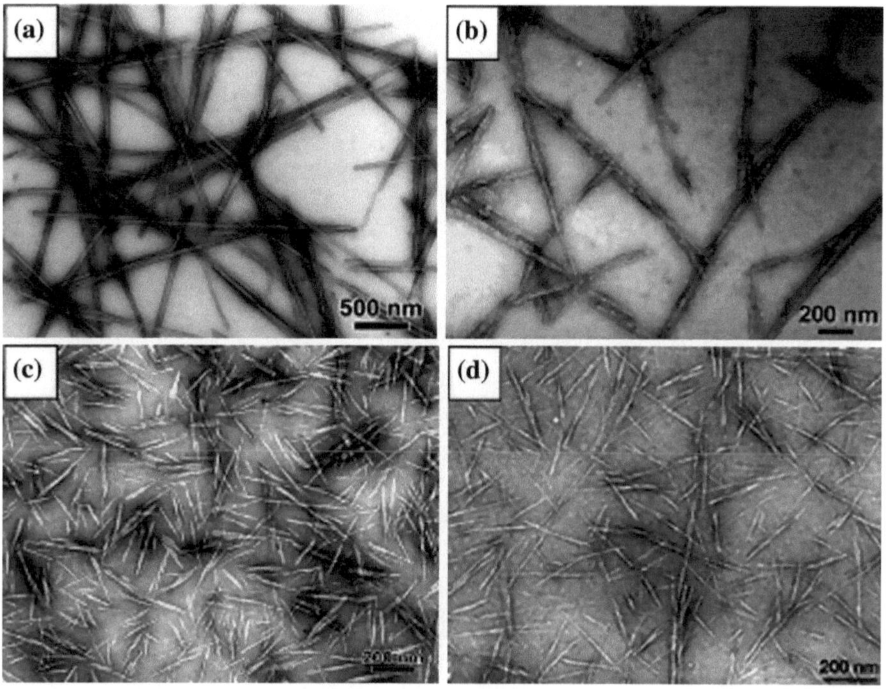

Fig. 5.4 TEM micrographs of cellulose nanowhsikers (CNWs) extracted from **a** tunicate, **b** bacterial, **c** ramie, and **d** sisal. Reproduced with permission from Ref. [50]

Thermal Properties

CNWs are semicrystalline polymers with high crystalline index. Thermal behaviors of CNWs are dependent on the sources, isolating method, and heating rate. Two endothermic peaks appear in the differential scanning calorimetry (DSC) curves, indicating the evaporation of water and melting of crystalline cellulose, respectively. The sharp melting peak usually appears at temperature ranging 210–350 °C, indicating its crystalline structure and a narrow range of molecule weight. Commercial microcrystalline cellulose (MCC) shows higher melting peak temperature of 358 °C due to its mild processing conditions [55]. The melting peak temperature (211 °C) of CNWs extracted from flax fibers by sulfuric acid method is much lower than that of commercial microcrystalline cellulose, as shown in Fig. 5.5a [56]. This low temperature is ascribed to the smaller size and lower molecule weight of cellulose because of their dependence on the extraction methods such as acidic concentration, hydrolysis time, and temperature.

Thermogravimetric analysis (TGA) was carried out to investigate the decomposition and thermal stability through the sample mass change during the heating process. The TGA and derivative TG curves are presented in Fig. 5.5b. A small weight loss of 5 % was seen below 150 °C, which is attributed to the evaporation of

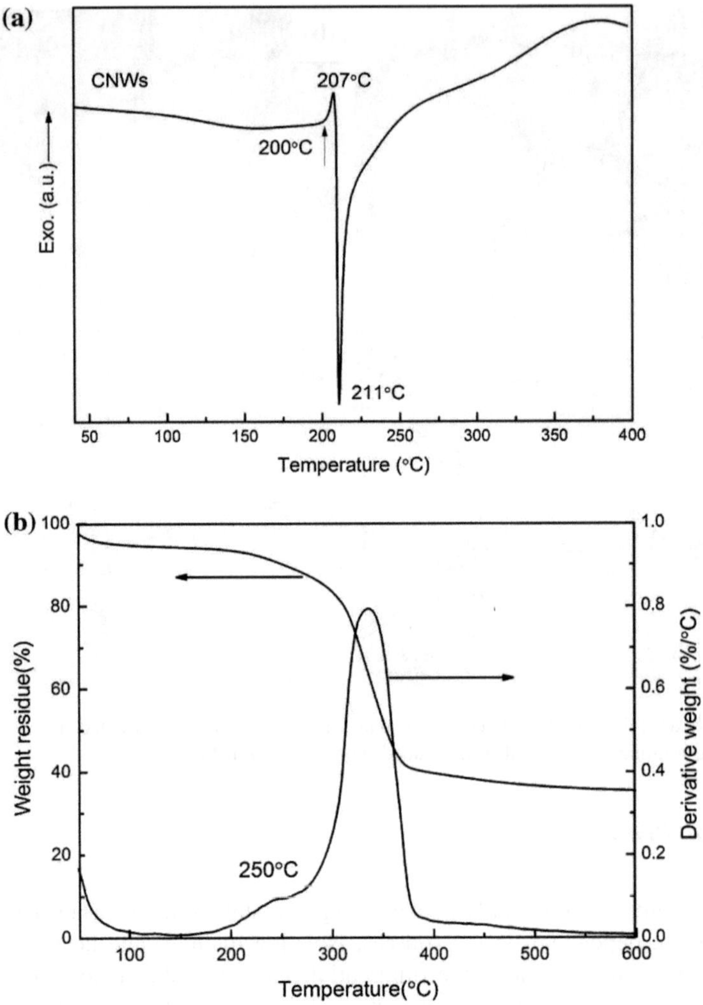

Fig. 5.5 TGA diagrams of CNWs from flax: **a** DSC curve and **b** TGA/DTGA curves. Reproduced from Ref. [56]

loosely bound moisture on the CNW surface [57]. A more pronounced weight loss starting in the range of 200–380 °C, indicates the cleavage of glycosidic linkages of cellulose including depolymerization, dehydration, and decomposition of glycosyl units, which was followed by the formation of CO_2, H_2O and a variety of hydrocarbon derivatives [58, 59]. This initial decomposition temperature of CNWs at 200 °C is much lower than those of cellulose fibers and microfibrillated cellulose due to the inclusion of acid sulfate ($O–SO_3H$) groups during sulfuric acid hydrolysis process [60, 61]. Cellulose nanocrystals extracted from corn stover also show a low decomposition temperature of 211 °C [62]. CNWs isolated from coconut fibers by 30 wt% sulfuric acid have an initial decomposition temperature of 270 °C [25]. This finding indicates the strong effect of sulfuric acid on the decomposition temperature of CNWs. Above 380 °C, the weight loss resulted from the oxidation and breakdown of charred residue to lower the molecule weight of gaseous products.

As seen in DTGA curves, the DTGA peaks positioned at 338 °C, which implies the maximum degradation rate of CNWs. A weak and broad shoulder centered at around 250 °C was observed prior to the main peak, resulting from the decomposition of CNWs with sulfate groups [63].

Elastic Modulus

Mechanical properties of nanofillers are very important for understanding their reinforcement in polymer composites. Unfortunately, there are very few direct and accurate characterization techniques that have been used to test CNWs due to their small size. Instead several methods have been used for characterizing the modulus of cellulose crystalline region. Sakurada et al. [64] reported a modulus value of 138 GPa for the crystalline region of cellulose determined by means of X-ray diffraction of deformed fiber bundles. Rusli and Eichhorn [65] measured the stiffness of CNWs by dispersing nanowhiskers into epoxy resin. The stiffness of cellulose nanowhiskers varied from 57 to 105 GPa, which was derived using theoretical models of their dispersion and matrix reinforcement through the molecular deformation of the whiskers from Raman spectroscopy. A different stiffness value of 143 GPa was measured by Sturcova et al. [66] with the same technique. Using the computational simulation method, Eichhorn and Davies [67] obtained the value from 149 to 155 GPa. In particular, tunicate cellulose nanowhisker shows a high modulus of 150.7 ± 28.8 GPa measured by AFM with three-point bending mode [68]. This method needs the whiskers with high longitude size to cover the grooves. Crystal structure, crystallinity index, testing method, and defects always affect the mechanical properties of CNWs. The key is to search for a direct and accurate method to measure the modulus of CNWs.

Production and Properties of CNWs Reinforced Polymer Composites

Cellulose is the most abundant natural polymer on Earth. Researchers have shown a growing interest in cellulose nanowhiskers due to its numerous advantages including biocompability, biodegradability, and unique chemical and reactive surface properties. Cellulose nanowhiskers have been used to reinforce a number of polymers by conventional methods, such as solvent casting, electrospinning, melt compounding, and freeze-drying technologies. CNWs reinforced polymer composites were early reported by Favier et al. [69] in 1995, which was followed by numerous investigations on the fabrication and properties of such composites [70, 71].

Solvent Casting

Solvent casting method is a simple and mostly used for producing polymer composites at the laboratory scale. CNWs are able to be disperses in water and various organic solvents [72, 73]. The good dispersion ability of CNWs is an important prerequisite for getting polymer composites with well dispersed nanofillers. Composite films can be obtained by casting the mixtures onto the substrate with smooth surfaces prior to the solvent evaporation. Composite films can maintain good transparency for the use in food packaging. Figure 5.6 showed good the transparent starch/CNWs composite film despite the inclusion of 20 wt% CNWs where the words underneath films still could be clearly seen. Additionally, the crystallization rate was strongly increased by only 1 wt% modified CNWs addition owing to the important role of CNWs as a nucleating agent [74]. The mechanical properties of polymer composites reinforced with CNWs can be also significantly improved. The tensile strength and storage modulus of poly (propylene carbonate)/CNWs composite at the CNW loading of 10 wt% exhibited a 130 % increase from 11.6 to 26.9 MPa, and a 76 % increase from 1680 to 2960 MPa, respectively [75]. The addition of 10 wt% CNWs into cellulose acetate butyrate (CAB) almost doubled the modulus with an increase from 0.9 GPa for neat CAB film to 1.75 GPa for composites [76]. The starch-based composite films reinforced with flax CNWs exhibited improved tensile strength and Young's modulus from 3.9 to 11.9 MPa,

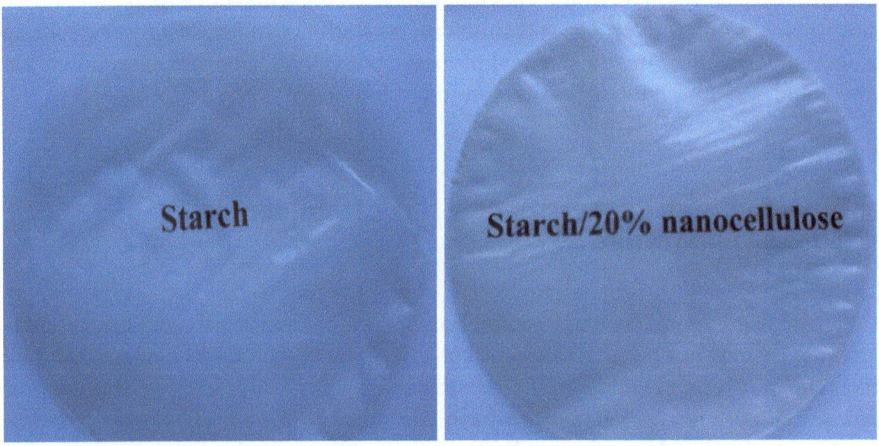

Fig. 5.6 Digital images of neat starch and starch composite films with 20 % CNWs

and from 31.9 to 498.2 MPa, respectively, as opposed to neat starch films [77]. The barrier properties of polymer matrices can be improved with the addition of nanoparticles due to an increase in tortuosity provided by dispersed nanofillers. The incorporation of cellulose nanowhiskers into polymer matrices inevitably provides a physical barrier through the creation of tortuous path. In addition to mechanical properties, permeabilities of oxygen and water decreased by 82 and 90 % when CNWs were incorporated into PLA matrices [78]. The addition of 1 wt% modified CNWs can achieve the reduction of 34 % for the water permeability of cast films. Good oxygen barrier properties were detected for nano-biocomposites with both 1 and 5 wt% of modified and unmodified cellulose nanocrystals. Moreover, the migration level of studied nano-biocomposites was below the overall migration limits required by the current standard for food packaging materials. Biodegradable polymers reinforced with CNWs with improved properties are promising materials for food packaging [79]. Nanocomposites with 10 wt% CNWs decreased the water vapor transmission rate (WVTR) of xylan/sorbit films from 304 to 174 g h^{-1} m^{-2} [80]. Porous composites can also be prepared by freeze-drying the mixtures of CNWs and polymer solutions [81, 82]. Solution casting technology is unlikely to be used as a large industrial scale process owing to the existence of volatile solvents.

Electrospinning

Electrospinning is a versatile and simple technology to prepare nanofibers from a polymer melt or solution [83]. The charged polymer melt or solution deforms due to the repulsion of electric charges at the surface. Thin fibers are generated on the collecting substrate after the charged polymer jet travels to the collector at a high speed. The setup of electrospinning and resulting electrospun mat are shown in Fig. 5.7. The nanofibrous mats demonstrate particular characteristics such as large

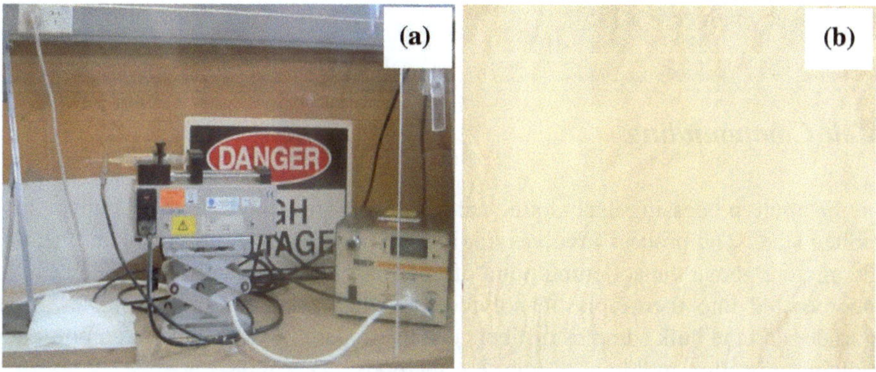

Fig. 5.7 **a** Electrospinning setup and **b** electrospun polylactic acid (PLA)/cellulose nanowhiskers (CNWs) nanocomposite fibrous mat

surface area-to-volume ratio and high porosity with small pore size, leading to a variety of applications including filtration, sensors, electrode materials, and tissue engineering scaffolds [84–87]. However, electrospun nanofibrous mat is not stiff enough and sometimes becomes difficult to handle. The incorporation of stiff nano-reinforcements, including carbon nanotubes, nanoclays, and cellulose nanowhiskers with high aspect ratios, has been regarded as an effective approach to enhance mechanical, electrical, and magnetic properties of electrospun fibers [88–91].

Cellulose nanowhiskers have been incorporated into various polymer solutions, such as polyethylene oxide (PEO) [92], polylactic acid (PLA) [93–97], poly(methyl methacrylate) (PMMA) [98], polystyrene (PS) [99], polycaprolactone (PCL) [100], poly(vinyl alcohol) (PVA) [101], cellulose acetate (CA) [102]. CNWs are cellulose short fibers with the length of several hundred nanometers. The incorporation of rigid CNWs into electrospun polymer nanofibers can greatly improve mechanical properties of matrix fibers. PLA, as one of mostly used biopolymer, has been widely used as a popular material candidate for electrospinning because of its good mechanical properties, excellent biodegradability, and biocompatibility. The strength and modulus of polymeric nanofibers can be enhanced by the small addition of cellulose nanowhiskers. Tensile strength and modulus of PCL nanofibers increased by 68 and 37 % with the addition of 2.5 wt% cellulose nanowhiskers [99]. Shi et al. [97] reported that tensile strength and tensile modulus of PLA nanofibers were improved by 5 and 22 times with the addition of 5 wt% CNWs. The tensile strength and Young's modulus of PCL nanofibrous webs with 2.5 wt% CNWs have been reported to be 6.54 and 1.51 MPa, which are higher than the values of neat PCL counterpart at 3.89 and 1.1 MPa, respectively [100]. These reinforcing effects are ascribed to the flow-induced and charge-induced orientations during electrospinning process. It allows the CNW distribution along the axis of polymer fibrers (i.e., the arrangement direction of molecular chains of polymers). Compared with solution cast samples, the electrospun semicrystalline polymers usually showed amorphous structure and lower cold crystallization temperatures because of the rapid solvent evaporation rate [96]. Biocompatible polymers reinforced with CNWs with nontoxicity are capable of supporting cell proliferation in bone tissue engineering [95].

Melt Compounding

In the melt processing technique, nanoparticles are mixed with polymers in a molten state. The process involves mixing the particles with polymers and heating the mixture above the softening point of polymers under a shear force. They can be incorporated into thermoplastic polymers by melt compounding using an screw extruder and the bulk samples or fibers can be prepared by subsequent compression molding, injection molding or spinning methods.

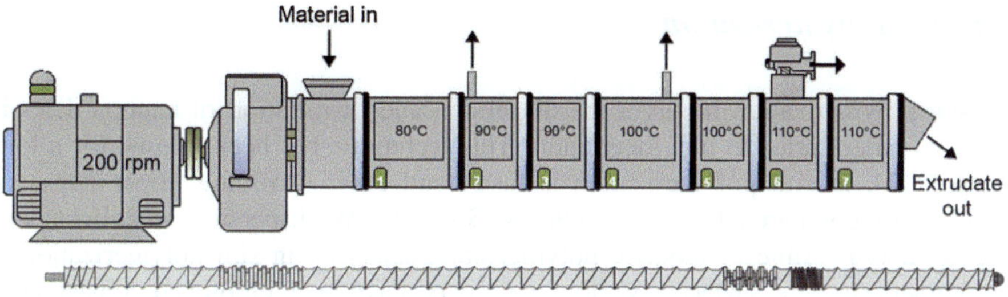

Fig. 5.8 Extrusion setup used in the compounding of thermoplastic (TPS)/cellulose nanofibers (CNF) composites. Reproduced from Ref. [102]

Cellulose nanowhiskers with reasonable thermal stability below 200 °C can be compounding with several polymers [62]. These polymers usually possess low melting points, such as polyethylene (PE) [103], PLA [104, 105], PCL [106] and starch [107]. The extrusion setup including the screw configuration and the temperature profile is shown in Fig. 5.8. The temperature setup is determined by polymer matrices that should be fully melted. The CNWs can be blended with the polymer melt in a corotating twin-screw extruder under the shearing force. The solvent-free thermal processing method is potential for the industrial application. The biodegradable polymers such as PLA and starch are the promising polymer matrices reinforced with CNWs for packaging applications.

The weakness of the thermal blending method is the poor CNW dispersion in polymer matrices because of the high viscosity of polymer melt, as opposed to solution casting method. The CNWs have a very high surface area and have a tendency to aggregate when dried. This problem can be avoided by first mixing them in a suitable medium, either in a liquid, which is then fed into the extruder, or first dried and then extruded. As a result, the master batch of polymer/CNWs composites is generally prepared by solution casting prior to their melt compounding [108]. The tensile modulus and strength of nanocomposites increased from 2.9 to 3.6 GPa and from 58 to 71 MPa, respectively, with 5 wt% CNWs when compared to those of neat polymer. CNWs surface modifying or grafting methods is applied to achieve better dispersion and good adhesion with polymer matrices [109–111]. The loading of 8 wt% CNWs-g-PCL resulted in increases of strength and elongation by approximately 1.9- and 10.7-fold, respectively, over those of pure PLA [112]. Melt compounding is an industrially feasible technology. However, Due to the thermal sensitivity of CNWs, the temperature should be low to avoid their degradation during thermal blending and subsequent shape-forming process. The modification of CNWs to get better dispersion and thermal stability is necessary for the further development of polymer nanocomposites using melt compounding method.

In Situ Polymerization

In situ polymerization involves the dispersion and distribution of nanoparticles in the monomers prior to polymerization. This technique has been known for a long time [113]. In situ polymerization is faster and easy, involving fewer steps and lower cost compared to other methods. Since CNW dispersion is uniform and stable, it is possible to produce polymer composites by in situ polymerization of monomers on their surfaces. Various polymers, have been employed consisting of polyurethane (PU) [114], PMMA [115], conducting polymers, polyaniline (Pani) [116], polypyrrole (PPy) [117]. CNWs can be uniformly dispersed in polymer matrices because of their surface coated with polymer nanoparticles, further avoiding their re-aggregation in the following process. PU/CNWs composites can enhance tensile modulus and strength without the loss of ductility. The composite film with the addition 0.5 wt% CNWs showed the modulus and strength of 31.2 and 13.5 MPa, which are higher than those of 19.8 and 10.9 MPa for pure PU film, respectively [114]. In the conducting polymer/CNWs composites, CNWs offer composites with good mechanical strength and film formability while conducting polymers contribute to remarkable functional properties for energy storage applications. The electrically conductive CNWs/Pani composite films are highly flexible and can be bended without breakage, as depicted in Fig. 5.9 [56]. Highly flexible PPy/CNWs nanocomposite paper with electrical conductivities as high as 10^{-3} S/cm were obtained [117]. Composite films prepared by in situ polymerization methods are promising because they are processed in water without the necessity of dissolving polymers in organic solvents.

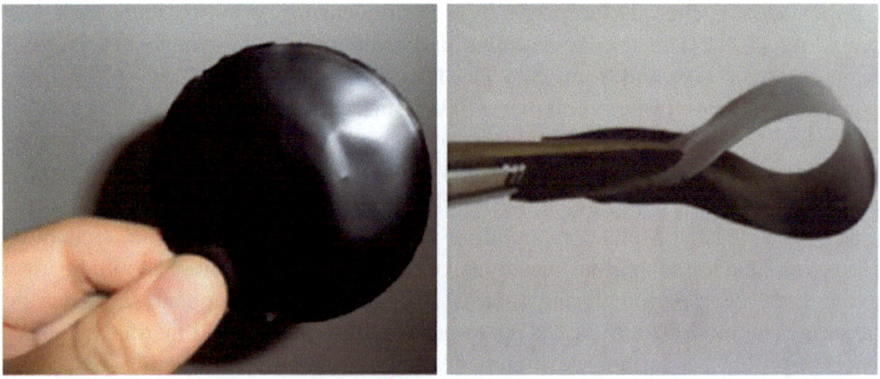

Fig. 5.9 Optical images of cellulose nanowhiskers (CNWs)-based composite film containing 20 wt% polyaniline (PANI) after bending by 18° (thickness of 50 m). Reproduced from Ref. [56]

Concluding Remarks and Future Trends

Cellulose nanowhiskers (CNWs) are promising nanomaterials due to their superior properties, high aspect ratio, crystallinity, mechanical modulus, surface area, and reactivity, together with the excellent biodegradability and renewability. They can be extracted from a variety of cellulosic materials by using chemical, mechanical and biological methods. However, these methods are still associated with environmental pollution, energy and time-consuming issues, respectively. The main challenges are to find an efficient CNW isolation process for environmental friendly mass production. The morphologies, physical and chemical structures, thermal behaviors of CNWs can be evaluated with the conventional instruments. Mechanical properties including strength and modulus of CNWs are essential to be measured in more direct and accurate manner.

CNWs as a natural reinforcing nanofillers, can be incorporated into numerous polymer matrices by solution casting, melt compounding, electrospinning, and in situ polymerizing methods. Although a better dispersion of nanofillers can be achieved using the solvent casting method, the process involves many chemical solvents, most of which are hazardous and hence impractical for the production upscaling at the industrial level. Electrospinning encounters the same problem due to the requirement to dissolve polymers with organic solvents. Thus water soluble polymers are preferable for the development of electrospun nanofibers. In situ polymerization tends to result in a better dispersion of nanoparticles than melt compounding, though it could not be as environmental friendly as the latter method at the industrial level. Unmodified and modified CNW reinforced polymer composite films showed improved mechanical properties, high transparency, and reduced gas permeability, which are promising material features to be used for food packaging. Electrospun nanofibrous composite mats exhibit great potential in tissue engineering and gas/liquid filtration. However, the biological toxicity is still of great concern for biomedical applications, conductive composite films reinforced with CNWs possess good mechanical properties and electrical conductivities, which are an ideal material candidate for energy storage devices. Despite the growing scientific production of nanocomposites, the transfer from laboratory scale to industrialization appears impractical and requires the development of advanced production technology of CNWs and their composites.

References

1. Ioelovich M (2008) Cellulose as a nanostructured polymer: a review. Bioresources 3:1403–1418
2. Klemm D, Heublein B, Fink HP, Bohn A (2005) Cellulose: fascinating biopolymer and sustainable raw material. Angew Chem Int Ed 44:3358–3393
3. Thomas S, Paul SA, Pothan LA, Deepa B (2011) In: Kalia S, Kaith BS, Kaur I (eds) Natural fibres: structure, properties and applications in cellulose fibers: bio- and nano-polymer composites. Springer, Berlin, pp 3–42

4. Kalia S, Dufresne A, Cherian BM, Kaith BS, Avérous L, Njuguna J, Nassiopoulos E (2011) Cellulose-based bio- and nanocomposites: a review. Int J Polym Sci 1–35. doi:10.1155/2011/837875
5. Eichhorn SJ (2011) Cellulose nanowhiskers: promising materials for advanced applications. Soft Matter 7:303–315
6. Lu HJ, Gui Y, Zheng LH, Liu X (2013) Morphological, crystalline, thermal and physicochemical properties of cellulose nanocrystals obtained from sweet potato residue. Food Res Int 50:121–128
7. Siro I, Plackett D (2010) Microfibrillated cellulose and new nanocomposite materials: a review. Cellulose 17:459–494
8. Chen W, Yu H, Liu Y (2011) Preparation of millimeter-long cellulose I nanofibers with diameters of 30–80 nm from bamboo fibers. Carbohd Polym 86:453–461
9. Brinchi L, Cotana F, Fortunati E, Kenny JM (2013) Production of nanocrystalline cellulose from lignocellulosic biomass: technology and applications. Carbohd Polym 94:154–169
10. Nishino T, Takano K, Nakamae K (1995) Elastic modulus of the crystalline regions of cellulose polymorphs. J Polym Sci Part B Polym Phys 33:1647–1651
11. Geyer U, Heinze T, Stein A, Klemm D, Marschs S, Schumann D (1994) Formation, derivation and applications of bacterial cellulose. Int J Biol Macromol 16:343–347
12. Mukherjee SM, Woods HJ (1953) X-ray and electron microscope studies of the degradation of cellulose by sulfuric acid. Biochim Biophys Acta 10:499–511
13. Strømme M, Mihranyan A, Ek R (2012) What to do with all these algae? Mater Lett 57:569–572
14. Nascimentoa JHO, Luz RF, Galvão FMF (2015) Extraction and characterization of cellulosic nanowhisker obtained from discarded cotton fibers. Mater Today Proc 2:1–7
15. Sheltami RM, Abdullah I, Ahmad I, Dufresne A, Kargarzdeh H (2012) Extraction of cellulose nanocrystals from mengkuang leaves (*Pandanus tectorius*). Carbohd Polym 88:772–779
16. Gelin K, Bodin A, Gatenholm P (2007) Characterization of water in bacterial cellulose using dielectric spectroscopy and electron microscopy. Polymer 48:7623–7631
17. Rusli R, Eichhorn SJ (2011) Interfacial energy dissipation in a cellulose nanowhisker composite. Nanotechnology 22:325706
18. Bettaieb F, Khiari R, Hassan ML, Belgacem MN, Bras J, Dufresne A, Mhenni MF (2015) Preparation and characterization of new cellulose nanocrystals from marine biomass Posidoniaoceanica. Ind Crops Prod 72:175–182
19. Zuluaga R, Putaux JL, Restrepo A, Mondragon I, Gañán P (2007) Cellulose microfibrils from banana farming resudues: isolation and characterization. Cellulose 14:585–592
20. Hiasa S, Iwamoto S, Endo T, Edashige Y (2014) Isolation of cellulose nanofibrils from mandarin (*Citrus unshiu*) peel waste. Ind Crop Prod 62:280–285
21. Alemdar A, Sain M (2008) Isolation and characterization of nanofibers from agricultural residues—Wheat straw and soy hulls. Bioresour Technol 99:1664–1671
22. Lu P, Hsieh YL (2012) Preparation and characterization of cellulose nanocrystals from rice straw. Carbohd Polym 87:564–573
23. Moran JI, Alvarez VA, Cyras V, Vázquez A (2008) Extraction of cellulose and preparation of nanocellulose from sisal fibers. Cellulose 15:149–159
24. Liu YY, Liu DY, Sui GX (2015) Effects of cellulose nano-whiskers on the properties of poly (vinyl alcohol)/graphene nanoplatelets nanocomposites. Polym Comp (under review)
25. Nascimento DM, Almeida JS, Dias AF, Figueirêdod MCB, Moraise JPS, Feitosa JPA, Rosa MF (2014) A novel green approach for the preparation of cellulose nanowhiskersfrom white coir. Carbohd Polym 110:456–463
26. Li R, Fei J, Cai Y, Li Y, Feng J, Yao J (2009) Cellulose whiskers extracted from mulberry: a novel biomass production. Carbohyd Polym 76:94–99
27. Teixeira EM, Bondancia TJ, Teodoro KBR, Corrêaa AC, José JM, Mattos LHC (2011) Sugarcane bagasse whiskers: extraction and characterization. Ind Crop Prod 33:63–66

28. Mandal A, Chakrabarty D (2011) Isolation of nanocellulose from waste sugarcane bagasse (SCB) and its characterization. Carbohd Polym 86:1291–1299
29. Sadeghifar H, Filpponen I, Clarke SP (2011) Production of cellulose nanocrystals using hydrobromic acid and click reactions on their surface. J Mater Sci 46:7344–7355
30. Cherian BM, Pothan LA, Nguyen-Chung T, Mcnnig G, Samy K, Thomas S (2008) A novel method for the synthesis of cellulose nanofibril whiskers from banana fibers and characterization. J Agr Food Chem 56:5617–5627
31. Yu H, Qin Z, Liang B, Liu N, Zhou Z, Chen L (2013) Facile extraction of thermally stable cellulose nanocrystals with a high yield of 93 % through hydrochloric acid hydrolysis under hydrothermal conditions. J Mater Chem A 1:3938–3944
32. Bradg PL, Bekkum HV, Besemer AC (2004) TEMPO-mediated oxidation of polysaccharides: survey of methods and applications. Top Catal 27:49–66
33. Okita Y, Saito T, Isogai A (2010) Entire surface oxidation of various cellulose microfibrils by TEMPO-mediated oxidation. Biomacromolecules 11:1696–1700
34. Saito T, Nishiyama Y, Putaux JL, Vignon M, Isogai A (2006) Homogeneous suspensions of individualized microfibrils from TEMPO-catalyzed oxidation of native cellulose. Biomacromolecules 7:1687–1691
35. Shinoda R, Saito T, Okita Y, Isogai A (2012) Relationship between length and degree of polymerization of TEMPO-oxidized cellulose nanofibrils. Biomacromolecules 13:842–849
36. Sun X, Wu Q, Ren S (2015) Lei T, Comparison of highly transparent all-cellulose nanopaper prepared using sulfuric acid and TEMPO-mediated oxidation methods. Cellulose 22:1123–1133
37. Satyamurthy P, Jain P, Balasubramanya RH, Vigneshwaran N (2011) Preparation and characterization of cellulose nanowhiskers from cotton fibres by controlled microbial hydrolysis. Carbohd Polym 83:122–129
38. Jiang F, Hsieh Y (2013) Chemically and mechanically isolated nanocellulose and their self-assembled structures. Carbohd Polym 95:32–40
39. Zhang J, Zhang B, Zhang J, Lin L, Liu S, Ouyang P (2010) Effect of phosphoric acid pretreatment on enzymatic hydrolysis of microcrystalline cellulose. Biotechnol Adv 28:613–619
40. Segal S, Creely JJ, Martin AE, Conrad C (1959) An empirical method for estimating the degree of crystallinity of native cellulose using the X-ray diffractometer. Textile Res J 29:786–794
41. Moon RJ, Martini A, Simonsen J (2011) Cellulose nanomaterials review: structure, properties and nanocomposites. Chem Soc Rev 40:3941–3994
42. Liu DY, Yuan XW, Bhattacharyya D, Easteal AJ (2010) Characterisation of solution cast cellulose nanofiber—Reinforced poly (lactic acid). Express Polym Lett 4:26–31
43. Mohamad Haafiz MK, Hassan A, Zakariac Z, Inuwa IM (2014) Isolation and characterization of cellulose nanowhiskers from oil palm biomass microcrystalline cellulose. Carbohd Polym 103:119–125
44. Fahma F, Iwamoto S, Hori N, Iwata T, Takemura A (2011) Effect of pre-acid-hydrolysis treatment on morphology and properties of cellulose nanowhiskers from conconut husk. Cellulose 18:443–450
45. Martins MA, Teixeira EM, Correa AC, Ferreika M, Mattoso LHC (2011) Extraction and characterization of cellulose whiskers from commercial cotton fibers. J Mater Sci 46: 7858–7864
46. Rosa MF, Medeiros ES, Malmonge JA, Gregorski KS, Wood DF, Mattoso LHC, Glenn G, Orts WJ, Imam SH (2010) Cellulose nanowhiskers from coconut husk fibers: effect of preparation conditions on their thermal and morphological behavior. Carbohd Polym 81: 83–92
47. Yoshiharu N, Paul L, Henri C (2002) Crystal structure and hydrogen-bonding system in cellulose Iβ from synchrotron X-ray and neutron fiber diffraction. J Am Chem Soc 124: 9074–9082
48. Trachea D, Donnot A, Khimeche K, Benelmir R, Brosse N (2014) Physico-chemical properties and thermal stability of microcrystallinecellulose isolated from Alfa fibres. Carbohd Polym 104:223–230

49. Pereira ALS, do Nacimento DM, Souza Filho MM, Morais JPS, Uasconcelos NF, Feitosa JPA, Brigida AI, Rosa MF (2014) Improvement of polyvinyl alcohol properties by adding nanocrystalline cellulose isolated from banana pseudostems. Carbohyd Polym 112:165–172
50. Habibi Y, Lucia LA, Rojas OJ (2010) Cellulose nanocrystals: chemistry, self-assembly, and applications. Chem Rev 110:3479–3500
51. Martins M, Teixeira E, Corrêa A, Ferreira M, Mattoso LHC (2011) Extraction and characterization of cellulose whiskers from commercial cotton fibers. J Mater Sci 46: 7858–7864
52. Roohani M, Habibi Y, Belgacem NM (2008) Cellulose whiskers reinforced polyvinyl alcohol copolymers nanocomposites. Eur Polym J 44:2489–2498
53. Angles MN, Dufresne A (2000) Plasticized starch/tunicin whiskers nanocomposites. 1. Structural analysis. Macromolecules 33:8344–8353
54. Zhou YM, Fu SY, Zheng LM, Zhan HY (2012) Effect of nanocellulose isolation techniques on the formation of reinforced poly (vinyl alcohol) nanocomposite films. eXPRESS Polym Lett 6:794–80
55. Trache D, Donnot A, Khimeche K, Benelmir R, Brosee N (2014) Physico-chemical properties and thermal stability of microcrystalline cellulose isolated from Alfa fibres. Carbohyd Polym 104:223–230
56. Liu DY, Sui SX, Bhattacharyya D (2014) Synthesis and characterisation of nanocellulose-based polyaniline conducting films. Comp Sci Technol 99:31–36
57. Yang H, Yan R, Chen H (2006) In-depth investigation of biomass pyrolysis based on three major components: hemicellulose, cellulose, and lignin. Energy Fuels 20:388–393
58. Bourbigot S, Chlebicki S, Mamleev V (2002) Thermal degradation of cotton under linear heating. Polym Degrad Stabil 78:57–62
59. Quievy N, Jacquet N, Sclavons M, Deroanne C, Paquot M, Devaux J (2010) Influence of homogenization and drying on thermal stability of microfibrillated cellulose. Polym Degrad Stabil 95:306–314
60. Johar N, Ahmad I, Dufresne A (2012) Extraction, preparation and characterization of cellulose fibres and nanocrystals from rice husk. Ind Crop Prod 37:93–99
61. Roman M, Winter WT (2004) Effect of sulfate groups from sulfuric acid hydrolysis on the thermal degradation behavior of bacterial cellulose. Biomacromolecules 5:1671–1677
62. DES, Costa LA, Fonseca AF, Pereira FV, Druzian J (2015) Extraction and characterization of cellulose nanocrystals from corn stover. Cellulose Chem Technol 49:127–133
63. Martinez-Sanz M, Lopez-Rubio A, Lagron JM (2011) Optimization of the nanofabrication by acid hydrolysis of bacterial cellulose nanowhiskers. Carbohd Polym 85:228–236
64. Sakurada I, Nukushina Y, Ito T (1962) Experimental determination of the elastic modulus of crystalline regions in oriented polymers. J Polym Sci 57:651–660
65. Rusli R, Eichhorn SJ (2008) Determination of the stiffness of cellulose nanowhiskers and the fiber-matrix interface in a nanocomposite using Raman spectroscopy. Appl Phys Lett 93:033111
66. Sturcova A, Davies GR, Eichhorn SJ (2005) Elastic modulus and stress-transfer properties of tunicate cellulose whiskers. Biomacromolecules 6:1055–1061
67. Eichhorn SJ, Davies GR (2006) Modelling the crystalline deformation of native and regenerated cellulose. Cellulose 13:291–307
68. Iwamoto S, Kai W, Isogai A, Iwata T (2009) Elastic modulus of single cellulose whiskers from tunicate measured by atomic force microscopy. Biomacromolecules 10:2571–2576
69. Favier V, Chanzy H, Cavaille JY (1996) Polymer nanocomposites reinforced by cellulose whiskers. Macromolecules 28:6365–6367
70. Eichhorn SJ, Dufresne A, Aranguren M, Marcocich NE, Capadona JR, Rowan SJ (2010) Review: current international research into cellulose nanofibres and nanocomposites. J Mater Sci 45:1–33
71. Dufresne A (2010) Processing of polymer nanocomposites reinforced with polysaccharide nanocrystals. Molecules 15:4111–4128

72. Beck S, Bouchard J, Berry R (2012) Dispersibility in water of dried nanocrystalline cellulose. Biomacromolecules 13:1486–1494
73. van den Berg O, Capadona JR, Weder C (2007) Preparation of homogeneous dispersions of tunicate cellulose whiskers in organic solvents. Biomacromolecules 8:1353–1357
74. Pei Aihua, Zhou Qi, Berglund Lars A (2010) Functionalized cellulose nanocrystals as biobased nucleation agents in poly(L-lactide) (PLLA)—Crystallization and mechanical property effects. Comp Sci Technol 70:815–821
75. Wang D, Yu J, Zhang J, He J, Zhang J (2010) Transparent bionanocomposites with improved properties from poly (propylene carbonate)(PPC) and cellulose nanowhiskers (CNWs). Comp Sci Technol 85:83–89
76. Gruner M, Winter WT (2002) Nanocomposites of cellulose acetate butyrate reinforced with cellulose nanocrystals. J Polym Environ 10:27–30
77. Cao X, Chen Y, Chang PR, Muir AD, Falk G (2008) Starch-based nanocomposites reinforced with flax cellulose nanocrystals. Express Polym Lett 2:502–510
78. Sanchez-Garcia MD, Lagaron JM (2010) On the use of plant cellulose nanowhiskers to enhance the barrier properties of polylactic acid. Cellulose 17:987–1004
79. Olsson RT, Fogelström L, Martínez-Sanz M, Henriksson M (2011) Cellulose nanofillers for food packaging. In: Lagaroân JM (ed) Multifunctional and nanoreinforced polymers for food packaging. Woodhead Publishing Limited, pp 86–107
80. Saxena A, Ragauskas AJ (2009) Water transmission barrier properties of biodegradable films based on cellulosic whiskers and xylan. Carbohyd Polym 78:357–360
81. Schyrr B, Pasche S, Voirin G, Weder C, Simon YC, Foster EJ (2014) Biosensors based on porous cellulose nanocrystals–poly(vinyl alcohol) scaffolds. Appl Mater Interf 6:12674–12683
82. Blaker JJ, Lee KY, Mantalaris A, Bismarck A (2010) Ice-microsphere templating to produce highly porous nanocomposite PLA matrix scaffolds with pores selectively lined by bacterial cellulose nano-whiskers. Comp Sci Technol 70:1879–1888
83. He JH, Liu Y, Mo LF, Wan YQ, Xu L (2008) In: He JH, Liu Y, Mo LF, Wan YQ, Xu L (ed) Electrospun nanofibers and their applications. iSmithers, UK, pp 6–16
84. Ding B, Wang M, Yu J, Sun G (2009) Gas sensors based on electrospun nanofibers. Sensors 9:1609–1624
85. Bhattarai SR, Bhattarai N, Yi HK, Hwang PH, Cha DI, Kim HY (2004) Novel biodegradable electrospun membrane: scaffold for tissue engineering. Biomaterials 25:2595–2602
86. Xu L, Wu Y, Liu Y (2010) Electrospun nanoporous materials: reality, potential, and challenges. Mater Sci Technol 26:1304–1308
87. Ramakrishna S, Jose R, Archana PS, Nair AS, Balamurugan R, Venugopal J, Teo WE (2010) Science and engineering of electrospun nanofibers for advances in clear energy, water filtration, and regenerative medicine. J Mater Sci 45:6283–6312
88. Hou H, Ge JJ, Zeng J, Li Q, Reneker DH, Greiner A (2005) Electrospun polyacrylonitrile nanofibers containing a high concentration of well-aligned multiwall carbon nanotubes. Chem Mater 17:967–973
89. Ji J, Sui G, Yu Y, Liu Y, Lin Y, Du Z (2009) Significant improvement of mechanical properties observed in highly aligned carbon-nanotube-reinforced nanofibers. J Phys Chem C 113:4779–4785
90. Hong JH, Jeong EH, Lee HS, Baik DH, Seo SW, Youk JH (2005) Electrospinning of polyurethane-organically modified Montmorillonite nanocomposites. J Polym Sci Part B Polym Phys 43:3171–3177
91. Peresin MS, Habibi Y, Zoppe JO (2010) Nanofiber composites of polyvinyl alcohol and cellulose nanocrystals: manufacture and characterization. Biomacromolecules 11:674–681
92. Park WI, Kang M, Kim HS (2007) Electrospinning of poly(ethylene oxide) with bacterial cellulose whiskers. Macromol Symp 249–250:289–294
93. Xiang C, Taylor AG, Hinestroza JP, Frey MW (2013) Controlled release of nonionic compounds from poly (lactic acid)/cellulose nanocrystal nanocomposite fibers. J Appl Polym Sci 127:79–86

94. Shi QF, Zhou C, Yue YY, Guo WH, Wu YQ, Wu QL (2012) Mechanical properties and in vitro degradation of elcetrospun bio-nanocomposite mats from PLA and cellulose nanocrystals. Carbohyd Polym 90:301–308
95. Zhou C, Shi Q, Guo W, Terrel L, Qureshi AT, Hayes DJ, Wu Q (2013) Electrospun bio-nanocomposites scaffolds for bone tissue engineering by cellulose nanocrystals reinforcing maleic anhydride grafted PLA. Appl Mater Interf 5:3847–3854
96. Liu DY, Yuan XW, Bhattacharyya D (2012) The effects of cellulose nanowhiskers on electrospun poly(lactic acid) nanofibers. J Mater Sci 47:3159–3165
97. Shi QF, Zhou CJ, Yue YY, Guo W, Wu Y, Wu Q (2012) Mechanical properties and in vitro degradation of electrospun bio-nanocomposite mats from PLA and cellulose nanocrystals. Carbohyd Polym 90:301–308
98. Dong H, Strawhecker KE, Snyder JF, Orlicki JA, Reiner RS, Rudie AW (2012) Cellulose nanocrystals as a reinforcing material for electrospun poly(methyl methacrylate) fibers: formation, properties and nanomechanical characterization. Carbohyd Polym 2012:2488–2495
99. Rojas OJ, Montero GA, Habibi Y (2009) Electrospun nanocomposites from polystyrene loaded with cellulose nanowhiskers. J Appl Polym Sci 113:927–935
100. Zoppe JO, Peresin MS, Habibi Y, Venditti RA, Rojas OJ (2009) Reinforcing poly (ε-caprolactone) nanofibers with cellulose nanocrystals. Appl Mater Interf 9:1996–2004
101. Lee J, Deng Y (2013) Nanoindentation study of individual cellulose nanowhisker-reinforced PVA electrospun fiber. Polym Bull 70:1205–1219
102. Vallejos ME, Peresin M, Rojas OJ (2012) All-cellulose composite fibers obtained by electrospinning dispersions of cellulose acetate and cellulose nanocrystals. J Polym Environ 20:1075–1083
103. de Menezes AJ, Siqueir G, Curvelo AAS, Dufresne A (2009) Extrusion and characterization of functionalized cellulose nanowhiskers reinforced polyethylene nanocomposites. Polymer 50:4552–4563
104. Raquez JM, Murena Y, Goffin AL, Habibi Y, Ruelle B, DeBuyl F, Dubois P (2012) Surface-modification of cellulose nanowhiskers and their use as nanoreinforcers into polylactide: a sustainably-integrated approach. Comp Sci Technol 72:544–549
105. Blaker JJ, Lee KY, Walters M, Drouet M, Bismarck A (2014) Aligned unidirectional PLA/bacterial cellulose nanocomposite fibre reinforced PDLLA composites. React Funct Polym 85:185–192
106. Goffin AL, Raquez JM, Duquesne E, Siqueira G, Habibi Y, Dufresne A, Dubois Ph (2011) Poly(e-caprolactone) based nanocomposites reinforced by surface-grafted cellulose nanowhiskers via extrusion processing: morphology, rheology, and thermo-mechanical properties. Polymer 52:1532–1538
107. Hietala M, Mathew AP, Oksman K (2013) Bionanocomposites of thermoplastic starch and cellulose nanofibers manufactured using twin-screw extrusion. Eur Polym J 49:950–956
108. Jonoobi M, Harun J, Mathew AP, Oksman K (2010) Mechanical properties of cellulose nanofiber (CNF) reinforced polylactic acid (PLA) prepared by twin screw extrusion. Comp Sci Technol 70:1742–1747
109. Bondeson D, Oksman K (2007) Dispersion and characteristics of surfactant modified cellulose whiskers nanocomposites. Comp Interf 14:617–630
110. Lin N, Huang J, Chang PR, Feng J, Yu J (2011) Surface acetylation of cellulose nanocrystals and its reinforcing function in poly(lactic acid). Carbohydr Polym 83:1834–1842
111. Chen G, Dufresne A, Huang J, Chang PR (2009) A novel thermoformable bionanocomposite based cellulose nanocrystal-graft-poly(ε-caprolactone). Macromol Mater Eng 294:59–67
112. Lin N, Chen G, Huang J, Dufresne A, Chang PR (2009) Effects of polymer-grafted natural nanocrystals on the structure and mechanical properties of poly(lactic acid): a case of cellulose whisker-*graft*-polycaprolactone. J Appl Polym Sci 113:3417–3425
113. Theng BKG (1979) In: Theng BKG (ed) Formation and properties of clay-polymer complexed. Elsevier, Amsterdam, pp 201–241

114. Rueda L, Saralegi A, Fernández-d'Arlas B, Zhou Q, Alonso-Varona A, Berglund LA, Mondragon I, Corcuera MA, Eceiza A (2013) In situ polymerization and characterization of elastomeric polyurethane-cellulose nanocrystal nanocomposites. Cell response evaluation. Cellulose 20:1819–1828
115. Maiti S, Sain S, Ray D, Mitr D (2013) Biodegradation behaviour of PMMA/cellulose nanocomposites prepared by in-situ polymerization and ex-situ dispersion methods. Polym Degrad Stabil 98:635–642
116. Luong ND, Korhonen JT, Soininen AJ, Ruokolainen J, Johansson LS, Seppälä J (2013) Processable polyaniline suspensions through in situ polymerization. Eur Polym J 49:335–344
117. Zhang D, Zhang Q, Gao X, Piao G (2013) A nanocellulose polypyrrole composite based on tunicate cellulose. Int J Polym Sci 2013:175609

Chapter 6
From Polymer Blends to Nano-size Materials with Controlled Nanomorphology

Stoyko Fakirov

Introduction

Although nanoparticles are generally considered a discovery of modern science, they actually have a very long history. Nanoparticles were used by artisans as far back as the ninth century in Mesopotamia for generating a glittering effect on the surface of pots. The peculiarities of nanomaterials arise mainly from their sizes and for this reason the search of methods for their preparation is of increasing importance.

Nanomaterials are materials with morphological features on the nanoscale, and especially those that have special properties stemming from their nanoscale dimensions. Nanoscale is usually defined as smaller than a one-tenth of a micrometer in at least one dimension [1]. In 2011 the European Commission adopted the following definition of a nanomaterial: "… 50 % or more of the particles in the number size distribution, one or more external dimensions is in the size range 1–100 nm…." [2].

An important aspect of nanotechnology is the vastly increased ratio of surface area to volume present in many nanoscale materials. For example, 1 kg of particles of 1 mm^3 has the same surface area as 1 mg of particles of 1 nm^3. The interesting and sometimes unexpected properties of nanoparticles are, therefore, largely due to the large surface area of the material, which dominates the contributions made by the small bulk of the material.

Nanoparticles may or may not exhibit size-related properties that differ significantly from those observed in fine particles or bulk materials. Although the size of most molecules would fit into the previous outline, individual molecules are usually not referred to as nanoparticles.

S. Fakirov (✉)
Department of Mechanical Engineering, Centre for Advanced Composite Materials,
The University of Auckland, Private Bag 92019, Auckland, New Zealand
e-mail: s.fakirov@auckland.ac.nz

S. Fakirov (ed.), *Nano-size Polymers*,
DOI 10.1007/978-3-319-39715-3_6

Nanoparticles of usually yellow gold and gray silicon are red in color. Gold nanoparticles melt at much lower temperatures (300 °C for 2.5 nm size) than the gold slabs (1064 °C) [3]. Also, the absorption of solar radiation in photovoltaic cells is much higher in materials composed of nanoparticles than it is in thin films of continuous sheets of material (i.e., the smaller the particles, the greater the solar absorption).

Another example being closer to the topic of the present study is the polymer nanocomposites. Clay nanoparticles when incorporated into polymer matrices increase reinforcement, leading to stronger plastics, verifiable by a higher glass transition temperature and other mechanical property tests. These nanoparticles are hard, and impart their properties to the polymer. For example, with tensile moduli in the tera-pascal range and lengths exceeding 10 μ of carbon nanotubes (CNTs), simple composite models predict order-of-magnitude enhancement in modulus at loadings less than 1 %. For this reason, a decade ago it was believed that the most common polymer composites comprising about 30 % glass fibers will be replaced by the nanocomposites having only 2–5 % nano-size filler as reinforcement! Unfortunately, with the exception of reinforced elastomers, nanocomposites have not lived up to expectations [4].

Nanoparticle research is currently an area of intense scientific interest due to the wide variety of potential applications in biomedical, optical, and electronic fields.

There are several methods for creating nanoparticles, including both attrition and pyrolysis. In attrition, macro- or microscale particles are ground in a ball mill, a planetary ball mill, or other size reducing mechanism. The resulting particles are air classified to recover nanoparticles. In pyrolysis, a vaporous precursor (liquid or gas) is forced through an orifice at high pressure and burned. The final solid is air classified to recover oxide particles from by-product gases. Pyrolysis often results in aggregates and agglomerates rather than single primary particles.

Another method is the sol–gel process (also known as chemical solution deposition) widely used recently in the fields of materials science. Such methods are used primarily for the fabrication of materials (typically a metal oxide) starting from a chemical solution, which acts as the precursor for an integrated network (or gel) of either discrete particles or network polymers.

As a matter of fact, just for polymers since the beginning of the last century exists a technique for converting many polymers into nanofibrillar state using their solutions or melts. Electrospinning was first observed by Rayleigh in 1897 [5], studied further by Zeleny in 1914 [6] and patented by Formhals in 1934 [7]. Electrospinning has been widely used in the past decade to produce nanofibers from a variety of different polymers [8, 9] with a particular application in the regenerative medicine [10].

Nowadays, the electrospinning is used in hundreds of labs worldwide because of its elegancy and the fascinating electron microscopic images of the nanofibers obtained which can be seen in some 1000 papers published per year [9]. In the same time, this attractive technique has the disadvantage that not so much can be done practically with the fine nanofibers because of the not easy handling of the spun material [8–10]. In addition, this technique has another peculiarity—among the

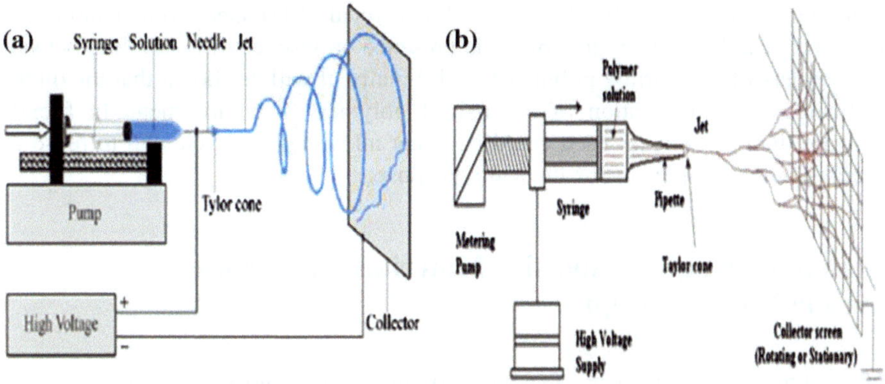

Fig. 6.1 Schematic the electrospinning setup illustrating the two different mechanisms of nanofiber formation. Reproduced with permission from Ref. [64]; Copyright 2015 Elsevier

electrospinning community there is still no consensus regarding the mechanism of nanofibre formation. As depicted in Fig. 6.1, there are two rather different mechanisms for nanowires formation. According to the first one (Fig. 6.1a), the nanowire represents a single filament, while according to the second one (Fig. 6.1b), the nanofibers are interconnected similarly to a spider network. Possibly, the answer could be found in the comparison of electrospinning with the natural silk formation and production.

After growing enough for a month, each caterpillar begins spinning a cocoon by moving its head in a pattern. Two glands produce liquid silk and force it through openings in the head (spinnerets). By the way, the misleading interpretation of the filament formation mechanism, i.e., pressing instead of drawing as later established, resulted some 150 years ago in the creation of the industry for artificial silk (rayon, viscose) [11]. Liquid silk is coated in sericin, a water-soluble protein, and solidifies on contact with air. Within 2–3 days, the caterpillar spins about 1.5 km of filament and is completely encased in a cocoon. Harvested cocoons are then soaked in boiling water to soften sericin holding the silk fibers together in a cocoon shape. The fibers are then unwound to produce a continuous tread. Since a single filament is too fine and fragile for commercial use, anywhere from three to ten strands are spun together to form a single tread of silk [12].

It seems important to mention an interesting detail regarding the structure of the cocoon and that of the electrospun material—from textile point of view both of them represent a nonwoven textile. The fact that only from a cocoon it is possible to unwound single filament is in favor of the second mechanism of nanofiber formation (Fig. 6.1b), i.e., the final electrospun product represents interconnected nanofibers similarly to the case of spider network.

The main target of this chapter is to describe a relatively new technique for converting of bulk polymers into nano-size materials with controlled nanomorphology, which does not suffer from the disadvantages of the electrospinning. The second target is to demonstrate that the final nanomorphology—individual non-interconnected nanofibrils or three dimensional (3-D) nanoporous nanofibrillar network—can be reliably governed via presence or absence of H-bonding between the partners of the starting polymer blend. Finally, it will be shown that the offered technology for preparation of nano-sized polymers is environmentally friendly because the only solvent used could be water allowing to recycle and to reuse the second blend component for the same purpose.

Manufacturing of Nano-size Materials and Articles via the MFC Concept

Opportunities for converting the bulk polymers into nano-size materials are also offered by the developed during the last two decades new type of polymer–polymer composite, the microfibril- or nanofibril reinforced composites (MFCs or NFCs) [13–28]. MFCs and NFCs are comprised of a polymer with a lower melting temperature (T_m) as a matrix, reinforced with extremely thin fibrils of a higher T_m polymer. They are manufactured from polymer blends. The two basic requirements to the blend partners are: (i) the two polymers should not be thermodynamically miscible, and (ii) they should be easily convertable in a highly oriented state [22]. After melt blending the polymer blend is extruded, and the resulting material is then drawn, thus transforming the minor component into a fibrillar state.

In case the final target is the preparation of nanofibrillar polymer–polymer composite, a thermal treatment between the melting points of the two polymers converts the lower T_m component (the matrix) into isotropic state, while leaving the reinforcing fibrils intact. In such cases an additional requirement to the blend partners exists—a difference in their melting temperatures of at least 40 °C. MFCs and NFCs created in this way have several advantages over composites with macro-size reinforcements, the most important of which is the fact that the dispergation stage (introducing and homogenation of the reinforcing component in the matrix) is missing since the reinforcing element (fibrils) are created during the manufacturing process.

The manufacturing process for converting the bulk polymers into nano-size material can be divided into two distinct steps:

- *Mixing and extrusion*: the matrix and reinforcing polymers are dried and mixed, before being compounded and extruded. This forms an isotropic, continuous blend filament (Fig. 6.2a).
- *Drawing with fibrillation*: the blend filament is drawn through pairs of rollers (Fig. 6.2b). This step creates highly oriented micro- or nanofibrils with properties biased predominantly along a linear dimension or symmetry axis. The drawing ratio is defined as the ratio of the linear speeds (S_2/S_1) of the two sets of rollers used to draw the filament (Fig. 6.2) and gives an indication as to the amount of alignment imparted to the blend. Next, the filament is either collected on a spool or pelletized in dependence of the further type of processing, i.e., compression molding or injection molding, respectively (Fig. 6.2a).

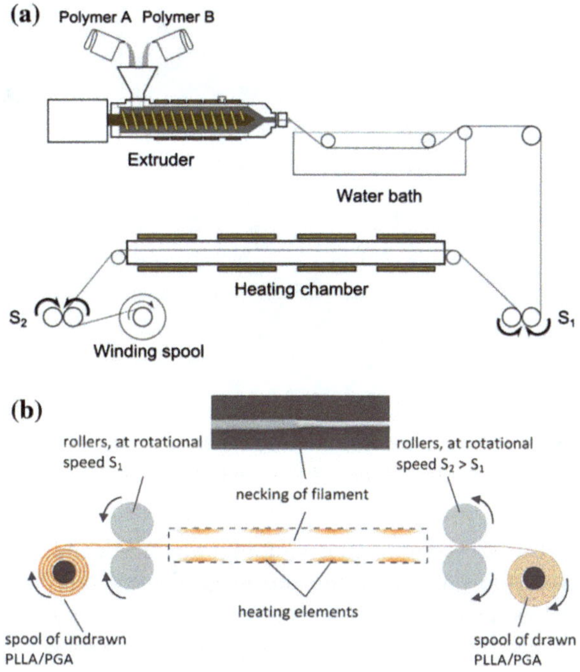

Fig. 6.2 Schematic the setup for manufacturing of MFCs (**a**) and the drawing device in more details demonstrating the neck formation (**b**); S_1 and S_2 are the rotating velocities of the pairs of rolls, where $S_1 < S_2$ and PLLA/PGA means blend of poly(L-lactic acid) and poly(glycolic acid)

Two different in situ fibril formation techniques can be employed to create the reinforcing microfibrils: (i) *cold drawing* of the solidified filament at a temperature far below the melting temperature of each blend constituent but still above their glass transition temperatures (T_g) [13–30], or (ii) *hot stretching* directly from the melt at a temperature far above both polymers' glass transition temperatures [31–39]. Cold drawing generally results in a greater molecular alignment within the microfibrils, but hot stretching can produce much higher draw ratios (although these ratios are not analogous with better molecular orientation). The drawing at temperatures around T_g takes place via the *necking phenomenon* as demonstrated in Fig. 6.2b, thus contrasting the case of hot stretching where the elongation does not require fundamental conformational changes resulting in stretching and parallel alignment of macromolecules.

The drawing performed slightly above the T_g of the reinforcing component (e.g., around 70 °C for poly(ethylene terephthalate) (PET) of the most studied polypropylene (PP)/PET blend) results in a perfect molecular orientation of the two blend partners as can be concluded from the wide-angle X-ray scattering (WAXS) patterns shown in Fig. 6.3.

At this point an important detail needs a clarification. For the cases when the target is converting a bulk polymer into nano-size material, i.e., isolation of neat nanofibrils, it is not necessary to perform the last (isotropization) step, which leads to preparation of polymer–polymer composite. In case this has to be done one has

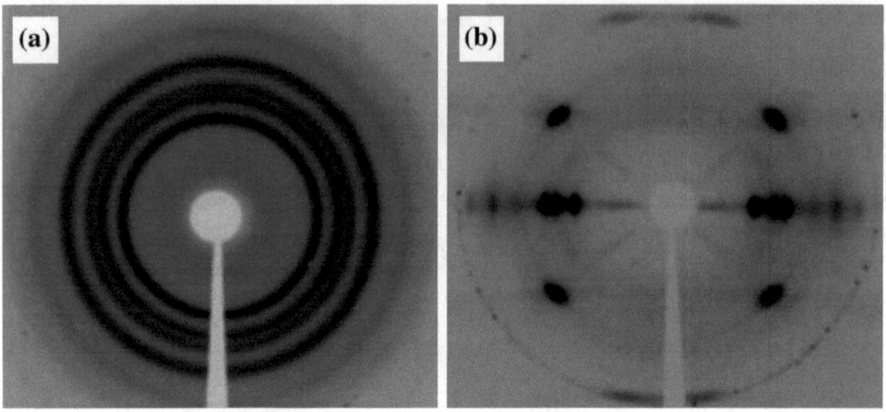

Fig. 6.3 WAXS patterns of PP/PET blends: **a** after extrusion (no drawing) and **b** after drawing. Reproduced with permission from Ref. [64]; Copyright 2015 Elsevier

to use the compression molding where aligned highly stretched drawn filaments from the polymer blend are used. During this heat treatment the lower melting component melts but the fibrils remain unchanged, even highly aligned.

For the isolation of the neat nanofibrils one can use the two rather different materials—the highly drawn polymer blend or the sheet (film) after compression molding. Either of these two materials has to be subjected to extraction of the major component using selective solvent. The rest represents parallel aligned neat nano- or microfibrils.

The extraction is a very important step of the technology. If organic solvent is used as selective solvent, the extraction has to be done at elevated temperature for many hours. The extraction time can be drastically reduced if one uses the recently modified Soxhlet apparatus [40] allowing extraction with boiling solvent.

The isolated in the described way neat micro- or nanofibrils can be further used as a starting material for manufacturing of micro-/nanofibrillar single polymer composites, as scaffolds in the regenerative medicine, as carriers for controlled drug delivery, as nanofilter and others.

Effect of Hydrogen Bonding in Polymer Blends on Nanomorphology

Non-hydrogen Bonding Polymers

To the best studied blends of this group of polymers belong the blends of polyolefins (not comprising functional groups) with condensation polymers (e.g., polyesters). In order to obtain a final product in the form of neat nanofibrils two

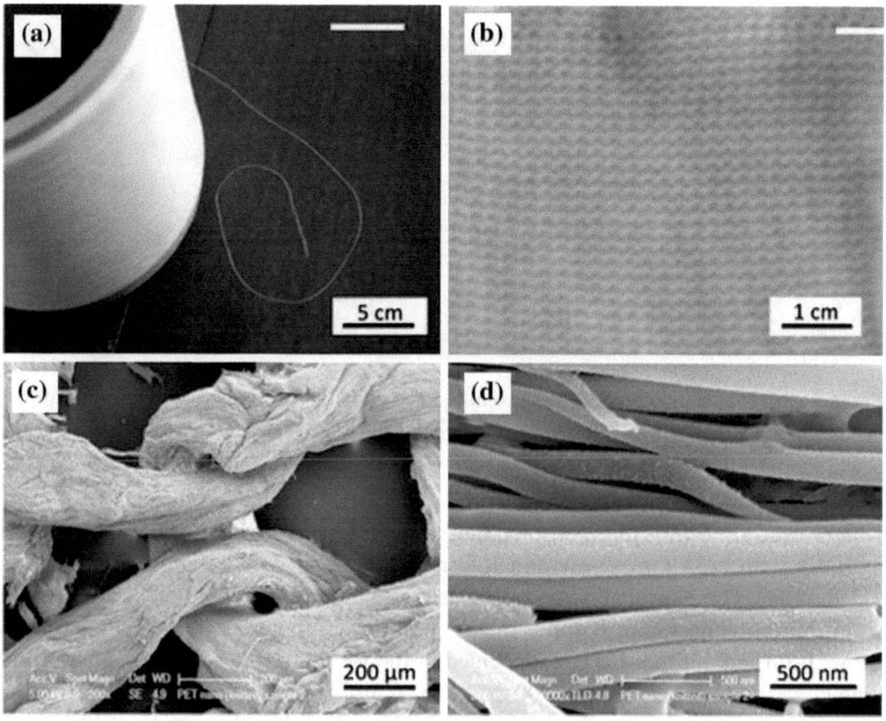

Fig. 6.4 Photograph of a PP/PET yarn of filaments with diameters of 30 μm (**a**), knitted fabrics of PP/PET yarn (**b**) and SEM micrographs of the same fabrics after removing of PP with xylene, i.e., fabrics comprising PET nanofibrils only at lower (**c**) and higher (**d**) magnifications. Reproduced with permission from Ref. [80]; Copyright 2015 Taylor & Francis

practical approaches have to be taken into account, namely: (i) intensive melt mixing leading to fine dispersion of the minor blend component, and/or (ii) using drawing conditions resulting in perfect molecular orientation. A good example for the effect of orientation could be the preparation of PET nanofibrils (Fig. 6.4).

After drying, melt blending and palletizing the PP/PET blend (80/20 wt%) was subjected to spinning using commercial equipment for manufacturing of synthetic fibers (Fig. 6.4a) thus preparing yarn of filaments with diameters of 30 μm. The final nanomaterial represents smooth not interconnected nanofibrils with thickness in the range of 50–150 nm, Fig. 6.4d. Figure 6.4 demonstrates also another peculiarity of this method of converting bulk polymers into nano-size materials—its applicability to any textile technique for manufacturing of the final articles comprising nanofibrils only. As a matter of fact, the article (Fig. 6.4b) is prepared from highly drawn polymer blend (textile yarn) and after removing of the dominating blend component (matrix) remain the nanofibrils only (Fig. 6.4d) organized in the desired way (Fig. 6.4d).

The importance of the good blend homogenization is demonstrated by the next PP/poly(butylene terephthalate) (PBT) blend. After drying and mixing (PP/PBT in

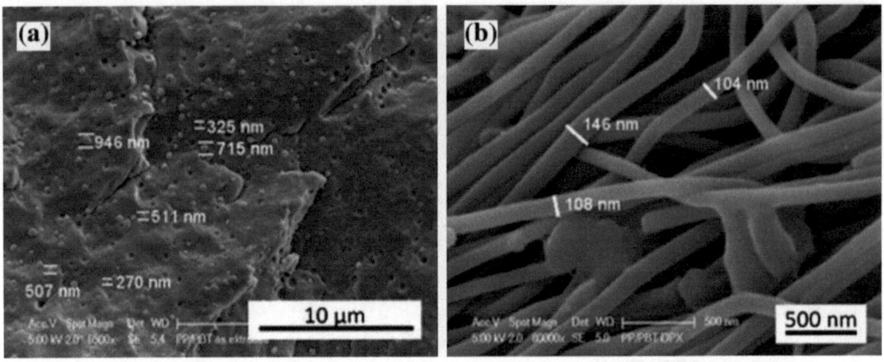

Fig. 6.5 SEM micrographs of PP/PBT blend **a** after exit from die of extruder (no drawing), and **b** after removal of PP from cold-drawn PP/PBT blend. Reproduced with permission from Ref. [80]; Copyright 2015 Taylor & Francis

weight ratio of 70/30) the melt blending was performed using a Brabender DSE20 extruder with a 25 mm screw and L/D ratio of 40, at 250 °C and 5 rpm, using a 1.3 mm die. The extrudate was cooled using a water bath immediately after its exit from the die (Fig. 6.2a). Sample of this non-drawn material was inspected in the SEM. It turned out that PBT is dispersed in the PP matrix as very fine spheres with diameters in the range of 300–700 nm (Fig. 6.5a). It is worthwhile to notice that in this particular case the smaller particles dominate. This fine dispersion of PBT in PP enhances drastically the preparation of rather thin (diameters around 100 nm) nanofibrils (Fig. 6.5b).

The extruded blend was subjected to cold drawing at 80 °C, to a ratio of about 5, using two pairs of rollers and a 2 m long heated chamber (Fig. 6.2b). This completes the fibrillation stage in the process of nanofibril formation. In order to separate the PBT neat nanofibrils, the drawn bristle was wound uniaxially on a wire frame and wrapped tightly in steel gauze. This frame was then inserted into the container of the modified Soxhlet apparatus [40], using xylene as a solvent and running it for at least 9 h, so as to completely remove PP. The isolated nanofibrils retained enough structural integrity to be unwound from the frame. The SEM observation showed that they are very smooth cylindrical formations with rather uniform diameters of about 100 nm, Fig. 6.5b. The neat PBT nanofibrils were further used for preparation of single polymer composite via winding on a metal plate, and hot compaction at temperature nearly 10 °C below the melting peak temperature [41].

The next system used for isolation of neat nanofibrils without formation of hydrogen bonding between the blend partner is the blend of linear low density polyethylene (LLDPE) (as a matrix) and poly(vinylidene fluoride) (PVDF) (as a reinforcement) in a weight ratio of 70/30 [42]. After the melt blending the extruded bristle was drawn up to a draw ratio of approximately 6. The drawing process was performed without intermittent breaking at room temperature due to high stretching

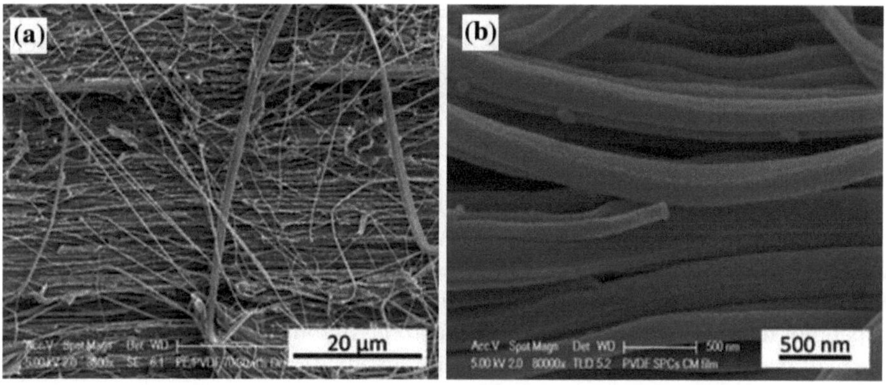

Fig. 6.6 SEM micrographs of nanofibrillar polymer–polymer composite based on LLDPE/PVDF and their reinforcing component: **a** cryofractured surface of a NFC made parallel to the draw direction, **b** neat PVDF nanofibrils after removing of LLDPE from the drawn LLDPE/PVDF blend

ability of the LLDPE. The highly drawn LLDPE/PVDF blend was wound onto a stainless steel plate and subjected to a hot pressing at 133 °C to produce a film with a NFC structure. For isolation of the PVDF neat nanofibrils the prepared nanofibrillar polymer–polymer composite (Fig. 6.6a) was subjected to extraction. The same final material (Fig. 6.6b) could be prepared if from the highly drawn blend the LLDPE will be removed using again a selective solvent.

Figure 6.6 shows SEM images of the described manufacturing stages. Figure 6.6a illustrates the perfect alignment of the reinforcing PVDF nanofibrils in the isotropic matrix of LLDPE. The same picture offers an idea about the aspect ratio of the reinforcement—nanofibrils of at least 50 μm in the length can be easily observed. Assuming an average thickness of the same nanofibrils of 100 nm (Fig. 6.6b), the aspect ratio will be of 500. Similarly to the previous SEM micrographs [12, 21, 26–28], these ones also demonstrate the very important fact—each nanofibril is surrounded individually by matrix material thus assuring a perfect dispersion of the nanoreinforcement and, what is more important, also making possible isolation of neat not agglomerated nanofibrils (Fig. 6.6).

Hydrogen Bonding Polymers

These blends are characterized by two peculiarities arising from the dominating second blend component, namely poly(vinyl alcohol) (PVA) contrasting the former dominating component (PP): (i) PVA is capable to form H-bonds because of the presence of OH groups in its molecule, and (ii) PVA belongs to the group of the few water-soluble polymers, and thus, for its extraction from the drawn blend it is not necessary to use organic solvent. To this rather new development of the method one has to come because of the application of micro- and nanofibrillar materials in medicine (e.g., tissue engineering) where even traces of organic solvents in the scaffolds have a negative effect on the cell growth. By replacing of organic solvents with water the method became environmentally friendly, cost effective (the water-soluble polymer can be regenerated and reused for the same purpose), and, last but not least, the final nanoarticles became more attractive for biomedical applications.

The first blend of this group is that of PVA/PETG. PETG being a glycol containing copolymer of PET is characterized by lower T_m as compared with PET and thus making possible the melt blending with the temperature sensitive PVA. Further, the most important characteristic feature of these two polymers is their capability to form H-bonds between the carbonyl group =CO of the polyester and the hydroxyl group –OH of PVA. Using the already developed technique for drying (being particularly essential for the polyesters), melt blending (in a wt ratio 70/30 = PVA/PETG) with extrusion, cold drawing and extraction of PVA with water, a nano-size material is prepared, as shown in Fig. 6.7.

Figure 6.7a shows that the final material looks as a thin continuous film, and only at high magnifications it became clear that one deals with a 3-D nanofibrillar nanoporous network (Fig. 6.7b), i.e., we observe a completely different nanomorphology as compared with the cases when no H-bonding is possible (Figs. 6.4, 6.5 and 6.6).

Before considering the reasons for formation of these two rather different types of nanomorphology, namely not interconnected smooth individual nanofibrils (Figs. 6.4, 6.5 and 6.6) or 3-D nanofibrillar nanoporous network (Fig. 6.7b) let notice that the second type of morphology has been also observed for other polyesters blended with PVA (Fig. 6.8).

Starting from the respective blends, the PVA/poly(lactic acid) (PLA), PVA/poly (capro lacton) (PCL), and PVA/poly(hydroxybutyrate) (PHB), and applying the standard treatment including extraction of PVA with water, the 3-D structures have been observed (Fig. 6.8).

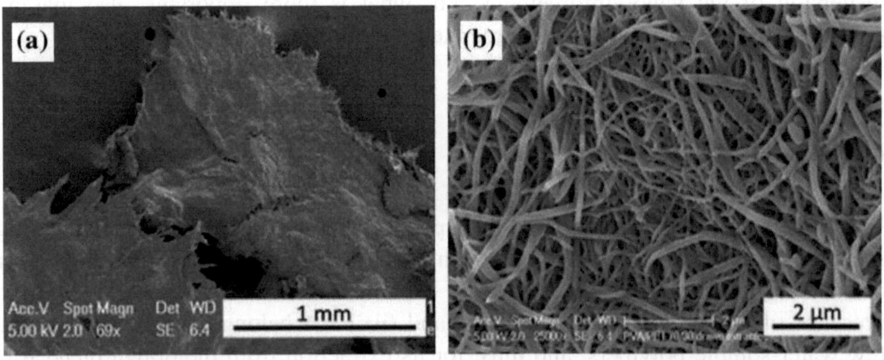

Fig. 6.7 SEM micrographs of 3-D nanofibrillar nanoporous network of PVA/PETG (70/30) blend after extrusion, cold drawing and extraction with WATER at **a** low and **b** high magnification

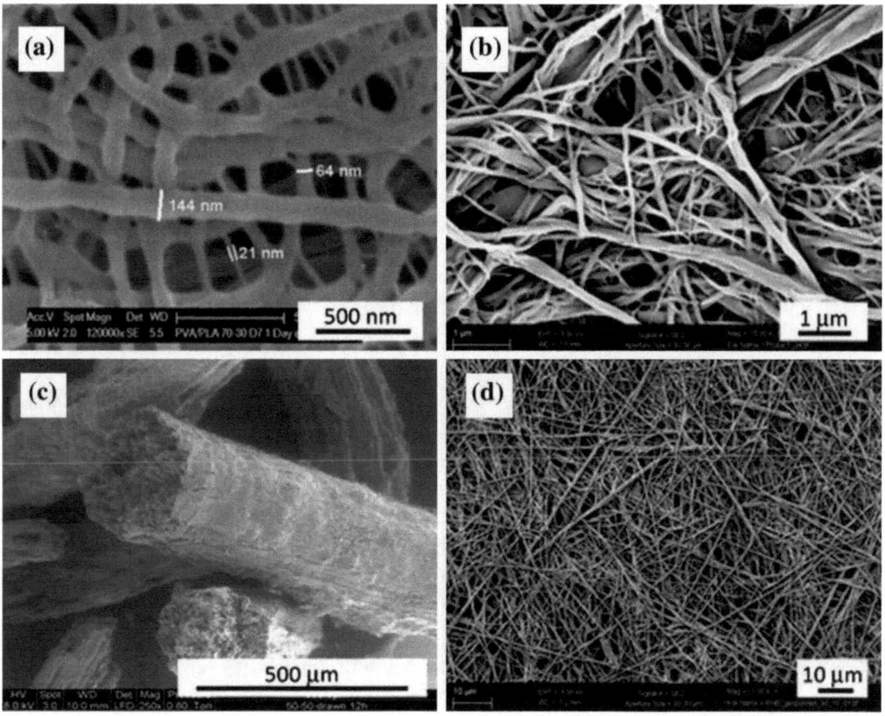

Fig. 6.8 SEM micrographs of biocompatible biodegradable polyesters after their melt blending with PVA with extrusion, cold drawing and extraction with water: **a** PLA in amount of the blend 20 wt%, **b** PCL of 20 wt%, **c** PHB after melt spinning from a PVA/PHB blend 90/10 by wt, and **d** the same at higher magnification. Reproduced with permission from Ref. [80]; Copyright 2015 Taylor & Francis

Comparing the results of the above described two types of experiments an interesting question arises: what could be the reason for obtaining of two completely different nanomorphologies of the nano-size polyesters when blended with PVA as a second blend component or with polyolefines (PP or PE)? The further systematic studies led to the conclusion that the main factor determining the type of the final nanomorphology is the possibility for formation of hydrogen bonds between the blend partners. In the cases when no hydrogen bonds exist the isolated neat nano-size materials represent individual non-interconnected fibrils and if hydrogen bonds are formed between the two blend partners the final morphology looks as a three-dimensional nanofibrillar nanoporous network.

Figure 6.9 demonstrates in the best way the crucial importance of the hydrogen bonding in polymer blends for obtaining of one or another nanomorphology because in this particular case the same polymer (PBT) has been blended with a H-bonding partner (PVA) and later with a non-hydrogen bonding partner (PP).

When H-bonding is possible as in the case of the blend PVA/PETG (Fig. 6.7) as well as in the blends of PLA, PHB, and PCL with PVA (Fig. 6.8) the final material

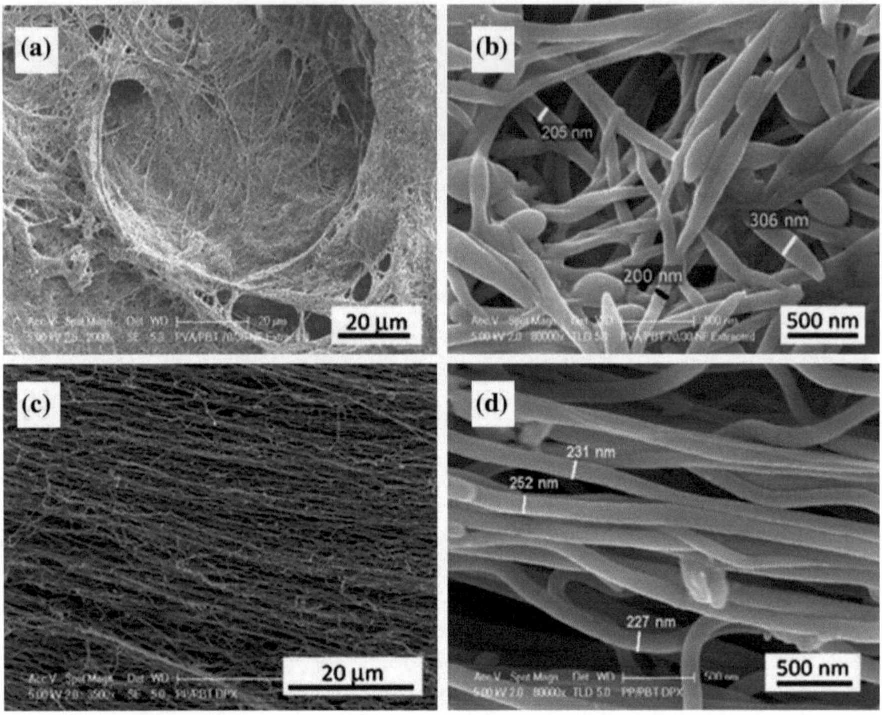

Fig. 6.9 SEM micrographs of PBT nanomorphology after removing of the second blend component: **a** and **b** PVA, **c** and **d** PP. Reproduced with permission from Ref. [80]; Copyright 2015 Taylor & Francis

is characterized by a 3-D structure (Fig. 6.9a, b) comprising nanopores and interconnected nanofibrils with diameters of about 200 nm (Fig. 6.9b). If the two blend partners are not capable to form hydrogen bonds as the case of the majority of studied blends (Figs. 6.4, 6.5 and 6.6) the final material is in the form of individual non-interconnected nanofibrils. Just the same situation one observes with the blend of PBT with PP (Fig. 6.9c, d), where the final nanomorphology represents continuous, not interconnected individual nanofibrils with diameter of about 250 nm (Fig. 6.9d). It should be noticed also that Fig. 6.9 demonstrates not only the importance of H-bonding for the formation of the final nanomorphology but also the potentials of the suggested method for converting of bulk polymers into nano-size materials with controlled nanomorphology. As a matter of fact, starting from the same bulk polymer (PBT) we are able to prepare two nano-size materials characterized by rather different nanomorphology, namely as individual not interconnected nanofibrils or as 3-D nanofibrillar nanoporous network.

Figure 6.10 shows schematically the formation of hydrogen bonds between polyesters and PVA.

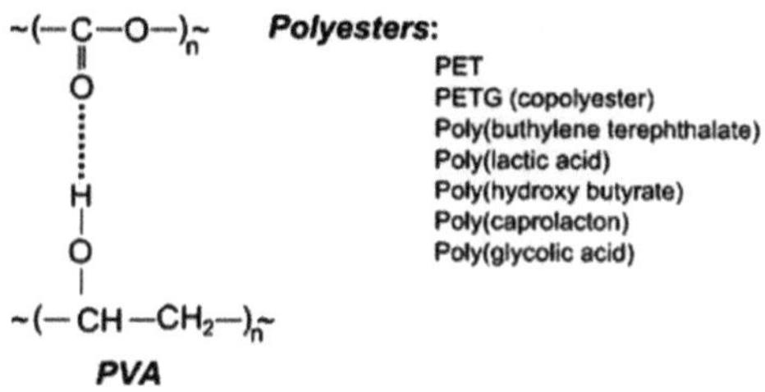

Fig. 6.10 Hydrogen bonding between polyesters and poly (vinyl alcohol)

Hydrogen bonding in polymer blends is a topic of great interest to polymer scientists because such systems have many potential applications [43–48]. For example, introducing functional groups to one component to make it capable of forming hydrogen bonds to another, thereby enhancing miscibility of otherwise immiscible blends, is one of the major achievements during the past 20 years of polymer science, as stated in a recent review on hydrogen bonding in polymer blends [49].

Coming back to our particular systems, the blends of PVA with various polyesters, one can assume that one deals with partial solubility leading to a good compatibilization of the two polymers due to the formation of H-bonds between them (Fig. 6.10). Existence of a complete solubility (thermodynamic miscibility) is excluded because in such a case one will obtain a one-phase melt. On the contrary, the partial solubility provoked by the H-bonding is the driving force for penetrating a small amount of the dominating component (PVA) in the dispersed particles of the minor component (e.g., PLA), forming a structure consisting of two co-continuous phases. After the extraction of this small amount finely dispersed PVA, a 3-D network of micro- or nanopores is formed (Figs. 6.7b, 6.8 and 6.9a, b). This assumption [50] is supported by the observation that in a ternary blend of poly (vinyl butyral)/PVA/PA 6, a thermodynamic miscibility of 0.4–0.6 volume fraction of vinyl alcohol has been found [51]. The SEM inspection of this blend after selective extraction of the dissolved component revealed a network of micropores [51, 52].

For a system much closer chemically and compositionally to PVA/PLA, Park and Im [53] reported that PLA/poly(vinyl acetate) (PVAc) blends were miscible systems for the entire composition range, but for the blends with even 10 % hydrolyzed PVAc copolymer, the phase separation and double glass transition could be observed. Another thorough study [54] on miscibility and phase structure of binary blends of poly(L-lactic acid) (PLLA) and PVA indicated that PLLA and PVA were immiscible in the amorphous regions. However, the data of the differential scanning calorimetry analysis still demonstrated that some degree of compatibility related to block composition existed in the blend systems. Furthermore, the formation of interpolymer hydrogen bonding in the amorphous region which is regarded as the driving force leading to some degree of component compatibility in these immiscible systems, has been confirmed by FTIR and further studied by ^{13}C solid-state NMR analysis [54].

Obviously, the hydrogen bonding is a powerful tool for controlling the properties of polymer blends, and more specifically the nanomorphology when converting the bulk polymers into nano-size materials. Blending completely immiscible polymers and applying the MFC concept makes possible the isolation of nano-size material in the form of individual not interconnected nanofibrils. In contrast to this situation, dealing with blend partners inclined to formation of hydrogen bonds and thus converting the blend in a partially miscible one the final nano-size material is a nanofibrillar nanoporous 3-D network.

Mechanism of Nanomorphology Formation in Polymer Blends Without and with Hydrogen Bonding

In addition to the outlined morphological difference between the two types of polymer blends, without and with hydrogen bonding between the blend partners, it turned out that the mechanism of formation of the nano-size materials is completely different for the one or the other case. Detailed studies on the mechanism of formation of the individual micro- and nanofibrils led to the conclusion that it takes place during the cold drawing via *coalescence* of the elongated droplets [55], as schematically illustrated in Fig. 6.11.

The rare statements (e.g., [56]) that each final fibril originates from a single spherical particle could hardly be correct for the following reasons: the comparison of the volumes of a starting sphere with that of the final fibril shows difference of many tens in favor of the fibril; further on, the draw ratio is typically around 5 and never higher than 10 (i.e., the starting spheres will be converted in particles with

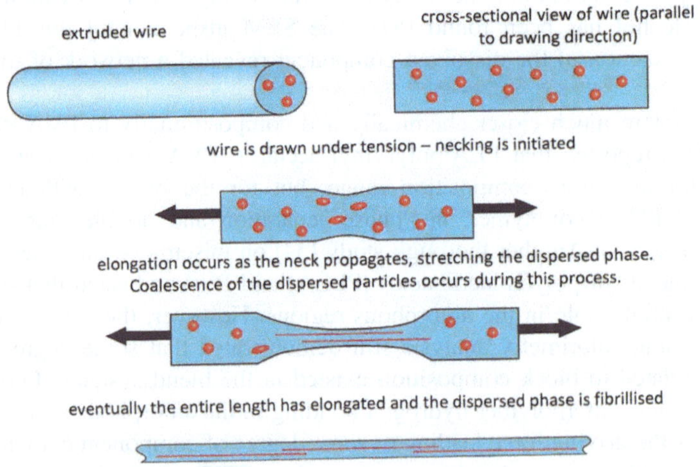

Fig. 6.11 Schematic the microfibril formation mechanism in polymer blends without hydrogen bonding during cold drawing (transformation of the spherical particles into microfibrils via coalescence under transverse contraction)

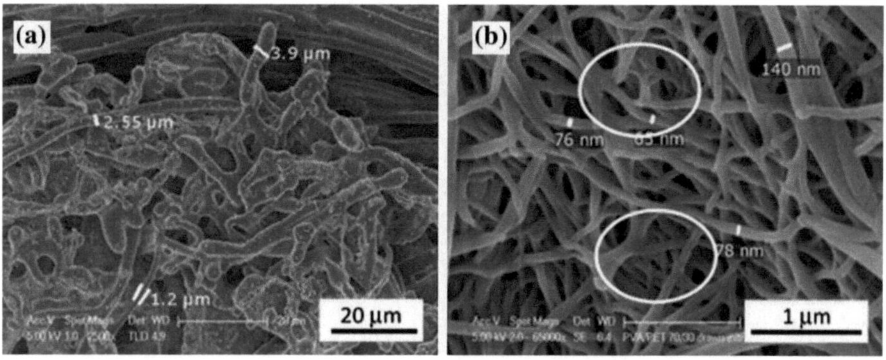

Fig. 6.12 SEM micrographs of PVA/PETG blend (70/30 wt%) taken after melt blending and extraction of PVA with water: **a** the sample is taken just after the die (no cold drawing) and **b** the sample is taken after the cold drawing. Reproduced with permission from Ref. [64]; Copyright 2015 Elsevier

maximum 10 times larger length but not 100 times as it follows from the final length of fibrils).

Systematic study of the mechanism of formation of the 3-D network in the case of polymer blends with H-bonding demonstrated that this process takes place in the melt *before* the drawing step. The subsequent cold drawing results in drastic reduction of the diameters of the network's fibrillar elements. This conclusion was proved by SEM observation of melt blended samples taken immediately after the extruder die. They were treated with water in order to extract the PVA and analyzed by SEM. The results are displayed in Fig. 6.12.

Figure 6.12a shows that the formation of the basic structure takes place in the extruder where the two partners are in a molten state, which favors the formation of H-bonds for the following reasons. In the melt the polymer chains are more flexible and mobile and additionally agitated by the rotating screw thus contributing to the intimated mixing of the two blend components and establishing of maximum H-bonds. The arising structure of the blend is of the type of two co-continuous phases, which is stabilized by the established H-bonds and further fixed by the subsequent cooling to room temperature. An important characteristic of this 3-D network is the thickness of its elements—their diameters are in the range of a couple of microns (Fig. 6.12a).

What happens during the cold drawing? First of all, the character of the 3-D structure arising in the molten blend is completely preserved. The only change, which takes place, is the conversion of the *microfibrillar* 3-D network into *nanofibrillar* 3-D network (Fig. 6.12b). With the progress of cold drawing the sizes of the constituting fibrils of the 3-D network become finer approaching the nanorange as can be concluded from Fig. 6.12b. This microphotograph, taken at higher magnification (65,000×) demonstrates that the majority of nanofibrils have a diameter around 70 nm and the nanopores are typically between 50 and 200 nm in size. The same images indicate on another peculiarity of the system—the really

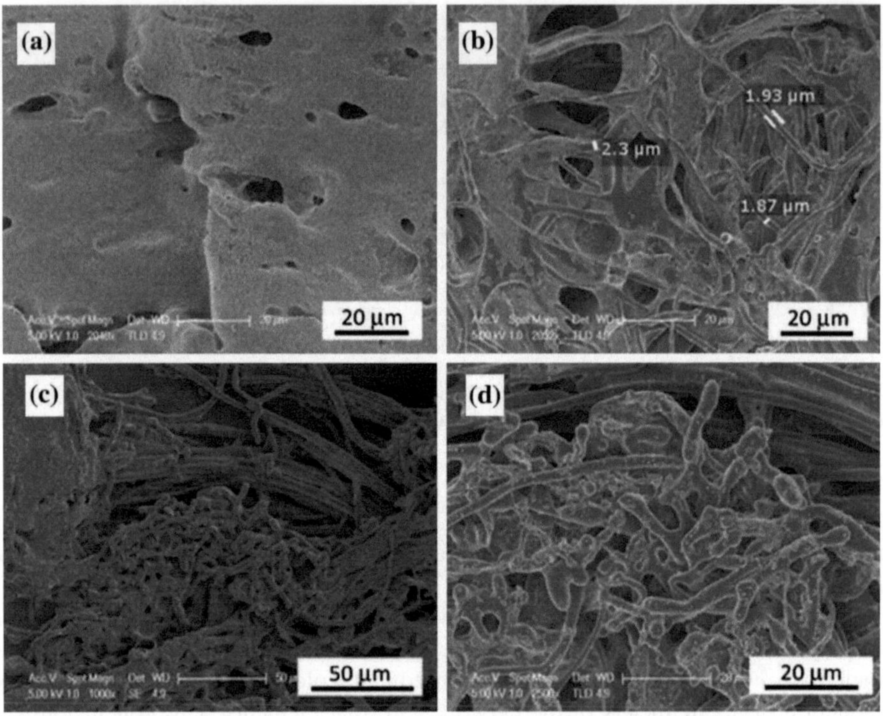

Fig. 6.13 SEM micrographs of PVA/PETG blend samples taken immediately after the die and subjected to PVA extraction with water with PVA/PETG wt ratio: **a** 30/70, **b** 50/50, **c** and **d** 70/30. Reproduced with permission from Ref. [80]; Copyright 2015 Taylor & Francis

branched character of the structure formed. Some of the branching "points" are highlighted in Fig. 6.12b using white ellipses for a better visualization.

The effect of concentration ratio of the blend components on the formation of 3-D network was also studied. For this purpose again samples just after the extruder die were taken and subjected to extraction of PVA followed by analysis using SEM. Blends of PVA/PETG in three different ratios were prepared: 30/70, 50/50, and 70/30. The respective results of SEM examination are shown in Fig. 6.13.

The microphotographs in Fig. 6.13 demonstrate that (i) the formation of a polymer blend with a co-continuous structure takes place during the melt blending, i.e., before drawing the extrudate, and (ii) the effect of the ratio of the two blend components for the formation of co-continuous phases is essential. The blend with the highest amount of PVA (70 %, Fig. 6.13c, d) is characterized by the best mutual penetrating structures with the finest fibrils as compared with the blends with less PVA (50 %, Fig. 6.13b) and particularly the case with the lowest PVA content (30 %, Fig. 6.13a). For the last case of blend composition the lack of fibrillar structures is understandable—PETG is the dominating component (70 %) in which PVA is dispersed, and using a selective for the PETG solvent it would be possible to isolate a PVA fibrillar structure (as a 3-D network in the present case).

Application Opportunities of Nano-size Polymers

Possibly, the main application of polymer nanofibrils is their use as reinforcing component of polymer nanocomposites. The most important advantage of this type of nanocomposites in comparison to the common polymer nanocomposites (prepared via blending of nano-size reinforcement with the matrix polymer) is the lack of dispersion step in their manufacturing process [28, 57–59, 61], similarly to their microfibrillar analogs [13, 24–26, 28, 30, 41, 42, 60, 61].

Quite different application opportunities are offered after the isolation of neat nanofibrils. At the first place is their usage for preparation of nanofibrillar single polymer composites (i.e., the case when the matrix and the reinforcement are of the same chemical composition). Regardless of some controversy from terminological point of view [62–64], this rather novel type of polymer composites represents possibly the best solution of the problem regarding the adhesion quality between matrix and reinforcement. In addition, recyclability after disposal is other main merit of SPCs [65].

The most recent development in the area of SPCs is the creation of micro- or nanofibrillar single polymer composites, where the starting material represents neat microfibrils or nanofibrils [21, 66–71] instead of commonly used highly oriented textile fibers and strips or their products [72–77]. This was possible due to the fact that the MFC and NFC concept allows not only to convert the dispersed polymer into highly oriented fibrils but also to isolate them as a neat material using selective solvent for removing of the second blend component [40].

Another important opportunity for the application of the polymer nano-size materials is their use for biomedical purposes. Organ transplantation nowadays practically has no technical problems. The problem is the lack of donors—only in USA alone a quarter of patients in need die while waiting for suitable donor. The solution was found in the creation of a new science, the tissue engineering, i.e., the use of a combination of cells, engineering, and materials methods to improve or replace biological functions (e.g., tissue, organs).

Cells are often implanted or "seeded" into an artificial structure capable of supporting three-dimensional tissue formation. These structures, typically called *scaffolds*, usually serve at least one of the following purposes: (i) allow cell attachment and migration, (ii) deliver and retain cells and biochemical factors, (iii) enable diffusion of vital cell nutrients and expressed products, and (iv) exert certain mechanical and biological influences to modify the behavior of the cell phase. Taking into account one of the basic requirements to the scaffolds, namely the high specific surface, which can be achieved using fibrillar and/or porous materials, it looked challenging to apply for the same purpose the concept of converting of bulk polymers into nano-size materials. Particularly attractive seemed to be the nano-size biodegradable biocompatible polymers with a 3-D network structure because of their nanoporosity and extremely high specific surface. Additional advantage of these materials is the fact that they are manufactured without any use of organic solvents since the only solvent used is water.

The results of the biomedical testing with living cells are quite promising—the cells attach rather well to the scaffold surface proliferate and grow further [52, 59, 78, 79].

Conclusions and Outlook

The peculiar properties of nanomaterials arise mainly from their sizes and for this reason the search of methods for converting the known materials into nano-size ones is of paramount importance. The electrospinning used for this purpose is simple and cost-effective method but the final product always represents a non-woven textile from nanofibers with a quite limited application potential. Contrasting to the electrospinning, the *concept of nanofibrillar composites* makes possible the application of any textile technique for preparation the respective articles starting from textile yearn of polymer blend, followed by removing of the second blend component in order to prepare articles comprising nanofibrils only (with diameters of 50–250 nm). The neat nanofibrils and articles thereof can be used as scaffolds in tissue engineering, micro- and nanofilters in industry, as starting materials for single polymer composites, as carriers for controlled drug delivery, and others. In this way it is possible to convert any bulk polymer into nano-size material. Recently, this approach was essentially improved be excluding the use of organic solvents—the only solvent used is water. In this way the method became environmentally friendly and cost-effective (the water-soluble polymer can be regenerated and reused for the same purpose), and, last but not least, the final nanoarticles became more attractive for biomedical applications.

Further development of the same method was the finding that the final nanomorphology, being of two basic types, can be controlled using hydrogen bonding as a tool for governing. If hydrogen bonding between the blend partners is missing, the observed morphology is of individual not interconnected nanofibrils. In case hydrogen bonds are present the nanomorphology represents a nanofibrillar nanoporous 3-D network.

Regarding the future trends in the area of described studies two topics of the main interest have to be mentioned: (i) elaboration of effective methods for suppressing the H-bonding of the polymer of interest in order to avoid formation of nanomorphology of network type and thus to prepare neat nano-size material with superior mechanical properties, and (ii) widening the application opportunities of the new nanomaterials and attempting their commercialization.

Acknowledgments The author would like to thank the Foundation for Research Science and Technology of New Zealand for the financial support (Grant No. UOAX 0406). He acknowledges also the hospitality of the Centre for Advanced Composite Materials at the University of Auckland where this study was completed.

References

1. Buzea C, Pacheco II, Robbie K (2007) Nanomaterials and nanoparticles: sources and toxicity. Biointerphases 2:MR17–MR71
2. http://ec.europa.eu/environment/chemicals/nanotech/faq/definition_en.htm. Accessed 20 Apr 2014
3. Buffat P, Borel J (1976) Size effect on melting temperature of gold particles. Phys Rev A 13:2287–2298
4. Schaefer DW, Justice RS (2007) How nano are nanocomposites? Macromolecules 40 (24):8501–8517
5. Bhardwaj N, Kundu SC (2010) Electrospinning: a fascinating fiber fabrication technique. Biotechnol Adv 28:325–347
6. Zeleny J (1914) The electrical discharge from liquid points, and a hydrostatic method of measuring the electric intensity at their surfaces. J Phys Rev 3(2):69–91
7. Formhals A (1934) Process and apparatus for preparing artificial threads, U.S. Patent No. 1,975,504
8. Huang ZM, Zhang YZ, Kotaki M, Ramakrishna S (2003) A review on polymer nanofibers by electrospinning and their applications in nanocomposites. Compos Sci Technol 63(15): 2223–2253
9. Greiner A, Wendorff JH (2007) Electrospinning: a fascinating method for the preparation of ultrathin fiber. Angew Chem Int Ed 46(30):5670–5703
10. Agarwal S, Wendorff JH, Greiner A (2008) Polymer 49(26):5603–5621
11. Editors (1991) Time-life, inventive genius. Time-Life Books, New York, p 52
12. Glaeson C (2007) The biography of silk. Crabtree Publishing Company, p 12
13. Fakirov S (2012) The concept of micro- or nanofibrils reinforced polymer–polymer composites. In: Bhattacharyya D, Fakirov S (eds) Synthetic polymer–polymer composites. Hanser Publisher, Munich, pp 353–400
14. Fakirov S (2013) H-bonding—a chance for novel nano-sized polymers. Expr Polym Lett 7 (6):494–494
15. Fakirov S, Evstatiev M, Schultz JM (1993) Microfibrillar reinfporced composites from drawn poly(ethylene terephthalate)/nylon blends. Macromolecules 34(22):4669–4679
16. Evstatiev M, Fakirov S, Krasteva B, Friedrich K, Covas J, Cunha A (2002) Recycling of PET as polymer–polymer composites. Polym Eng Sci 42(4):826–835
17. Fakirov S, Kamo H, Evstatiev M, Friedrich K (2004) Microfibrillar reinforced composites from PET/LDPE blends: morphology and mechanical properties. J Macromol Sci B Phys B43 (4):775–789
18. Friedrich K, Ueda E, Kamo H, Evstatiev M, Fakirov S, Krasteva B (2002) Direct electron microscopic observation of transcrystalline layers in microfibrillar reinforced polymer–polymer composites. J Mater Sci 37(20):4299–4305
19. Krumova M, Michler GH, Evstatiev M, Friedrich K, Stribeck N, Fakirov S (2005) Transcrystallization with reorientation of polypropylene in drawn PET/PP and PA66/PP blends. Part 2 Electron microscopic observations on the PET/PP blend. Prog Colloid Polym Sci 130:167–173
20. Sapoundjieva D, Denchev Z, Evstatiev M, Fakirov S, Stribeck N, Stamm M (1999) Transcrystallization with reorientation in drawn PET/PA12 blend as revealed by WAXS from synchrotron radiation. J Mater Sci 34(13):3063–3067
21. Bhattacharyya D, Maitrot P, Fakirov S (2009) Polyamide 6 single polymer composites. Expr Polym Lett 3(8):525–532
22. Evstatiev M, Fakirov S, Bechtold G, Friedrich K (2000) Structure—property relationships of injection- and compression-molded microfibrillar reinforced PET/PA-6 composites. Adv Polym Technol 19(4):249–259
23. Fakirov S, Sarac Z, Anbar T, Boz B, Bahar I, Evstatiev M, Apostolov A, Mark J, Kloczkowski A (1996) Mechanical properties and transition temperatures of crosslinked

oriented gelatin. 1. Static and dynamic mechanical properties of cross-linked gelatin. Colloid Polym Sci 274(4):334–341
24. Fakirov S, Evstatiev M (1990) New routes to poly(ethylene terephthalate) with improved mechanical properties. Polymer 31(3):431–434
25. Kargin VA, Bakeev NF, Fakirov SK (1964) New direct observation technique of structure of polymer solutions with aid of electron microscope. Dokl Acad Nauk SSSR 159(4):885
26. Fuchs C, Bhattacharyya D, Fakirov S (2006) Microfibril reinforced polymer–polymer composites: application of Tsai-Hill equation to PP/PET composite. Compos Sci Technol 66 (16):3161–3171
27. Fuchs C, Bhattacharyya D, Friedrich K, Fakirov S (2006) Application of Halpin–Tsai equation to microfibril reinforced polypropylene/poly(ethylene terephthalate) composites. Compos Interfaces 13(4–6):331–344
28. Fakirov S, Bhattacharyya D, Shields RJ (2008) Nanofibril reinforced composites from polymer blends. Coll Surf A Physicochem Eng Aspects 313–314:2–8
29. Fakirov S, Evstatiev M, Schultz JM (1993) Microfibrillar reinforced composite from drawn poly(ethylene terephthalate) nylon blends. Polymer 34(22):4669–4679
30. Fakirov S, Evstatiev M, Petrovich S (1993) Microfibrillar reinforced composites. Macromolecules 26(19):5219–5226
31. Li ZM, Huang CG, Yang W, Yang MB, Huang R (2004) Morphology dependent double yielding in injection molded polycarbonate/polyethylene blend. Macromol Mater Eng 289 (11):1004–1011
32. Li ZM, Lu A, Lu ZY, Shen KZ, Li LB, Yang MB (2005) In-situ microfibrillar PET/iPP blend via a slit die extrusion, hot stretching and quenching process: influences of PET concentration on morphology and crystallization of iPP at a fixed hot stretching ratio. J Macromol Sci Phys B44(2):203–216
33. Li ZM, Yang MB, Lu A, Feng JM, Huang R (2002) Tensile properties of poly(ethylene terephthalate) and polyethylene in-situ microfiber reinforced composite formed via slit die extrusion and hot stretching. Mater Lett 56(5):756–762
34. Li ZM, Yang MB, Xie BH, Lu A, Feng JM, Huang R (2003) In-situ microfiber reinforced composite based on PET and PE via slit die extrusion and hot stretching. Influences of hot stretching ratio on morphology and tensile properties at a fixed composition. Polym Eng Sci 43(3):615–628
35. Li ZM, Li LB, Shen KZ, Yang MB, Huang R (2004) In-situ microfibrillar PET/iPP blend via slit die extrusion, hot stretching, and quenching: influence of hot stretch ratio on morphology, crystallization, and crystal structure of iPP at a fixed PET concentration. J Polym Sci B Polym Phys 42(22):4095–4106
36. Zhong GJ, Li LB, Mendes E, Byelov D, Fu Q, Li ZM (2006) Suppression of skin-core structure in injection-molded polymer parts by in-situ incorporation of a microfibrillar network. Macromolecules 39(19):6771–6775
37. Zhong GJ, Li ZM, Li LB, Shen KZ (2008) Crystallization of oriented isotactic polypropylene (iPP) in the presence of in situ poly(ethylene terephthalate) (PET) microfibrils. Polymer 49 (19):4271–4278
38. Yi X, Xu LK, Wang YL, Zhong GJ, Ji X, Li ZM (2010) Morphology and properties of isotactic polypropylene/poly(ethylene terephthalate) in situ microfibrillar reinforced blends: influence of viscosity ratio. Eur Polym J 46(4):719–730
39. Yi X, Chen C, Zhong GJ, Xu L, Tang JH, Ji X, Li ZM (2011) Suppressing the skin–core structure of injection-molded isotactic polypropylene via combination of an in situ microfibrillar network and an interfacial compatibilizer. J Phys Chem B 115(23):7497–7504
40. Fakirov S (2006) Modified Soxhlet apparatus for high-temperature extraction. J Appl Polym Sci 102(2):2013–2014
41. Panamoottil SM, Bhattacharyya D, Fakirov S (2013) Nanofibrillar polymer–polymer and single polymer composite involving poly(butylene terephthalate): preparation and mechanical properties. Polym Plast Technol Eng 52(11):1106–1112

42. Kim NK, Bhattacharyya D, Fakirov S (2014) Polymer–polymer and single polymer composites involving nanofibrillar poly(vinylidene fluoride): manufacturing and mechanical properties. J Macromol Sci Phys B53(7):1168–1181
43. Kotek R, Tonelli A, Vasanthan N (2003) Lewis acid-base complexation of polyamides and the effect of hydrogen bonding on structure development, M01-NS03, NTC Project
44. Kotek R, Jung D, Tonelli A, Vasanthan N (2005) Novel methods for obtaining high modulus aliphatic polyamide fibers. J Macromol Sci Polym Rev C45(3):201–230
45. Kotek R, Pang K, Schmidt B, Tonelli A (2004) Synthesis and gas barrier characterization of poly(ethylene isophthalate). J Polym Sci Part B Polym Phys 42(23):4247–4254
46. Jung DW, Kotek R, Vasanthan N, Tonelli A (2004) High modulus nylon 66 fibers through Lewis acid-base complexation to control hydrogen bonding and enhance drawing behavior. 228th ACS National Meeting, Philadelphia, PA
47. Vasanthan N, Kotek R, Jung DW, Shin D, Tonelli AE, Salem DR (2004) Lewis acid-base complexation of polyamide 66 to control hydrogen bonding, extensibility and crystallinity. Polymer 45(12):4077–4085
48. Vasanthan N, Kotek R, Jung DW, Salem DR, Tonelli AE (2004) Lewis acid-base complexation of polyamide 66 as a means to control hydrogen bonding to form high strength fibers and films. 227th ACS National Meeting, Anaheim, CA
49. Kuo SW (2008) Hydrogen-bonding in polymer blends. J Polym Res 15(6):459–486
50. Fakirov S, Bhattacharyya D, Hutmacher D (2008) Applications of microfibrillar polymer–polymer composites concept for biomedical purposes. In: Bhatnagar N, Srivatsan TS (eds) Processing and fabrication of advanced materials—XVII, vol 2. IIIK International, New Delhi, pp 794–803
51. Shuai X, He Y, Asakawa N, Inoue YJ (2001) Miscibility and phase structure of binary blends of poly(L-lactide) and poly(vinyl alcohol). J Appl Polym Sci 81(3):762–772
52. Bini T, Gao S, Wang S, Ramakrishna S (2006) Poly(l-lactide-co-glycolide) biodegradable microfibers and electrospun nanofibers for nerve tissue engineering: an in vitro study. J Mater Sci 41(19):6453–6459
53. Park JW, Im SS (2003) Miscibility and morphology in blends of poly(L-lactic acid) and poly (vinyl acetate-co-vinyl alcohol). Polymer 44(15):4341–4354
54. Chiu JB, Luu YK, Fang D, Hsiao BS, Chu B, Hadjiargyrou M (2005) Electrospun nanofibrous scaffolds for biomedical applications. J Biomed Nanotechnol 1(2):115–132
55. Fakirov S, Bhattacharya D, Lin R, Fuchs C, Friedrich K (2007) Contribution of coalescence to microfibrils formation in polymer blends during cold drawing. J Macromol Sci Phys B46 (1):183–194
56. Denchev Z, Dencheva N (2012) Preparation, mechanical properties and structural characterization of microfibrillar composites based on polyethylene/polyamide blends. In: Bhattacharyya D, Fakirov S (eds) Synthetic polymer–polymer composites. Hanser Publisher, Munich, pp 465–524
57. Bhattacharyya D, Fakirov S, Organoclay (2009) Particulate and nanofibril reinforced polymer-polymer composites: manufacturing, modeling and applications. In: Karger-Kocsis, Fakirov S (eds) Nano- and micro-mechanics of polymer blends and composites. Hanser, Munich, pp 167–208
58. Shields R, Bhattacharyya D, Fakirov S (2008) Fibrillar polymer-polymer composites: morphology, properties and applications. J Mater Sci 43(20):6758–6770
59. Shields RJ, Bhattacharyya D, Fakirov S (2012) Application opportunities of the microfibril reinforced composite concept. In: Bhattacharyya D, Fakirov S (eds) Synthetic polymer–polymer composites. Hanser, Munich, pp 589–626
60. Evstatiev M, Fakirov S, Friedrich K (2000) Microfibrillar reinforced composite: another approach to polymer blends processing. In: Fakirov S (ed) Structure development during polymer processing. Kluwer Academic, Dordrecht, The Netherlands, pp 311–325
61. Evstatiev M, Fakirov S, Friedrich K (2005) Manufacturing and characterization of microfibrillar reinforced composites from polymer blends. In: Friedrich K, Fakirov S, Zhang Z (eds) Polymer composites: from nano- to macro-scale. Springer, Boston, pp 149–167

62. Fakirov S (2015) Is the use of correct terms and definitions important in creation of new materials? Expr Polym Lett 9(8):671
63. Fakirov S (2015) Composite materials—is the use of proper definitions important? Mater Today 18(10):529
64. Fakirov S (2013) Nano-/microfibrillar polymer-polymer and single polymer composites: the converting instead of adding concept. Comp Sci Technol 89:211–225
65. Matabola K, De Vries A, Moolman F, Luyt A (2009) Single polymer composites: a review. J Mater Sci 44(23):6213–6222
66. Duhovic M, Fakirov S, Holschuh R, Mitschang P, Bhattacharyya D (2012) Micro- and nanofibrillar single polymer composites. In: Bhattacharyya D, Fakirov S (eds) Synthetic polymer–polymer composites. Hanser, Munich, pp 643–672
67. Karger-Kocsis J, Fakirov S (2012) Polymorphism- and stereoregularity-based single polymer composites. In: Bhattacharyya D, Fakirov S (eds) Synthetic polymer–polymer composites. Hanser, Munich, pp 673–698
68. Fakirov S (2013) Nano- and microfibrillar single-polymer composites: a review. Macromol Mater Eng 298(1):9–32
69. Duhovic M, Maitrot P, Fakirov S (2009) Polyamide 66 polymorphic single polymer composites. Open Macromol J 3:37–40
70. Duhovic M, Bhattacharyya D, Fakirov S (2010) Nanofibrillar single polymer composites of poly(ethylene terephthalate). Macromol Mater Eng 295(2):95–99
71. Fakirov S, Duhovic M, Maitrot P, Bhattacharyya D (2010) From PET nanofibrils to nanofibrillar single-polymer composites. Macromol Mater Eng 295(6):515–518
72. Capiati NJ, Porter RS (1975) Concept of one polymer composites modeled with high-density polyethylene. J Mater Sci 10(10):1671–1677
73. Hine P, Ward I, Olley R, Bassett D (1993) The hot compaction of high modulus melt-spun polyethylene fibers. J Mater Sci 28(2):316–324
74. Hine P, Olley R, Ward I (2008) The use of interleaved films for optimising the production and properties of hot compacted, self-reinforced polymer composites. Comp Sci Technol 68 (6):1413–1421
75. Cabrera NO, Alcock B, Klompen BET, Peijs T (2008) Filament winding of co-extruded polypropylene tapes for fully recyclable all-polypropylene composite products. Appl Compos Mater 15(1):27–45
76. Li R, Yao D (2008) Preparation of single poly(lactic acid) composites. J Appl Polym Sci 10:2909–2916
77. Barkoula NM, Alcock B, Cabrera NO, Peijs T (2008) Fatigue properties of highly oriented polypropylene tapes and all-polypropylene composites. Polym Polym Compos 16(2):101–113
78. Lin STC, Bhattacharyya D, Fakirov S, Cornish J (2014) Novel organic solvent free micro-/nano-fibrillar, nanoporous scaffolds for tissue engineering. Int J Polym Mater Polym Biomater 63(8):416–423
79. Lin STC, Bhattacharyya D, Fakirov S, Matthews BG, Cornish J (2014) A novel microfibrillar composite approach towards manufacturing nanoporous tissue scaffolds. Mech Adv Mater Struct 21(3):237–243
80. Fakirov S, Bhattacharyya D, Panamoottil SM (2014) Converting of bulk polymers into nanosized materials with controlled nanomorphology. Int J Polym Mater Polym Biomater 63 (15):777–793

Chapter 7
Electrospun Polymer Nanofiber Separators and Electrolyte Membranes for Energy Storage and Conversion Applications

Yue-E Miao and Tianxi Liu

Introduction

With severe stress from energy crisis and global environmental concerns, development of high-performance energy storage and conversion systems, such as lithium-ion batteries (LIBs) and proton exchange membrane fuel cells (PEMFCs), has evolved into an urgent task [1, 2]. As vital components in most energy storage and conversion systems, the polymer separator and electrolyte membrane, which respectively refer to the porous polymer membrane placed between LIB electrodes, and nonporous ion-conducting polymer film that is employed in moderate-temperature fuel cells, play the key role of physically separating anode and cathode to prevent electrical shorting as well as regulate cell kinetics [3, 4]. Therefore, polymer separators and electrolyte membranes with rationally designed structures and architectures are essentially needed for high-performance energy storage and conversion devices.

One-dimensional (1-D) nanomaterials have been extremely appealing in terms of their high specific surface area, effective ion diffusion/electron transfer pathways, and good contact with electrodes or current collectors compared to bulk materials [5]. Therefore, 1-D nanomaterials with a variety of morphologies, such as nanowires, nanorods, nanotubes, and nanofibers, have been proposed as prospective separators or electroactive materials in energy storage and conversion applications

Y.-E. Miao · T. Liu (✉)
State Key Laboratory for Modification of Chemical Fibers and Polymer Materials,
College of Materials Science and Engineering, Donghua University, Shanghai 201620,
People's Republic of China
e-mail: txliu@dhu.edu.cn

T. Liu
State Key Laboratory of Molecular Engineering of Polymers,
Department of Macromolecular Science, Fudan University,
Shanghai 200433, People's Republic of China

S. Fakirov (ed.), *Nano-size Polymers*,
DOI 10.1007/978-3-319-39715-3_7

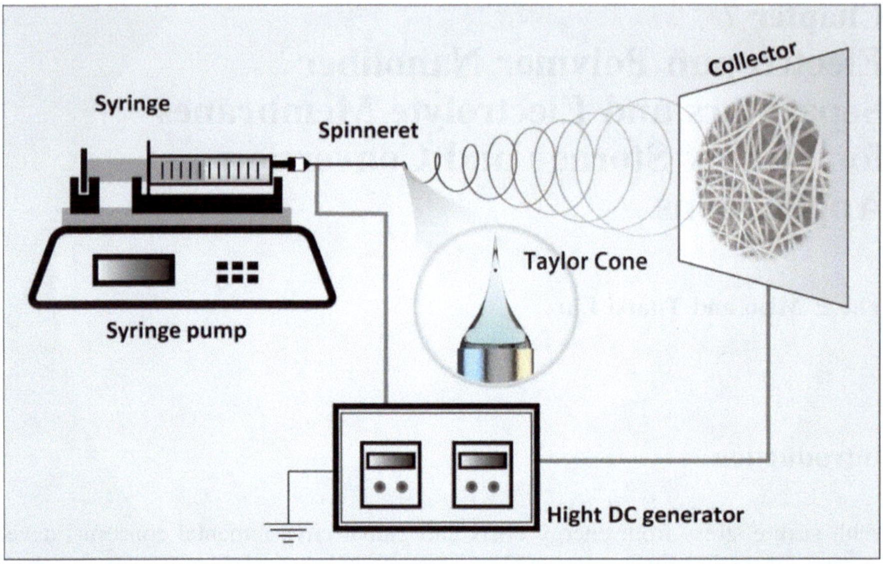

Fig. 7.1 Schematic illustration of the basic electrospinning setup. Reproduced with permission from Ref. [10]; Copyright 2014 Wiley Periodicals

[4, 6, 7]. Electrospinning is a versatile technique for fabricating 1-D polymer nanofibers with controllable diameter, morphology, and compositions [8, 9]. The reproducibility and simplicity make this technique very attractive to prepare 1-D architectures for multi applications in filtration, catalysis, sensors, as well as, energy storage/conversion areas. Figure 7.1 shows a schematic illustration of the basic setup for electrospinning, which consists of three major modules: a high voltage power generator, a syringe pump (syringe included), and a collector [10]. Polymer solution is hosted in the syringe which can be continuously fed through the spinneret. Under a high voltage, an evenly distributed electric field is formed between the collector and polymer fluid stream, resulting in charged liquid surface at the end of the needle. Once the electric field intensity exceeds liquid surface tension, the continuous jet flow of charged solution undergoes stretching and elongation while the solvent evaporates, thus making diameter of the jet significantly reduced to hundreds of nanometers to form nanofibers.

The outstanding properties of high surface area to volume ratio, large porosity with fully interconnected pore structures greatly facilitate high electrolyte uptake and easy transport of ions, thus making electrospun polymer nanofiber separators and electrolyte membranes particularly suitable for high energy density rechargeable batteries and high-power output fuel cells [11, 12]. Moreover, complex nanostructures with controllable hierarchical features, such as random three-dimensional (3-D) structures, aligned/patterned fibers, and nonwovens could be facilely obtained by modifying the electrospinning solution/processing parameters, and set-up geometries (e.g., co-electrospinning technique using specially designed spinnerets) [13]. In situ or surface modifications also can introduce multifunctional molecules or nanoparticles in polymer nanofibers to construct inorganic nanofiller/polymer nanocomposites, thus achieving significantly improved electrochemical properties [14]. Therefore, the aim of this chapter concerns the recent advances in preparation of different kinds of polymer nanofiber separators and electrolyte membranes, including single-component, multi-component polymer nanofibers, and inorganic nanofiller/polymer composite nanofibers, which can be successfully achieved by conventional electrospinning, coaxial electrospinning and surface modification methods. Moreover, physical and electrochemical properties of these electrospun nanofiber-based separators and electrolyte membranes have been thoroughly investigated for energy storage and conversion applications in LIBs and PEMFCs.

Electrospun Polymer Nanofiber Separators for LIBs

LIBs have been considered as one of the most promising power sources for electric vehicles due to their high energy density, low gravimetric density, long cycle life, and flexible design [15, 16]. As a critical component, the separator behaves as a physical barrier between positive and negative electrodes as shown in Fig. 7.2, serving as a medium for ion transport during charge/discharge processes as well. Nevertheless, conventional polyolefin microporous membranes are still need to be largely improved regarding their weak electrochemical contacts with electrodes and poor wettability toward electrolytes [17, 18]. Therefore, separators which can be easily swollen by electrolytes to achieve good electrode–electrolyte contacts, high ionic conductivity and fast ion transport is very important to greatly improve the electrochemical performance of lithium-ion batteries.

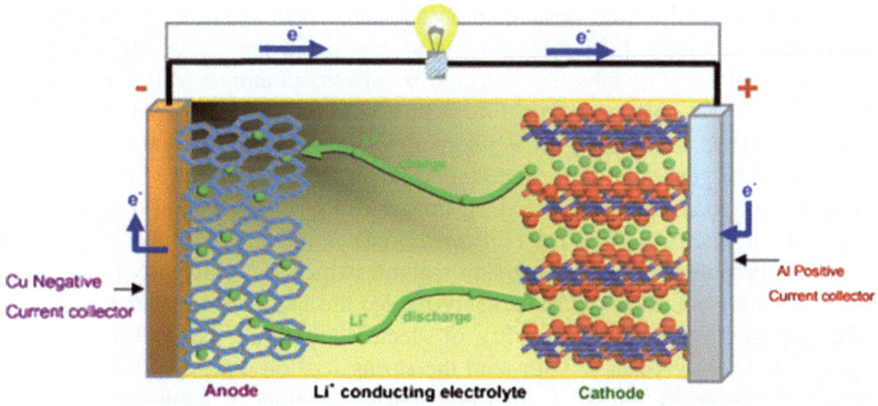

Fig. 7.2 Schematic of a lithium-ion battery which consists of an anode and a cathode separated by electrolyte containing dissociated lithium salts, enabling transfer of lithium ions between the two electrodes. Reproduced with permission from Ref. [16]; Copyright 2007 Elsevier

Nanofiber membranes obtained via electrospinning exhibit high porosity and interconnected open pore structure to guarantee good electrolyte encapsulation and retention, thus leading to high ionic conductivity. Moreover, polymer nanofiber membranes could be facilely prepared through the controllable tuning of electrospinning solution/processing parameters, thus achieving high-performance nanofiber separators with different compositions for LIBs, including single-component polymer, multi-component polymers and inorganic nanofiller/polymer nanocomposites.

Single-Component Polymer Nanofiber Separators

Polymer electrolytes have received considerable attentions for higher-performance rechargeable battery applications due to their advantages of safer, lighter, and more flexible in shape compared with their liquid counterparts. Many kinds of polymers, such as poly(vinylidene fluoride) (PVDF), poly(vinylidenefluoride-co-hexafluoropropylene) (PVDF-HFP), polyacrylonitrile (PAN), and poly(methyl methacrylate) (PMMA), have been widely studied as host polymers for preparation of electrospun fibrous membranes as separators for lithium-ion batteries [19–21].

Because of the good electrochemical stability and affinity to electrolyte solutions, PVDF has been considered as one of the most attractive host polymers for separators of lithium-ion batteries. Choi et al. first reported electrospun PVDF membrane as a microporous matrix for polymer electrolyte [22], which exhibits excellent electrochemical properties with high ionic conductivity, low interfacial resistance, and good mechanical strength. However, the highly crystalline part of PVDF homopolymer largely hinders Li^+ ion migration, thus leading to low ionic conductivity and electrochemical stability. To enhance the ionic conductivity of PVDF-based polymer electrolytes, electrospun PVDF-HFP membranes have been developed, in which the amorphous HFP phase effectively captures large amount of liquid electrolyte while the crystalline PVDF phase acts as the mechanically supported polymer matrix [23]. Raghavan et al. compared the electrochemical properties of PVDF-HFP membranes prepared by electrospinning and phase inversion methods [24]. As observed from Fig. 7.3a–d, the membrane prepared by phase inversion method exhibits highly nonhomogeneous structure while the electrospun fiber membrane generates uniform microporous structures with fully interconnected pores. Atomic force microscopy (AFM) images in Fig. 7.3e and f also indicate that roughness of electrospun PVDF-HFP membrane is much higher than that of the membrane prepared by phase inversion method, which can efficiently enhance the adhesion with electrodes and remarkably reduce the interface resistance.

Nevertheless, mechanical stability of electrospun PVDF-based fibrous membranes is inherently weak to withstand the strong tension during battery assembly process, which severely restricts their practical applications. Chemical modification by grafting or crosslinking with functional reagents has been proved as an effective approach to improve the mechanical property of PVDF [25]. Kader et al. reported the preparation of electrospun poly(vinylidene fluoride)-graft-poly(tert-butyl

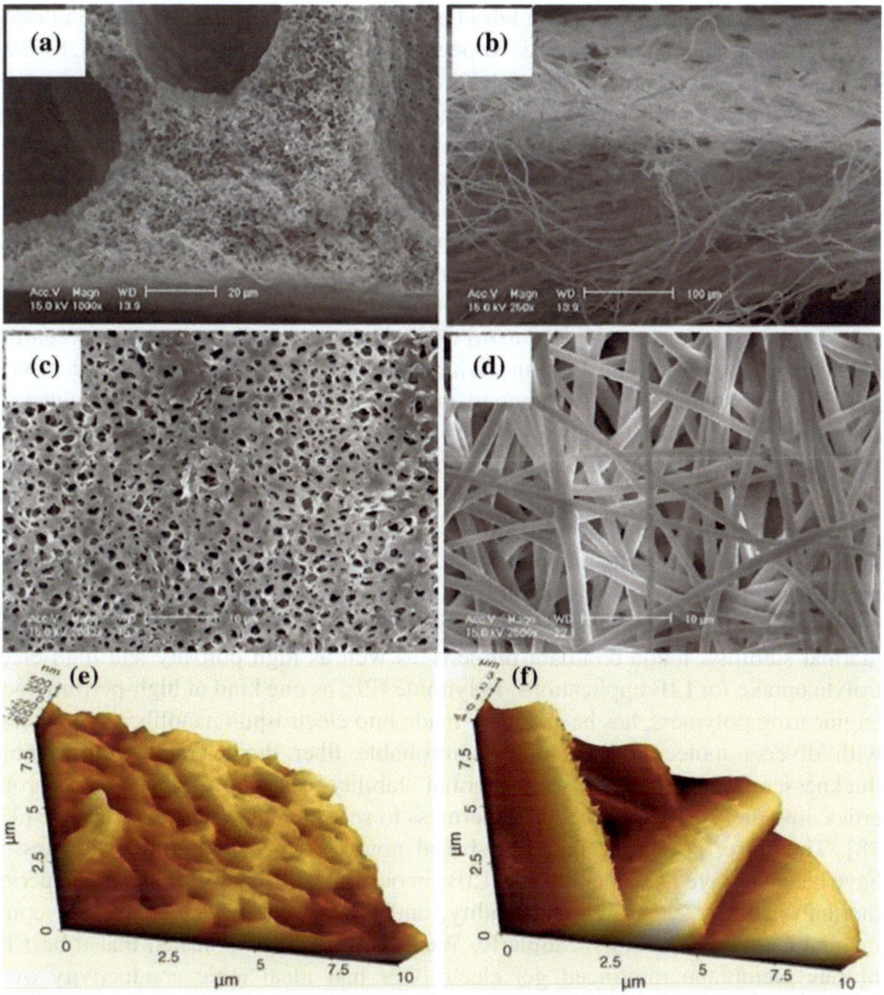

Fig. 7.3 The cross-section (**a**, **b**) and *top-view* (**c**, **d**) field emission scanning electron microscopy (FESEM), and AFM (**e**, **f**) images of PVDF-HFP membranes, respectively prepared by phase inversion methods (**a**, **c**, **e**) and electrospinning (**b**, **d**, **f**) methods. Reproduced with permission from Ref. [24]; Copyright 2009 Elsevier

acrylate) microporous membrane by grafting tert-butyl acrylate onto PVDF chains using a ultraviolet irradiation technique [26], showing obvious improvement in ionic conductivity, electrochemical stability and cyclic performance. Liang et al. found that heat treatment of PVDF fibrous membranes could significantly improve the tensile strength and modulus as well as elongation at break [27], which results from the formation of strong interfiber bonding, increased fiber diameter and crystallinity after high-temperature treatment. Although chemical modification can inherently improve the mechanical property of electrospun polymer nanofiber membranes, long-time soaking in liquid electrolyte solutions easily leads to internal short circuits and safety hazards especially at elevated temperatures [28, 29]. Therefore, an alternative approach is developed by electrospinning polymer nanofibers onto mechanically strong microporous supporting substrates. Zhang et al. systematically prepared PVDF copolymer nanofiber-coated microporous membranes using single-nozzle and nozzle-less electrospinning techniques, indicating that the nanofiber coating via nozzle-less electrospinning method exhibits better adhesion to the microporous separator membrane and higher electrolyte uptake capacity than that obtained by single nozzle electrospinning [30–32].

Among various polymer hosts employed to prepare polymer electrolytes, PAN is also demonstrated as an efficient matrix due to the strong interactions between the nitrile groups of PAN and Li^+ ions, which can greatly improve ionic conductivity [33, 34]. PAN nanofiber-based nonwoven separators have been facilely obtained by electrospinning for LIB applications [35]. However, low flexibility of PAN main chain makes the electrospun nanofiber membranes suffer from severe brittleness. Therefore, Carol et al. prepared porous PAN membranes with a sufficiently high molecular weight of 4×10^5 [36], which consists of uniformly dispersed bead-free fibers with diameter in the range of 880–1260 nm to act as a host matrix with good mechanical strength and porosity for gel polymer electrolyte in LIBs.

It is always an important task in developing new polymer matrix with superior thermal stability, flame retardant property as well as high porosity and high electrolyte uptake for LIB applications. Polyimide (PI), as one kind of high-performance engineering polymers, has been widely made into electrospun nanofiber membranes with diverse molecular structures, controllable fiber diameters, and membrane thicknesses due to their excellent thermal stability, outstanding mechanical properties, low dielectric constants, and inertness to solvent and radiation resistance [37, 38]. Therefore, polyimide nanofiber-based nonwovens with different thicknesses have been deployed as separators for LIBs in our work [39], which indicates superior thermal stability, electrolyte wettability, and battery performance over the commercial Celgard separators. Similarly, Wang et al. further indicated that robust PI fibrous membrane reinforced gel electrolytes had ideal ionic conductivity over 2.79×10^{-3} S cm^{-1} at 25 °C, excellent electrochemical stability up to 5.0 V (vs. Li/Li^+), and appropriate tensile strength of 7.22 MPa [40, 41]. Moreover, a direct flame on PI and Celgard shows quick contraction and bright flame for Celgard whereas PI membrane still holds excellent flame retardancy until 25 s as shown in Fig. 7.4a and b, providing great potentials in safe, reliable, and long cycle life LIB applications.

A novel poly(phthalazinone ether sulfone ketone) (PPESK) membrane for lithium-ion battery was also prepared using electrospinning technique [42], which indicates high porosity of 92 % and high electrolyte uptake of 1210 %, thus leading to high ionic conductivity of 3.79×10^{-3} S cm^{-1} at 25 °C. Moreover, the cell assembled with PPESK membrane displays better electrochemical stability and higher cycling discharge capacity than that assembled with PVDF membrane, thus making it promising for high-performance lithium-ion battery applications.

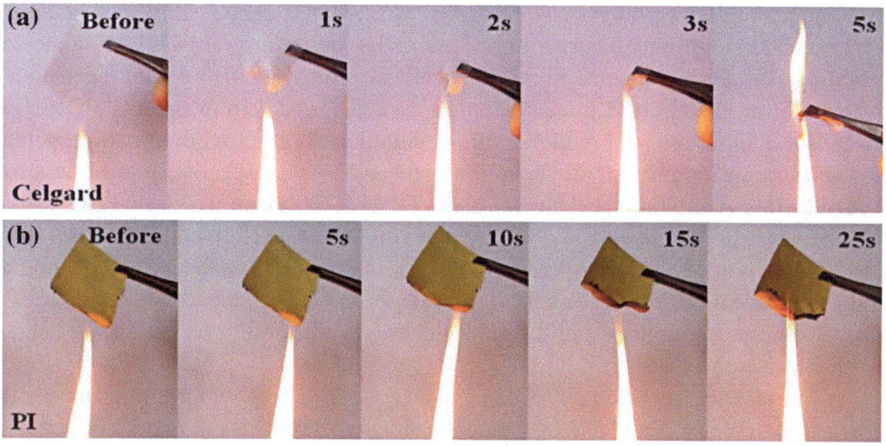

Fig. 7.4 Photograph of flammability: **a** Celgard and **b** PI fibrous membrane. Reproduced with permission from Ref. [41]; Copyright 2014 Elsevier

Multi-component Polymer Nanofiber Separators

Despite many kinds of polymers, including PVDF, PVDF-HFP, PAN, and PI, have already been applied in electrospun fibrous membrane separators for lithium-ion batteries, most of the above mentioned single-component polymer nanofiber separators are still far from satisfying for their inherent weak mechanical strength, poor thermal stability or electrolyte wettability. Exactly, multi-component polymer nanofiber separators could be facilely obtained through single-nozzle electrospinning technique [30], which can effectively combine the advantages of different components meanwhile avoid their respective disadvantages, thus leading to synergistic improvements in mechanical property, electrolyte uptake efficiency, and battery performances.

Thermoplastic polyurethane (TPU) has two-phase microstructure in which the hard segments play the critical role of keeping dimensional stability while the soft segments easily dissolve alkali metal to provide good ionic conductivity. Therefore, nonwoven composites of TPU with different proportions of PVDF (80, 50 and 20 %, w/w) have been prepared by electrospinning, which was then soaked in 1 M $LiClO_4$/ ethylene carbonate/propylene carbonate to obtain gel polymer electrolytes with improved ionic conductivity of 3.2×10^{-3} S cm^{-1} at room temperature [43–45]. The same group also prepared TPU reinforced PVDF-HFP gel polymer electrolyte [46], which exhibits a maximum ionic conductivity of 4.1×10^{-3} S cm^{-1} with electrochemical stability up to 5.5 V (vs. Li/Li^+) at room temperature. Moreover, TPU/PVDF-HFP membrane keeps high tensile strength of 8.4 MPa and elongation at break of 118.7 %, making it suitable for application in polymer lithium-ion batteries.

Nanocrystalline cellulose (NCC) with whisker-like shape shows outstanding advantages of high tensile strength and modulus, being a promising candidate for modification of electrochemical and mechanical properties of polymer electrolytes [47]. Lalia et al. reported the preparation of 2 wt% NCC-reinforced PVDF-HFP composite membrane [48], which exhibits significantly improved electrolyte retention and storage modulus of the separator by 63 and 15 %, respectively. Additionally, PVDF/PMMA/cellulose acetate [49], PVDF/poly(ethylene oxide) (PEO) [50], PAN/PMMA/polystyrene [51], PMMA/polyvinyl chloride [52], cellulose/PEO [53], PVDF-HFP/polyethylene glycol/polyethylene glycol dimethacrylate oligomer composite nanofiber membranes [54], have been facilely obtained via single-nozzle electrospinning to achieve good liquid electrolyte wettability as well as excellent mechanical properties.

As previously reported [55, 56], ideal reproduction and strong interfacial interactions can be achieved by using in situ reinforcement strategy for making high-performance polymer composites. Electrospun PVDF fibrous membrane embedded in a poly(4-vinylpyridine) (P4VP) matrix were successfully fabricated as newly proposed polymer electrolyte membranes in order to improve their mechanical/dimensional stabilities and ionic conductivity for lithium rechargeable battery applications [57]. Results show that the electrolyte uptakes of electrospun PVDF and PVDF/P4VP composite membranes were higher than that of PVDF cast film. Moreover, the tensile properties and ionic conductivity of PVDF/P4VP composite membrane were considerably improved compared to those of the traditional PVDF cast film.

Attributed to the various setups of electrospinning technology, multi-component polymer nanofiber separators can be facilely obtained through the fine tuning of electrospinning processes. Chen et al. developed a bicomponent cross-electrospinning method to prepare PVDF-HFP/PI composite nanofiber membranes [58]. Comparing with traditional single-component membranes, excellent thermal stability and mechanical property were simultaneously achieved for enhancing the safety performance of lithium-ion battery. Liu et al. also obtained composite PI@PVDF-HFP core@sheath nanofibrous separators through a facile coaxial electrospinning technique [59], in which the core material of high-performance polyimide provides superior thermal stability while PVDF-HFP shell guarantees good affinity with liquid electrolyte, thus leading to synergistic characteristics of thermal, mechanical and electrochemical properties. A core–shell structured polysulfonamide (PSA)@PVDF-HFP composite nonwoven separator was exploited by the same group [60], which manifested higher porosity, better electrolyte wettability, superior thermal stability, rate capability, and better cycling capacity retention for high power lithium-ion battery applications. Recently, Huang et al. reported an eco-friendly approach for extracting cellulose acetate from waste cigarette filter to construct cellulose-based membrane separator via coaxial electrospinning as shown in Fig. 7.5 [61], which not only effectively alleviates environmental pressure but also develops high additional-value products. Through the

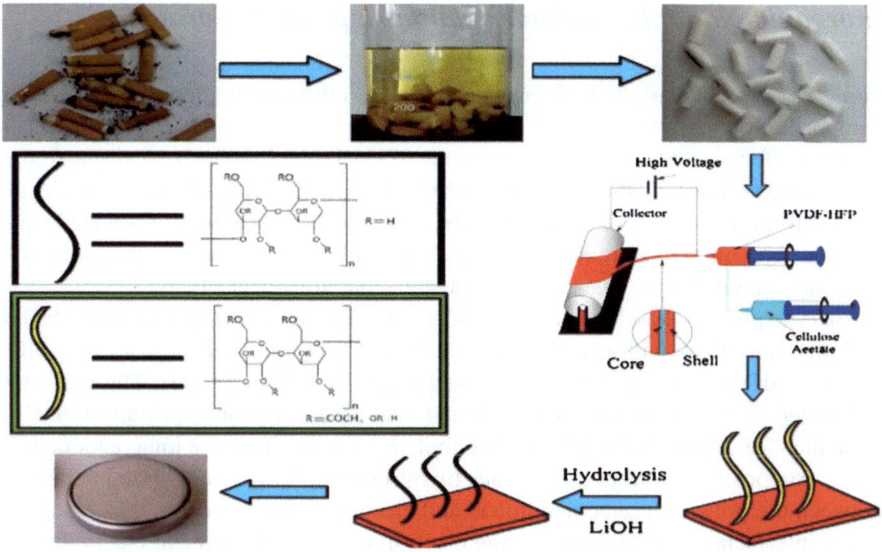

Fig. 7.5 Preparation of cellulose-based coaxial nanofiber separators for lithium-ion battery. Reproduced with permission from Ref. [61]; Copyright 2015 American Chemical Society

simple hydrolysis of cellulose acetate to cellulose in LiOH, cellulose-core/PVDF-HFP-shell fibrous membrane shows good tensile strength (34.1 MPa), excellent thermal stability (over 200 °C), as well as lower interfacial resistance (98.5 Ω) and higher ionic conductivity (6.16×10^{-3} S cm^{-1}) than those of commercial separators (280.0 Ω and 0.88×10^{-3} S cm^{-1}).

Sequential electrospinning technique is proved to be an effective approach to prepare multi-layered porous nanofiber membranes. Angulakshmi et al. obtained PVDF-HFP/poly (vinyl chloride)/PVDF-HFP based-trilayer porous polymeric membranes by sequential electrospinning [62], achieving highly improved mechanical integrity compared with the single layer PVDF-HFP membrane. Similarly, novel sandwich-structured PVDF/poly(m-phenylene isophthalamide) (PMIA)/PVDF nanofibrous battery separators with robust mechanical strength and thermal stability are fabricated via a sequential electrospinning technique [63]. PVDF and PMIA nanofiber layers are found to be strongly bonded and interconnected with each other at the interface boundary without any polymer binder or posttreatment, thus resulting in higher ionic conductivity of 2.3 times than that of Celgard membrane, robust tensile strength of 13.96 MPa, excellent thermal stability, and capacity retention of Li/$LiCoO_2$ cell. Therefore, it provides new insight into design and development of high-performance rechargeable lithium-ion battery separators.

Inorganic Nanofiller/Polymer Nanocomposite Separators

SiO_2, as a particulate nanofiller, has been an attractive material for preparing composite polymer electrolytes due to its large surface area, high conductivity, and good interfacial contact with the electrode [21]. To be emphasized, dispersion of SiO_2 nanoparticles in the polymer matrix plays a key role in the final electrochemical performance. Consequentially, how to achieve well dispersed SiO_2 nanoparticles in polymer membranes is a critical issue during the preparation of inorganic nanofiller/polymer nanocomposite separators/electrolytes.

Electrospinning has been employed as an effective technique to produce fine polymer nanofibers with excellent distribution of SiO_2 nanoparticles, such as SiO_2/nylon 6, SiO_2/PVDF, and SiO_2/PAN nanocomposites [20, 21, 64]. The uniformly embedded SiO_2 nanoparticles in polymer nanofibers not only provide tunneling structures for liquid electrolyte holding, but also greatly improve ionic conductivity of the composite nanofiber membranes by introducing Lewis acid–base interactions between the polar inorganic nanofillers and electrolyte ionic species, thus leading to better interfacial stability between the electrolyte and lithium metal electrode [21]. Yanilmaz et al. reported the preparation of nanoparticle-on-nanofiber hybrid membranes by electrospraying of SiO_2 dispersions and electrospinning of PVDF solution simultaneously, resulting in high-loading efficiency of SiO_2 nanoparticle decorated PVDF nanofibers [65]. The unique hybrid nanostructure with increased surface area exhibits greatly improved electrolyte wettability, ionic conductivity and electrochemical properties compared with commercial microporous polyolefin membranes.

PVDF doped with ion-complex (SiO_2-PAALi) was prepared by electrospinning technique [66], with the corresponding composite gel polymer electrolytes (CGPEs) obtained after being activated in liquid electrolyte. With the help of SiO_2-PAALi, CGPEs present higher ionic conductivity compared with pure PVDF and PVDF/SiO_2 membranes as shown in Fig. 7.6a. This can be attributed to the unique nanoporous structure of CGPEs as shown in the scheme of Fig. 7.6b–d, in which PVDF effectively provides mechanical support with fully interconnected network structure, while the ion-complexes uniformly absorbed on the surface of PVDF nanofibers can absorb sufficient liquid electrolyte, creating a favorable lithium ion conduction pathway in the vicinity of nanoparticles. Moreover, the amorphous phase of SiO_2-PAALi efficiently prevents the leakage of liquid electrolyte by tightly holding electrolyte solution. Therefore, incorporation of SiO_2-PAALi is proved to be an effective method to achieve high-performance CGPEs for lithium-ion battery applications.

Nevertheless, formation of lithium dendrites and deteriorated battery performance still exist owing to uneven current distribution derived from the relatively large pore size of electrospun membranes. To construct nanofibrous separators with controllable pore structure, Zhai et al. prepared hierarchically structured SiO_2 nanoparticles coated polyetherimidepolyurethane (PEI-PU) composite nanofibrous membranes by combining the advantages of electrospun nanofibers with surface

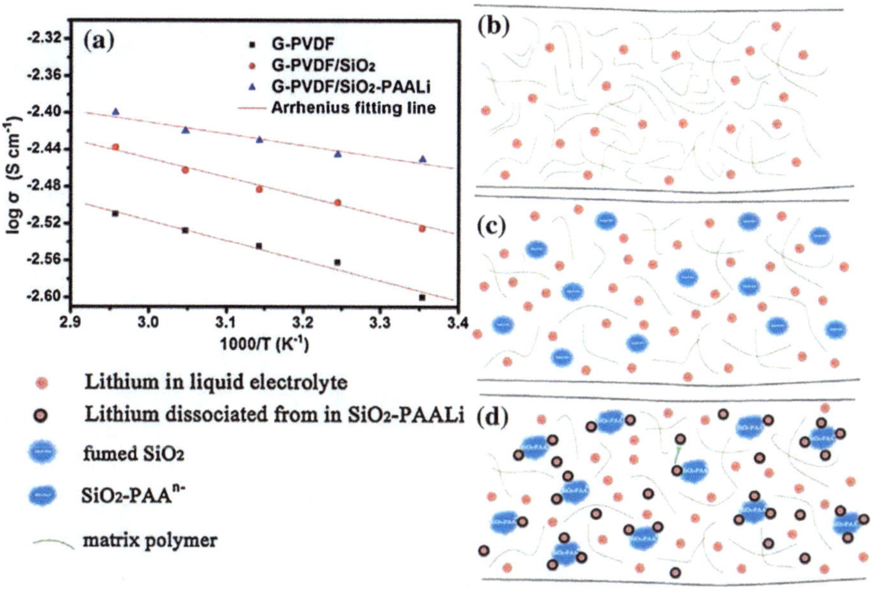

Fig. 7.6 **a** Arrhenius fitting curves about reciprocal temperature dependence of ionic conductivity of different composite gel polymer electrolytes (frequency range from 0.1 to 100 kHz, amplitude of 5 mV); Schematic representation of ionic conduction channel in CGPEs (or GPE): **b** pure PVDF, **c** PVDF/SiO_2, and **d** PVDF/SiO_2-PAALi. Reproduced with permission from Ref. [66]; Copyright 2014 Elsevier

modification [67]. Decreased average pore size is achieved for SiO_2/PEI-PU compared with PEI-PU, as shown in Fig. 7.7a, which is beneficial for resolving thermal abuse tolerance and restraining the micro-shorting caused by lithium dendrites. Due to the effective bonding structure between adjacent layers, tensile strength of SiO_2/PEI-PU composite membranes (15.65 MPa) is also much higher than that of PEI-PU membranes (8.74 MPa) (Fig. 7.7b). Moreover, severe shrinking of 92 % with color change from white to transparent is observed for Celgard while PEI-PU and SiO_2/PEI-PU membranes respectively shrink only by 9 and 5 % (Fig. 7.7c and d), thus making SiO_2/PEI-PU composite membrane promising separator candidates for next-generation LIBs. Similarly, through the combination of electrospinning and a dip-coating process, heat-resistant SiO_2 nanoparticle enhanced polysulfonamide nonwoven separator has been successfully explored [68], further promoting the significant step toward separators for LIBs that can address the persisting challenges in current battery technologies. PI nanofibers coated with thin Al_2O_3 over-layers are also obtained via a dip-coating process [69], exhibiting greatly enhanced rate capability (78.91 %) compared to bare PI separator (68.65 %) and commercial polypropylene separator (18.25 %) at a current density of 10 C.

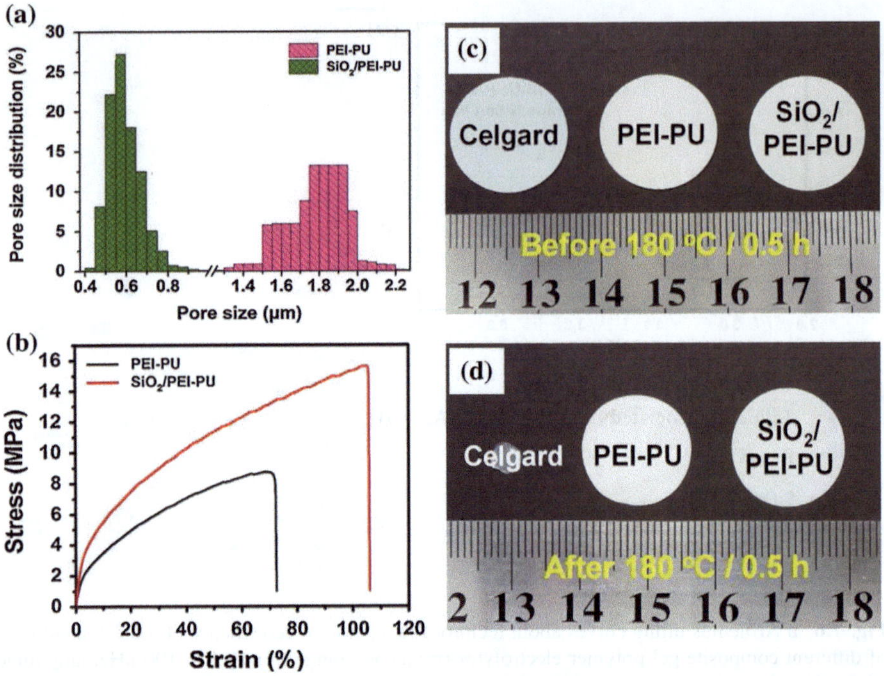

Fig. 7.7 Pore size distributions **a** and stress-strain curves **b** of as-prepared PEI-PU and SiO_2/PEI-PU composite membranes; Photographs of as-prepared PEI-PU and SiO_2/PEI-PU composite membranes **b** before and **c** after exposure to 180 °C for 0.5 h. Reproduced with permission from Ref. [67]; Copyright 2014 Elsevier

Due to the lacking of sufficient interfiber bonding, electrospun nanofiber membranes still suffer from poor physical properties. Therefore, more kinds of inorganic nanoparticles like TiO_2, La_2O_3, and SnO_2 have been introduced into electrospun nanofiber composite electrolytes [70, 71]. Electrospinning from poly (vinyl alcohol)/TiO_2 blend solutions results in composite nanofiber membranes with different amounts of TiO_2 nanoparticles [72], which exhibited high electrolyte uptake of 659 % to guarantee fully reversible charge/discharge processes. Similarly, PVDF/PMMA gel electrolytes with in situ formed TiO_2 nanoparticles embedded in the composite nanofibers were obtained, achieving highest ionic conductivity of 3.9×10^{-3} S cm^{-1} with electrochemical stability up to 5.1 V versus Li^+/Li at room temperature [73].

However, severe aggregation attributed to large specific surface area and high surface energy is an inevitable drawback when nanoparticles serve as modification additives, which would largely hinder the efficient improvement of battery performance [74]. Thus, PMMA grafted titanium dioxide (PMMA-g-TiO_2) was synthesized via atom transfer radical polymerization (ATRP) [75], which was then incorporated into PVDF membranes during electrospinning process to obtain composite nanofibrous membranes with TiO_2 nanoparticles more uniformly distributed in polymer nanofiber matrix. Chen et al. also reported the functionalization of TiO_2 with 2-hydroxyethyl methacrylate through a same ATRP method to obtain functionalized TiO_2 (f-TiO_2), which significantly reduced the agglomeration of TiO_2 nanoparticles [76]. Therefore, the composite separator with 2 % f-TiO_2 nanoparticle was found to show smaller fiber diameter, higher porosity, larger electrolyte uptake, more excellent thermal dimensional stability and electrochemical properties. Wu et al. further compared properties of TPU-PVDF-based gel polymer electrolytes with in situ ceramic fillers of SiO_2 and TiO_2, respectively [77], founding that the gel polymer electrolyte with in situ formed TiO_2 nanoparticles displays a wider electrochemical stability up to 5.4 V versus Li^+/Li at 25 °C. Moreover, better mechanical stability with high tensile strength (8.7 ± 0.3 MPa) and elongation at break (110.3 ± 0.2 %), and good ionic conductivity as high as 4.8×10^{-3} S cm^{-1} suggest that incorporation of in situ generated TiO_2 is more efficient in improving the properties of gel polymer electrolyte for practical applications.

$MgAl_2O_4$, as a bimetallic ceramic filler, shows strong Lewis acid character which could form complexes with the strong Lewis base of PVDF-HFP polymer chains, thus resulting in easy pathways for Li^+ ions and improved ionic conductivity [78]. Therefore, PVDF-HFP/$MgAl_2O_4$ hybrid fibrous nanocomposites were newly prepared by electrospinning for high-performance polymer electrolyte membranes [79]. Novel electrospun nanocomposite electrolyte membrane of PVDF-HFP/PMMA/$MgAl_2O_4$/$LiPF_6$ further shows low crystallinity and average fiber diameter, high thermal stability, electrolyte uptake, and conductivity of 2.6×10^{-3} S cm^{-1} at room temperature for CR 2032 cell lithium battery [80]. Padmaraj et al. even studied the effect of $ZnAl_2O_4$ nanofiller concentration on physical and electrochemical properties of electrospun PVDF-HFP based nanocomposite electrolyte membranes [81], indicating that optimized reduction of crystallinity, good thermal stability, and high ionic conductivity can be achieved with 5 wt% $ZnAl_2O_4$ fillers.

Electrospun Polymer Nanofiber Electrolyte Membranes for PEMFCs

PEMFCs have attracted much attention due to their high power density and high efficiency with low greenhouse gas emission [82, 83]. In PEMFCs, proton exchange membranes (PEMs) are the key components which efficiently transport protons from anode to cathode. Typically, Nafion-based membranes are the most widely studied PEMs, which possess good thermal, mechanical, and chemical stabilities, as well as relatively high proton conductivity [84]. Nevertheless, high production cost, environmental incompatibility, and low operation temperature of Nafion largely hinder the widespread commercialization of PEMFCs [85]. Therefore, considerable interest has been focused on designing chemically and mechanically stable proton exchange membranes with high ionic conductivity.

A new direction for Nafion has been persuaded via electrospinning to obtain high-purity Nafion nanofibers with the addition of only 0.1 wt% carrier polymer of PEO as shown in Fig. 7.8a and b [86]. Proton conductivity of single Nafion nanofiber with diameter of 400 nm reaches as high as 1.5 S cm^{-1} (Fig. 7.8c and d), being an order of magnitude higher than that (0.1 S cm^{-1}) of the bulk film. Strong anisotropic scattering at smaller angles (appears white in Fig. 7.8e) and the scattering intensity as a function of azimuthal angle (Fig. 7.8g) further indicate that ionic aggregates are isotropic in the cast film whereas anisotropic in electrospun nanofibers, thus making electrospinning a promising technique to produce anisotropic polymer nanofiber electrolyte membranes for fuel cell applications. However, Nafion itself is hard to be produced into electrospun nanofibers due to its high molecular weight and poor solubility in most common solvents. Therefore, Nafion/PVA and Nafion/poly(vinyl pyrrolidone) (PVP) composite nanofiber electrolytes are also obtained as promising low-cost proton exchange membranes [87, 88]. Moreover, Nafion-filled sulfonated poly(ether sulfone) (SPES) nanofiber-based bilayer membranes has been fabricated to show significantly decreased methanol permeability and improved membrane selectivity compared with commercial Nafion112 and Nafion117 membranes [89], which can be ascribed to the reinforcing effect, swelling reduction of Nafion caused by the key role of SPES membranes.

Tamura and Kawakami synthesized uniaxially-aligned sulfonated polyimide nanofiber membrane electrolyte by parallel electrospinning [90], showing ultra-high proton conductivities above 1 S cm^{-1} due to the formation of effective proton conduction multi-pathways by molecular orientation in nanofibers. Polybenzoxazine (PBz)-modified polybenzimidazole (PBI) nanofibers also have been prepared by electrospinning [91], which was further crosslinked through a ring-opening addition reaction of PBz benzoxazine groups. Thus, crosslinked PBI nanofibers-based composite membranes exhibit significantly improved mechanical properties, acid uptakes, and dimensional stability upon acid doping. Liu et al. developed highly conducting poly(diallyl dimethyl ammonium chloride)/sulfonated poly(2,6-dimethyl 1,4-phenylene oxide) layer-by-layer (LBL) films on a highly controllable electrospun fiber scaffold through the combination of electrospinning and LBL assembly processes [92], in which the electrospun membranes provide mechanical support and the LBL assembly produces highly conductive films that provide ionic conductivity and fuel blocking characteristics. Therefore, mechanically stable membranes with high selectivity are produced for potential use in fuel cell applications.

High operation temperature (over 100 °C) of PEMFEs has pushed forward the development of novel polymer electrolyte membranes with excellent thermal stability, mechanical strength as well as good proton conductivity. Therefore, Kim et al. prepared inorganic silicon carbide fibers by electrospinning [93], which were further functionalized by hydroxyl groups and phosphoric acid to achieve an ion-exchange capability of 70 % higher than the conventional cast electrolyte

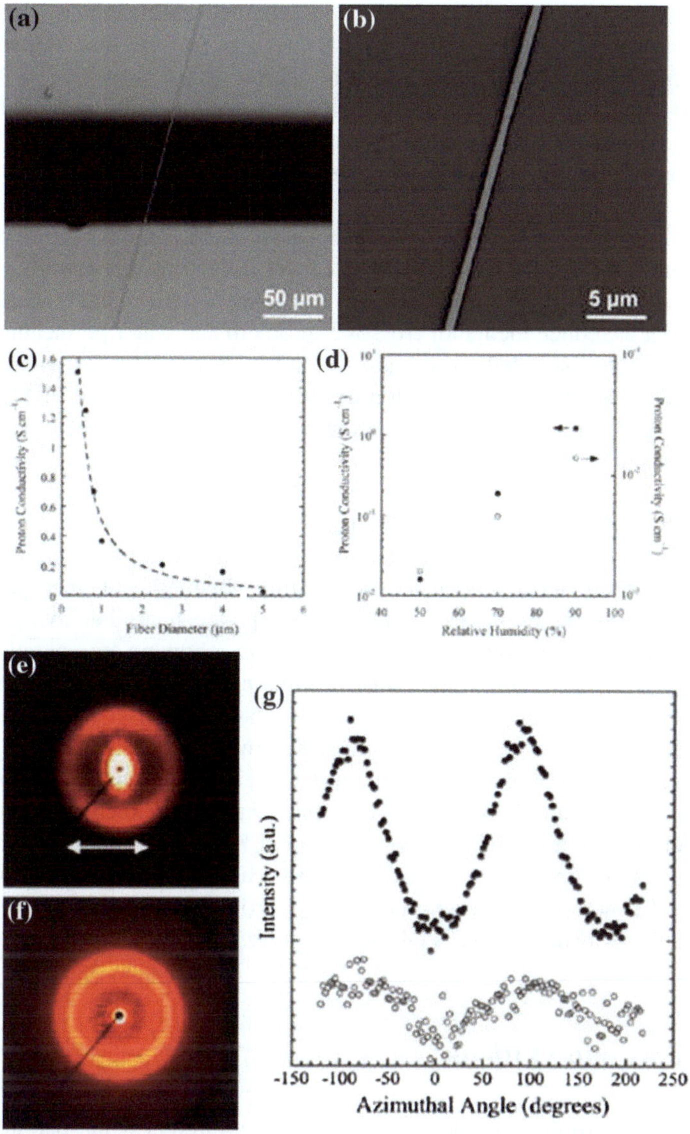

Fig. 7.8 **a** SEM image of a single high-purity Nafion nanofiber (99.9 wt% Nafion) bridging two electrodes; **b** enlarged image of **a**; **c** proton conductivity (at 30 °C, 90 % RH) versus fiber diameter for high-purity Nafion nanofibers measured on individual nanofibers; **d** humidity-dependent proton conductivity (30 °C) of a single high-purity Nafion nanofiber (*solid symbols*) and a cast Nafion film (*open symbols*); 2D X-ray scattering patterns (at 86 % RH, 25 °C) of **e** macroscopically aligned high-purity Nafion nanofibers (*arrows* indicating longitudinal direction of fibers) and **f** Nafion cast film; **g** X-ray scattering intensity as a function of azimuthal angle for macroscopically aligned high-purity Nafion nanofibers (*solid symbols*) and a cast Nafion film (*open symbols*). Reproduced with permission from Ref. [86]; Copyright 2010 American Chemical Society

membrane. Moreover, solid acids have been paid much attention in fuel cells due to their high proton conductivity (10^{-2}–10^{-3} S cm^{-1}) at the intermediate temperature range of 150–300 °C [94]. Hence, pure inorganic CsH_2PO_4 fibers (CDPf) were synthesized by electrospinning of a viscous solution of partially polymerized CDP [95], showing a maximum proton conductivity of 8×10^{-3} S cm^{-1} at 250 °C for potential applications in an intermediate temperature electrochemical device.

In spite of excellent thermal stability and high proton conductivity, the fiber integrity and mechanical strength of pure inorganic nanofibers are still needed to be greatly improved after the elimination of carrier polymers. Meanwhile, inorganic nanofillers are always considered as effective reinforcement agent in polymeric PEMs in order to reduce methanol crossover problem and water permeability. Thus, novel conductive sulfated zirconia (S-ZrO_2) fiber-filled crosslinked poly (2-acrylamido-2-meathylpropane-sulfonic acid) (C-PAMPS) hybrid PEMs has been developed via electrospinning and posttreatment (Fig. 7.9a and b) [96], in which the continuous inorganic fibers not only served as continuous hopping pathways for protons, but also significantly improved the proton conductivity of PEMs. Moreover, hybrid membranes with thinner fiber diameters exhibit higher proton conductivities which can be attributed to the increased surface area and ionic channels of S-ZrO_2 fibers. Lee et al. successfully obtained SiO_2-incorporated sulfonated poly (ether ether ketone) (SPEEK) composite nanofiber membrane as a support for dip-coating with Nafion solution [97], showing comparative proton conductivity with cast Nafion and SPEEK films and much higher power density (170 mW cm^{-2}) than that (71 mW cm^{-2}) of cast Nafion. Choi et al. further reported the preparation of sulfonated SPEEK/directly spinnable carbon nanotube (dsCNT) composite proton exchange membranes via combining electrospinning of SPEEK and direct spinning of CNT forests [98]. Due to the excellent alignment and interconnectivity of thin CNT forest web, SPEEK/dsCNT composite PEM possesses significantly improved physical and electrical performance over cast SPEEK and Nafion212 membrane, which undoubtedly provides a new strategy for preparation of high-performance dsCNT-based polymer composite membranes.

Conclusions and Outlook

Through various tuning of electrospinning solution/processing parameters and set-up geometries, as well as combining facile in situ or surface modification approaches, one-dimensional electrospun nanofiber-based nanocomposites with hierarchical structures (such as random three-dimensional structures, aligned or patterned fibers) and multi-components (including single-component, multi-component, and inorganic nanofiller/polymer composite) have been successfully

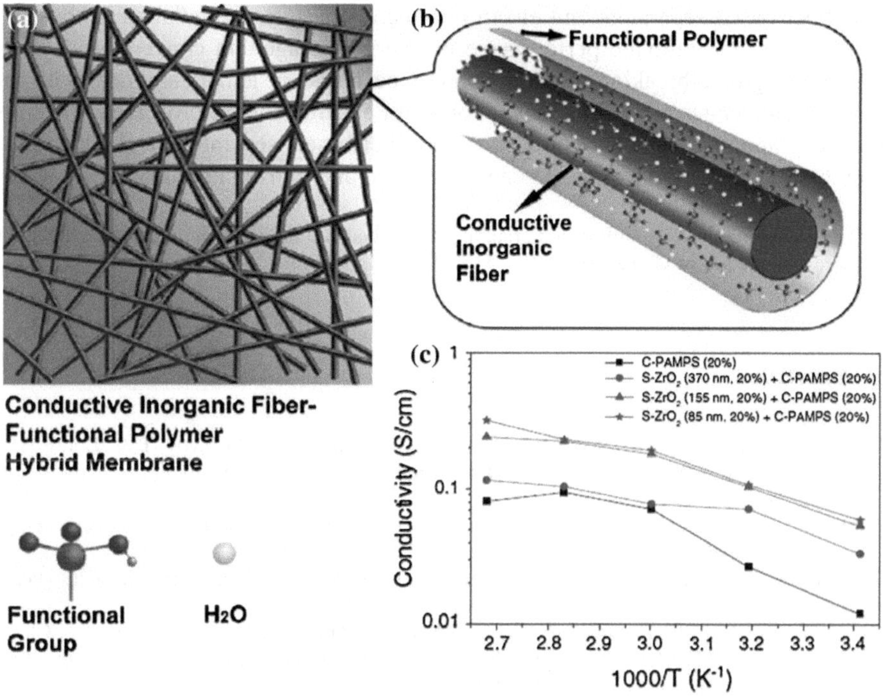

Fig. 7.9 Schematic diagrams of **a** conductive inorganic fiber/polymer hybrid membrane; **b** the long-range ionic channel between conductive fiber and functional polymer matrix; **c** proton conductivities of C-PAMPS (crosslinking degree = 20 %) and hybrid PEMs (fiber volume fraction = 20 %) with different fiber diameters. Reproduced with permission from Ref. [96]; Copyright 2011 Elsevier

achieved. Due to the high surface area to volume ratio, large porosity with fully interconnected pore structures for high electrolyte uptake and easy transport of ions, these electrospun polymer nanofiber separators and electrolyte membranes exhibit outstanding physical and electrochemical properties for high energy density rechargeable batteries and high power output fuel cell applications. Therefore, electrospinning technique has been playing an increasingly important role in the development of new clean energy areas.

Nevertheless, several technical and theoretical issues are still urgently needed to be addressed before the practical commercialization of electrospun nanofiber-based separators and electrolyte membranes. First, electrospinning processing parameters are not fully controllable, which may inevitably lead to defects such as beads and inhomogeneous nanofibers during large scale productions. Second, although multi-component and inorganic nanofiller/polymer nanocomposites have been obtained to improve mechanical properties, how to maintain good fiber integrity and mechanical strength during industrial cell assembly processes still keeps a great challenge. Moreover, the large scale-up and down-stream processing of electrospun nanofiber-based separators and electrolyte membranes are essential steps that definitely affect the ultimate performance of energy storage and conversion devices. Therefore, groundbreaking progresses are expected to systematically deal with the existing challenges and difficulties involved in electrospinning processes for future practical applications of electrospun nanofiber-based separators and electrolyte membranes in high-performance energy storage and conversion areas.

Acknowledgements The authors are grateful for the financial support from the National Natural Science Foundation of China (51125011, 51373037, 51433001).

References

1. Rolison DR, Long JW, Lytle JC, Fischer AE, Rhodes CP, McEvoy TM, Bourg ME, Lubers AM (2009) Multifunctional 3D nanoarchitectures for energy storage and conversion. Chem Soc Rev 38(1):226–252
2. Guo YG, Hu JS, Wan LJ (2008) Nanostructured materials for electrochemical energy conversion and storage devices. Adv Mater 20(15):2878–2887
3. Pintauro PN (2015) Perspectives on membranes and separators for electrochemical energy conversion and storage devices. Polym Rev 55(2):201–207
4. Choi NS, Chen ZH, Freunberger SA, Ji X, Sun YK, Amine K, Yushin G, Nazar LF, Cho J, Bruce PG (2012) Challenges facing lithium batteries and electrical double-layer capacitors. Angew Chem Int Ed 51(40):9994–10024
5. Bruce PG, Scrosati B, Tarascon JM (2008) Nanomaterials for rechargeable lithium batteries. Angew Chem Int Ed 47(16):2930–2946
6. Lee KT, Cho J (2011) Roles of nanosize in lithium reactive nanomaterials for lithium ion batteries. Nano Today 6(1):28–41
7. Zhang XW, Ji L, Toprakci O, Liang YZ, Alcoutlabi M (2011) Electrospun nanofiber-based anodes, cathodes, and separators for advanced lithium-ion batteries. Polym Rev 51(3):239–264
8. Doshi J, Reneker DH (1995) Electrospinning process and applications of electrospun fibers. J Electrostat 35(2–3):151–160
9. Tan SH, Inai R, Kotaki M, Ramakrishna S (2005) Systematic parameter study for ultra-fine fiber fabrication via electrospinning process. Polymer 46(16):6128–6134
10. Huang YP, Miao YE, Liu TX (2014) Electrospun fibrous membranes for efficient heavy metal removal. J Appl Polym Sci 131(19)
11. Aravindan V, Sundaramurthy J, Kumar PS, Shubha N, Ling WC, Ramakrishna S, Madhavi S (2013) A novel strategy to construct high performance lithium-ion cells using one dimensional electrospun nanofibers, electrodes and separators. Nanoscale 5(21):10636–10645
12. Dong Z, Kennedy SJ, Wu Y (2011) Electrospinning materials for energy-related applications and devices. J Power Sources 196(11):4886–4904
13. Aravindan V, Sundaramurthy J, Suresh Kumar P, Ramakrishna Y-S, Ramakrishna S, Madhavi S (2015) Electrospun nanofibers: a prospective electro-active material for constructing high performance Li-ion batteries. Chem Commun 51(12):2225–2234
14. Cavaliere S, Subianto S, Savych I, Jones DJ, Roziere J (2011) Electrospinning: designed architectures for energy conversion and storage devices. Energy Environ Sci 4(12):4761–4785
15. Kalluri S, Seng KH, Guo ZP, Liu HK, Dou SX (2013) Electrospun lithium metal oxide cathode materials for lithium-ion batteries. RSC Adv 3(48):25576–25601
16. Patil A, Patil V, Wook Shin D, Choi JW, Paik DS, Yoon SJ (2008) Issue and challenges facing rechargeable thin film lithium batteries. Mater Res Bull 43(8–9):1913–1942

17. Djian D, Alloin F, Martinet S, Lignier H, Sanchez JY (2007) Lithium-ion batteries with high charge rate capacity: influence of the porous separator. J Power Sources 172(1):416–421
18. Venugopal G, Moore J, Howard J, Pendalwar S (1999) Characterization of microporous separators for lithium-ion batteries. J Power Sources 77(1):34–41
19. Liang YZ, Ji LW, Guo BK, Lin Z, Yao YF, Li Y, Alcoutlabi M, Qiu YP, Zhang XW (2011) Preparation and electrochemical characterization of ionic-conducting lithium lanthanum titanate oxide/polyacrylonitrile submicron composite fiber-based lithium-ion battery separators. J Power Sources 196(1):436–441
20. Raghavan P, Choi J-W, Ahn J-H, Cheruvally G, Chauhan GS, Ahn H-J, Nah C (2008) Novel electrospun poly(vinylidene fluoride-co-hexafluoropropylene)-in situ SiO_2 composite membrane-based polymer electrolyte for lithium batteries. J Power Sources 184(2):437–443
21. Jung HR, Ju DH, Lee WJ, Zhang XW, Kotek R (2009) Electrospun hydrophilic fumed silica/polyacrylonitrile nanofiber-based composite electrolyte membranes. Electrochim Acta 54(13):3630–3637
22. Choi SW, Jo SM, Lee WS, Kim YR (2003) An electrospun poly(vinylidene fluoride) nanofibrous membrane and its battery applications. Adv Mater 15(23):2027–2032
23. Li X, Cheruvally G, Kim JK, Choi JW, Ahn JH, Kim KW, Ahn HJ (2007) Polymer electrolytes based on an electrospun poly(vinylidene fluoride-co-hexafluoropropylene) membrane for lithium batteries. J Power Sources 167(2):491–498
24. Raghavan P, Zhao X, Manuel J, Shin C, Heo MY, Ahn JH, Ryu HS, Ahn HJ, Noh JP, Cho GB (2010) Electrochemical studies on polymer electrolytes based on poly(vinylidene fluoride-co-hexafluoropropylene) membranes prepared by electrospinning and phase inversion-A comparative study. Mater Res Bull 45(3):362–366
25. Kim HS, Shin JH, Doh CH, Moon SI, Kim SP (2002) Preparation and electrochemical performance of gel polymer electrolytes using tri(ethylene glycol) dimethacrylate. J Power Sources 112(2):469–476
26. Kader MA, Kwak SK, Kang SL, Ahn JH, Nah C (2008) Novel microporous poly(vinylidene fluoride)-graft-poly(tert-butyl acrylate) electrolytes for secondary lithium batteries. Polym Int 57(11):1199–1205
27. Liang YZ, Cheng SC, Zhao JM, Zhang CH, Sun SY, Zhou NT, Qiu YP, Zhang XW (2013) Heat treatment of electrospun Polyvinylidene fluoride fibrous membrane separators for rechargeable lithium-ion batteries. J Power Sources 240:204–211
28. Bansal D, Meyer B, Salomon M (2008) Gelled membranes for Li and Li-ion batteries prepared by electrospinning. J Power Sources 178(2):848–851
29. Kritzer P (2006) Nonwoven support material for improved separators in Li–polymer batteries. J Power Sources 161(2):1335–1340
30. Alcoutlabi M, Lee H, Watson J, Zhang XW (2013) Preparation and properties of nanofiber-coated composite membranes as battery separators via electrospinning. J Mater Sci 48(6):2690–2700
31. Lee H, Alcoutlabi M, Watson JV, Zhang XW (2013) Electrospun nanofiber-coated separator membranes for lithium-ion rechargeable batteries. J Appl Polym Sci 129(4):1939–1951
32. Lee H, Alcoutlabi M, Toprakci O, Xu G, Watson J, Zhang XW (2014) Preparation and characterization of electrospun nanofiber-coated membrane separators for lithium-ion batteries. J Solid State Electrochem 18(9):2451–2458
33. Huang BY, Wang ZX, Li GB, Huang H, Xue RJ, Chen LQ, Wang F (1996) Lithium ion conduction in polymer electrolytes based on PAN. Solid State Ionics 85(1–4):79–84
34. Yang CR, Perng JT, Wang YY, Wan CC (1996) Conductive behaviour of lithium ions in polyacrylonitrile. J Power Sources 62(1):89–93
35. Cho TH, Sakai T, Tanase S, Kimura K, Kondo Y, Tarao T, Tanaka M (2007) Electrochemical performances of polyacrylonitrile nanofiber-based nonwoven separator for lithium-ion battery. Electrochem Solid-State Lett 10(7):A159–A162
36. Carol P, Ramakrishnan P, John B, Cheruvally G (2011) Preparation and characterization of electrospun poly(acrylonitrile) fibrous membrane based gel polymer electrolytes for lithium-ion batteries. J Power Sources 196(23):10156–10162

37. Chen D, Liu TX, Zhou XP, Tjiu WW, Hou HQ (2009) Electrospinning fabrication of high strength and toughness polyimide nanofiber membranes containing multiwalled carbon nanotubes. J Phys Chem B 113(29):9741–9748
38. Chen D, Miao YE, Liu TX (2013) Electrically conductive polyaniline/polyimide nanofiber membranes prepared via a combination of electrospinning and subsequent in situ polymerization growth. ACS Appl Mater Interfaces 5(4):1206–1212
39. Miao YE, Zhu GN, Hou HQ, Xia YY, Liu TX (2013) Electrospun polyimide nanofiber-based nonwoven separators for lithium-ion batteries. J Power Sources 226:82–86
40. Wang QJ, Song WL, Wang LN, Song Y, Shi Q, Fan LZ (2014) Electrospun polyimide-based fiber membranes as polymer electrolytes for lithium-ion batteries. Electrochim Acta 132:538–544
41. Wang QJ, Jian ZX, Song WL, Zhang SC, Fan LZ (2014) Facile fabrication of safe and robust polyimide fibrous membrane based on triethylene glycol diacetate-2-propenoic acid butyl ester gel electrolytes for lithium-ion batteries. Electrochim Acta 149:176–185
42. Qi W, Lu C, Chen P, Han L, Yu Q, Xu RQ (2012) Electrochemical performances and thermal properties of electrospun Poly(phthalazinone ether sulfone ketone) membrane for lithium-ion battery. Mater Lett 66(1):239–241
43. Wu N, Cao Q, Wang XY, Li XY, Deng HY (2011) A novel high-performance gel polymer electrolyte membrane basing on electrospinning technique for lithium rechargeable batteries. J Power Sources 196(20):8638–8643
44. Wu N, Cao Q, Wang XY, Chen QQ (2011) Study of a novel porous gel polymer electrolyte based on TPU/PVdF by electrospinning technique. Solid State Ionics 203(1):42–46
45. Wu N, Jing B, Cao Q, Wang XY, Kuang H, Wang Q (2012) A novel electrospun TPU/PVdF porous fibrous polymer electrolyte for lithium ion batteries. J Appl Polym Sci 125 (4):2556–2563
46. Zhou L, Cao Q, Jing B, Wang XY, Tang XL, Wu N (2014) Study of a novel porous gel polymer electrolyte based on thermoplastic polyurethane/poly(vinylidene fluoride-co-hexafluoropropylene) by electrospinning technique. J Power Sources 263:118–124
47. Lu P, Hsieh Y-L (2010) Preparation and properties of cellulose nanocrystals: Rods, spheres, and network. Carbohyd Polym 82(2):329–336
48. Lalia B, Samad Y, Hashaikeh R (2013) Nanocrystalline cellulose-reinforced composite mats for lithium-ion batteries: electrochemical and thermomechanical performance. J Solid State Electrochem 17(3):575–581
49. Yvonne T, Zhang CY, Zhang CH, Omollo E, Ncube S (2014) Properties of electrospun PVDF/PMMA/CA membrane as lithium based battery separator. Cellulose 21(4):2811–2818
50. Prasanth R, Shubha N, Hng HH, Srinivasan M (2014) Effect of poly(ethylene oxide) on ionic conductivity and electrochemical properties of poly(vinylidenefluoride) based polymer gel electrolytes prepared by electrospinning for lithium ion batteries. J Power Sources 245:283–291
51. Prasanth R, Aravindan V, Srinivasan M (2012) Novel polymer electrolyte based on cob-web electrospun multi component polymer blend of polyacrylonitrile/poly(methyl methacrylate)/ polystyrene for lithium ion batteries-Preparation and electrochemical characterization. J Power Sources 202:299–307
52. Jung HR, Lee WJ (2011) Electrochemical characteristics of electrospun poly(methyl methacrylate)/polyvinyl chloride as gel polymer electrolytes for lithium ion battery. Electrochim Acta 58:674–680
53. Samad YA, Asghar A, Hashaikeh R (2013) Electrospun cellulose/PEO fiber mats as a solid polymer electrolytes for Li ion batteries. Renew Energy 56:90–95
54. Kimura N, Sakumoto T, Mori Y, Wei K, Kim BS, Song KH, Kim IS (2014) Fabrication and characterization of reinforced electrospun poly(vinylidene fluoride-co-hexafluoropropylene) nanofiber membranes. Compos Sci Technol 92:120–125

55. Chen D, Wang RY, Tjiu WW, Liu TX (2011) High performance polyimide composite films prepared by homogeneity reinforcement of electrospun nanofibers. Compos Sci Technol 71 (13):1556–1562
56. Miao YE, Yan JJ, Huang YP, Fan W, Liu TX (2015) Electrospun polymer nanofiber membrane electrodes and an electrolyte for highly flexible and foldable all-solid-state supercapacitors. RSC Adv 5(33):26189–26196
57. Nah C, Jeong KU, Lee YS, Lee SH, Kader MMA, Lee HK, Ahn JH (2013) Polymer electrolyte membranes composed of an electrospun poly(vinylidene fluoride) fibrous mat in a poly(4-vinylpyridine) matrix. Polym Int 62(3):375–381
58. Chen WY, Liu YB, Ma Y, Liu JZ, Liu XR (2014) Improved performance of PVdF-HFP/PI nanofiber membrane for lithium ion battery separator prepared by a bicomponent cross-electrospinning method. Mater Lett 133:67–70
59. Liu ZH, Jiang W, Kong QS, Zhang CJ, Han PX, Wang XJ, Yao JH, Cui GL (2013) A core@sheath nanofibrous separator for lithium ion batteries obtained by coaxial electrospinning. Macromol Mater Eng 298(7):806–813
60. Zhou XH, Yue LP, Zhang JJ, Kong QS, Liu ZH, Yao JH, Cui GL (2013) A core-shell structured polysulfonamide-based composite nonwoven towards high power lithium ion battery separator. J Electrochem Soc 160(9):A1341–A1347
61. Huang FL, Xu YF, Peng B, Su YF, Jiang F, Hsieh YL, Wei QF (2015) Coaxial electrospun cellulose-core fluoropolymer-shell fibrous membrane from recycled cigarette filter as separator for high performance lithium-ion battery. ACS Sustain Chem Eng 3(5):932–940
62. Angulakshmi N, Stephan AM (2014) Electrospun trilayer polymeric membranes as separator for lithium-ion batteries. Electrochim Acta 127:167–172
63. Zhai YY, Wang N, Mao X, Si Y, Yu JY, Al-Deyab SS, El-Newehy M, Ding B (2014) Sandwich-structured PVdF/PMIA/PVdF nanofibrous separators with robust mechanical strength and thermal stability for lithium ion batteries. J Mater Chem A 2(35):14511–14518
64. Yanilmaz M, Dirican M, Zhang XW (2014) Evaluation of electrospun SiO_2/nylon 6,6 nanofiber membranes as a thermally-stable separator for lithium-ion batteries. Electrochim Acta 133:501–508
65. Yanilmaz M, Lu Y, Dirican M, Fu K, Zhang XW (2014) Nanoparticle-on-nanofiber hybrid membrane separators for lithium-ion batteries via combining electrospraying and electrospinning techniques. J Membrane Sci 456:57–65
66. Li WL, Xing YJ, Wu YH, Wang JW, Chen LZ, Yang G, Tang BZ (2015) Study the effect of ion-complex on the properties of composite gel polymer electrolyte based on Electrospun PVdF nanofibrous membrane. Electrochim Acta 151:289–296
67. Zhai YY, Xiao K, Yu JY, Ding B (2015) Fabrication of hierarchical structured SiO_2/polyetherimide-polyurethane nanofibrous separators with high performance for lithium ion batteries. Electrochim Acta 154:219–226
68. Zhang JJ, Yue L, Kong QS, Liu ZH, Zhou XH, Zhang CJ, Pang SP, Wang XJ, Yao JH, Cui GL (2013) A heat-resistant silica nanoparticle enhanced polysulfonamide nonwoven separator for high-performance lithium ion battery. J Electrochem Soc 160(6):A769–A774
69. Lee J, Lee C-L, Park K, Kim I-D (2014) Synthesis of an Al_2O_3-coated polyimide nanofiber mat and its electrochemical characteristics as a separator for lithium ion batteries. J Power Sources 248:1211–1217
70. Wootthikanokkhan J, Phiriyawirut M, Pongchumpon O (2014) Effects of electrospinning parameters and nanofiller content on morphology and gel electrolyte properties of composite nanofibers based on La_2O_3-filled PVDF-HFP. Int J Polym Mater 64(8):416–426
71. Zaccaria M, Fabiani D, Cannucciari G, Gualandi C, Focarete ML, Arbizzani C, De Giorgio F, Mastragostino M (2015) Effect of silica and tin oxide nanoparticles on properties of nanofibrous electrospun separators. J Electrochem Soc 162(6):A915–A920
72. Shokrollahi M, Morshed M, Semnani D, Rezaei B (2013) Development of electro-spun poly (vinyl alcohol)/titanium dioxide membrane-based polymer electrolytes for lithium-ion batteries. Int J Polym Mater 63(3):161–171

73. Zhou L, Wu N, Cao Q, Jing B, Wang XY, Wang Q, Kuang H (2013) A novel electrospun PVDF/PMMA gel polymer electrolyte with in situ TiO_2 for Li-ion batteries. Solid State Ionics 249–250:93–97
74. Hojjati B, Sui R, Charpentier PA (2007) Synthesis of TiO_2/PAA nanocomposite by RAFT polymerization. Polymer 48(20):5850–5858
75. Cui WW, Tang DY, Gong ZL (2013) Electrospun poly(vinylidene fluoride)/poly(methyl methacrylate) grafted TiO_2 composite nanofibrous membrane as polymer electrolyte for lithium-ion batteries. J Power Sources 223:206–213
76. Chen WY, Liu YB, Ma Y, Yang WX (2015) Improved performance of lithium ion battery separator enabled by co-electrospinnig polyimide/poly(vinylidene fluoride-co-hexafluoropropylene) and the incorporation of TiO_2-(2-hydroxyethyl methacrylate). J Power Sources 273:1127–1135
77. Wu N, Cao Q, Wang XY, Li S, Li XY, Deng HY (2011) In situ ceramic fillers of electrospun thermoplastic polyurethane/poly(vinylidene fluoride) based gel polymer electrolytes for Li-ion batteries. J Power Sources 196(22):9751–9756
78. Shi BL, He MJ (2008) Lewis acid-base property of P(VDF-co-HFP) measured by inverse gas chromatography. J Appl Polym Sci 107(3):1642–1646
79. Padmaraj O (2014) Nageswara Rao B, Jena P, Venkateswarlu M, Satyanarayana N. Electrochemical studies of electrospun organic/inorganic hybrid nanocomposite fibrous polymer electrolyte for lithium battery. Polymer 55(5):1136–1142
80. Padmaraj O, Rao BN, Venkateswarlu M, Satyanarayana N (2015) Electrochemical characterization of electrospun nanocomposite polymer blend electrolyte fibrous membrane for lithium battery. J Phys Chem B 119(16):5299–5308
81. Padmaraj O, Venkateswarlu M, Satyanarayana N (2014) Characterization and electrochemical properties of P(VdF-co-HFP) based electrospun nanocomposite fibrous polymer electrolyte membrane for lithium battery applications. Electroanalysis 26(11):2373–2379
82. Steele BCH, Heinzel A (2001) Materials for fuel-cell technologies. Nature 414(6861):345–352
83. Bauen A, Hart D (2000) Assessment of the environmental benefits of transport and stationary fuel cells. J Power Sources 86(1–2):482–494
84. Hickner MA, Pivovar BS (2005) The chemical and structural nature of proton exchange membrane fuel cell properties. Fuel Cells 5(2):213–229
85. Li QF, He RH, Jensen JO, Bjerrum NJ (2003) Approaches and recent development of polymer electrolyte membranes for fuel cells operating above 100 °C. Chem Mater 15(26):4896–4915
86. Dong B, Gwee L, Salas-de la Cruz D, Winey KI, Elabd YA (2010) Super proton conductive high-purity Nafion nanofibers. Nano Lett 10(9):3785–3790
87. Mollá S, Compañ V, Gimenez E, Blazquez A, Urdanpilleta I (2011) Novel ultrathin composite membranes of Nafion/PVA for PEMFCs. Int J Hydrogen Energy 36(16):9886–9895
88. Pan CF, Wu H, Wang C, Wang B, Zhang L, Cheng ZD, Hu P, Pan W, Zhou ZY, Yang X, Zhu J (2008) Nanowire-based high-performance "micro fuel cells": one nanowire, one fuel cell. Adv Mater 20(9):1644–1648
89. Shabani I, Hasani-Sadrabadi MM, Haddadi-Asl V, Soleimanid M (2011) Nanofiber-based polyelectrolytes as novel membranes for fuel cell applications. J Membrane Sci 368:233–240
90. Takemori R, Ito G, Tanaka M, Kawakami H (2014) Ultra-high proton conduction in electrospun sulfonated polyimide nanofibers. RSC Adv 4(38):20005–20009
91. Li HY, Liu YL (2013) Polyelectrolyte composite membranes of polybenzimidazole and crosslinked polybenzimidazole-polybenzoxazine electrospun nanofibers for proton exchange membrane fuel cells. J Mater Chem A 1(4):1171–1178
92. Liu DS, Ashcraft JN, Mannarino MM, Silberstein MN, Argun AA, Rutledge GC, Boyce MC, Hammond PT (2013) Spray layer-by-layer electrospun composite proton exchange membranes. Adv Funct Mater 23(24):3087–3095
93. Kim TE, Juon SM, Park JH, Shul YG, Cho KY (2014) Silicon carbide fiber-reinforced composite membrane for high-temperature and low-humidity polymer exchange membrane fuel cells. Int J Hydrogen Energy 39(29):16474–16485

94. Haile SM, Boysen DA, Chisholm CRI, Merle RB (2001) Solid acids as fuel cell electrolytes. Nature 410(6831):910–913
95. Goni-Urtiaga A, Scott K, Cavaliere S, Jones DJ, Roziere J (2013) A new fabrication method of an intermediate temperature proton exchange membrane by the electrospinning of CsH_2PO_4. J Mater Chem A 1(36):10875–10880
96. Yao YF, Guo BK, Ji L, Jung KH, Lin Z, Alcoutlabi M, Hamouda H, Zhang XW (2011) Highly proton conductive electrolyte membranes: fiber-induced long-range ionic channels. Electrochem Commun 13(9):1005–1008
97. Lee C, Jo SM, Choi J, Baek KY, Truong YB, Kyratzis IL, Shul YG (2013) SiO_2/sulfonated poly ether ether ketone (SPEEK) composite nanofiber mat supported proton exchange membranes for fuel cells. J Mater Sci 48:3665–3671
98. Choi J, Lee C, Hawkins SC, Huynh CP, Park J, Jeon Y, Truong YB, Kyratzis IL, Shul Y-G, Caruso RA (2014) Direct spun aligned carbon nanotube web-reinforced proton exchange membranes for fuel cells. RSC Adv 4(62):32787–32790

Chapter 8
Converting of Bulk Polymers into Nanofibrils via Hot Stretching of Polymer Blends

Ben Niu, Gan-Ji Zhong, Jia-Zhuang Xu, Huan Xu, Lan Xie and Zhong-Ming Li

Introduction

Blending modification of various polymers is widely considered as an economically viable and versatile approach to design polymeric materials with a desirable combination of properties, thus satisfying the strict needs for specific applications in polymer industry. Nevertheless, since most of the polymer pairs are thermodynamically immiscible and technologically incompatible, a multicomponent system with a deformable minor component always forms during melt processing. Under appropriate conditions, the dispersed component can be in situ deformed into diverse structures, such as spheres, ellipsoids, fibers, and plates/ribbons [1–4]. The factors determining the above morphology formation can be summarized as blend composition, viscosity ratio, thermo/mechanical history, etc. [5–7]. Furthermore, it has been well revealed that some morphologies such as fibers and plates, are beneficial to improving the specific properties, even endowing newly functional properties to the blend. Typical examples involve the capability of the plates to improve the gas barrier, and fibers enhance the strength and stiffness of the blends [8–10].

Inspired by the intrinsic deformation nature of polymer melt, various proposals have been put forward to construct particular structure in polymer blend, aiming at the achievement of excellent performance. One of the typical examples is the methodology of "in situ micro/nanofibrillation," in which, under the process of "melt extrusion-hot stretching-quenching," the spherical dispersion domains in situ deform into micro/nanofibrils, realizing the scalable achievement of in situ micro/nanofibrillar composites, accompanied with greatly improved mechanical

B. Niu · G.-J. Zhong · J.-Z. Xu · H. Xu · L. Xie · Z.-M. Li (✉)
State Key Laboratory of Polymer Materials Engineering,
College of Polymer Science and Engineering, Sichuan University,
Chengdu 610065, Sichuan, People's Republic of China
e-mail: zmil@scu.edu.cn

S. Fakirov (ed.), *Nano-size Polymers*,
DOI 10.1007/978-3-319-39715-3_8

properties[4, 10–13]. Compared with the traditional mineral filler/polymer systems, the in situ micro/nanofibrillar composites perform a number of unique advantages, such as, easy processing, reduction in weight, complete recycling, etc. [14]. Additionally, regardless of the difference between the various reinforcing fibrils, two factors are generally considered to determine the final reinforcement effects: (i) the adhesion quality between the matrix and fibrils, and (ii) the aspect ratio of the fibrils [15–17]. With these respects, in situ nanofibrillar composites present eminent superiority in contrast with the microfibrillar counterpart. In this chapter, the recent results of attempts to prepare in situ nanofibrillar composites starting from polymer blends, including low density polyethylene/isotactic polypropylene and the fully biodegradable poly(lactic acid)/poly(butylene succinate) blends, are presented.

Nanofibrillation of Polymer Blends via "Melt Extrusion-Hot Stretching-Quenching"

Detailed process of "in situ nanofibrillation" can be well illustrated by Fig. 8.1 [18, 19]. First, the polymer blend is melt blended in an extruder. Particularly, two extrusion parameters are crucial to achieve nano-sized dispersed particles: a high feeding rate to minimize the staying in the extruder, and a high screw speed to strengthen the mechanical stress and shearing action imposed on the blends and at the same time to reduce the residence time. Second, after exiting from the die, the extrudate is hot stretched by a take-up device with two pinching rolls to fulfill the in situ fibrillation of dispersed component. Moreover, by controlling the extrusion and stretching speed, different hot stretching ratio, which is defined as the area of the transverse section of the die to the area of the transverse section of the extrudate, can be obtained. After hot stretching, the extrudate is immediately quenched in the cooling unit, which is full of cold water, to freeze the formed nanofibrils in the composites and finally, the ribbons of the in situ nanofibrillar composites with adjustable thickness, are collected by the pelletizer or wind-up apparatus.

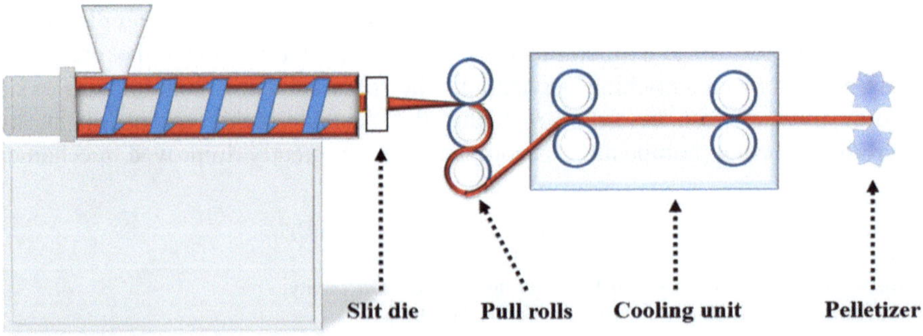

Fig. 8.1 Schematic diagram depicting the "melt extrusion-hot stretching-quenching" process

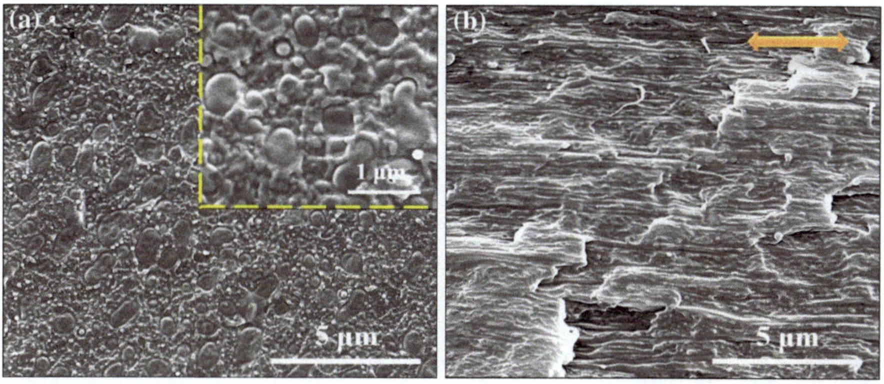

Fig. 8.2 Scanning electron micrograph (SEM) images of LDPE/iPP **a** common blend, higher magnification image is inserted; **b** in situ nanofibrillar composite; (the flow direction is indicated by the *double arrows*)

Via the above methodology of "melt extrusion-hot stretching-quenching," well-defined nanofibrils are successfully achieved in many polymer blends, such as classic polyethylene/polypropylene, fully degradable polymer systems, etc. One of the typical examples is the low density polyethylene (LDPE)/isotactic polypropylene (iPP) system. As shown in Fig. 8.2a, a classic "sea-island" structure and poor interactions, in which submicron droplets of iPP component are uniformly dispersed in the LDPE matrix without obvious interfacial bonding, are evidently presented for the LDPE/iPP (weight ratio 80/20) blend. It essentially arises from the thermodynamic incompatibility between LDPE and iPP, which will heavily deteriorate the mechanical properties of the blend. After stretching, it is visible to find that iPP particles deform into nanofibrils along the flow direction (Fig. 8.2b).

Moreover, as depicted in Fig. 8.3, synchrotron small-angle X-ray scattering (SAXS), wide angle X-ray diffraction (WAXD), and polarized Fourier transform infrared spectroscopy (FTIR) are conducted to learn more information about iPP nanofibrils' lamellar and molecular orientation. Obviously, the 2D-SAXS image inserted in Fig. 8.3a displays a pair of symmetrical bulb-shaped lobes and needle-like streaks around the meridian and equator direction, respectively. Moreover, the corresponding azimuthal scan denotes that most lamellae (kebabs) are arranged perpendicular to the flow direction, while signals of shish are parallel with the flow direction (Fig. 8.3a). Molecular orientation and crystalline morphology of LDPE/iPP in situ nanofibrillar composite are followed utilizing 2D-WAXD as demonstrated in Fig. 8.3b. Arc-like diffraction patterns of iPP (110)/(040)/(130) planes suggest highly oriented α-form iPP along the flow direction, which is well coincident with SEM and SAXS results. Figure 8.3c represents the polarized FTIR spectra of LDPE/iPP in situ nanofibrillar composites with the electron vector perpendicular and parallel to the stretching direction in the wavenumber range from 800 to 1200 cm^{-1}. Among them, 841, 998, 1167 cm^{-1} represent the 3/1 helical conformation bands of iPP, while 973 cm^{-1} represents the

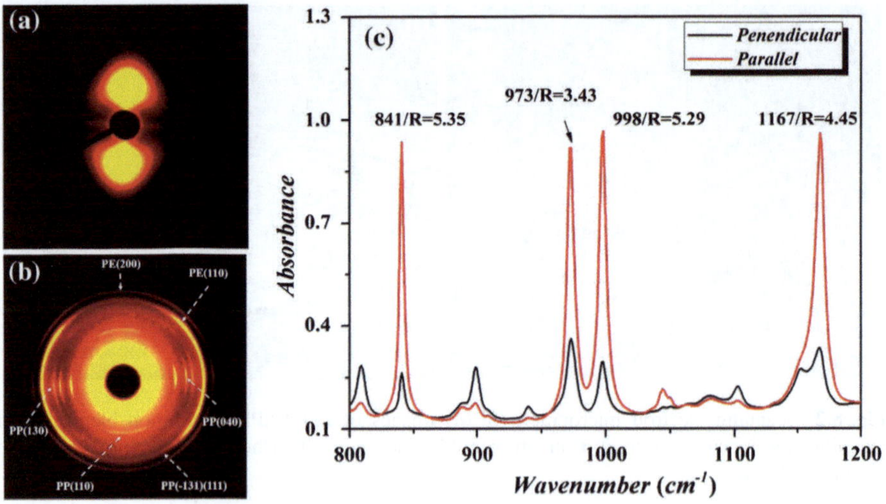

Fig. 8.3 **a** 2D-SAXS pattern; **b** 2D-WAXD pattern and **c** Polarized FTIR spectra of LDPE/iPP in situ nanofibrillar composite in the 800–1200 cm^{-1} region with the electron vector perpendicular and parallel to the stretching direction

isotactic band, respectively. Both the 3/1 helical conformation bands and isotactic band are related to the critical lengths of regular isotactic sequence [20–22]. Moreover, it should be noted that the transition moments of the above 4 characteristic absorption bands are parallel to the axis of iPP chains. In Fig. 8.3c, obvious different absorbance intensities are observed in the parallel- and perpendicular-polarized FTIR spectra, implying high molecular orientation of iPP nanofibrils. Characterized by dichroic ratio (*R*), which is calculated as Eq. (8.1):

$$R = \frac{I_{//}}{I_{\perp}} \quad (8.1)$$

where $I_{//}$ and $I_{\perp}$ are the infrared-absorbance intensity which is parallel and perpendicular to the stretching direction, respectively, *R* values of the 4 characteristic absorption bands are 5.35 for 841 cm^{-1}, 3.43 for 973 cm^{-1}, 5.29 for 998 cm^{-1}, and 4.45 for 1167 cm^{-1}, respectively. This is, no doubt, indicative of highly oriented iPP molecules along the stretching direction in iPP nanofibrils.

With a green concept in recent years, synthetic biopolymers based on renewable resources hold the basis for the twenty-first-century portfolio of eco-efficient, biodegradable, and sustainable plastics. As the front runner in the emerging bioplastics market, poly (lactic acid) (PLA) has long been attractive as a versatile model thanks to the renowned merits of high strength and modulus, coupled with excellent biodegradability, biocompatibility, and renewability. This property combination predestines PLA to be the appealing candidate for tissue engineering, biomedical, and packaging applications [23–25]. However, PLA finds limited usages in practical application primarily due to the low heat resistance, poor gas barrier, and intrinsic brittleness [26–29]. Generally, to obtain ductile PLA, adding flexible biopolymers and plasticizers is a natural and frequently used technique, and the toughness of PLA is improved greatly, but unfortunately on the huge sacrifice of strength and modulus [30, 31]. Herein, via in situ nanofibrillation, ductile poly (butylene succinate) (PBS) nanofibrils are firstly introduced into brittle PLA matrix. Due to the formation of highly oriented PBS nanofibrils and strong interfacial crystallization, strong and stiff PLA, simultaneously coupled with ductility, is highly expected.

The nanofibrillation of PBS component after the application of in situ fibrillation technique is evidenced in Fig. 8.4. Fully extended, compact nanofibrils embedded in the matrix are clearly observed for all in situ fibrillar composites, showing an extremely high aspect ratio (over 50) and orientation degree along the stretching direction. After selective etching of the PLA component, the exposed nanofibrils

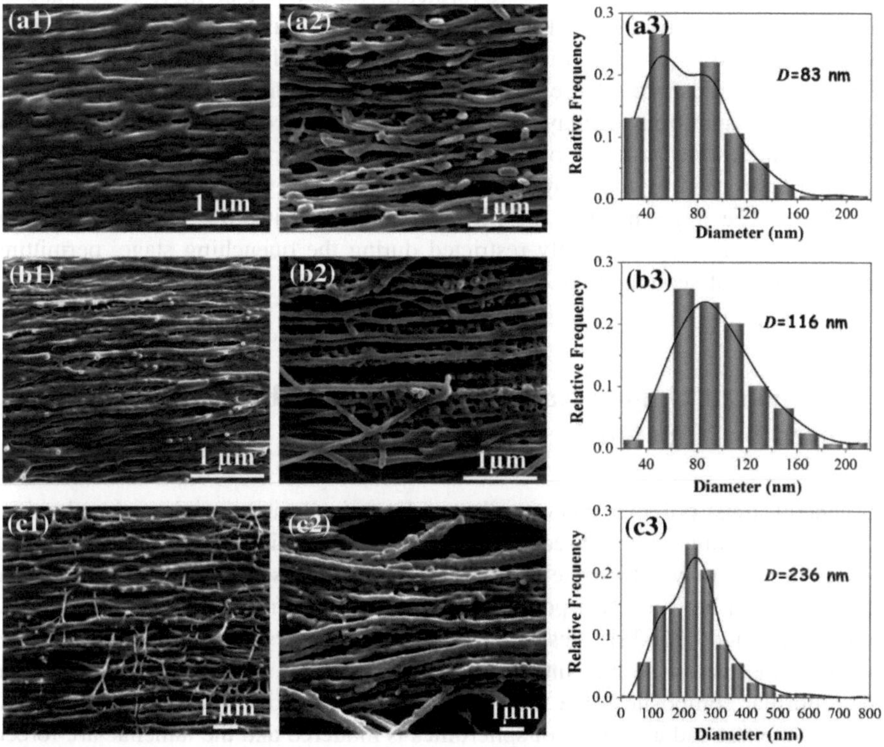

Fig. 8.4 Morphological observation of in situ fibrils in the nanofibrillar composites. SEM images of cryofracture surfaces of **a1** PLA/PBS (90/10), **b1** PLA/PBS (80/20), and **c1** PLA/PBS (60/40) composite samples. **a2**, **b2**, and **c2** present the fibrillar morphology of PBS after etching the PLA matrix of **a1**, **b1**, and **c1**, which produce the quantitative analyses in terms of the distribution of PBS nanofibrils as shown in **a3**, **b3**, and **c3**, respectively. The average diameter (D) is marked in the *right up corner* of **a3**, **b3**, and **c3**

are clearly observed in Fig. 8.4a2–c2, indicating the fact that the nanofibrils originate from the PBS component. Of particular interest are the "kebab-like" filaments strung by the nanofibrils, especially for PLA/PBS (80/20) as shown in Fig. 8.4b2, which may stem from the crystalline entities of PLA matrix that show stronger ability to resist the etching agent. The formation of interfacial ligaments offers great opportunity in enhancing interactions between the PBS nanofibrils and PLA matrix with respect to the interfacial enhancement mechanism that arises from the interfacial crystallization [32–34]. Figure 8.4a3–c3 shows the strong dependence of nanofibril diameter on the PBS content despite the diameter distribution is slightly varied. The average diameter rapidly grows with increasing PBS addition, rising from the minimum value of 83 nm for PLA/PBS (90/10) to 116 and 236 nm for the nanofibrillar composites penetrated 20 and 40 wt% PBS nanofibrils, respectively. It may be a result of the enhanced collision and agglomeration possibility in the existence of higher concentrations of PBS nanofibrils.

SEM observations of the section surfaces of nanofibrillar composites after the selective etching of PLA matrix were further performed to gain more understanding on the fibrillar morphology (Fig. 8.5). Without the support of interlinked PLA backbone, plenty of nanofibrils tend to lean on the surfaces. Specifically, Fig. 8.5c1, c2 illustrate a large quantity of nanofibril bundles consisting of tens to hundreds of nanofibrils, displaying the unique appearance of seaweed rooted in the sea. The interesting structure probably stems from the integration of separate PBS nanofibrils aroused by the extremely high surface energy at the absence of matrix backbone. The agglomeration of PBS nanofibrils during the in situ fibrillation process, nevertheless, is largely restricted during the quenching stage, permitting the formation of separated nanofibrils.

Formation of Unique Nanohybrid Shish Kebabs Induced by Nanofibrils

In general, most polymer pairs are thermodynamically immiscible and technologically incompatible, and consequently producing a multicomponent system during processing that may deteriorate the mechanical properties of the blends. To enhance the interaction between two polymer components, adding compatibilizer is a traditional method [35–38]. Besides that, interfacial crystallization offers another possible means to strong interface [32, 33, 39]. As previously reported in microfibrillar composites, due to the high density of active nuclei on microfibrils' surface, the free radial growth of spherulites is hindered and the lamellae are forced to grow in one direction, leading to the typical transcrystallinity [11, 13, 14]. Then, whether similar transcrystallinity will also form in nanofibrillar composites or not? If not, what is the case for the blend reinforced with nanofibrils?

To reveal the origin of the interfacial filaments, the crystalline morphology is tracked by etching the amorphous LDPE component as manifested in Fig. 8.6.

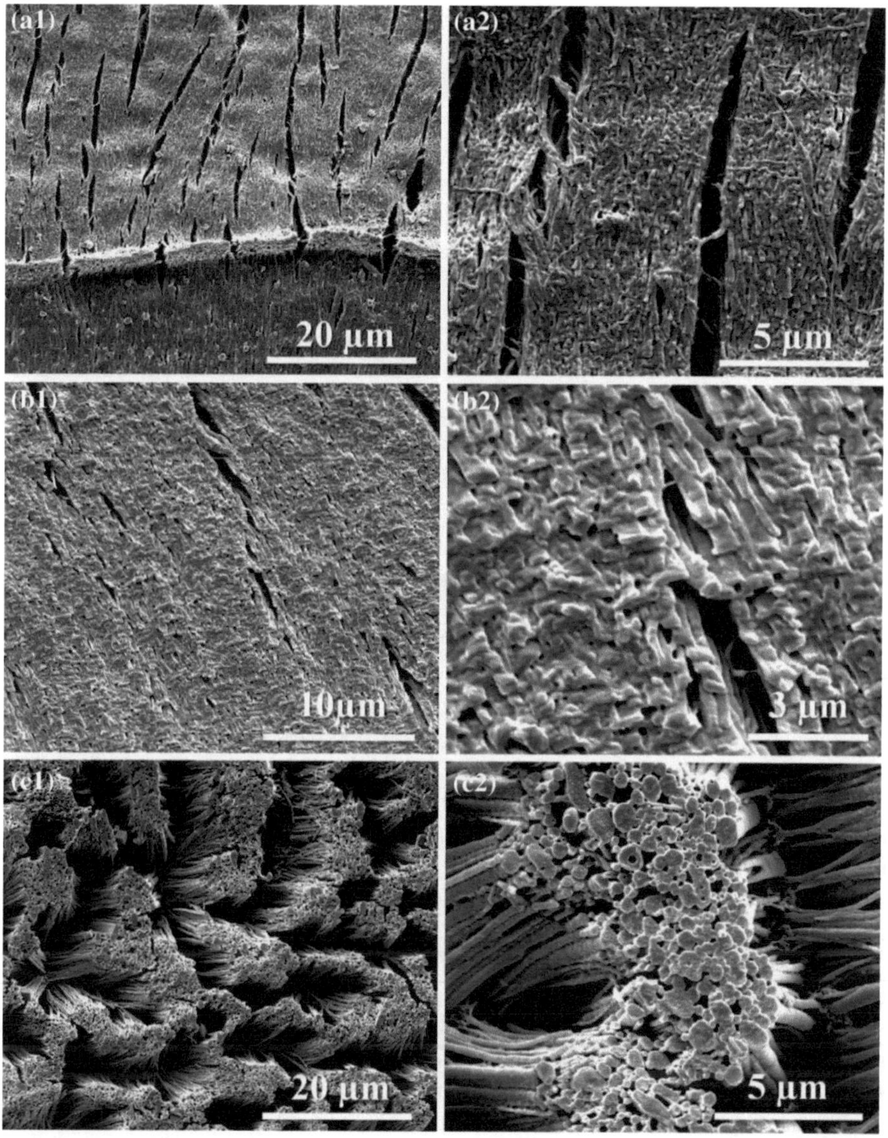

Fig. 8.5 SEM micrographs of section cryofracture surfaces of in situ nanofibrillar composites after etching the PLA matrix for **a1** PLA/PBS (90/10), **b1** PLA/PBS (80/20), and **c1** PLA/PBS (60/40) composite samples, and **a2**, **b2**, and **c2** present their nanofibrillar morphology in higher magnitude, respectively

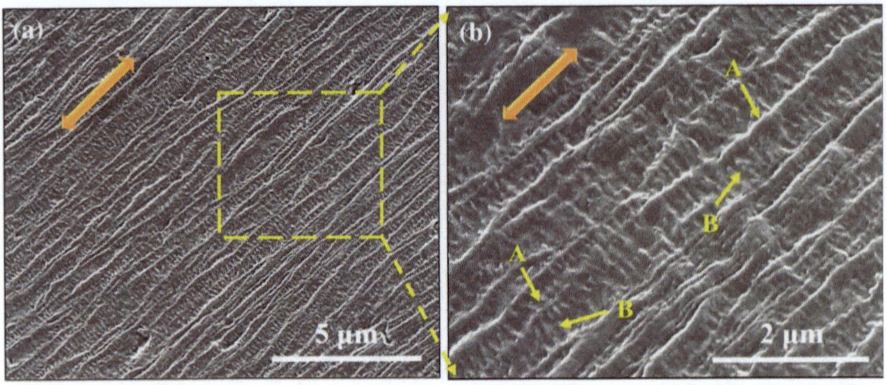

Fig. 8.6 SEM images of acid etched LDPE/iPP in situ nanofibrillar composite with different magnifications. **b** Is the magnification of *yellow rectangle* region of (**a**), showing the detailed structure of NHSK, where *A* and *B* inside image **b** represent iPP nanofibrils and LDPE kebabs induced by iPP nanofibrils, respectively (The flow direction is indicated by the *double arrows*)

What is immediately noticeable from Fig. 8.1c is that iPP fibrils, with an average diameter of 85 nm, are well aligned along the flow direction, and meanwhile they are wrapped by regularly aligned lamellae perpendicular to the nanofibrils axis, serving as the central shish to elicit the ordered arrangement of brush-like LDPE crystal lamellae, i.e., kebab structure. Based on the structural features, the unique superstructure can be termed as nanohybrid shish kebabs (NHSK) consisting of the central polymer nanohybrid shish and ordered lamellar arrangement, which is similar to the classical NHSK structure induced by CNTs [40–43]. The generation of NHSK structure is of crucial significance, for it can desirably bridge the LDPE matrix and iPP nanofibrils, and cooperate with the highly oriented iPP nanofibrils to make significant contributions to the mechanical properties of LDPE, desirably allowing its larger scale application.

Furthermore, the crystallization of LDPE on iPP nanofibrils substrate under quiescent conditions is conducted utilizing time-resolved SAXS, in which LDPE is selectively melted by heating up to 150 °C, reserving only iPP nanofibrils to induce the recrystallization of LDPE under quiescent conditions. Figure 8.7 represents some typical 2D-SAXS patterns, revealing the evolution of crystalline structure during heating and cooling. On heating, it is obvious to observe the gradually weakening scattering signals in the meridian due to the melting of LDPE, and when heating up to 150 °C, LDPE is completely molten, the remaining 2D-SAXS pattern displays a characteristic shish kebab pattern, denoting that iPP nanofibrils are, in fact, composed of shish kebabs structure of themselves, just as illustrated by the right schematic in Fig. 8.7. For the recrystallization process under quiescent condition, there occur four symmetrical bulb-shaped lobes, displaying a beautiful flower-like pattern. Undoubtedly, the new emerging signals spring from the recrystallization of LDPE induced by iPP nanofibrils. Azimuthal scan result in Fig. 8.8a demonstrates that the regenerated LDPE lamellae under quiescent

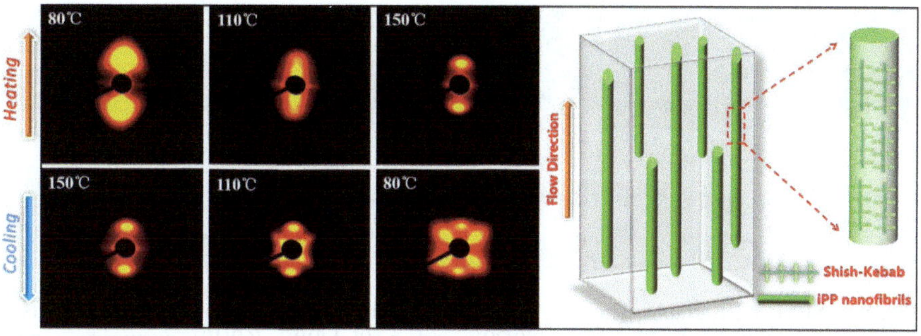

Fig. 8.7 Selected 2D-SAXS patterns of LDPE/iPP in situ nanofibrillar composite during heating from room temperature to 150 °C and cooling from 150 °C to room temperature. The right model illustrates the crystalline structure of iPP nanofibrils (The flow direction is *vertical*)

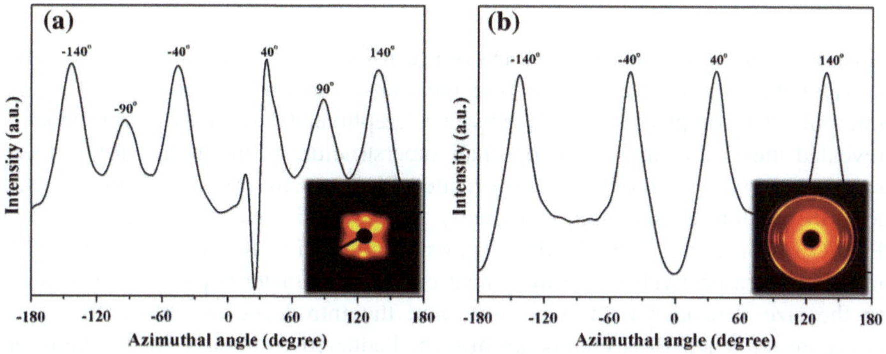

Fig. 8.8 **a** Azimuthal scan of 2D-SAXS; **b** Azimuthal scan of LDPE (110) plane in 2D-WAXD of LDPE/iPP in situ nanofibrillar composite after recrystallization at 150 °C under quiescent condition. 2D-SAXS and 2D-WAXD patterns are inserted (The flow direction is *vertical*)

conditions are ±40° with respect to the equator (axis of iPP nanofibrils). For the molecular orientation of LDPE, the WAXD result depicted in Fig. 8.8b signifies that LDPE chains are oriented ±40° apart from the equator, viz., ±50° off the axis of iPP nanofibrils, showing a characteristic heteroepitaxy for PE/iPP system, mainly based on the chain-row matching by parallel alignment of PE chains in the (100) lattice plane along the methyl group rows in the [101] direction of iPP with a mismatch of 2 % [33, 44–48]. In summary, under quiescent condition, LDPE preferentially overgrows on iPP nanofibrils substrate in the classic epitaxy way, implying that the flow field does have significant effect on the overgrowth of LDPE on iPP nanofibrils substrate, finally resulting in the formation of exclusive NHSK.

Similarly, to reveal the origin of the interfacial filaments between PLA and PBS nanofibrils, the matrix crystalline morphology is tracked by etching the amorphous PLA component as manifested in Fig. 8.9, where PBS nanofibrils are wrapped by regularly aligned lamellae perpendicular to the nanofibril axis, serving as the central

Fig. 8.9 SEM micrographs showing crystalline filaments of PLA at PBS nanofibrils for **a** PLA/PBS (90/10), **b** PLA/PBS (80/20), and **c** PLA/PBS (60/40) composite samples. Cryofracture surfaces of in situ nanofibrillar composites after etching the amorphous PLA component are employed for SEM observation. The existence of 40 wt% PBS makes it difficult to clearly expose the lamellar structure in **c** during the same etching processing. The *big arrows* indicate the stretching direction, while the *small brown* and *yellow arrows* refer to the nanofibrils serving as hybrid shish and oriented PLA lamellae, respectively

shish to elicit the ordered arrangement of brush-like PLA crystal lamellae (so-called kebabs). In an earlier case involving the transcrystalline morphology of in situ microfibrillar composites of poly(ethylene terephthalate) (PET)/iPP, we evidently revealed the similar hybrid shish kebab superstructure in the atomic force microscopic observation. Herein, strictly oriented PLA kebabs are periodically attached to PBS nanofibrils, showing a gradually increased diameter with the addition of PBS (1.6, 2.1, and 3.2 μm for the composites containing 10, 20, and 40 wt% PBS nanofibrils, respectively). The increment of kebab diameter is probably a response to the size dependence of nanofibrils and the introduced flexible PBS chains: (1) more stretched PLA chains are absorbed/adhered to involve in the nanofibril entities due to the increased nanofibril size, holding the chance to trigger the subsequent lamellar growth of adjacent PLA chains; (2) the incorporation of flexible PBS chains allows the enhancement of chain mobility PLA and thus the crystallization kinetics [49]. The NHSK can serve as a favorable self-reinforced element and desirably bridge the nanofibrils and PLA matrix, both of which are predestined to make significant contributions to the mechanical and barrier performances in favor of packaging applications [28, 50].

As a complement for the SEM observation, two-dimensional small-angle X-ray scattering (2D-SAXS) was performed to offer quantitative insights into the NHSK induced by the PBS nanofibrils, as demonstrated in Fig. 8.10. For PLA/PBS (90/10) nanofibrillar composite, Fig. 8.10a1 apparently displays a pair of asymmetrical triangular streaks in the equatorial direction and a pair of symmetrical bulb-shaped lobes in the meridional direction, showing high similarity with those of PLA shish kebabs and PLA hybrid shish kebabs induced by nanowhiskers or natural nanofibers [27, 28]. It therefore evidences the presence of stretch-aligned shish and oriented lamellae decorated at the shish. Coupled with the crystalline morphology observation (Fig. 8.9a), it can be fairly concluded that the PBS nanofibrils serving as the hybrid shish string the PLA kebabs. Figure 8.10a2 illustrates the extracted

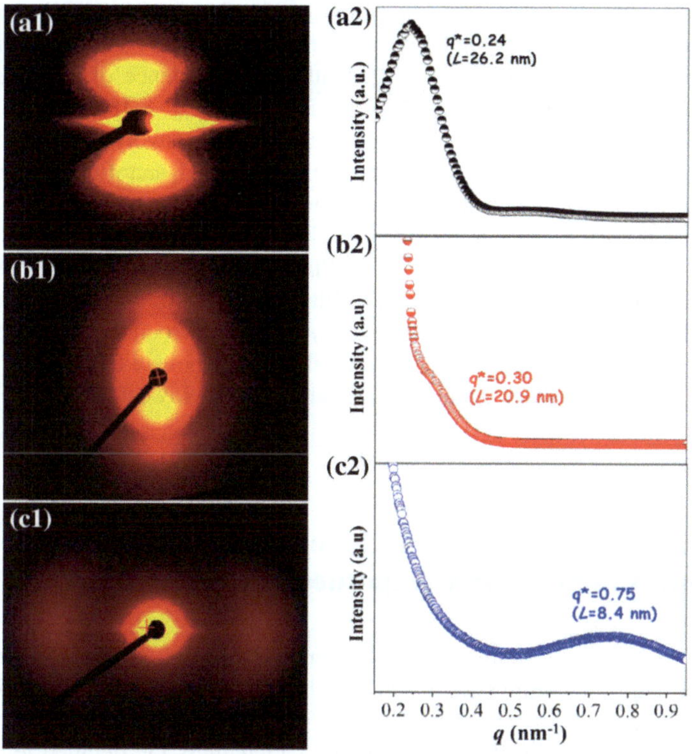

Fig. 8.10 **a1–c1** Representative 2D-SAXS patterns and (**a2–c2**) 1D-SAXS intensity profiles of in situ nanofibrillar composites. The peak position (q^*) and corresponding long spacing (L) are marked on the intensity curves. **a** PLA/PBS (90/10), **b** PLA/PBS (80/20), and **c** PLA/PBS (60/40)

1D-SAXS intensity curve for PLA/PBS (90/10), in which the maxima (q^*) at $q = 0.24\ \text{nm}^{-1}$ is observed. The long period (L) regarding the lamellar structure is calculated to be 26.2 nm, using the Bragg equation, $L = 2\pi/q^*$. This L value is moderately higher than that of pure PLA shish kebabs (22.8 nm) [27], an indicative of the larger lamellar spacing between adjacent regular lamellae. With regard to the nucleation ability, PBS nanofibrils cannot match the extended nanostructured PLA shish in providing favorable geometrical lattice matching and available nucleating sites to absorb the folded-chain lamellae. It should give rise to the increment of L. The similar observations were also observed in shear-induced hybrid shish kebabs of PLA [34] and iPP [51].

With the increase of PBS content, the two-point SAXS signal representing the oriented PBS lamellae is observed as suggested in Fig. 8.10b, c (note that the PBS component cannot be etched during the remove of amorphous PLA). Figure 8.10b1 shows a pair of symmetrical bulb-shaped lobes in the meridional direction with strong scattering intensity, accompanied by the weak scattering presenting a pair of arc-like lobes probably assigned to the formation of well-aligned PBS lamellae parallel to the PLA kebabs. The incorporation of ordered PBS lamellae pushes the maxima q^* higher to 0.30 nm^{-1}, because PBS crystals arouse an intrinsic peak position at around $q = 0.75\ nm^{-1}$ [52]. With the coexistence of PLA kebabs and PBS lamellae, the long spacing falls to 20.9 nm (Fig. 8.10b2). Figure 8.10c1 illustrates the weak arc-like scattering patterns exclusively for oriented PBS lamellae along the stretching direction (in the extended nanofibrils or epitaxially grown lamellae), and the central homogeneous scattering patterns are probably attributed to the PLA crystals confined by the PBS nanofibrils [34]. In contrast to the poor crystallization kinetics of PLA, PBS is characterized by high molecular mobility and favorable crystallization ability, resulting in the scattering patterns dominated by the PBS lamellae [31]. The q^* mainly associated with the nanofibrillar structure and crystal structure of PBS shifts enormously to $q = 0.75\ nm^{-1}$ for PLA/PBS (60/40) fibrillar composite, producing the lowest L value at 8.4 nm (Fig. 8.10c2).

Excellent and Comprehensive Mechanical Properties of In Situ Nanofibrillar Composites

Since exclusively well-defined NHSK superstructure with high concentration has been successfully accomplished by the methodology of in situ nanofibrillar composite. It is imperative to verify whether NHSK can lead to notable improvement in mechanical properties for the blend. Figure 8.11 gathers the detailed tensile properties regarding the tensile strength, Young modulus, and elongation at break of the samples. Compared to the initial values of 15.6 MPa, 154.8 MPa, and 879.1 % of pure LDPE, iPP is characterized by higher strength (29.2 MPa) and modulus (510.2 MPa), but poorer ductility (11.9 %). With the introduction of 20 wt% iPP, i.e., LDPE/iPP common blend, dramatic mechanical deterioration is presented for the blend, dropping from 15.6 MPa and 879.1 % to 10.9 MPa and 9.3 % for tensile strength and elongation at break, respectively. The above phenomena arise from, in principle, the excessive separation caused by inherent immiscibility between LDPE and iPP. In clear contrast, the creation of iPP nanofibrils and attendant improved interfacial adhesion with epitaxy or NHSK, enormously benefit the mechanical properties of the blend. As depicted in Fig. 8.11, due to the formation of epitaxial interface, greatly enhanced tensile strength (from 10.9 to 27.0 MPa) and elongation at break (from 9.3 to 138.0 %) are presented compared with the common blend, indicating that interfacial epitaxy structure is actually effective at improving the interfacial adhesion between LDPE and iPP, finally leading to greatly enhanced mechanical properties [45, 47]. In addition, compared with epitaxy, it can be concluded that the superstructure of NHSK is more effective at transferring load from LDPE matrix to iPP component because of the higher orientation degree of PE lamellae, which can be clearly verified by Fig. 8.11, where the tensile strength, Young modulus, and elongation at break further increase from 27.0 MPa,

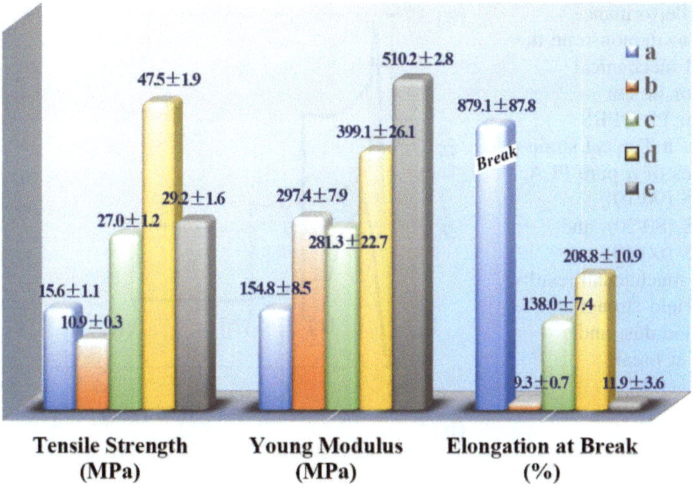

Fig. 8.11 Tensile strength, Young modulus, and elongation at break of **a** pure LDPE; **b** LDPE/iPP common blend; **c** LDPE/iPP in situ nanofibrillar composite with epitaxy structure; **d** LDPE/iPP in situ nanofibrillar composite with NHSK structure; **e** pure iPP

281.3 MPa, and 138.0 % of epitaxy-samples to 47.5 MPa, 399.1 MPa, and 208.8 % of NHSK-samples. In summary, the generation of NHSK structure is of crucial significance, for it can desirably bridge the LDPE matrix and iPP nanofibrils, and cooperate with the highly oriented iPP nanofibrils to make significant contributions to the mechanical properties of LDPE, desirably allowing its larger scale application. Moreover, of particular importance is the common polymer processing techniques involved in the preparation of NHSK in polymer blend, desirably allowing an industrially feasible fabrication in the future.

The introduction of highly flexible biopolymers, such as PBS, poly[(butylene succinate)-co-adipate], and poly(3-hydroxybutyrate-co-hydroxyvalerate), to fabricate ductile PLA blends has been the subject of much recent interest [31, 53, 54]. As yet, it is unfortunate to observe the sacrifice of strength and stiffness. In clear contrast, the creation of PBS nanofibrils enormously benefits the mechanical performances in terms of strength, modulus, and ductility (Fig. 8.12). Figure 8.12a describes representative stress–strain curves for pure PLA and PLA/PBS nanofibrillar composites. We see from Fig. 8.12a that the stress–strain curves of PLA containing PBS nanofibrils evidently tower over that of pure PLA. Figure 8.12b gathers the detailed tensile properties regarding yield strength, Young's modulus, and elongation at break. Compared to the initial values of 43.1 and 1130 MPa of pure PLA, the composites loaded 10 and 20 wt% PBS nanofibrils obtains an unexpected promotion of yield strength and Young's modulus (58.7 and 2533 MPa, 78.7 and 3171 MPa, respectively), although the ductility drops to some extent. Moreover, all the nanofibrillar composites present higher yield strength, ultimate strength, and Young's modulus compared to those of pure PBS (17.7 MPa,

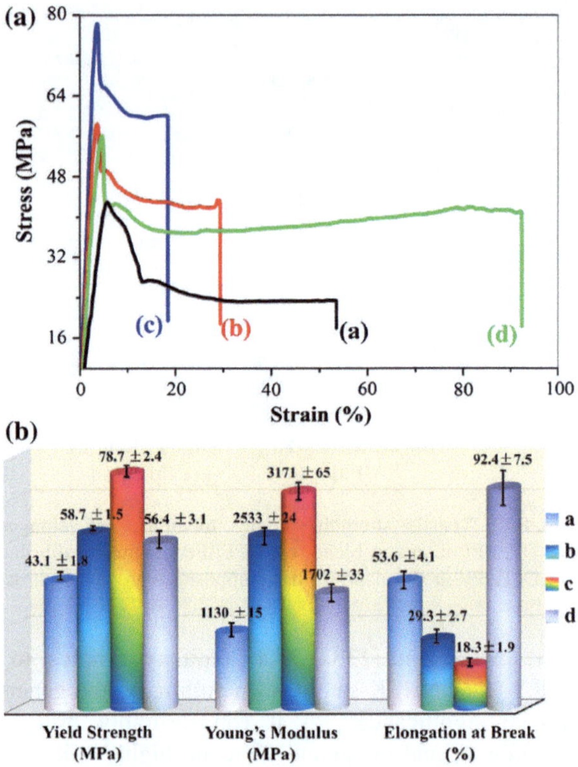

Fig. 8.12 Performance evaluation to demonstrate the exceptional mechanical properties of in situ nanofibrillar PLA/PBS composites. **a** Typical strain–stress curves of *a* pure PLA, *b* PLA/PBS (90/10), *c* PLA/PBS (80/20), and *d* PLA/PBS (60/40). **b** Detailed mechanical results regarding yield strength, Young's modulus, and elongation at break

33.8 and 340.5 MPa). The unusual combination of strength and stiffness, in principal, lies in the excellent performance of stiff nanofibrils with sufficiently extended chains and substantially strengthened interfacial interactions due to the nanofibril-induced NHSK. Desirable promotion of ductility is observed for the composite constituting 40 wt% PBS nanofibrils, achieving a nearly twofold elongation at break (92.4 %) compared to that of pure PLA (53.6 %). Notably, its strength and stiffness (56.4 and 1702 MPa) are favorably far above those of pure PLA although 40 wt% PBS is added. It is clearly seen that, the creation of stiff nanofibrils desirably permits the effective transfer of applied stress and external deformation through the greatly enhanced interfacial interactions.

Figure 8.13 depicts the major structural features for the in situ nanofibrillar composites, leading to the functional definition of two unique superstructures: plenty of in situ PBS nanofibrils containing stretched chains and partially folded lamellae, and highly oriented PLA kebabs strung by nanofibrils that act as the hybrid shish. With respect to the function role, the well-aligned PLA kebabs attached onto the nanofibrils render a significant self-reinforcement of the PLA matrix, and create strong interfacial ligaments for the immiscible blends. The preferred generation of strong interfacial bonding is of crucial importance, bridging the stress transfer from the PLA matrix to the PBS nanofibrils. The PBS nanofibrils

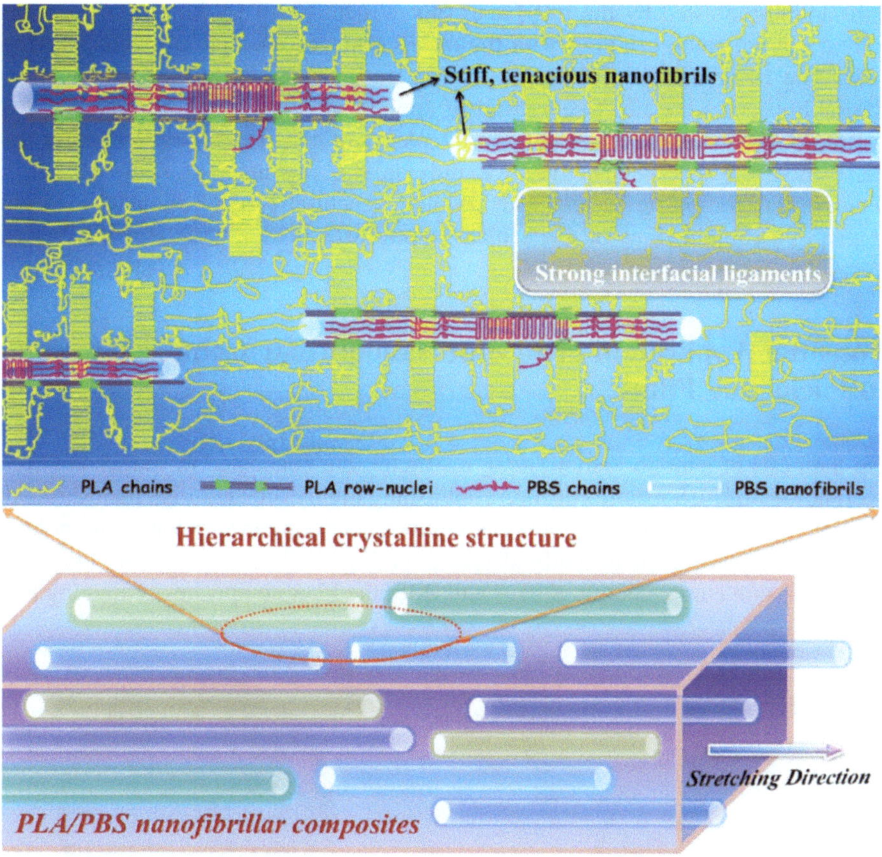

Fig. 8.13 Schematic diagram demonstrating the structural features in the PLA/PBS (60/40) nanofibrillar composites that are responsible for the exceptionally high strength and ductility. Extended chains and ordered lamellae of PBS existing in the nanofibrils render high stiffness and strength for the flexible component, while highly oriented PLA kebabs strung by the PBS nanofibrils serve as the self-reinforcing elements and strong ligaments. The well-aligned PBS nanofibrils consist of flexible amorphous chains and strong oriented lamellae. The PBS nanofibrils, which present large surface energy, may provide pinning points to the surrounding PLA row-nuclei. Such anchoring interactions can contribute to the retention of the molecular stretch and orientation after flow, conferring the formation of PLA row-nuclei attached onto the nanofibrils

possess the unique combination of strength and tenacity: (1) high strength and stiffness attributed to the sufficiently stretched chains and orderly folded lamellae; (2) they are also resilient due to the intrinsic flexibility of amorphous PBS chains. When encountered, the external stress deformation, the tight nanofibrils, and PLA kebabs show strong retardation of crack propagation, and the applied stress is prone to traverse along the length of them quite easily rather than the conventional stress concentration, leading to the unusual observation of increased strength and modulus with the existence of 40 wt% PBS component (Fig. 8.12b). This mechanism can be substantially assisted by the interfacial superstructures, like a kind of sheath, that may construct crack bridging with the dissipation and absorption of much energy [28]. The combination of multiple reinforcement and strong interfacial bonding desirably permits the effective transfer of applied stress and impact load from PLA matrix to PBS nanofibrils, resulting in the plenty of plastic deformation and energy dissipation. In this perspective, we provide the organisms for the PLA/PBS nanofibrillar composite system with peculiar structural configurations that exhibit impressive combinations of mechanical response.

Potential Application of In Situ Nanofibrillar Composites in Packaging Films

On the basis of nanofibrillation, more researches are conducted to further tailor the morphology of PBS nanofibrils, i.e., establishing highly oriented PBS nanosheets by confined flaking of in situ nanofibrillar composites during the compression molding process, which is schematically depicted in Fig. 8.14. Starting from the direct melt extrusion compounding of PLA and PBS, the PLA composites containing PBS

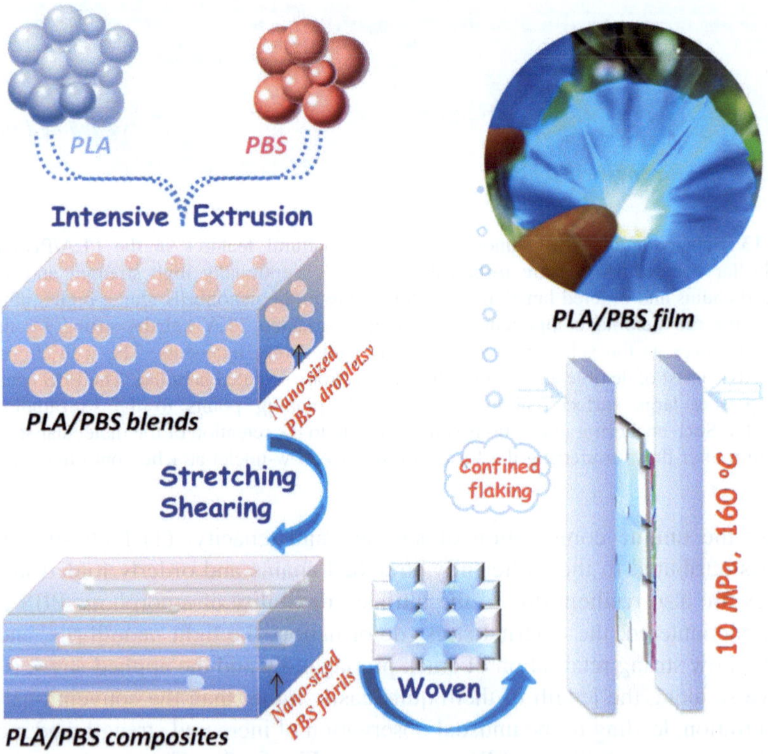

Fig. 8.14 Schematic representation describing the processing approach for structuring the PLA/PBS films

nanofibrils are manufactured employing the "melt extrusion-hot stretching-quenching" technique. Subsequently, the composite sheets are woven into ordered platelets and finally, PLA/PBS composite films with a thickness of $\sim$100 μm are obtained by confined flaking at the specific temperature of 160 °C. It is worth noting that a woven structure is applied to preserve the oriented units by minimizing the random flow during confined flaking and to achieve the isotropic structure and performances, while the molding temperature is fixed at 160 °C to allow the sufficient thermal deformation of PBS nanofibrils but partial melting of PLA matrix. The digital photograph presents the appearance for the PLA/PBS (80/20) film, demonstrating high transparence and desirable toughness. The key element in this methodology is the application of appropriate molding temperature (160 °C), allowing the preservation of oriented units and sufficient thermal deformation of PBS nanofibrils but partial melting of PLA matrix. Moreover, the woven design may help build biaxial structures for the composite films rather than the anisotropic alignment of functional units, which is in great need for packaging, energy, and transport-related applications [19].

Combining the in situ fibrillation and confined flaking techniques, we aim at creating and orderly organizing nanolaminar structure for PBS component, potentially rendering the construction of compact "nano-barrier walls," as well as the generation of effective reinforcing elements [14]. Figure 8.15 offers the direct evidence of this. Figure 8.15a1 evidently suggests the formation of laminar PBS after confined flaking in the composite film penetrated 10 wt% PBS, in clear contrast to the nanofibrillar structure in the in situ nanofibrillar composite. Figure 8.15a1, 2 reveals some important features in the laminar structure of PBS. First, extremely high density and good alignment of PBS nanosheets are clearly observed, indicative of sufficient structural conversion from the rich, oriented PBS nanofibrils. Second, these PBS nanosheets are characterized by a large width of 1.5 ± 0.3 μm but a low thickness of several to tens of nanometers, demonstrating little variations of volume compared to that of PBS nanofibrils with an average diameter of 83 nm. Such a high width/thickness ratio (428 ± 13.1) of PBS nanosheets definitely may be on a par with or even exceed most 2D nanofillers such as nanoclay and few-layered graphene sheets [55, 56]. Third, the PBS nanosheets are interconnected to each other to form an enormous laminar network without visible ends, primarily arising from the branched agglomeration of molten PBS nanofibrils during the confined flaking. The strong linkages created among the PBS nanosheets may present evident advantages over normal nanofillers that are individually dispersed in the matrix in terms of enhancing permeability resistance and mechanical performance [57].

The significantly increased density of PBS nanosheets is clearly observed for the PLA/PBS (80/20) film in Fig. 8.15b1, b2, accompanied by the decreased width (0.9 ± 0.2 μm). The reduction in sheet width can be explained by the decreased viscosity of composite system in the presence of higher contents of PBS, resulting in effective flow and extension of PBS fibrillar melts along the channels in PLA backbone. This restricts the lateral extension of PBS component during the confined flaking, leading to the formation of orderly individual nanosheets rather than

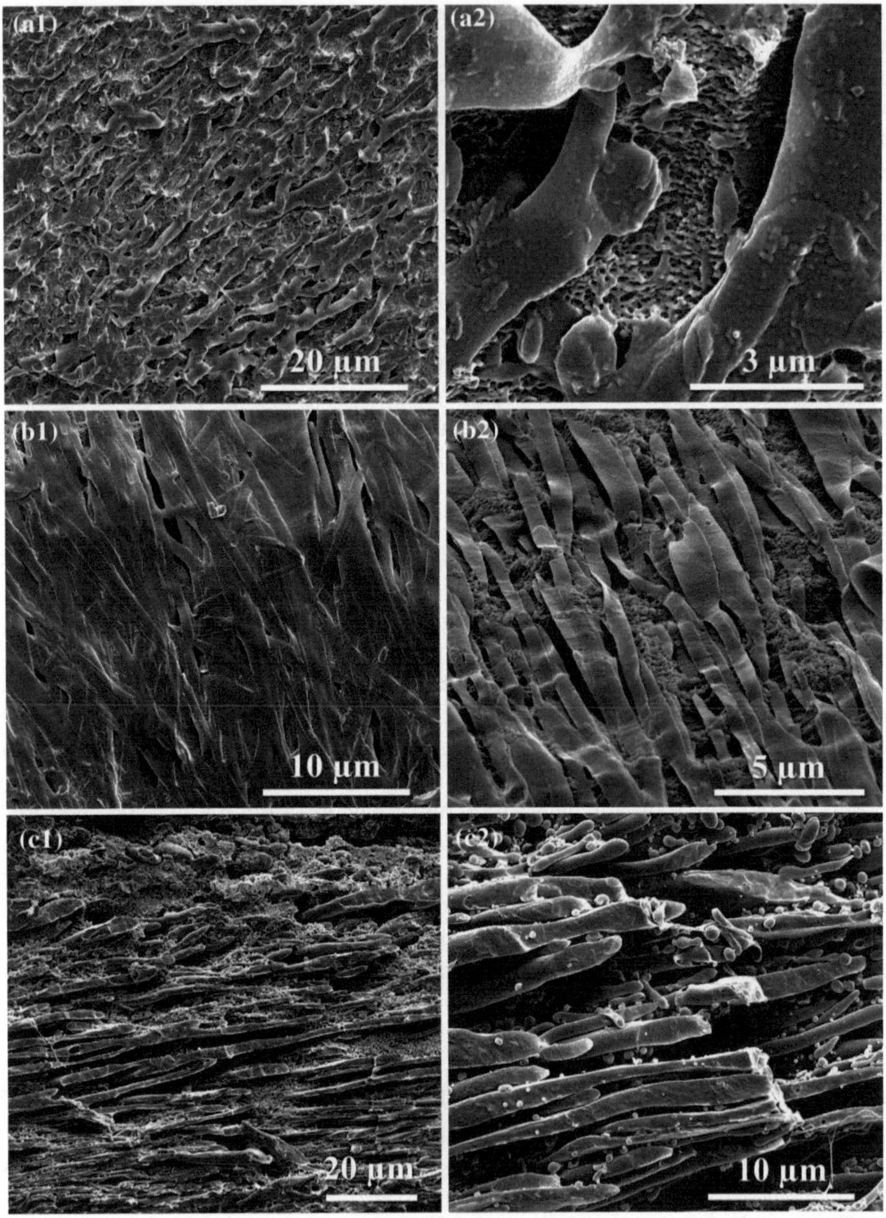

Fig. 8.15 SEM images of section cryofracture surfaces of PLA/PBS composite films after etching the PLA matrix for **a1** PLA/PBS (90/10), **b1** PLA/PBS (80/20) and **c1** PLA/PBS (60/40); **a2–c2** show the local morphology of PBS nanosheets in higher magnitude, respectively

interconnected nanosheets in the PLA/PBS (90/10) film. Figure 8.15c1, c2 suggests PBS nanofibrils in the PLA/PBS (60/40) sample cannot be effectively converted to nanosheets, displaying tabular ribbons with a width of 1.6 ± 0.3 μm are generated. The agglomeration of molten PBS component and decreased channels of PLA framework are, to different degrees, responsible for it.

Figure 8.16a reveals the large changes in oxygen permeability coefficient (P_{O2}) for the composite films. Evidently, the highest permeability (1.4×10^{-14} cm^3 cm cm^{-2} s^{-1} Pa^{-1}) occurs in pure PLA film, and pure PBS film shows a much lower P_{O2} at 0.6×10^{-14} cm^3 cm cm^{-2} s^{-1} Pa^{-1}. It is a surprise to find the substantial decrease in P_{O2} for all composite films (even lower than that of pure PBS). In particular, PLA/PBS (80/20) is characterized by the lowest P_{O2} of 0.2×10^{-14} cm^3 cm cm^{-2} s^{-1} Pa^{-1}, whereas moderately increased P_{O2} values are observed for composite films with 20 and 40 wt% PBS (0.5 and 0.4×10^{-14} cm^3 cm cm^{-2} s^{-1} Pa^{-1}). Apparently, the barrier properties of composite films are closely associated with the morphology of

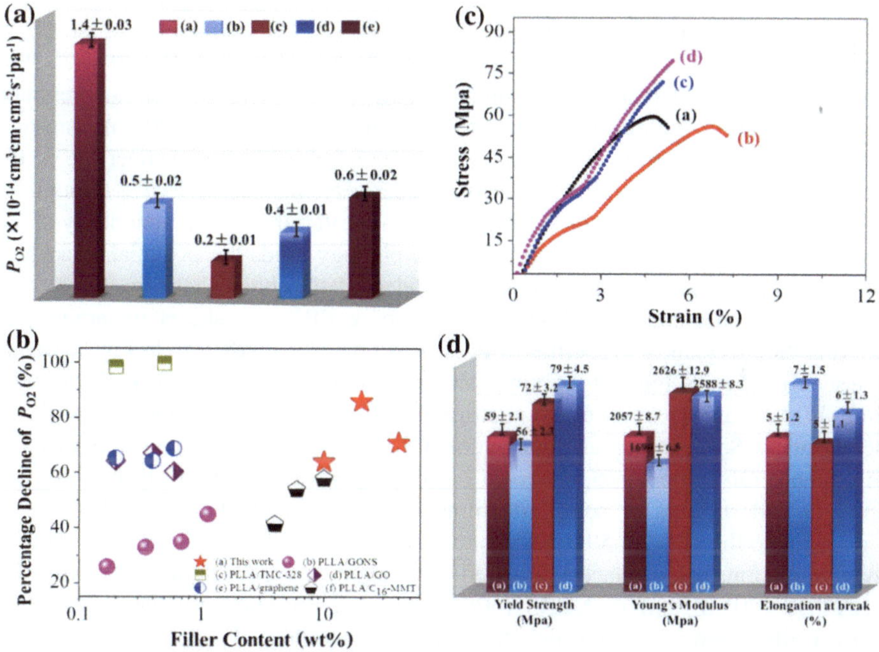

Fig. 8.16 Performance evaluation to demonstrate the impressive permeability resistance to oxygen of the PLA films containing PBS nanosheets, as well as the exceptional mechanical properties. **a** *Bar chart* showing the values of oxygen permeability coefficient (P_{O2}) with standard deviations for *a* pure PLA, *b* PLA/PBS (90/10), *c* PLA/PBS (80/20), *d* PLA/PBS (60/40), and *e* pure PBS. **b** Comparison of percentage decline of P_{O2} between *a* this work and other PLLA composite systems based on *b* GO nanosheets [55], *c* *N,N′,N″*-Tricyclohexyl-1,3,5-benzenetricarboxylamide (trade name TMC-328, a nucleating agent of PLLA) [50], *d* GO [69], *e* graphene [69] and *f* hexadecylamine–montmorillonite (C16-MMT) [70], the percentage decline is defined as the decrease in P_{O2} to the initial value of pure matrix. **c** Typical strain–stress curves and **d** detailed tensile results regarding yield strength, Young's modulus, elongation at break for *a* pure PLA, *b* PLA/PBS (90/10), *c* PLA/PBS (80/20) and *d* PLA/PBS (60/40)

incorporated PBS component. For PLA/PBS (80/20) composite film, an array of compact, well-organized PBS nanosheets render the construction of robust barriers to the diffusing gas molecules, conferring the highest promotion in the permeability resistance [58]. This mechanism can be further enhanced by the increased concentration of crystalline PLA component due to the enhancement of crystallization kinetics with the aid of the existing PBS nanosheets, forcing a tortuous penetration path for the gas molecules by offering compact crystalline lamellae [50, 59, 60]. As such, the PLA/PBS (60/40) shows moderately higher permeability in the absence of sufficiently flaked PBS nanosheets, suggesting that 1D structural units (or fillers) fail to provide effective resistance to gas diffusion and permeation compared to the 2D ones. It is therefore instructive to convert 1D structural units into 2D ones to advanced barrier properties, instead of traditional preparation of nanocomposites based on nanotubes and nano-whiskers [29, 61–64]. Although unique interconnected nanosheets are created for PLA/PBS (90/10), the low concentration of "barrier walls" slightly pushes up the oxygen permeability. It should be emphasized that PLA/PBS blends manufactured by common processing methods are normally characterized by poor barrier performance, instead of translating the relatively high barrier ability of PBS to the PLA products. As an example, Gupta et al. demonstrated the PLA/PBS blends loaded 20 wt% PBS prepared by melt extrusion compounding followed by confined flaking presented almost the same oxygen permeability with pure PLA (around 20 cm^3 mm m^{-2} day^{-1} atm) [65]. It seems that uncontrollable morphology and poor interface in the immiscible blends cause the easy diffusion and permeation of gas molecules through the PLA matrix and interfaces [66, 67]. Figure 8.16b clarifies the distinct superiority of our proposed methodology, i.e., tailoring the morphology holds the potential to achieve superior oxygen-barrier properties in fully biodegradable composites film.

Next, the enormous possibility of developing PBS nanosheets in largely improving the mechanical properties of PLA is examined, as indicated in Fig. 8.16c, d. We see from Fig. 8.16c that the composite films present similar tensile behavior with pure PLA, starting from elastic deformation and fracturing with slight plastic deformation. Some important parameters of the tensile properties are summarized in Fig. 8.16d. Of particular significance is the proportional relationship between the strength and stiffness of composite films and the content of oriented PBS units. Compared to the value of 59.5 MPa for pure PLA, the incorporation of PBS is found to give an unusual enhancement of yield strength up to 72.1 and 79.1 MPa for PLA/PBS (80/20) and PLA/PBS (60/40), respectively. Note that pure PBS film intrinsically shows low tensile strength and modulus of 33.7 and 623 MPa, respectively, well below those of pure PLA. The highly ordered PBS nanosheets can serve as the effective reinforcing elements instead producing the simple plasticization for the common PLA/PBS blends [31, 68]. This is the reason for the slightly decreased strength observed in PLA/PBS (90/10) film containing the lowest content of PBS nanosheets (56.1 MPa). Similar results are found in the Young's modulus of composite films, achieving the profoundly increased modulus for PLA/PBS (80/20) and PLA/PBS (60/40) (2626 and 2588 MPa) in comparison to low modulus of pure PLA (2057 MPa). Moreover, PLA/PBS (90/10) presents the highest elongation at

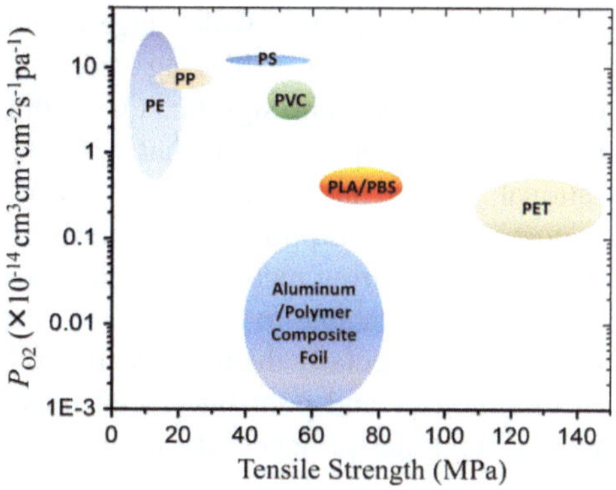

Fig. 8.17 Comparison of P_{O2} and tensile strength for a wide range of packaging film materials, showing that PLA films containing PBS nanosheets present the distinct combination of low P_{O2} and high strength even compared to the common packaging materials existing in the films market

break (7.1 %), which can be appraised from the formation of interconnected nanosheets that may show strong retardation of crack propagation with the dissipation and absorption of much energy when encountering the stress deformation.

In addition to the distinct superiority in offering high resistance to oxygen permeation compared to the traditional PLA-based composing systems, our PLA/PBS composite films featuring the combination of high mechanical properties and low gas permeation can even challenge the common plastic films existing in the market, as illustrated in Fig. 8.17. Although making up a huge share of the films market, PE, PP, polystyrene (PS), and poly(vinyl chloride) (PVC) are characterized by the relatively poor barrier properties compared to our PLA/PBS composite films, by for example showing a P_{O2} of 5–10 × 10^{-14} cm^3 cm cm^{-2} s^{-1} Pa^{-1} for PP films [70]. More importantly, these traditional films normally fall into the category of poor mechanical properties, as demonstrated by the tensile strength of 50–60 MPa for PVC films far below that of our PLA/PBS films (up to nearly 80 MPa). It is apparent that our PLA/PBS composite films are in the comparable level with PET with respect to the ability to resist gas molecules diffusion, both of which present the extremely low values of P_{O2} (0.1–0.6 × 10^{-14} cm^3 cm cm^{-2} s^{-1} Pa^{-1}) [70]. The property elucidation evidently brings to light the possibility to substitute the fossil fuel films with biobased materials. As growing industrial development drastically reduces the production costs of bioplastics, scaled-up biorefineries can ultimately compete with those traditional plastics coming only from crude oil [71].

Conclusions and Outlook

The exceptional combination of strength, stiffness, and ductility suggests interesting generalizations concerning the role of in situ nanofibrillation in creating evolutionary innovations and adaptive radiation for the fabrication of high-performance polymeric products. Moreover, the strength, stiffness, and ductility can be sufficiently tailored in a wide range by adjusting nanofibrils contents, representing enormous promise in achieving specific properties and thus expanding its application. Of paramount significance is the optimized morphology control method (i.e., in situ nanofibrillation) that will inspire and provide design principles for the rational design and reproducible construction of nanofibrillar composites with nanostructured morphology for multifunctional integration. We anticipate the proposed structuring methodology for the structural design of blend systems should be broadly applicable to the industrial manufacturing of various immiscible polymer blends, especially the biomass-based ones, because uncontrollable optimization of morphology and poor interfacial interactions are commonly encountered problems in high-performance blends preparation.

Acknowledgments The authors gratefully acknowledge the financial support to this work by the National Natural Science Foundation of China (Nos. 21276168, 51273131, 51227801, and 51473101), China Postdoctoral Science Foundation (Grant No. 2014T70868), the Specialized Research Fund for the Doctoral Program of Higher Education (Grant No. 20120181120101) and the Fundamental Research Funds for the Central Universities. We would also like to express our heartfelt thanks to the Shanghai Synchrotron Radiation Facility (SSRF, Shanghai, China), for the kind help on WAXD and SAXS measurements.

References

1. Boyaud M-F, Aıt-Kadi A, Bousmina M, Michel A, Cassagnau P (2001) Organic short fibre/thermoplastic composites: morphology and thermorheological analysis. Polymer 42 (15):6515–6526
2. Li ZM, Lu A, Lu ZY, Shen KZ, Li LB, Yang MB (2005) In-situ microfibrillar PET/iPP Blend via a slit die extrusion, hot stretching and quenching process: influences of PET concentration on morphology crystallization of iPP at a fixed hot stretching ratio. J Macromol Sci Part B Phys 44(2):203–216
3. Robeson L (1984) Applications of polymer blends: emphasis on recent advances. Polym Eng Sci 24(8):587–597
4. Shields R, Bhattacharyya D, Fakirov S (2008) Fibrillar polymer–polymer composites: morphology, properties and applications. J Mater Sci 43(20):6758–6770
5. Miles IS, Zurek A (1988) Preparation, structure, and properties of two-phase co-continuous polymer blends. Polym Eng Sci 28(12):796–805
6. Huneault M, Shi Z, Utracki L (1995) Development of polymer blend morphology during compounding in a twin-screw extruder. Part IV: a new computational model with coalescence. Polym Eng Sci 35(1):115–127
7. Gonzalez-Nunez R, De Kee D, Favis B (1996) The influence of coalescence on the morphology of the minor phase in melt-drawn polyamide-6/HDPE blends. Polymer 37 (21):4689–4693
8. Subramanian P (1985) Permeability barriers by controlled morphology of polymer blends. Polym Eng Sci 25(8):483–487
9. Monticciolo A, Cassagnau P, Michel A (1998) Fibrillar morphology development of PE/PBT blends: rheology and solvent permeability. Polym Eng Sci 38(11):1882–1889
10. Fakirov S, Evstatiev M (1994) Microfibrillar reinforced composites—new materials from polymer blends. Adv Mater 6(5):395–398

11. Dencheva N, Denchev Z, Oliveira MJ, Funari SS (2010) Microstructure studies of in situ composites based on polyethylene/polyamide 12 blends. Macromolecules 43(10):4715–4726
12. Zhong G-J, Li L, Mendes E, Byelov D, Fu Q, Li Z-M (2006) Suppression of skin-core structure in injection-molded polymer parts by in situ incorporation of a microfibrillar network. Macromolecules 39(19):6771–6775
13. Li ZM, Li LB, Shen KZ, Yang W, Huang R, Yang MB (2004) Transcrystalline morphology of an in situ microfibrillar poly (ethylene terephthalate)/poly (propylene) blend fabricated through a slit extrusion hot stretching-quenching process. Macromol Rapid Commun 25(4):553–558
14. Fakirov S (2013) Nano-/microfibrillar polymer–polymer and single polymer composites: The converting instead of adding concept. Compos Sci and Technol 89:211–225
15. Fakirov S, Bhattacharyya D, Shields R (2008) Nanofibril reinforced composites from polymer blends. Colloids Surf, A 313:2–8
16. Fakirov S (2013) Nano-and microfibrillar single-polymer composites: a review. Macromol Mater Eng 298(1):9–32
17. Li J-X, Wu J, Chan C-M (2000) Thermoplastic nanocomposites. Polymer 41(18):6935–6937
18. Xie L, Xu H, Niu B, Ji X, Chen J, Li Z-M, Hsiao BS, Zhong G-J (2014) Unprecedented access to strong and ductile poly (lactic acid) by introducing in situ nanofibrillar poly (butylene succinate) for green packaging. Biomacromolecules 15(11):4054–4064
19. Xie L, Xu H, Chen J-B, Zhang Z-J, Hsiao BS, Zhong G-J, Chen J, Li Z-M (2015) From nanofibrillar to nanolaminar poly (butylene succinate): paving the way to robust barrier and mechanical properties for full-biodegradable poly (lactic acid) films. ACS Appl Mater Interfaces 7(15):8023–8032
20. Zhu X, Yan D, Fang Y (2001) In situ FTIR spectroscopic study of the conformational change of isotactic polypropylene during the crystallization process. J Phys Chem B 105(50):12461–12463
21. Konishi T, Nishida K, Kanaya T, Kaji K (2005) Effect of isotacticity on formation of mesomorphic phase of isotactic polypropylene. Macromolecules 38(21):8749–8754
22. Kissin YV, Rishina L (1976) Regularity bands in the ir spectra of C_3H_6-C_3D_6 copolymers. Eur Polym J 12(10):757–759
23. Gupta B, Revagade N, Hilborn J (2007) Poly (lactic acid) fiber: an overview. Prog Polym Sci 32(4):455–482
24. Raquez J-M, Habibi Y, Murariu M, Dubois P (2013) Polylactide (PLA)-based nanocomposites. Prog Polym Sci 38(10):1504–1542
25. Fukushima K, Kimura Y (2006) Stereocomplexed polylactides (Neo-PLA) as high-performance bio-based polymers: their formation, properties, and application. Polym Int 55(6):626–642
26. Tang H, Chen J-B, Wang Y, Xu J-Z, Hsiao BS, Zhong G-J, Li Z-M (2012) Shear flow and carbon nanotubes synergistically induced nonisothermal crystallization of poly (lactic acid) and its application in injection molding. Biomacromolecules 13(11):3858–3867
27. Xu H, Zhong G-J, Fu Q, Lei J, Jiang W, Hsiao BS, Li Z-M (2012) Formation of shish-kebabs in injection-molded poly (l-lactic acid) by application of an intense flow field. ACS Appl Mater Interfaces 4(12):6774–6784
28. Xu H, Xie L, Chen J-B, Jiang X, Hsiao BS, Zhong G-J, Fu Q, Li Z-M (2014) Strong and tough micro/nanostructured poly (lactic acid) by mimicking the multifunctional hierarchy of shell. Mater Horiz 1(5):546–552
29. Martínez-Sanz M, Lopez-Rubio A, Lagaron JM (2012) Optimization of the dispersion of unmodified bacterial cellulose nanowhiskers into polylactide via melt compounding to significantly enhance barrier and mechanical properties. Biomacromolecules 13(11):3887–3899
30. Jiang L, Wolcott MP, Zhang J (2006) Study of biodegradable polylactide/poly (butylene adipate-co-terephthalate) blends. Biomacromolecules 7(1):199–207
31. Zhang K, Mohanty AK, Misra M (2012) Fully biodegradable and biorenewable ternary blends from polylactide, poly (3-hydroxybutyrate-co-hydroxyvalerate) and poly (butylene succinate) with balanced properties. ACS Appl Mater Interfaces 4(6):3091–3101

32. Ning N, Fu S, Zhang W, Chen F, Wang K, Deng H, Zhang Q, Fu Q (2012) Realizing the enhancement of interfacial interaction in semicrystalline polymer/filler composites via interfacial crystallization. Prog Polym Sci 37(10):1425–1455
33. Li H, Yan S (2011) Surface-induced polymer crystallization and the resultant structures and morphologies. Macromolecules 44(3):417–428
34. Xu H, Xie L, Chen Y-H, Huang H-D, Xu J-Z, Zhong G-J, Hsiao BS, Li Z-M (2013) Strong shear flow-driven simultaneous formation of classic shish-kebab, hybrid shish-kebab, and transcrystallinity in poly (lactic acid)/natural fiber biocomposites. ACS Sustain Chem Eng 1 (12):1619–1629
35. Kim H-S, Lee B-H, Choi S-W, Kim S, Kim H-J (2007) The effect of types of maleic anhydride-grafted polypropylene (MAPP) on the interfacial adhesion properties of bio-flour-filled polypropylene composites. Compos A Appl Sci Manuf 38(6):1473–1482
36. Yi X, Xu L, Wang Y-L, Zhong G-J, Ji X, Li Z-M (2010) Morphology and properties of isotactic polypropylene/poly (ethylene terephthalate) in situ microfibrillar reinforced blends: influence of viscosity ratio. Eur Polymer J 46(4):719–730
37. Friedrich K, Evstatiev M, Fakirov S, Evstatiev O, Ishii M, Harrass M (2005) Microfibrillar reinforced composites from PET/PP blends: processing, morphology and mechanical properties. Compos Sci Technol 65(1):107–116
38. Pracella M, Chionna D, Pawlak A, Galeski A (2005) Reactive mixing of PET and PET/PP blends with glycidyl methacrylate–modified styrene-b-(ethylene-co-olefin) block copolymers. J Appl Polym Sci 98(5):2201–2211
39. Quan H, Li Z-M, Yang M-B, Huang R (2005) On transcrystallinity in semi-crystalline polymer composites. Compos Sci Technol 65(7):999–1021
40. Laird ED, Li CY (2013) Structure and morphology control in crystalline polymer–carbon nanotube nanocomposites. Macromolecules 46(8):2877–2891
41. Li L, Li CY, Ni C (2006) Polymer crystallization-driven, periodic patterning on carbon nanotubes. J Am Chem Soc 128(5):1692–1699
42. Li CY, Li L, Cai W, Kodjie SL, Tenneti KK (2005) Nanohybrid shish-kebabs: periodically functionalized carbon nanotubes. Adv Mater 17(9):1198–1202
43. Yang J, Wang C, Wang K, Zhang Q, Chen F, Du R, Fu Q (2009) Direct formation of nanohybrid shish-kebab in the injection molded bar of polyethylene/multiwalled carbon nanotubes composite. Macromolecules 42(18):7016–7023
44. Wittmann J, Lotz B (1990) Epitaxial crystallization of polymers on organic and polymeric substrates. Prog Polym Sci 15(6):909–948
45. Petermann J, Broza G, Rieck U, Kawaguchi A (1987) Epitaxial interfaces in semi-crystalline polymers and their applications. J Mater Sci 22(4):1477–1481
46. Yan S, Lin J, Yang D, Petermann J (1994) Critical epitaxial layers of different kinds of polyethylene on highly oriented isotactic poly (propylene) substrates. Macromol Chem Phys 195(1):195–201
47. Sherman ES (1984) Reinforcement of polyethylene with polypropylene by a blending and deformation process. J Mater Sci 19(12):4014–4020
48. Kojima M, Satake H (1984) Morphological and structural features of heat-treated drawn polypropylene/high-density polyethylene blends. J Polym Sci: Polym Phys Ed 22(2):285–294
49. Yokohara T, Yamaguchi M (2008) Structure and properties for biomass-based polyester blends of PLA and PBS. Eur Polym J 44(3):677–685
50. Bai H, Huang C, Xiu H, Zhang Q, Deng H, Wang K, Chen F, Fu Q (2014) Significantly improving oxygen barrier properties of polylactide via constructing parallel-aligned shish-kebab-like crystals with well-interlocked boundaries. Biomacromolecules 15(4):1507–1514
51. Larin B, Avila-Orta CA, Somani RH, Hsiao BS, Marom G (2008) Combined effect of shear and fibrous fillers on orientation-induced crystallization in discontinuous aramid fiber/isotactic polypropylene composites. Polymer 49(1):295–302
52. Wang K, Jiao T, Wang Y, Li M, Li Q, Shen C (2013) The microstructures of extrusion cast biodegradable poly (butylene succinate) films investigated by X-ray diffraction. Mater Lett 92:334–337

53. Ojijo V, Sinha Ray S, Sadiku R (2012) Role of specific interfacial area in controlling properties of immiscible blends of biodegradable polylactide and poly [(butylene succinate)-co-adipate]. ACS Appl Mater Interfaces 4(12):6690–6701
54. Ojijo V, Sinha Ray S, Sadiku R (2013) Toughening of biodegradable polylactide/poly (butylene succinate-co-adipate) blends via in situ reactive compatibilization. ACS Appl Mater Interfaces 5(10):4266–4276
55. Huang H-D, Ren P-G, Xu J-Z, Xu L, Zhong G-J, Hsiao BS, Li Z-M (2014) Improved barrier properties of poly (lactic acid) with randomly dispersed graphene oxide nanosheets. J Membr Sci 464:110–118
56. Carosio F, Colonna S, Fina A, Rydzek G, Hemmerle J, Jierry L, Schaaf P, Boulmedais F (2014) Efficient gas and water vapor barrier properties of thin poly (lactic acid) packaging films: Functionalization with moisture resistant nafion and clay multilayers. Chem Mater 26 (19):5459–5466
57. Wu Q, Xu Y, Yao Z, Liu A, Shi G (2010) Supercapacitors based on flexible graphene/polyaniline nanofiber composite films. ACS Nano 4(4):1963–1970
58. Kim H, Miura Y, Macosko CW (2010) Graphene/polyurethane nanocomposites for improved gas barrier and electrical conductivity. Chem Mater 22(11):3441–3450
59. Drieskens M, Peeters R, Mullens J, Franco D, Lemstra PJ, Hristova-Bogaerds DG (2009) Structure versus properties relationship of poly (lactic acid). I. effect of crystallinity on barrier properties. J Polym Sci, Part B: Polym Phys 47(22):2247–2258
60. Evstatiev M, Simeonova S, Friedrich K, Pei X-Q, Formanek P (2013) MFC-structured biodegradable poly (l-lactide)/poly (butylene adipate-co-terephatalate) blends with improved mechanical and barrier properties. J Mater Sci 48(18):6312–6330
61. Thielemans W, Warbey CR, Walsh DA (2009) Permselective nanostructured membranes based on cellulose nanowhiskers. Green Chem 11(4):531–537
62. Sánchez-García MD, Hilliou L, Lagarón JM (2010) Morphology and water barrier properties of nanobiocomposites of κ/ι-hybrid carrageenan and cellulose nanowhiskers. J Agric Food Chem 58(24):12847–12857
63. Kim S, Jinschek JR, Chen H, Sholl DS, Marand E (2007) Scalable fabrication of carbon nanotube/polymer nanocomposite membranes for high flux gas transport. Nano Lett 7 (9):2806–2811
64. Ma H, Burger C, Hsiao BS, Chu B (2012) Highly permeable polymer membranes containing directed channels for water purification. ACS Macro Lett 1(6):723–726
65. Bhatia A, Gupta R, Bhattacharya S, Choi H (2010) Effect of clay on thermal, mechanical and gas barrier properties of biodegradable poly (lactic acid)/poly (butylene succinate)(PLA/PBS) nanocomposites. Int Polym Proc 25(1):5–14
66. Wu D, Yuan L, Laredo E, Zhang M, Zhou W (2012) Interfacial properties, viscoelasticity, and thermal behaviors of poly (butylene succinate)/polylactide blend. Ind Eng Chem Res 51 (5):2290–2298
67. Di Lorenzo ML, Rubino P, Cocca M (2013) Miscibility and properties of poly (L-lactic acid)/ poly (butylene terephthalate) blends. Eur Polym J 49(10):3309–3317
68. Ojijo V, Sinha Ray S, Sadiku R (2012) Effect of nanoclay loading on the thermal and mechanical properties of biodegradable polylactide/poly [(butylene succinate)-co-adipate] blend composites. ACS Appl Mater Interfaces 4(5):2395–2405
69. Pinto AM, Cabral J, Tanaka DAP, Mendes AM, Magalhães FD (2013) Effect of incorporation of graphene oxide and graphene nanoplatelets on mechanical and gas permeability properties of poly (lactic acid) films. Polym Int 62(1):33–40
70. Lange J, Wyser Y (2003) Recent innovations in barrier technologies for plastic packaging—a review. Packag Technol Sci 16(4):149–158
71. Xu H, Wu D, Yang X, Xie L, Hakkarainen M (2015) Thermostable and impermeable "nano-barrier walls" constructed by poly (lactic acid) stereocomplex crystal decorated graphene oxide nanosheets. Macromolecules 48(7):2127–2137

Chapter 9
Nano-size Polymers via Precipitation of Polymer Solutions

Prabhakar Dwivedi, Kariappa M. Karumbaiah and Raj Das

Introduction

Polymer nanoparticles (PNPs) are gaining wide recognition, owing to their unique properties, such as the formation of suspensions, UV blocking properties, quantum confinement, surface plasmon resonance, and dimensionality. They nowadays play pivotal roles in many applications, such as drug-loaded carriers, sensors in biomedical technology, scaffolds in tissue engineering, pollution control in environmental technology, photonics, and data storage devices in electronics engineering. The development of nanoparticles has received significant consideration in the past three decades. Figure 9.1 shows the increasing trends, associated with the number of researches conducted to determine their properties, manufacturing methods, and applications in the past decade (as cited in Science Direct database for 2000–2014 [1]). Increase in demands of PNPs has been a good motivation to develop advanced production techniques to increase the quality and yield of nanoparticles [2]. Nanoparticles are solid, colloidal particles, with size ranging from 5 to 1000 nm. Despite this usually defined range, PNPs are mostly obtained and accepted in the range of 100–500 nm. Size of the particles plays a vital role in the adsorption of the opsonins (plasma proteins), which results in the use of nanoparticles in intravenous drug administration [3]. Systems made of nanoparticles are promising for efficient drug delivery, owing to their ability to release drugs [4]. They are synthesized from materials which are either biodegradable or biocompatible; so the material can be used with tissues and cells without inducing toxicity. The nanoparticles can be classified into nanospheres and nanocapsules. Nanospheres are particles, which are solid and are used as matrix particles. Specific chemical molecules can be either encapsulated in the particle or absorbed on the sphere surface.

P. Dwivedi · K.M. Karumbaiah · R. Das (✉)
Department of Mechanical Engineering,
University of Auckland, Auckland 1010, New Zealand
e-mail: r.das@auckland.ac.nz

S. Fakirov (ed.), *Nano-size Polymers*,
DOI 10.1007/978-3-319-39715-3_9

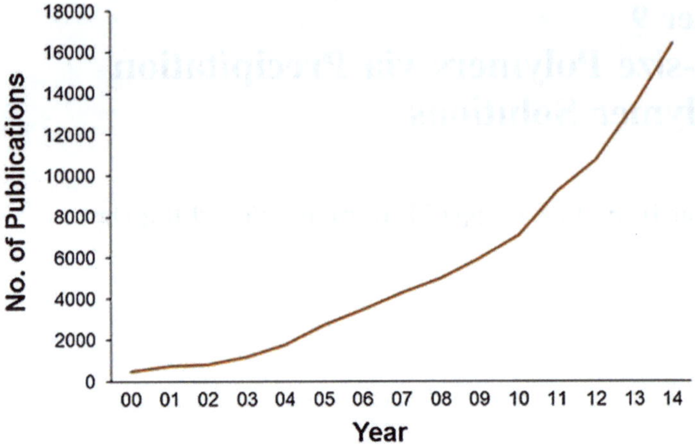

Fig. 9.1 Graphical representation of number of publications on polymer nanoparticles during the period of 2000–2014 [1]

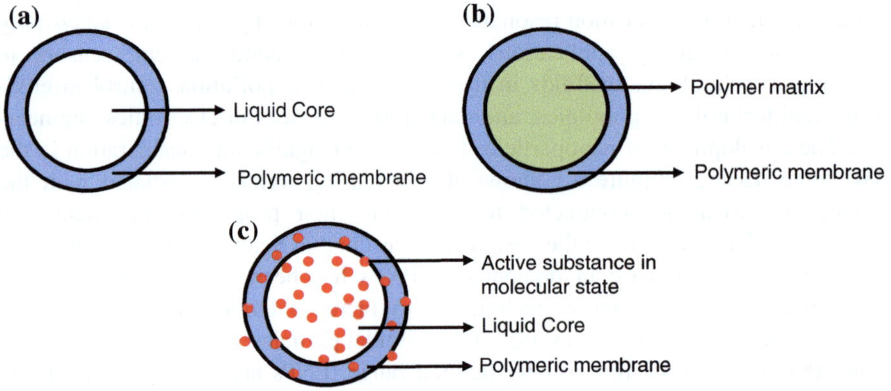

Fig. 9.2 Structures of different nanocapsules: **a** liquid core, **b** polymer matrix and **c** molecularly dispersed active substance [5]

Nanocapsules act like a reservoir, which are called vesicular systems. They carry the active substance entrapped in the solid polymeric membrane or on their surfaces. The cavity inside contains either oil or water. A schematic diagram of Polymer Nanocapsules is shown in Fig. 9.2 [5]. There are different methods that are used nowadays to prepare polymeric nanoparticles, such as nanoprecipitation (also termed as the solvent diffusion and solvent displacement method), solvent evaporation, dialysis, microemulsion, surfactant-free emulsion, salting-out, supercritical fluid technology, and interfacial polymerization [2]. Among these methods, nanoprecipitation is a fast and simple process, which does not require a pre-prepared polymer emulsion before the nanoparticle preparation. It produces a dispersion of nanoparticles by precipitation of preformed hydrophobic polymer solution. Under

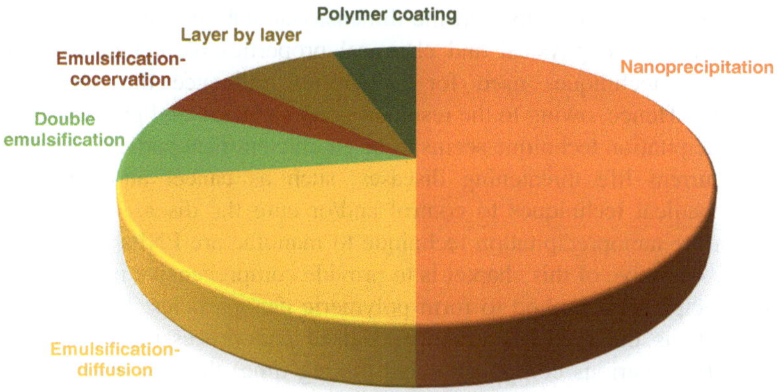

Fig. 9.3 Trends of method selection in nanocapsule research (data source from [5])

appropriate condition, this method spontaneously produces nanoparticles in the range of 50–300 nm with narrow distribution range [6]. In addition, this method has distinct advantages, (i) the use of potentially toxic materials is minimized, (ii) a reproducible particle size with low polydispersity index can be obtained, (iii) the setup is comparatively easy and cheap, and (iv) the yield of particles is high [5]. Nanoprecipitation has emerged as an important method, and has been consistently yielding desirable results over a long period of time. Nanoprecipitation is one of the commonly used methods to produce PNPs as shown in Fig. 9.3. It is considered to be a reliable method, where yield efficiency and particle size are major concerns for producing PNPs. Approximately half of the research conducted to date is based on PNPs using nanoprecipitation. The technical advancements in nanoprecipitation assist the manufacturers to produce nanoparticles economically [7]. Nanoprecipitation is also known as the solvent displacement method, because the technique involves mixing of a solution, containing a biodegradable polymer and a water soluble organic solvent. The nanoparticles are produced due to the turbulences created by the mixing of two liquids [8]. The preparation of nanoparticles using nanoprecipitation was first developed by H. Fessi and his colleagues in 1989 [9]. Indomethacin-loaded nanocapsules were the first nanoparticles prepared using the nanoprecipitation technique. The method was developed to meet the increasing demand of nanopolymers in the drug delivery system. It involved poly-(D, L-lactide) depositing on the interface after displacement of acetone with water [9].

A range of PNPs were used for biomedical applications comprising of drug-loaded carriers, biosensors, and anticancer therapies [10]. In the drug delivery application, these particles protect the drugs from degradation, target drugs to specific sites, and transfer biological molecules. They can be used to reduce the toxicity effects of drugs used and increase the therapeutic efficiency [11]. Overall, the efficiency depends on the chemical structure of the drug, structure of the carrier, its surface properties and polydispersity index (PDI). Thus, the nanoparticles used should have the capability to hold the dimensional stability and last for sufficient time. In general, the physical and chemical properties of nanoparticles and the manufacturing techniques used for preparation influence the applications of nanoparticles. Hence, owing to the resultant good chemical and physical properties, the nanoprecipitation technique seems to be an efficient nanoparticle manufacturing method. Current life threatening diseases, such as cancer and AIDS, require advanced medical techniques to control and/or cure the diseases. Hence, further research in the nanoprecipitation technique to manufacture PNPs is essential.

The key objective of this chapter is to provide comprehensive information about the nanoprecipitation method to form polymeric nanoparticles. It provides crucial factors, such as the type of components used and processing conditions (e.g., addition rate of various components and mixing time) that are involved in nanoprecipitation. The chapter comprises a brief discussion on the parameters affecting the size of the nanoparticles and the mechanism of nanoparticle formations. Additionally, the different setups used to mix the phases and the recent advancements in the manufacturing techniques are discussed. Finally, the biomedical applications of the nanoprecipitation technique are conversed.

Nanoprecipitation of Polymers

Nanoprecipitation uses preformed polymers to prepare PNPs. Hence, preparation of the emulsion before the initiation of the process is not necessary in nanoprecipitation. Precipitation of polymer occurs under spontaneous dispersion of the polymer solution. Use of preformed polymers rather than monomers provides simplicity to the process and minimizes the probability of introducing foreign particles in the nanoparticles. Nanoprecipitation is considered as the simple, reproducible, economical, less toxic, and efficient process in encapsulation [12–14].

The two key phases involved in nanoprecipitation are the solvent and non-solvent phases, which synthesize nanoparticles by interfacial deposition following rapid solvent displacement. The solvent phase consists of the polymer dissolved in a water miscible organic solvent. The non-solvent phase has a solvent or a mixture of solvents, in which the polymer is insoluble. The solvent phase is added to a large amount of the non-solvent phase, as the polar solvent is completely miscible in the non-solvent phase. Consequently, the polymer becomes completely insoluble in the mixture leading to phase separation, resulting in the formation of PNPs [2]. Generally, the solvent phase is an organic solvent (organic phase), and the non-solvent phase is water (aqueous phase). A surfactant can also be used in the process to control the size of the nanoparticles produced. In this process, the condition of miscibility and solubility between the two phases should be satisfied for the process to occur. If these conditions are satisfied, there is a scope to experiment on different materials to be used as the organic or aqueous phase, resulting in more diverse processes in terms of the selection of materials [7]. Role of mixing of both the phases is important, as it accounts for the driving force of the

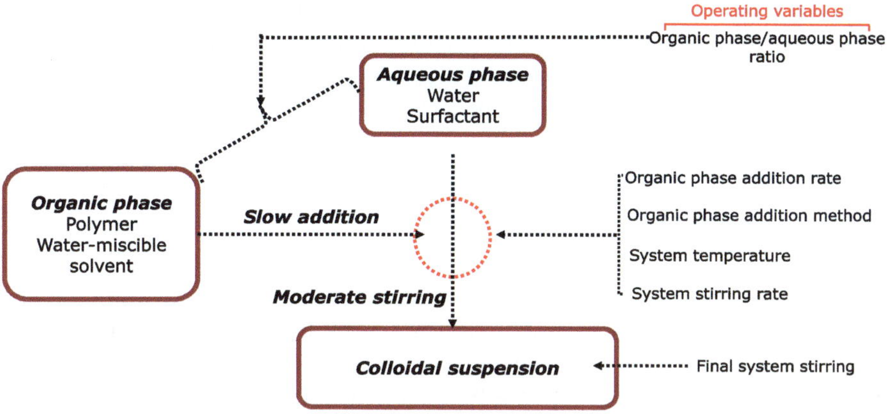

Fig. 9.4 Schematic diagram showing procedure and operating parameters of nanoprecipitation (data source from [7])

process. Generally, the organic phase is added to the aqueous phase dropwise, or at once. Different mixers and setups are also used to mix both the phases, which help in controlling the operating parameters of nanoprecipitation, e.g., polymer flow rate, aqueous phase stirring rate, concentration of the polymer, and concentration of the surfactant added [14, 15]. The two phases contain four basic components, polymer, solvent, non-solvent, and surfactant. Since the nanoparticles produced are affected by many parameters, such as the type of components used, rate of their addition, and rate of their stirring. It is critical to choose suitable components and control the operating parameters to produce nanoparticles. The following sections describe the various components involved in the nanoprecipitation (Fig. 9.4).

Polymers

Nanoprecipitation is a flexible method, because a wide range of PNPs can be produced. It can be used with many hydrophobic biodegradable polymers, but it is mainly used to produce nanoparticles of poly-(lactic-co-glycolic) acid (PLGA) and polycaprolactone (PCL) particles. PLGA and PCL are commonly used in drug delivery [16, 17]. The main advantage of using this method is its ability to encapsulate a hydrophobic drug, and increase its water solubility with nanoparticles [18]. Encapsulation of a hydrophilic drug can also be accomplished with the aid of nanoprecipitation [12]. The choice of polymer depends on the application of the nanoparticles. In biomedical applications, a drug carrier hydrophobic polymer is used with a surfactant or with a (di)block copolymer that provides stronger interaction with the carrier polymer [17]. PLGA, PCL, and poly (D, L-lactic acid) (PDLLA) are the most common polymers used in nanoprecipitation to produce polymeric nanoparticles, but other polymers are also being explored to make the method more robust.

Polymers like starch and cellulose derivatives, poly(methyl vinyl ether-co maleic anhydride), hydrophobic dextranes, poly(3-hydroxybutyrate-co-hydroxyvalerate), polyethylene sebacate, polycyanoacrylates, and poly(methyl methacrylates) have been investigated as alternative polymers [19–25].

Solvents and Non-solvents

Solvent phase is generally an organic solvent, which can dissolve the polymer completely. It should also be water miscible, i.e., polar; so that when the polymer solution is added to the non-solvent, it can rapidly interact, leading to the formation of nanoparticles. Solvent or non-solvent is generally opposite in nature to the polymer. Water is the most common non-solvent used in nanoprecipitation [6]. The choice of organic solvent also depends upon the requirement of the process and the nanoparticles that need to be produced. Additionally, solvent phase is not the only parameter responsible for the final size of the nanoparticles. Since nanoprecipitation is governed by the diffusion of the solvent into the non-solvent, the nanoparticle size depends on the combination of the solvent and non-solvent used. One solvent used in this process is acetone, owing to its polar molecular structure. It can dissolve the polymer and is soluble in water. Other solvents, such as acetonitrile, dimethyl sulfoxide, tetrahydrofuran, methanol, ethanol, isopropanol, dimethyl formamide, and ethyl lactate are also used to produce nanoparticles [5, 7, 12, 26].

One of the important parameters associated with the solvent is its dielectric constant. The difference in the dielectric constants between the solvent and the non-solvent affects the efficiency of nanoprecipitation. If the difference is very high, nanoprecipitation may fail. It has been seen that smaller nanoparticles are formed when dielectric constant of the solvent is high [25]. Nanoprecipitation depends on various phenomena that affect the diffusion of the solvents, in addition to the dielectric constant. Hence, the effectiveness of the technique is dependent on combined properties of the solvent and the non-solvent, such as their solubility values and volumes ratios [12] (which will be discussed in Section "Controlling Nanoparticle Size").

Surfactants

In nanoprecipitation, it is necessary to control the size of nanoparticles, so that a colloidal suspension of PNPs is obtained, rather than a macroscopic solid. When the polymer solution is added to the aqueous phase, polymer particles become insoluble. Hence, they start colliding and sticking to each other. The aggregation will continue to grow to a macroscopic size unless a surfactant is added to control it. Surfactants may get absorbed on the surface of the growing polymer with the help of their hydrophobic tail, hence delaying the adherence among polymer particles. Therefore, the aggregates covered with the surfactant do not recombine with each other, leading to the formation of stable suspension of PNPs. There are two regimes for the use of surfactants, (a) at low concentration of the surfactant, growing particles absorb all surfactant molecules, which is the efficient regime. (b) At high concentration of the surfactant, all surfactant molecules are not used and the excess is left in water. The problem associated with the use of surfactant is that the amount of surfactant molecules needed is more, which may result in decrease in efficiency of the nanocarrier, and can cause unwanted toxicity in nanoparticles [27].

Controlling Nanoparticle Size

PNPs have remained of great interest because of their functionality in different applications. Mainly PNPs are used in electronics, food, photonics, and biomedical applications. The application of PNPs depends on the physicochemical properties of the nanoparticles. The pharmacokinetics and bio-administration of drugs can be varied by several parameters, e.g., size, composition [28], and surface morphology of the nanoparticle [29]. The nanoparticle size is a pivotal characteristic, which determines the functionality of the nanoparticles, cellular uptake [7], and is directly linked to the stability of the particle dispersion [17]. Nanoparticle size plays a vital role in vivo administration of drugs, as it affects the adsorption of the plasma proteins, leading to their recognition by the macrophages of the endothelial system and fast removal from the bloodstream [6]. Since the first use of nanoprecipitation to produce nanoparticles, many experimental and theoretical studies have been conducted on the processing parameters, working conditions, stabilizers and starting materials of nanoprecipitation. These parameters affect the final size of PNPs, PDI, and production yield. The following sections will explain the effects of operating conditions, and type and volume of polymer, organic solvent, and surfactant used on the final size of nanoparticles formed.

Effects of Various Operating Conditions on the Final Size

Operating conditions during particle formation play a vital role on the final size of PNPs. Organic phase addition rate, type of setup/stirring system (method of mixing), stirring rate, interfacial tension, ratio of non-solvent to solvent flow rate, and final mixing time are the various factors that influence the particle size. The research conducted till date has focused on varying these parameters to produce nanoparticles with the required size [12, 17].

There are some contradictory results among the experiments conducted by different researchers, which may be attributed to the differences in the starting materials, working conditions, and post-processing treatments for purification and stabilization [7]. The mean size of nanoparticles depends on various operating factors simultaneously. In the following subsections, the effects of various operating conditions on PNP size will be discussed.

Organic Phase Addition Rate

When nanoprecipitation of polymer is performed by adding an organic phase to an aqueous phase, the addition rate of the organic phase has an effect on the nanoparticles produced, particularly it affects the nanoparticle size. However, the influence changes with the organic to aqueous phase ratio. Different behaviors were encountered in different researches. A controlled study needs to be conducted to investigate nanoprecipitation, and to address the contradictory behaviors observed, due to the difference in materials and operating conditions. The change in the mean particle size with the organic phase injection rate, considering the total organic phase volume simultaneously, is discussed below.

Smallest nanoparticles are formed by lowest injection rate of the organic phase (Fig. 9.5), particularly when the ratio of organic to aqueous phase is high. The nature of nanoparticle formation suggests that when organic phase addition rate is rapid, formed particles may continue to grow, because of insufficient diffusion time [6]. The behavior suggests that the method of nanoprecipitation to form polymeric nanoparticles is time-dependent. To obtain nanoparticles of small size, proper phase mixing should be ensured first before the formation of particles. The stirring rate of the system should be increased with the increase of organic phase injection rate so that mixing of the phases happens faster than formation of particles. Nanoparticles with size ranging from 160 to 190 nm can be produced with controlled stirring and the organic phase addition rates [7].

Method of Mixing of Phases

Different methods of mixing are employed to mix the organic phase with the aqueous phase, which will be discussed in Section "Manufacturing Techniques Used in Nanoprecipitation". The technique used to mix the phases could also have an effect on the nanoparticle size. Nanoprecipitation is a non-equilibrium process, which initiates with the mixing of the two phases, which suggests that the variation in the particle size can be obtained by change in mixing methods. Nanoparticle size

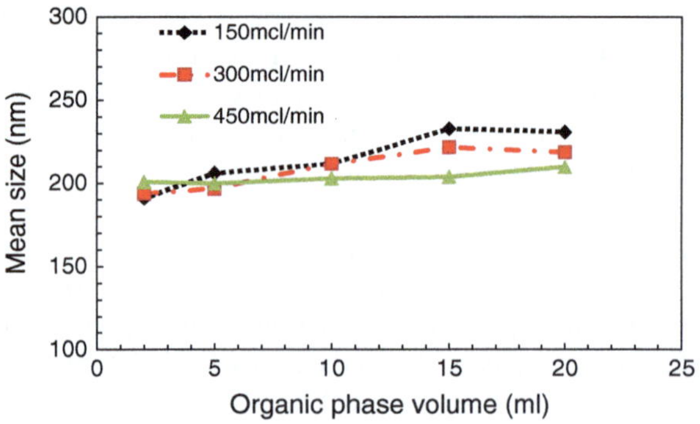

Fig. 9.5 Change in the mean size of nanoparticle with organic phase injection rate (data source from [7])

is also dependent on the order of phase mixing, i.e., aqueous phase into organic phase or vice versa, and the behavior of the organic solvent used. To evaluate the effects of mixing methods, four different methods were compared [15]: (a) Aqueous phase of largest volume was added into the organic phase in one shot, which produced the most rapid mixing of the phases. (b) Aqueous phase was added slowly dropwise to the organic solution. This method showed almost equilibrium phase separation due to a continuous decrease in the solubility of the polymer used. (c) Organic solution was added dropwise to the aqueous solution. This method is the commonly used technique to produce nanoparticles by nanoprecipitation. (d) This method is same as (a), but 10 % of the solvent was added to the aqueous solution instead of being used in the starting polymer solution.

All the above methods were compared by taking same parameters of polymer concentration, polymer mass fraction, and two sets of solvent, and solvent/non-solvent ratios. Method (a) is a vigorous process, which produced metastable dispersions, i.e., nanoparticle dispersions having colloidal stability, on the condition that the aqueous phase was made of perfectly pure water containing NaOH, or a surfactant, or both. Nanoparticle size was strongly affected by the properties of the aqueous phase, e.g., its pH and ionic strength. Methods (a), (c), and (d) produced nanoparticles of almost similar sizes and PDI, but method (b) produced much larger number of nanoparticles. Since method (b) is dropwise addition of aqueous phase into the organic solution, this difference can be expected. Dropwise addition of the aqueous phase causes a gradual increase in the supersaturation, which leads to uneven nucleation at low supersaturation levels. Precipitation at low polymer concentrations is slower, thus the mixing mode did not affect initially, and all the processes produced nanoparticles of approximately the same size [15].

The nanoparticle size obtained was small, when the organic phase was added dropwise into the aqueous medium (dropwise-in method) [7]. In the experiments, it was found that the drop size was smaller compared to that produced by other methods. This occurred due to the stirring shear strength, leading to the nucleation of small particles and resulting in the formation of small nanoparticles. The change in the mean size of the nanoparticles can be achieved by varying the method of adding organic phase, considering the organic phase volume is shown in Fig. 9.6.

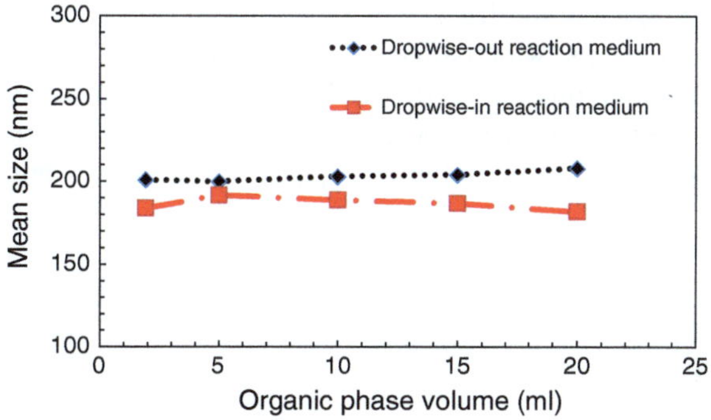

Fig. 9.6 Change in mean size of nanoparticles with the method of addition of organic phase (data source from [7])

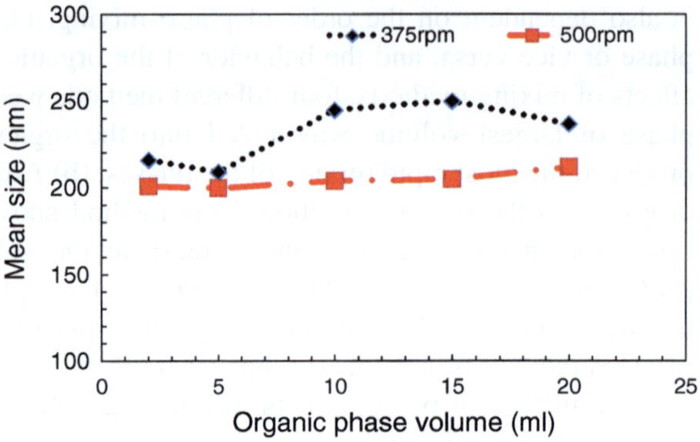

Fig. 9.7 Relationship between the nanoparticle size and system stirring rate (data source from [7])

System Stirring Rate

System stirring rate is a parameter, which affects nanoparticle size. Effects of stirring rate on the nanoparticle size are shown in Fig. 9.7 [7]. Its effect is also dependent on the volume and addition rate of the organic phase. Nanoparticle size reduces with an increase in the system stirring rate. Hence a higher stirring rate of the aqueous phase and organic phase produces smaller nanoparticles. The reason for this effect can be that high stirring rate of the phases improves the mixing performance, which promotes reduction of the mean size of the nanoparticles. Computational fluid dynamics (CFD) simulations show that the mixing at high stirring rate increases the nucleation rates, leading to production of small particles. Stirring rate potentially influences solvent migration, system micromixing, polymer supersaturation, and particle aggregation. Usually, terms like "gentle magnetic stirring" and "moderate stirring" are commonly used to refer the stirring rate of the system [7, 30].

Interfacial Tension

Nanoparticle formation by nanoprecipitation involves many steps, such as nucleation of particles, growth of particles, and aggregation. The kinetics of each step is responsible for the final nanoparticle size and PDI [31]. The driving force of each of these steps, which affects their kinetics is the supersaturation, which is caused due to rapid mixing of the solvent into the non-solvent. The supersaturation (S) is usually defined as the ratio of the concentration of the polymer ($c_{polymer}$) to the solubility of the polymer ($c_{polymer,\ eq}$) in the solvent, as given by Eq. 9.1 [31].

$$S = \frac{c_{polymer}}{c_{polymer,eq}} \tag{9.1}$$

Supersaturation plays an important role in nanoprecipitation, as it determines the nucleation rate of the particles (*J*). According to the classical theory of nucleation, the nucleation rate of spherical nuclei can be calculated by Eq. 9.2 [31]:

$$J = \frac{2D}{d^5}\exp\left(-\frac{16\pi\gamma^3\tilde{\upsilon}^2}{3k_B^3T^3[\ln(S)]^2}\right) \tag{9.2}$$

where d is the molecular diameter of the polymer molecule, D is its molecular diffusion rate, T and k_B are the absolute temperature and the Boltzmann constant, respectively, γ is the interfacial tension between the formed nanoparticles and the solution, and v is the polymer molecular volume. These are the factors on which the supersaturation depends. Additionally, local supersaturation is somewhat dependent on mixing and fluid dynamics. The interfacial tension depends on the local composition of the mixture (local fraction of solvent and non-solvent).

The effects of interfacial tension on the nanoparticle size are investigated [32]. In the experiments, PCL was chosen as the polymer, and acetone and water were chosen as the solvent and non-solvent, respectively. The interfacial tension between the polymer particles and the water–acetone mixture was calculated based on the composition of the mixture in the form of water molar fraction. A change in interfacial tension with an increase in the molar fraction of water was found, because of the difference in the affinity of the polymer with water and acetone. These effects can make a difference in the interfacial tension and in the nucleation rate (Eq. 9.2). Thus even a minute change in the interfacial tension can cause large change in the nucleation rate and affect the final particle size [32].

Ratio of Non-solvent to Solvent Flow Rates

The ratio of non-solvent to solvent flow rates has an effect on the nanoparticle size. The flow rates affect mixing, which in turn affects the nanoparticle size. The polymer is not soluble in the final mixture, which is the key mechanism in nanoprecipitation for preparation of nanoparticles. Hence, the influence of this ratio on the final nanoparticles size becomes prominent (Eq. 9.3) [31]:

$$R = \frac{\textit{Flow Rate of Water}}{\textit{Flow Rate of Polymer Solution}} \tag{9.3}$$

when R is increased, it leads to an increase in the supersaturation, which increases the number of nuclei formed, thus decreasing the final particle size. An increase in the potential interface due to the higher value of R increases the number of nanoparticles during phase separation. This results in the decrease in the local concentration of the polymer, leading to smaller particles. Because of the effects of the ratio R, influence of other parameters are analyzed, while taking R into consideration. The effect of R on the final particle size, taking different initial polymer

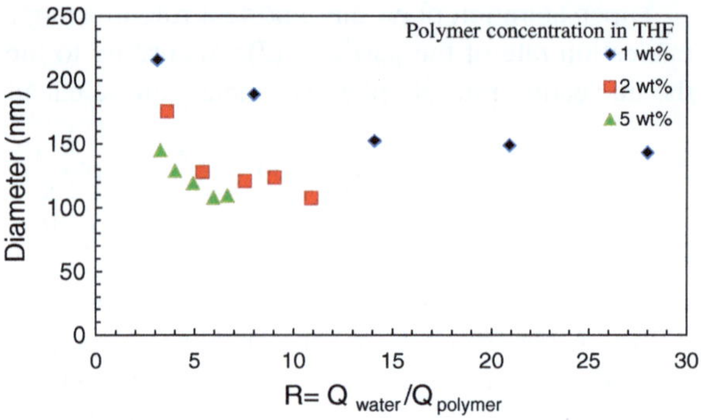

Fig. 9.8 Evolution in nanoparticle diameter with proportion of solvent and non-solvent (*R*) (data source from [31])

concentrations, with Tetrahydrofuran (THF) as the solvent, water as the non-solvent, and Poly(methyl methacrylate) (PMMA) as the polymer is shown in Fig. 9.8. Gradual decrease in the nanoparticle diameter with an increase in the value of *R* can be observed [31].

Mixing Time

The mixing time of the organic phase with the aqueous phase is a crucial factor affecting the nanoparticle size. Experiments conducted on nanoprecipitation showed that nanoparticle size increases with an increase in mixing time [17, 31, 33]. Thus, fast and efficient mixing is ensured between the phases so that the mixing time is minimum. Various approaches have been used till now to improve the efficiency of the mixing, which will be discussed in Section "Manufacturing Techniques Used in Nanoprecipitation". An experiment was conducted to study the influence of mixing time of the two phases on the size of nanoparticles produced by nanoprecipitation [34]. It used confined impinging jet mixture (Section "Confined Jet Impinging Mixers") to mix the solution of the polymer and the solvent with the aqueous medium, Jet velocities controlled the mixing time. The rapid dissolution of the solvent into water induces nucleation of the polymer and its growth as spherical nanoparticles. The results revealed that the size of the nanoparticles decreased with a decrease in the mixing time until a critical point, below which the size did not change. Additionally, when polymer concentration was considered, its effect on nanoparticle size was distinct for large mixing time, but was negligible in small mixing time as shown in Fig. 9.9.

Effects of the Materials on Final Size

The sections below discuss how the different types, properties, and concentrations of starting materials (polymer, solvents, and stabilizing agents) used in nanoprecipitation affect the final size of nanoparticles.

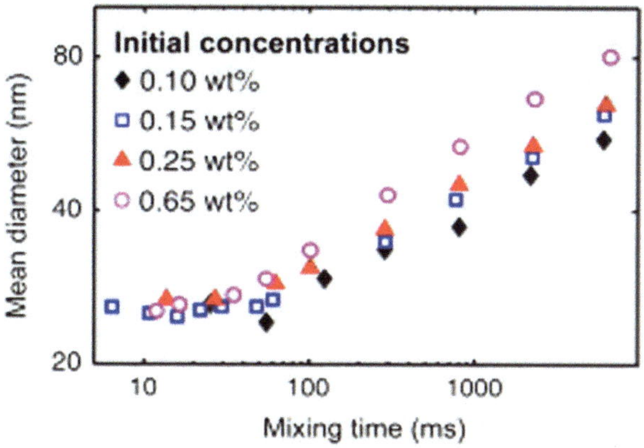

Fig. 9.9 Variation of the nanoparticle size with the mixing time of the phases [6]

Effect of Polymers

When nanoparticles are prepared, the polymer not only affects the size, but also the size is influenced by other parameters, such as mixing rate, solvent and non-solvent used, and stabilizing agent. So, it is necessary to determine the influence of the polymer on the nanoparticle size. Nanoparticle size is mainly affected by two properties of the polymer, nature and concentration. When the nature of the polymer is considered, the size of the nanoparticle is different based on the selection of the polymer, even though the same manufacturing technique is used for processing. An experiment was performed selecting PCL, PLGA, and PDLLA as the polymers [35]. Variation in the nanoparticle sizes can be observed using these polymers. Comparatively, the size of the PLGA nanoparticle is less than that of the PDLLA one, and the size of the PCL nanoparticle is greater than that of the PDLLA one (PLGA < PDLLA < PCL). In another recent study, similar results were obtained as given in [7]. The reason for such difference can be attributed to the crystalline and amorphous characteristics shown by the polymers during nanoprecipitation. The amorphous characteristics can be observed during precipitation of PLGA and PDLLA, whereas PCL exhibits both crystalline and amorphous characteristics under precipitation, which results in larger PCL particle size compared to those of PLGA and PDLLA [36]. In the case of PCL, a three-phase model was constructed, which consisted of the rigid and mobile amorphous, and the crystalline fractions, where mobile amorphous phase specific volume is less than that of the rigid amorphous phase. When the different molecular weights of PCL were used, there was no difference between the nanoparticle sizes (Fig. 9.10). Hence, the crystalline structure of the polymer affects largely the size of the particle produced using the nanoprecipitation technique [7, 30]. There is a considerable effect of polymer concentration on the size of the particle produced. The size of the nanoparticles is found to increase with an increase in the polymer concentration [17]. When the number of polymer chains increases per unit volume with an increase in the polymer concentration of the solvent, the formation of large nanoparticles can be observed. Moreover, when more

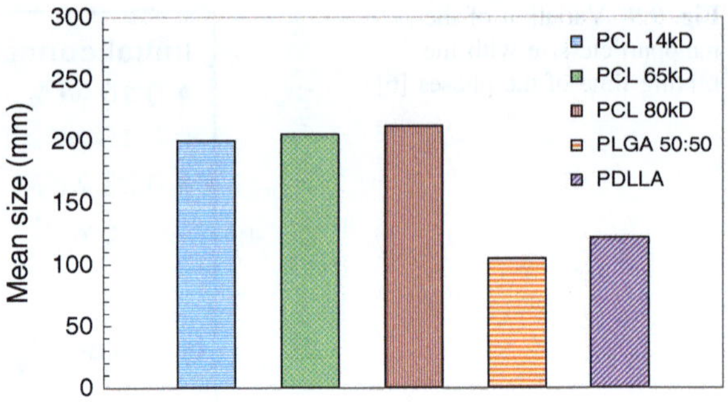

Fig. 9.10 Effect of different polymers on the mean size of nanoparticles (data source from [7])

polymer chains are associated during the diffusion process, an increase in the polymer–polymer interaction can be noticed.

Furthermore, an increase in polymer concentration produces a high viscous organic phase, which leads to resistance to mass transfer. Because of the high resistance, a decrease in diffusion rate of the organic phase into the aqueous phase can be observed, resulting in large nanoparticles. Hence, small sized nanoparticles can be obtained by reducing the polymer concentration. The change in nanoparticles size with the polymer concentration is shown in Fig. 9.11 [22]. A study has also been conducted to know the effect of initial polymer polydispersity on the final nanoparticle size. It was found that there is negligible effect of initial distribution of polymer particles on the final nanoparticle size; it only converges the molecular weight. Hence it can be concluded that the final particle size is independent of the initial PDI [17, 37].

The effect of polymer mass on the size has been experimentally tested. PCL polymers with different molar masses were used with acetone as the solvent. No effects were observed between the polymer mass and the final nanoparticle size. The particle size remained almost constant with different molar masses. A minor

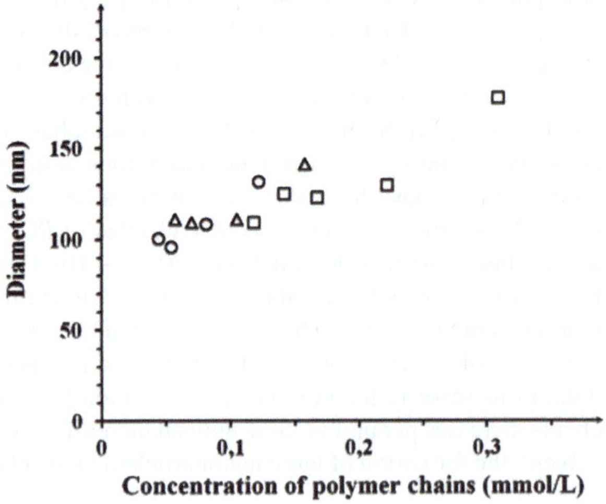

Fig. 9.11 Change in the nanoparticle size with the polymer concentration [31]

variation between the sizes of nanoparticles produced was encountered, which can be attributed to a small difference in the viscosities of the different polymer solutions [17]. When polymers of lower molar masses were used, influence of the polymer on viscosity was negligible, and thus no major effect on the nanoparticle size was noted. More pronounced effect on the final size can be observed when polymers of higher molar masses were used, because of their effect on the viscosity of the solution. At higher molar masses of the polymer, the size of nanoparticles increased with an increase in molar mass, which is attributed to the polar group attached at the end of the PLA chains, which gives them low amphiphilic characteristic [6, 38].

Effect of Solvents

Nanoprecipitation was also analyzed by varying the usage of solvent that affects the nature of the organic phase. Many solvents have been used for nanoprecipitation, such as ethanol, dimethyl sulfoxide, isopropyl alcohol, ethyl lactate, and most importantly acetone. The change in the final nanoparticle size in nanoprecipitation can be expected with the change of solvent, because of the differences created in the interaction of water-solvent and polymer-solvent. The effect of the solvent on the size of nanoparticles depends on its ability to dissolve the polymer. The mean size of the nanoparticle was found to increase in this order: ethanol < dimethyl sulfoxide < isopropyl alcohol < acetone < ethyl lactate.

To determine the ability of the polymer to dissolve in a particular solvent, it is important to know the solubility of the solvent used. For a solute to dissolve in a particular solvent, the solubility values between the solute and solvent should be less. Thus, the difference between the solubility values of the two components is the main criteria for solubility. A similar concept was used to determine the difference of solubility parameters of water and different solvents used in nanoprecipitation ($\Delta\delta_{s-w}$). The $\Delta\delta_{s-w}$ value increases in the following order: ethanol < dimethyl sulfoxide < isopropyl alcohol < ethyl lactate < acetone for Methacrylic acid copolymer as the polymer. Nanoparticles were prepared using these solvents, keeping the polymer concentration constant. The order of size obtained was as follows: ethanol < dimethyl sulfoxide < isopropyl alcohol < acetone < ethyl lactate. It can be seen that the order obtained is almost the same as the previous order of $\Delta\delta_{s-w.}$ This relationship suggests that when the affinity between the solvent and water is high, it has lower $\Delta\delta_{s-w}$ value, which means that the solvent diffuses into the aqueous phase more effectively resulting in the formation of smaller nanoparticles [22].

Similar results were observed when THF and acetone were compared as solvents. The nanoparticles formed by using THF were always larger than those formed by using acetone. Low viscosity acetone leads to proper mixing [38]. Four different solvents, i.e., Dimethylformamide (DMF), acetone, acetonitrile, and THF, were considered to analyze their effect on the size of the particle formed. It was observed that when miscibility of the solvent in water increased, the size of nanoparticles decreases [18]. Thus, it can be concluded that mixing between the solvent and the aqueous phase influences the size of the nanoparticles.

Effect of Stabilizers

Stabilizers are used to ensure the physical stability of the polymer dispersions formed in nanoprecipitation. This function depends on their physical properties and the role they play in the formation of nanoparticles. Thus, the concentration, molar mass, and nature of the stabilizer are the important factors that were used to determine the effects on the nanoparticle size.

The nature of stabilizer has a considerable effect on the nanoparticle size. The experiment was conducted; PCL as the polymer, dodecyltrimethylammonium bromide (DTAB) and sodium dodecyl sulfate (SDS) as the positively and negatively charged stabilizers, Poloxamer (Plx), polysorbate-80, and PVA (nonionic stabilizers) as the stabilizing agents, and acetone as the solvent. The stabilizing effect was found as there was no aggregation of the particles, without affecting their formation. The function and influence of the stabilizer were dependent on its steric, electrosteric, and electrostatic effect. The size of the nanoparticle was found to be increasing in this order: Polysorbate 80 = DTAB = SDS < PLX < PVA. DTAB and SDS have predominantly electrosteric effect, whereas the rest of the stabilizers exhibit steric effect. The results reveal that small nanoparticles are formed when a stabilizer with a strong electrosteric effect is used in nanoprecipitation. In addition, particle growth is favored by delay in the solvent diffusion, which may be due to the steric effect of the stabilizer [7].

The concentration and molar mass of the stabilizer are also considered to be important variables to study their effects on the nanoparticle size. The nanoparticles can be prepared without the use of stabilizer, as electrostatic repulsion between their surface charges stabilizes them. Still a stabilizing agent is used in the process because of its properties to stabilize the system and prevention of aggregation [39]. This makes concentration of the stabilizer a parameter of interest. It has been reported that the concentration of the stabilizer does not have any significant effect on the size of the nanoparticles in nanoprecipitation [7]. However, when the molar mass of the surfactant was concerned, it was reported that a surfactant with a high molar mass produced large nanoparticles. This is associated with the change in time scale induced by the surfactant. The adsorption time of the surfactant increases with the use of a surfactant of higher molar mass. Because of this effect, a minute increase in the nanoparticle size is observed. Also, nanoparticle size is increased with the use of a surfactant with a higher viscosity, which results in a longer mixing time [17].

Manufacturing Techniques Used in Nanoprecipitation

Nanoparticles are developed by mixing the components, i.e., polymer, solvent, non-solvent, and stabilizers. The central concept lies in the method used to mix them. Controlling the mixing gives the ability to control various physicochemical properties of the nanoparticles. Uncontrolled mixing process leads to properties which are undesirable and may be toxic in case of biomedical applications.

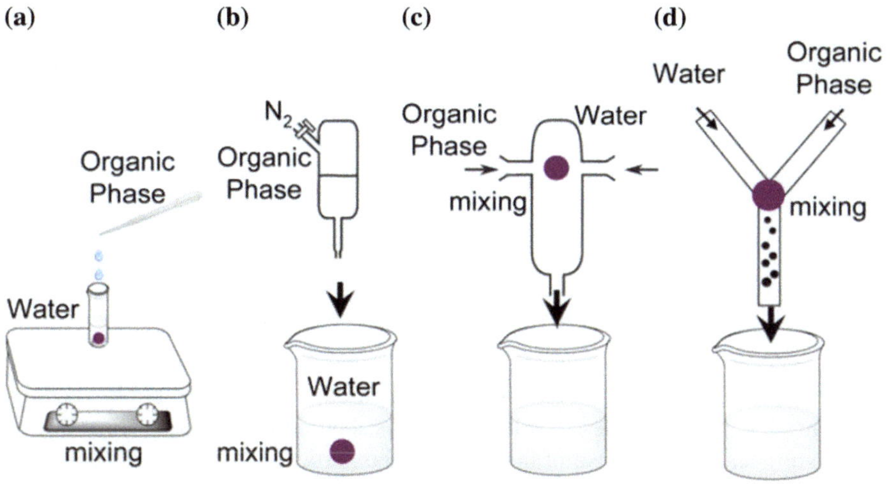

Fig. 9.12 Different nanoprecipitation devices: **a** bulk dropwise process [74], **b** pressure driven injection device [74], **c** confined impinging jet mixer [54], and **d** continuous flow setup [6]

Different methods have been implemented to mix and allow the interaction of the components. The simplest way to produce PNPs is to use "bulk" method to precipitate them by solvent exchange, the most conventional approach for production [18, 40]. But due to increased interest in nanoparticles, several efforts have been made to improve the mixing processes. Different types of methods allow manipulating the control parameters based on the application of nanoparticles. Basically, there are two types of configurations used. One is the stopped flow setup, where definite volumes of the phases are allowed to interact and they are transferred to a cell where nanoparticles grow. Other is the continuous flow setup, where two channels carrying the phases converge to mix at the junction. Nanoprecipitation occurs through the exit channel [6] (Fig. 9.12). To produce nanoparticles in a easy, fast, and reproducible way, different types of mixers were experimented upon [34, 41–44]. Some of the techniques which are used to mix the components are (a) Bulk method, (b) Hydrodynamic flow focusing, and (c) Confined impinging jet reactors (CIJR). These methods are discussed in the following subsections.

Bulk Method or Pot Pouring

Bulk method is the conventional, simple, and easy method to mix the components. Components are simply poured into one another, and after sometime polymeric nanodispersions are formed. Suitable volumes of phases are taken and are added either drop wise or simply poured. There are no conditions for mixing; hence parameters responsible for the process cannot be completely controlled. The first

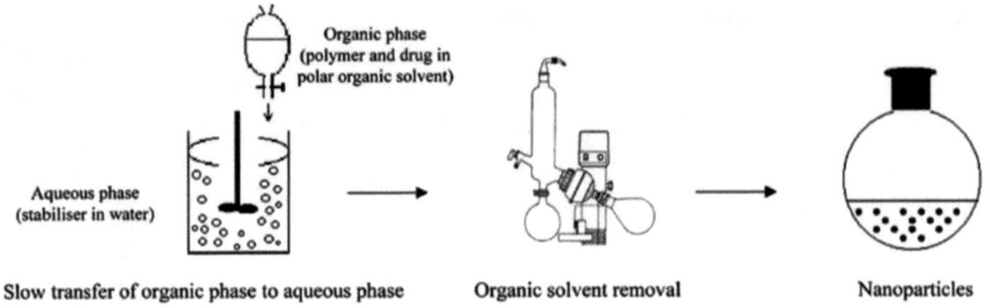

Fig. 9.13 Dropwise addition of organic phase in aqueous phase to produce nanoparticles [40]

nanoprecipitation was performed using this method, which successfully produced nanoparticles. Nanoparticles developed by this method cannot be always directly used in applications, because of high PDI of the particles produced. Generally, the organic phase is added drop wise to the aqueous solution [45–48]. But, other combinations of mixing have also been considered. Aqueous phase has always larger volume than organic phase to ensure rapid nanoprecipitation. Different mixing orders give different results and they are frequently tested [15]. Following mixing orders are the most frequently used: (a) The aqueous phase of large volume is added to the organic phase directly in one shot, (b) The aqueous phase is added drop wise to the organic phase, (c) The organic phase is added drop wise to the aqueous phase as discussed above. When this method is used to perform nanoprecipitation, the nucleation and growth of particles are induced following a continuous change in the composition of the added phases, i.e., organic phase and aqueous phase. As the organic phase is added, rapid diffusion of the organic solvent into the aqueous phase takes place with spontaneous reduction of the interfacial tension. Diffusion is followed by formation of nanospheres by polymer aggregation. After the process, the nanoparticles are subjected to heating to remove the organic solvent (Fig. 9.13) [14, 25].

Hydrodynamic Flow Focusing

Microfluidics is a useful technology, owing to its ability to mix different phases rapidly, maintaining homogenous reaction environments, changing reaction conditions, and introducing reagents at specified time intervals [49]. Since a decade, it has been successfully used in a variety of reactions by optimizing the quality of the reagents and conditions. Likewise, nanoparticles also attracted considerable attention, because of their potential applications in biomedical technology, photonics, and electronic technology. Hydrodynamic flow focusing harnesses the potential of microfluidics to exploit it in the production of nanoparticles by nanoprecipitation [50]. The bulk process of nanoprecipitation lacks control of the process parameters; so use of microfluidics may provide better control over the mixing process and

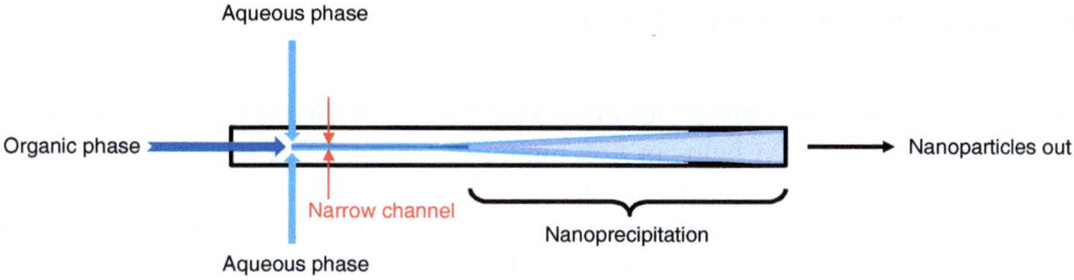

Fig. 9.14 2D microfluidic device using hydrodynamic flow focusing (data source from [33])

nanoparticle properties, such as size and surface morphology [33]. There are two kinds of microfluidic devices, which use hydrodynamic flow focusing, 2D channel [33, 51] and 3D channel [52]. They both have been proved to produce nanoparticles with controlled results.

The nanoparticles can be produced by 2D channel flow of the organic phase through the central channel, which meets the two adjacent streams of aqueous phase (Fig. 9.14). The narrow width of the central stream leads to rapid mixing of both the components. PLGA–PEG nanoparticles can be produced using this method. By using this method, a decrease in the PDI and particle size was observed [33].

A 3D microfluidic device is a more recent development of hydrodynamic flow focusing, which helps to remove clogging in a 2D device [52, 53]. Thus, a 3D channel is a development over a 2D channel to improve the productivity. It has different inlets in the form of horizontal nozzles and vertical chimneys as shown in Fig. 9.15. It used PLGA–PEG as polymer, Acetonitrile (ACN) as solvent, and water as non-solvent. It produced nanoparticles of a small size in a controlled manner.

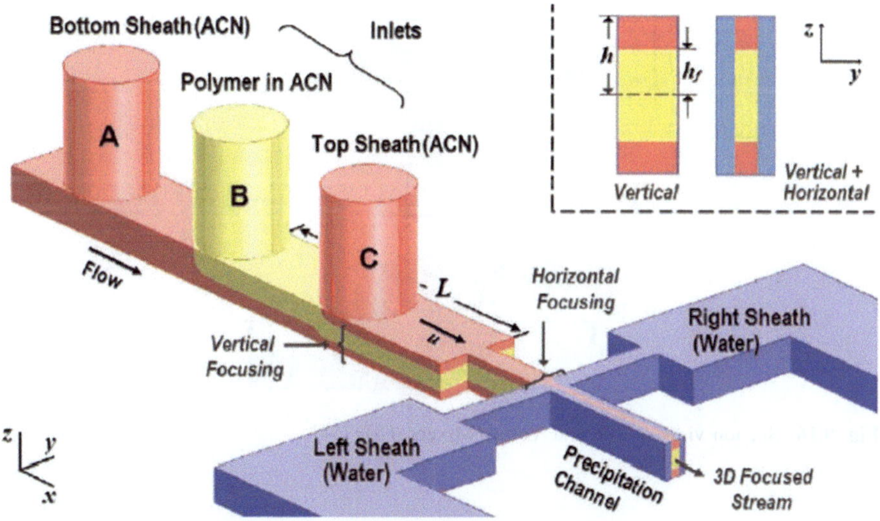

Fig. 9.15 A 3D microfluidic device using HFF [52]

Confined Jet Impinging Mixers

Nanoprecipitation in Confined Jet Impinging Mixers (CJIM) (Fig. 9.16) has shown promising results. CJIM helps in fine tuning of the final size of nanoparticles and their surface morphology. CJIM controls the process of nanoprecipitation by controlling the supersaturation efficiently. CJIM has two jets with high velocity fluids which collide and mix inside a chamber. These mixers have important properties of rapid homogenization and low mean residence time of feed streams [54]. The effects of CJIM construction on the size of nanoparticles is still in the research phase [32].

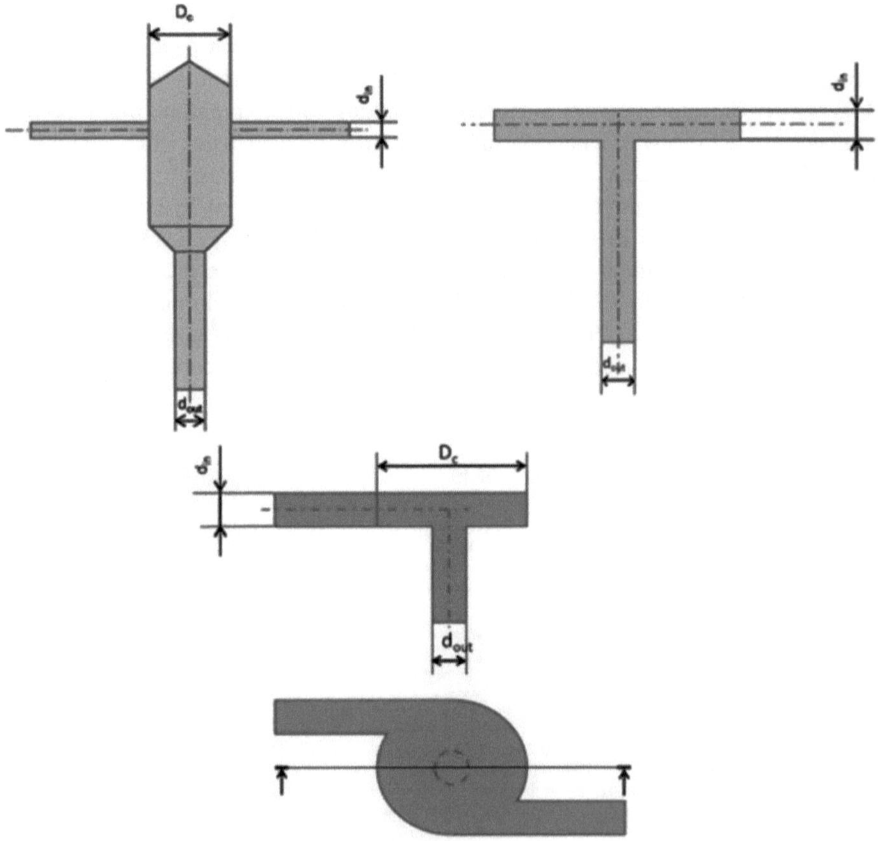

Fig. 9.16 Section view of Tee and Vortex Mixers (CIJ) [54]

Mechanism of Nanoparticle Growth

The mechanism of nanoparticle growth has always remained an intriguing concept. Many experiments have been carried out in the past to propose exact mechanisms of particle growth. The particle growth mechanisms are related to the basic theoretical model that governs the nanoprecipitation process. The preparation process may appear to be simple, which involves displacement of a semipolar solvent from a hydrophobic solution, followed by interfacial deposition of a biodegradable and preformed polymer. However, this involves an intricate hydrodynamic process. Initially, a bicomponent system of a dilute solution of a hydrophobic polymer is injected into an aqueous/surfactant solution. Brownian motion governs both the polymer and the surfactant. Both solvent and non-solvent, both are assumed to be properly miscible. Consequently, when the polymer solution is injected into the non-solvent, rapid mixing takes place. The formation process of nanodroplets is fast, due to unbidden emulsification and interfacial tension of the solution. The speedy diffusion of the solvent across the interface results in the decrease of the interfacial tension [9].

The method of preparation of nanoparticles via nanoprecipitation is experimentally well-known, but the mechanisms responsible for the formation of nanoparticles are not yet fully known, which need to be addressed [7]. Knowledge of the mechanisms that control the nanoprecipitation process can be useful. It can help us to understand how various controlling parameters, such as molar masses, concentration of polymer, surfactants, and mixing rates, affect the final particle size. This knowledge will help to develop an effective nanoprecipitation technique. There are two proposed mechanisms, which are supposed to be consistent with the experimental results: mechanical mechanism and classical nucleation mechanism. Both of these mechanisms abide by the fact that the formation of particles takes place in the metastable region of the phase diagram, i.e., between the binodal and spinodal curve [31].

The mechanical mechanism of particle formation in nanoprecipitation relies on a dispersion mechanism. The mechanism is attributed to turbulences or instabilities, which are formed at the interface of the two phases during solvent diffusion as shown in Fig. 9.17. These turbulences break up the polymer solution into small droplets and disperse them in the non-solvent. According to the assumption of complete phase miscibility, the continuous formation of eddies at the interface explains the molecular mechanism of interfacial turbulences. These eddies of the solvent can continue to break into smaller droplets. These eddies originate either due to the formation of drops or because of thermal inequalities of the system. This process takes place very rapidly, and it spontaneously stops, when the whole solvent flows away from the droplets, leading to polymer nanoprecipitation. It is vital to emphasize that the macromolecules, which are present in the droplets, aggregate to form PNPs. The nucleation and growth steps are not taken into consideration in this mechanism [8].

Classical mechanism of particle formation follows nucleation of nanoparticles in supersaturated regions created by solvent diffusion. It is an intricate process, which

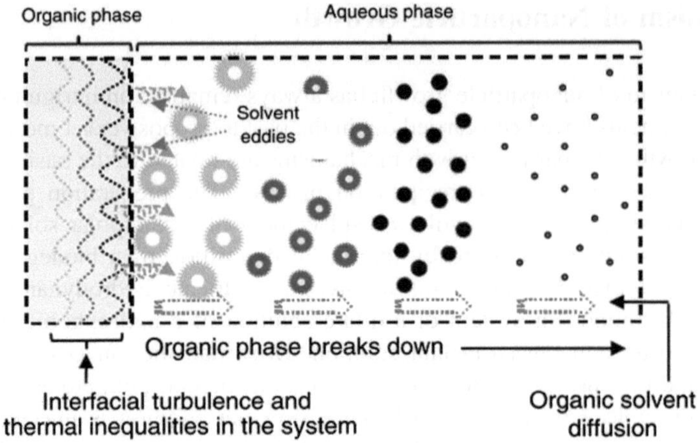

Fig. 9.17 Solvent displacement method of nanoparticle formation based on the interfacial disturbances [7]

includes nucleation of particles, their growth and aggregation. The rate of each steps determines the final nanoparticle size, PDI, and surface morphology. This mechanism is governed by classical theory of nucleation, which is followed in the case of the formation of spherical nuclei. The nucleation rate involved is determined by supersaturation, which is the driving process of the phenomena. The supersaturated states show the chemical instability of the system, which corresponds to the presence of nuclei. Supersaturation is generally defined by the ratio of the concentration of the polymer to the solubility of the polymer in the solution. After nucleation, molecular growth takes place. It occurs when a single polymer particle attaches itself into the particle matrix. It is generally exhibited as the continuous change in the particle size. After growth, aggregation also affects the final size and properties. It is dependent upon the stability of the particles and their frequency of collision [30].

Owing to the recent developments in the nanoprecipitation mechanism, the above mechanisms, mechanical and classical mechanisms, have been proved to be consistent with the experimental results. However, no study yet has concluded on a specific mechanism of nanoprecipitation process by exact experimental results. The two mechanisms can occur parallel to each other, or one can dominate over the other, depending upon the interrelationship and concentration of the materials used [31].

Post-Processing Treatments

Post-processing treatments refer to the processes, which are carried on the nanoparticles after they are produced by the nanoprecipitation method. These processes are carried out to purify and to increase the stability and concentration of the nanoparticles. Nanoparticles formed by nanoprecipitation may contain impurities in the form of traces of solvent, stabilizer, and other salts used in the process. It is important to remove these impurities, as nanoparticles are used in therapeutic and drug delivery applications, which require them to be highly pure and non-toxic. In the matter of concentration, nanoparticle dispersions produced by nanoprecipitation frequently has a low content of drug in it, which is a disadvantage when therapeutic application is considered. Concentration of the dispersions results in high variation of drug concentration in the nanoparticles [5].

Stable systems are those where the nanoparticles follow Brownian motion [7]. However, they can be destabilized by various factors such as polymer degradation and contamination of liquid systems. Stability of nanoparticles is one of the limiting factors in the large scale production for use in drug delivery system [5, 55]. Thus, post-processing treatments are required to ensure the stability of the nanoparticles.

To achieve the suitable concentration, purification and stabilization level of the nanoparticles developed by nanoprecipitation, nanoparticles are subjected to treatments after their formation. There are different techniques available to perform post-processing. The commonly used methods are shown in Fig. 9.18. These processes can be used on nanoparticles independently or consecutively. These processes are still in the research phase to determine the most efficient combination of the sequence of processes, which provides high yield and guarantees high purity of the nanoparticles. Different processes applied for solvent elimination, purification, and stabilization are discussed in the following sub-sections.

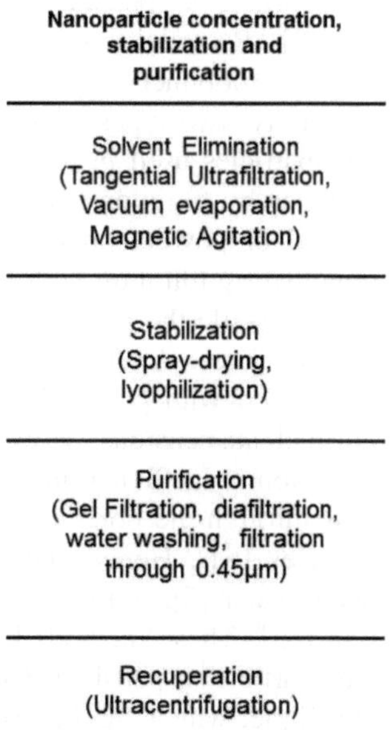

Fig. 9.18 Different techniques of post-processing treatments [5]

Solvent Elimination

Nanoparticles prepared by nanoprecipitation contain traces of the solvent used, as it is miscible with both the polymer solution and water. The remaining organic solvent either acts as a toxic impurity or decreases the nanoparticle stability against Ostwald ripening. Hence, it is necessary to remove the solvent. The technique for removing usual solvents like acetone, THF, and ethanol is evaporation under reduced pressure, because the boiling point of water is higher than that of the others. Crossflow microfiltration is used to remove surfactants and solvents efficiently [56]. Methods such as gel permeation chromatography are used for more efficient separation for small nanoparticles. Filtration, ultracentrifugation, and dialysis do not provide efficient solution for the separation of small nanoparticles (80–150 nm) [57]. The removal of solvent by flash evaporation and extraction process using supercritical carbon dioxide has been used for solvents like THF, acetone, and acetone/ethanol mixtures. In flash evaporation processes, preheated liquid stream is partially vaporized by spraying it inside a vacuum chamber. Concentration of THF was reduced from 10 wt% to less than 0.5 wt%. CO_2 extraction process also produced lower quantities of solvent in polymeric nanoparticle suspensions, compared to that in the normal evaporation process [58, 59].

Purification

The nanoparticles are administered via orally or parenteral route. Thus, the nanoparticle suspension formed should be pure to ensure biological tolerance [8, 60]. Potentially toxic materials, such as organic solvents, stabilizers, and residual polymer aggregates, are removed during the purification process. Generally, two types of approaches are implemented for purifying suspensions in nanoparticles used in pharmaceutical applications. (a) The separation of the solid particles by filtration from the liquid or fast-drying methods followed by washing the gel or powder obtained with deionised water. Large polymer aggregates are removed by filtration with the help of sintered glass filters [56]. (b) Centrifugation aims to settle the particles by activating their sedimentation by the action of centrifugal force. After many successive rounds, most of the supernatant liquid is exchanged with pure solvent with the help of the settling effect, which leads to purified suspensions. However, both of these processes have their limitations. Filtration is inefficient in the case of filtering nano-sized particles as they can also pass through the pores of filters, and centrifugation is not effective for nano-sized particles [61]. Hence, the purification of nanoparticles is performed by dialysis using water [5], filtration through 0.45 μm [62], and dialysis using polymer solution [63]. These methods have been successfully used to purify small batches of nanoparticle suspensions. Crossflow microfiltration has been suggested as a suitable technique, which can be used industrially to purify nanoparticle suspensions [8].

Stabilization

The stability of nanoparticle suspensions is an important factor, as it determines the efficacy of the nanocapsules used in drug delivery applications. The stability is determined by the balance between the attractive van der Waals force and repulsive electrostatic force caused by the double layer of the oppositely charged ions [64]. Attraction and aggregation of nanoparticles are due to induced dipole—dipole forces (London dispersion forces). Induced dipole force is a part of the van der Waals forces. These forces result from multipoles formed in molecules, caused by quantum induced instantaneous polarization. The formation of instantaneous dipoles occurs, because the electrons in adjacent molecules redistribute due to their correlated movements.

Electrostatic force depends on two parameters, Nernst potential and Zeta potential. Nernst potential has little effect on stabilization and is not practically assessable. Zeta potential is the potential difference between the electroneutral region and actual nanoparticle surface. The condition for stability of nanoparticles is that the electrostatic repulsive forces should be higher than the attractive van der Waals forces. Hence, a higher value of zeta potential will result in a more colloidal stability of nanoparticles [65]. The different strategies to increase the stability of nanoparticles for biomedical applications are as follows. The first is electrostatic stabilization, in which a electric double layer around nanoparticles is formed. The second is steric stabilization in which a steric stabilizer is chemically attached or adsorbed on the nanoparticle surface. In the third approach, both the above methods are combined. Electrosteric stabilization has many advantages, such as redispersibility of nanoparticles, ability to support high concentration of nanoparticle suspensions, and electrolytic insensitivity [65–68].

Applications

Novel PNPs have been a topic of interest since two decades, owing to their various applications in biotechnology, optoelectronics, photonics, and environmental technology. Nanoparticles have a larger surface area to volume ratio than that of bulk materials, which can be used to initiate surface interactions. The properties of PNPs are reformed according to different applications. The manufacturing technique of nanoparticles plays an important role to achieve the target properties, such as particle size and load carrying capacity.

Nanoparticles have some specific characteristics that are better than those of bulk materials, such as uniform geometry, increased binding capacity, chemical reactivity, kinetics, stability, and easy dispersive quality. Hence, PNPs are used in biomedical applications. They have shown potential in various aspects, discussed below.

Biomedical Applications

PNPs have provided promising solutions in biomedical applications due to their small size, biocompatibility, adaptability in formulations, and controlled release of drugs. They provide many advantages over traditional drug dosage routes. Hence, extensive research has been conducted on the vaccination and treatment of several diseases including tumor, cancer, and diabetes. The importance of size and parameters controlling nanoprecipitation has been discussed in this chapter. Liposomes have been conventional drug carriers, but polymeric nanoparticles have better advantages over liposomes. Some advantages of PNPs over conventional particles are: (a) they provide great stability in storage of biological fluids, (b) their small size allows parental administration easy and controlled release of drugs to a particular organ, and (c) increased efficiency in targeted drug release reduces the amount of drug use in therapeutic application, which reduces the unwanted toxicity. Use of polymeric nanoparticles reduces various problems, such as leakage of drugs in the presence of blood and poor encapsulation efficiency encountered, when lipid was used as drug carrier [11].

The characteristic applications of PNPs in the biomedical field are (i) PNPs loaded with corticoids are used for treating diseases like uveitis, (ii) targeting tumor by carrying and accumulating drugs to a specific site, (iii) carrying loaded drugs beyond blood brain barrier, and (iv) encapsulating genetic materials to protect them from degradation. These medical uses are of paramount importance in the current scenario [69].

The drug can be entrapped, attached, encapsulated or dissolved to a nanoparticle matrix depending on the form of nanosphere or nanocapsule. Drug can be loaded in either of the two ways: adsorbing the drug after the formation of nanoparticles or fuse the drug during nanoparticle formations. The hydrophobicity of the polymer and specific area of the nanoparticle determine the extent of adsorption of drugs. It has been observed that polymers with longer alkyl chain results in higher affinity for drugs [10].

Nanoprecipitation has played an important role in incorporating hydrophobic drugs into nanoparticles by precipitation of solution of drug-polymer in water. Hydrophobic drugs like docetaxel and paclitaxel have been incorporated into PLGA nanoparticles by employing nanoprecipitation. These antimitotic agents in nanoparticles show increased efficacy and lower toxicity compared to formulations, based on micelle using surfactants with low molecular weight [70]. To encapsulate more hydrophilic substances like protein by using an accurate combination of solvent, non-solvent and polymer, the nanoprecipitation has been used. PLGA and PLA nanoparticles incorporating protein can be obtained by precipitation using Dimethyl Sulfoxide (DMSO) as a solvent and alcohol as a non-solvent [12].

Nanoparticles are able to cross the Reticuloendothelial system (RES) due to their stealth characteristic that enhances the ‘Enhanced Permeability and Retention’ (EPR) effect. An increased EPR aids in treating cancers and tumors. The nanocarriers produced by the nanoprecipitation technique had PACA core, stealth property, and colloidal stability provided by PEG shell and active targeting provided by terminal ligands attached [71]. Paclitaxel was encapsulated in nanoparticles, which provided specific in vitro anticancer activity. Additionally, the nanoparticles developed by the nanoprecipitation method are used to treat Alzheimer's disease. Multifunctional nanoparticles (PACA core) carrying different hydrophobic drugs with appropriate ligands can be used to target different pathologies.

Moreover, nanoprecipitation has been used for the preparation of polymersomes and nanocapsules. Polymersomes are a new class of vesicles, which were inspired by phospholipids that self-assembled in liposomes. They have been obtained by nanoprecipitation using amphiphilic diblock copolymers. Polymersome showed higher stability and lower permeability of the shell compared to viral capsids. Polymersomes were loaded with doxorubicin using co-precipitation and were used to deliver drugs to breast cancer cells [71–73]. Nanocapsules were formed by dissolving the solution of polymer with a small amount of oil and an active compound in water. Precipitation of the hydrophobic polymer on the surface of oil droplets leads to the formation of core shell nanocapsule [9].

Summary

Nanoprecipitation is one of the commonly used processing methods to produce polymeric nanoparticles. Since the method is efficient, economical, and accurate, nanoprecipitation is better than the other methods, such as dialysis, solvent evaporation, and salting-out. It is a technique used to produce nanoparticles in the form of nanocapsules, nanospheres, and polymersomes. Polymeric nanoparticles have gained the interest, because of their applications, such as drug delivery, targeting specific organs, and reducing the problems associated with conventional particles. Therefore, proficient methods like nanoprecipitation are necessary to produce nanoparticles in an efficient way for many applications.

Nanoprecipitation is the process in which the polymer is first dissolved in a good solvent, generally a hydrophilic organic solvent, then the solution is mixed with aqueous phase, which leads to the formation of the nanoparticles. Surfactants can also be used in the process to control the size of the particles formed, by controlling the aggregation of the particles.

The nanoparticle size is affected by various operating parameters and materials used in nanoprecipitation. These effects have been discussed thoroughly in the chapter. The relationship of nanoparticle size with operating parameters, such as organic phase addition rate, flow rate ratio of solvent to antisolvent, stirring rate, configuration, and mixing time, has been discussed. By analyzing these different effects, it was found that proper mixing should be ensured between the two phases to obtain smaller nanoparticles. Also, the particle size is not dependent on a single operating parameter, rather it is affected by the combination of the various operating parameters mentioned. The influence of different properties of polymer, solvent, and surfactant on the nanoparticle size has also been discussed. The type of polymer used and its concentration, the type of solvent and the molar mass of surfactant were found to be critical factors to affect the nanoparticle size. The extent of the effect of the controlling parameters and material are still being researched. The final size is either influenced by the combination of all these parameters, or one of the parameters can have a dominating effect on the nanoparticle size over others depending upon the interrelationship between parameters.

There are different methods of nanoprecipitation, which are used to mix the two phases (organic and aqueous phases). To increase the yield and reduce the size of the nanoparticles, a proper method of mixing is necessary. The efficient methods, like hydrodynamic flow focusing and CJIM were discussed with a common method like bulk pouring. Continuous efforts have been made to develop more efficient configuration to precipitate nanoparticles and to improve the mixing with reduced mixing time.

Nanoprecipitation is a complex process; nanoparticles are precipitated in the regions of high supersaturation created by rapid mixing of the phases, where there is an instantaneous diffusion of the organic phase into the aqueous phase. Hence, a proper understanding of the mechanism of the particle growth is vital. Understanding the mechanism of the particle growth will aid in controlling the size and yield of the nanoparticles. Two mechanisms have been proposed for the growth: (a) the formation of nanoparticles results from the turbulences developed due to interfacial tension, and (b) particles are developed due to nucleations, which lead to the growth and aggregation of particles. The mechanism of particle growth has a significant influence on the size of the particle developed. During the particle growth it is possible that either both the mechanisms occur simultaneously, or one of the mechanism dominates over the other, depending upon the concentration of the organic phase.

Different post-processing treatments are carried on nanoparticles to stabilize, purify, and concentrate them. Nanoparticles are stabilized to make them suitable for long-term applications, purified to remove any impurity present in the nanoparticles, and concentrated to increase the drug loading capacity.

Different applications of nanoprecipitation were discussed in the field of biomedical technology. The advantages of nanoprecipitation, such as incorporating hydrophobic drugs into the nanoparticles and producing different particles like polymersomes, have been discussed.

Nanoprecipitation is an easy and reproducible technique. The rate of nanoparticle production can be high enough to meet the requirements of various applications. Development of polymeric nanoparticles by nanoprecipitation is a technology which requires selecting an appropriate polymer, solvent and surfactant, with an efficient mixing method. Different operating parameters, such as solvent used, mixing method, interfacial tension, and the type of polymer used, plays a vital role in generating nanoparticles with suitable size, good surface properties, and load carrying capacity. Specifically, the mechanism of particle formation has not been explored in detail; hence its effect on the final size of nanoparticles is yet to be understood. Moreover, there are several challenges that need to be addressed pertaining to loading drugs into nanoparticles, co-precipitation of different compounds, mechanisms of particle formation and low polymer concentrations. Future research needs to focus on these aspects to further improve the nanoprecipitation technique.

References

1. www.sciencedirect.com. Accessed 10 Jan 2015
2. Rao JP, Geckeler KE (2011) Polymer nanoparticles: preparation techniques and size-control parameters. Prog Polym Sci 36(7):887–913
3. Leite EA, Grabe-Guimarães A, Guimarães HN, Machado-Coelho GLL, Barratt G, Mosqueira VCF (2007) Cardiotoxicity reduction induced by halofantrine entrapped in nanocapsule devices. Life Sci 80(14):1327–1334
4. Cruz L, Soares LU, Costa TD, Mezzalira G, da Silveira NP, Guterres SS, Pohlmann AR (2006) Diffusion and mathematical modeling of release profiles from nanocarriers. Int J Pharm 313(1–2):198–205
5. Mora-Huertas CE, Fessi H, Elaissari A (2010) Polymer-based nanocapsules for drug delivery. Int J Pharm 385(1–2):113–142
6. Lepeltier E, Bourgaux C, Couvreur P (2014) Nanoprecipitation and the "Ouzo effect": application to drug delivery devices. Adv Drug Deliv Rev 71:86–97
7. Mora-Huertas CE, Fessi H, Elaissari A (2011) Influence of process and formulation parameters on the formation of submicron particles by solvent displacement and emulsification–diffusion methods: critical comparison. Adv Colloid Interface Sci 163(2): 90–122
8. Quintanar-Guerrero D, Allémann E, Fessi H, Doelker E (1998) Preparation techniques and mechanisms of formation of biodegradable nanoparticles from preformed polymers. Drug Dev Ind Pharm 24(12):1113–1128
9. Fessi H, Puisieux F, Devissaguet JP, Ammoury N, Benita S (1989) Nanocapsule formation by interfacial polymer deposition following solvent displacement. Int J Pharm 55(1):R1–R4
10. Soppimath KS, Aminabhavi TM, Kulkarni AR, Rudzinski WE (2001) Biodegradable polymeric nanoparticles as drug delivery devices. J Controlled Release 70(1):1–20
11. Pinto Reis C, Neufeld RJ, Ribeiro AJ, Veiga F (2006) Nanoencapsulation II. Biomedical applications and current status of peptide and protein nanoparticulate delivery systems. Nanomed Nanotechnol Biol Med 2(2):53–65
12. Bilati U, Allémann E, Doelker E (2005) Development of a nanoprecipitation method intended for the entrapment of hydrophilic drugs into nanoparticles. Eur J Pharm Sci 24(1):67–75
13. Chorny M, Fishbein I, Danenberg HD, Golomb G (2002) Lipophilic drug loaded nanospheres prepared by nanoprecipitation: effect of formulation variables on size, drug recovery and release kinetics. J Controlled Release 83(3):389–400
14. Gavory C, Durand A, Six JL, Nouvel C, Marie E, Leonard M (2011) Polysaccharide-covered nanoparticles prepared by nanoprecipitation. Carbohydr Polym 84(1):133–140
15. Aubry J, Ganachaud F, Cohen Addad JP, Cabane B (2009) Nanoprecipitation of polymethylmethacrylate by solvent shifting: 1. Boundaries. Langmuir 25(4):1970–1979
16. Becer CR, Babiuch K, Pilz D, Hornig S, Heinze T, Gottschaldt M, Schubert US (2009) Clicking pentafluorostyrene copolymers: synthesis, nanoprecipitation, and glycosylation. Macromolecules 42(7):2387–2394
17. Lebouille JGJL, Stepanyan R, Slot JJM, Cohen Stuart MA, Tuinier R (2014) Nanoprecipitation of polymers in a bad solvent. Colloids Surf A 460:225–235

18. Cheng J, Teply BA, Sherifi I, Sung J, Luther G, Gu FX, Levy-Nissenbaum E, Radovic-Moreno AF, Langer R, Farokhzad OC (2007) Formulation of functionalized PLGA–PEG nanoparticles for in vivo targeted drug delivery. Biomaterials 28(5):869–876
19. Ameller T, Marsaud V, Legrand P, Gref R, Barratt G, Renoir JM (2003) Polyester-poly (ethylene glycol) nanoparticles loaded with the pure antiestrogen RU 58668: physicochemical and opsonization properties. Pharm Res 20(7):1063–1070
20. Arbos P, Wirth M, Arangoa M, Gabor F, Irache J (2002) Gantrez® AN as a new polymer for the preparation of ligand–nanoparticle conjugates. J Controlled Release 83(3):321–330
21. Aumelas A, Serrero A, Durand A, Dellacherie E, Leonard M (2007) Nanoparticles of hydrophobically modified dextrans as potential drug carrier systems. Colloids Surf B 59(1):74–80
22. Galindo-Rodriguez S, Allémann E, Fessi H, Doelker E (2004) Physicochemical parameters associated with nanoparticle formation in the salting-out, emulsification-diffusion, and nanoprecipitation methods. Pharm Res 21(8):1428–1439
23. Guhagarkar SA, Malshe VC, Devarajan PV (2009) Nanoparticles of polyethylene sebacate: a new biodegradable polymer. Aaps Pharmscitech 10(3):935–942
24. Plasari E, Grisoni P, Villermaux J (1997) Influence of process parameters on the precipitation of organic nanoparticles by drowning-out. Chem Eng Res Des 75(2):237–244
25. Thioune O, Fessi H, Devissaguet JP, Puisieux F (1997) Preparation of pseudolatex by nanoprecipitation: influence of the solvent nature on intrinsic viscosity and interaction constant. Int J Pharm 146(2):233–238
26. Yoo HS, Choi HK, Park TG (2001) Protein–fatty acid complex for enhanced loading and stability within biodegradable nanoparticles. J Pharm Sci 90(2):194–201
27. Lannibois H, Hasmy A, Botet R, Chariol OA, Cabane B (1997) Surfactant limited aggregation of hydrophobic molecules in water. J Phys II 7(2):319–342
28. Lee H, Fonge H, Hoang B, Reilly RM, Allen C (2010) The effects of particle size and molecular targeting on the intratumoral and subcellular distribution of polymeric nanoparticles. Mol Pharm 7(4):1195–1208
29. Hillaireau H, Couvreur P (2009) Nanocarriers' entry into the cell: relevance to drug delivery. Cell Mol Life Sci 66(17):2873–2896
30. Lince F, Marchisio DL, Barresi AA (2008) Strategies to control the particle size distribution of poly-ε-caprolactone nanoparticles for pharmaceutical applications. J Colloid Interface Sci 322(2):505–515
31. Bally F, Garg DK, Serra CA, Hoarau Y, Anton N, Brochon C, Parida D, Vandamme T, Hadziioannou G (2012) Improved size-tunable preparation of polymeric nanoparticles by microfluidic nanoprecipitation. Polymer 53(22):5045–5051
32. Lince F, Marchisio DL, Barresi AA (2011) A comparative study for nanoparticle production with passive mixers via solvent-displacement: use of CFD models for optimization and design. Chem Eng Process 50(4):356–368
33. Karnik R, Gu F, Basto P, Cannizzaro C, Dean L, Kyei-Manu W, Langer R, Farokhzad OC (2008) Microfluidic platform for controlled synthesis of polymeric nanoparticles. Nano Lett 8 (9):2906–2912
34. Johnson BK, Prud'homme RK (2003) Mechanism for rapid self-assembly of block copolymer nanoparticles. Phys Rev Lett 91(11):118302
35. Lemoine D, Francois C, Kedzierewicz F, Preat V, Hoffman M, Maincent P (1996) Stability study of nanoparticles of poly (ε-caprolactone), poly (D, L-lactide) and poly (D, L-lactide-co-glycolide). Biomaterials 17(22):2191–2197
36. Leroueil-Le Verger M, Fluckiger L, Kim YI, Hoffman M, Maincent P (1998) Preparation and characterization of nanoparticles containing an antihypertensive agent. Eur J Pharm Biopharm 46(2):137–143
37. Whitesides TH, Ross DS (1995) Experimental and theoretical analysis of the limited coalescence process: stepwise limited coalescence. J Colloid Interface Sci 169(1):48–59

38. Legrand P, Lesieur S, Bochot A, Gref R, Raatjes W, Barratt G, Vauthier C (2007) Influence of polymer behaviour in organic solution on the production of polylactide nanoparticles by nanoprecipitation. Int J Pharm 344(1–2):33–43
39. Lourenco C, Teixeira M, Simões S, Gaspar R (1996) Steric stabilization of nanoparticles: size and surface properties. Int J Pharm 138(1):1–12
40. Avgoustakis K (2004) Pegylated poly (lactide) and poly (lactide-co-glycolide) nanoparticles: preparation, properties and possible applications in drug delivery. Curr Drug Deliv 1(4): 321–333
41. D'Addio SM, Prud'homme RK (2011) Controlling drug nanoparticle formation by rapid precipitation. Adv Drug Deliv Rev 63(6):417–426
42. Johnson BK, Prud'homme RK (2003) Chemical processing and micromixing in confined impinging jets. AIChE J 49(9):2264–2282
43. Jun H, Fabienne T, Florent M, Coulon PE, Nicolas M, Olivier S (2012) Understanding of the size control of biocompatible gold nanoparticles in millifluidic channels. Langmuir 28(45):15966–15974
44. Liu Y, Fox RO (2006) CFD predictions for chemical processing in a confined impinging-jets reactor. AIChE J 52(2):731–744
45. Bouchemal K, Briançon S, Perrier E, Fessi H (2004) Nano-emulsion formulation using spontaneous emulsification: solvent, oil and surfactant optimisation. Int J Pharm 280(1): 241–251
46. Brick MC, Palmer HJ, Whitesides TH (2003) Formation of colloidal dispersions of organic materials in aqueous media by solvent shifting. Langmuir 19(16):6367–6380
47. Stainmesse S, Orecchioni AM, Nakache E, Puisieux F, Fessi H (1995) Formation and stabilization of a biodegradable polymeric colloidal suspension of nanoparticles. Colloid Polym Sci 273(5):505–511
48. Yu W, do Egito E, Barratt G, Fessi H, Devissaguet JP, Puisieux F (1993) A novel approach to the preparation of injectable emulsions by a spontaneous emulsification process. Int J Pharm 89(2):139–146
49. Demello AJ (2006) Control and detection of chemical reactions in microfluidic systems. Nature 442(7101):394–402
50. Knight JB, Vishwanath A, Brody JP, Austin RH (1998) Hydrodynamic focusing on a silicon chip: mixing nanoliters in microseconds. Phys Rev Lett 80(17):3863
51. Valencia PM, Basto PA, Zhang L, Rhee M, Langer R, Farokhzad OC, Karnik R (2010) Single-step assembly of homogenous lipid–polymeric and lipid–quantum dot nanoparticles enabled by microfluidic rapid mixing. ACS Nano 4(3):1671–1679
52. Rhee M, Valencia PM, Rodriguez MI, Langer R, Farokhzad OC, Karnik R (2011) Synthesis of size-tunable polymeric nanoparticles enabled by 3D hydrodynamic flow focusing in single-layer microchannels. Adv Mater 23(12):H79–H83
53. Wong I, Ho CM (2009) Surface molecular property modifications for poly (dimethylsiloxane) (PDMS) based microfluidic devices. Microfluid Nanofluid 7(3):291–306
54. Valente I, Celasco E, Marchisio DL, Barresi AA (2012) Nanoprecipitation in confined impinging jets mixers: production, characterization and scale-up of pegylated nanospheres and nanocapsules for pharmaceutical use. Chem Eng Sci 77:217–227
55. Pohlmann AR, Mezzalira G, de Garcia Venturini C, Cruz L, Bernardi A, Jäger E, Battastini AMO, da Silveira NP, Guterres SS (2008) Determining the simultaneous presence of drug nanocrystals in drug-loaded polymeric nanocapsule aqueous suspensions: a relation between light scattering and drug content. Int J Pharm 359(1–2):288–293
56. Limayem I, Charcosset C, Fessi H (2004) Purification of nanoparticle suspensions by a concentration/diafiltration process. Sep Purif Technol 38(1):1–9
57. Ma J, Feng P, Ye C, Wang Y, Fan Y (2001) An improved interfacial coacervation technique to fabricate biodegradable nanocapsules of an aqueous peptide solution from polylactide and its block copolymers with poly (ethylene glycol). Colloid Polym Sci 279(4):387–392
58. Campardelli R, Della Porta G, Reverchon E (2012) Solvent elimination from polymer nanoparticle suspensions by continuous supercritical extraction. J Supercrit Fluids 70:100–105

59. Kumar V, Prud'homme RK (2009) Nanoparticle stability: processing pathways for solvent removal. Chem Eng Sci 64(6):1358–1361
60. Quintanar-Guerrero D, Fessi H, Allémann E, Doelker E (1996) Influence of stabilizing agents and preparative variables on the formation of poly(D, L-lactic acid) nanoparticles by an emulsification-diffusion technique. Int J Pharm 143(2):133–141
61. Al-kattan A, Dufour P, Drouet C (2011) Purification of biomimetic apatite-based hybrid colloids intended for biomedical applications: a dialysis study. Colloids Surf B 82(2):378–384
62. Schaffazick SR, Pohlmann AR, Dalla-Costa T, Guterres SS (2003) Freeze-drying polymeric colloidal suspensions: nanocapsules, nanospheres and nanodispersion. A comparative study. Eur J Pharm Biopharm 56(3):501–505
63. Stella B, Arpicco S, Rocco F, Marsaud V, Renoir JM, Cattel L, Couvreur P (2007) Encapsulation of gemcitabine lipophilic derivatives into polycyanoacrylate nanospheres and nanocapsules. Int J Pharm 344(1):71–77
64. Verwey EJW, Overbeek JTG, Van Nes K (1948) Theory of the stability of lyophobic colloids: the interaction of sol particles having an electric double layer. Elsevier, New York
65. Kulhari H, Kulhari DP, Singh MK, Sistla R (2014) Colloidal stability and physicochemical characterization of bombesin conjugated biodegradable nanoparticles. Colloids Surf A 443:459–466
66. Chen KL, Smith BA, Ball WP, Fairbrother DH (2010) Assessing the colloidal properties of engineered nanoparticles in water: case studies from fullerene C60 nanoparticles and carbon nanotubes. Environ Chem 7(1):10–27
67. Luckham PF (1996) Recent advances in polymers at surfaces: the steric effect. Curr Opin Colloid Interface Sci 1(1):39–47
68. Ortega-Vinuesa J, Martın-Rodrıguez A, Hidalgo-Alvarez R (1996) Colloidal stability of polymer colloids with different interfacial properties: mechanisms. J Colloid Interface Sci 184(1):259–267
69. Cismaru L, Popa M (2010) Polymeric nanoparticles with biomedical applications. Rev Roum Chim 55(8):433–442
70. Gaucher G, Marchessault RH, Leroux JC (2010) Polyester-based micelles and nanoparticles for the parenteral delivery of taxanes. J Controlled Release 143(1):2–12
71. Le Droumaguet B, Nicolas J, Brambilla D, Mura S, Maksimenko A, De Kimpe L, Salvati E, Zona C, Airoldi C, Canovi M (2012) Versatile and efficient targeting using a single nanoparticulate platform: application to cancer and Alzheimer's disease. ACS Nano 6(7): 5866–5879
72. Discher BM, Won YY, Ege DS, Lee JC, Bates FS, Discher DE, Hammer DA (1999) Polymersomes: tough vesicles made from diblock copolymers. Science 284(5417):1143–1146
73. Upadhyay KK, Bhatt AN, Mishra AK, Dwarakanath BS, Jain S, Schatz C, Le Meins JF, Farooque A, Chandraiah G, Jain AK (2010) The intracellular drug delivery and anti tumor activity of doxorubicin loaded poly (γ-benzyl L-glutamate)-*b*-hyaluronan polymersomes. Biomaterials 31(10):2882–2892
74. Molpeceres J, Guzman M, Aberturas MR, Chacon M, Berges L (1996) Application of central composite designs to the preparation of polycaprolactone nanoparticles by solvent displacement. J Pharm Sci 85(2):206–213

Chapter 10
Fabrication and Properties of Spin-Coated Polymer Films

Raj Das and Avishek Chanda

Introduction

Spin coating is a common technique to produce a thin, high quality and uniform polymer film on a plane substrate [1, 2]. In the recent years, owing to their exceptional properties, thin polymers films have received a great deal of attention [3]. Many studies have observed that thin polymer films exhibit unique properties when compared to bulk materials. The initial works date back to the time when Prest and Luca [4] noted a rise in the birefringence with decreasing film thickness. Such research paved the way for the evolution of thin polymer films. Reduction in the degree of crystallinity and crystallisation [5], reduction in glass transition temperature [6] and superior toughening mechanisms [7, 8] have been observed with the reduction in the thickness of the polymer films. Thin films are also very useful in the applications where surface properties are crucial for biocompatibility and adhesion of the films [3]. Polymer films as protective layers against ultraviolet light rays, corrosion, scratching and humidity are extensively used in the industries for the past few decades [9, 10].

There are many methods for producing polymer films with very thin thicknesses with each having its own advantages and limitations [10]. The different techniques are summarised in Fig. 10.1. The purpose of the basic working principle of all the techniques is to produce a film of the polymer solution on a substrate which can be in different orientations and can either be planar or even non-planar. The main difference between the techniques is the method of film formation and the controllability of each for film thickness.

R. Das (✉) · A. Chanda
Department of Mechanical Engineering, University of Auckland,
20 Symonds Street, 1010 Auckland, New Zealand
e-mail: r.das@auckland.ac.nz

A. Chanda
e-mail: acha553@aucklanduni.ac.nz

S. Fakirov (ed.), *Nano-size Polymers*,
DOI 10.1007/978-3-319-39715-3_10

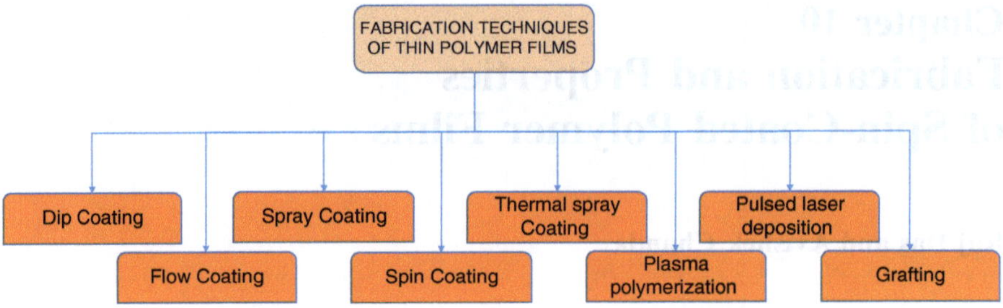

Fig. 10.1 Commonly available techniques for processing thin polymer films

The thickness of the polymer film, during dip-coating, is controlled by the withdrawal speed of the substrate immersed in the polymer solution [11–13]; whereas, during flow coating, the substrate's angle of inclination, on which the polymer solution is guided and the viscosity of the solution controls the final polymer thickness [14, 15]. The spray coating technique is comparably faster than the dip and flow coating methods and thus, is used widely, though, for non-planar substrates, the controllability of the film thickness is similar to that of the dip-coating method [16, 17]. The other techniques are improvised versions of the previous techniques. The thermal spray coating [17–19] and pulsed laser deposition [20] methods are solvent free; whereas the plasma polymerization [21] and grafting methods [22] depend on the chemical reaction between the polymer solution and the substrate, the former being used via developing a magnetic field and the latter grafting the polymers onto the substrate.

The spin coating process involves the deposition of thin and uniform polymer films on planar substrates. This method can be broadly divided into four stages or processes, namely, deposition, spin-up, spin-off and evaporation [23]. The substrate is mounted on a rotating platform called the spin coater or spinner. An excess amount of the polymer solution is first deposited on the centre of the substrate's surface, mounted on the spin-coater, when it is at rest or rotating at a very low speed (deposition stage). The substrate is then rotated, with the help of the coater, at a very high speed (1000–4000 rpm) [24, 25], resulting in the liquid (polymer solution) flowing radially outwards due to the centrifugal force (spin-up stage). The third stage incorporates the outward flow of the excess polymer solution towards the perimeter, forming accumulations and subsequently being ejected (spin-off stage). The final stage leads to further thinning of the polymer coating by evaporation of the polymer solution (evaporation stage), leaving behind a film in the solid or the liquid state [10]. The first three processes take place in sequence; whereas, evaporation occurs throughout the process [26]. The thickness of the final polymer film is insignificantly affected by the deposition and spin-up processes, the main deciding stage being the spin-off one. This method for polymer coating can be used more efficiently, compared to the other techniques, for coating a large area with a uniform polymer film [10] and this technique allows to maintain superior control on the thickness of the final film and thus is extremely effective, especially for planar or slightly curved substrates.

The highly uniform polymer coating is formed because of the resultant of two forces acting in the opposite directions, the centrifugal force acting outwards and the viscous force, which is the resisting force, acting inwards [26]. The thickness of the coated polymer is dependent mostly on the subtle transition process, experiencing evaporation, during the spin-off process, resulting in the unwanted volatile components being ejected and the unwanted non-volatile components being eliminated by the flow due to the centrifugal force. Meyerhofer [27] derived an expression for the thickness of the polymer coating at the end of the spin-off process given as:

$$h_{Spin-off} = \left(\frac{3\mu e}{2\rho_{AO}\omega^2}\right)^{1/3} \tag{10.1}$$

The time duration of the spin-off process to attain this thickness will be

$$t_{Spin-off} = \frac{3\mu}{4\pi\omega^2}\left(\frac{1}{h^2_{Spin-off}} - \frac{1}{h^2_O}\right) \tag{10.2}$$

In Eqs. (10.1) and (10.2), it is assumed that the layer of the polymer solution, applied initially, has a uniform thickness of h_O and contains volatile substances of density ρ_{AO}. The viscosity of the polymer solution is denoted by μ and e represents the evaporation rate, assumed to be constant. For the spin coating process, the angular velocity of the coater is represented by ω. It is assumed that evaporation commences when the rate of film thinning process is reduced below the threshold value of $e\rho_A/\rho_{AO}$, where ρ_A is the density of the volatile part of the solvent at that instance [26, 27]. In general, the higher the angular velocity of the coater, the thinner is the coating of the polymer film.

The final thickness of the polymer film and the total time elapsed, including the evaporation stage, was derived by Meyerhofer [27] as

$$\begin{aligned} h_{Final} &= \left(1 - \frac{\rho_{AO}}{\rho_A}\right)\left(\frac{3\mu e}{2\rho_{AO}\omega^2}\right)^{1/3} \\ t_{Final} &= t_{Spin-off} + h_{Spin-off}\frac{\rho_{AO}}{e\rho_A} \end{aligned} \tag{10.3}$$

The wide-spread use and the overwhelming interest in spin coating are because of its numerous possible applications in the field of science and engineering. The first ever application of spin coating can be found in the paint coatings, on various products, by the industries [28]. In the present times, spin coating has been used for fabricating thin layers in numerous microelectronic applications [29–43] including the production of photoresists [1]. Spin coating has also found excellent applications in the field of sensor coatings [44–47] and also optical coatings [10, 48].

Advances in the polymer coating technique has resulted in the polymerization or cross-linking of the spin-coated polymer membranes to form hydrogels [49] and also in the fabrication of porous structures by spin coating polymer membranes that are incompatible [50]. Spin coating has a few benefits over the other processes, which include reduced time for coating and superior uniformity over the planar substrate. Even surfaces with slight curvatures, such as lenses, can be efficiently and uniformly coated with minimal edge effects or thickness variation. Although, the process of spin coating is quite beneficial, it has a few significant limitations, such as, the substrate should always be essentially flat or with low curvature, the process fails for complex surfaces and also the material loss due to ejection of the excess material, during the spin-off stage, is quite high.

The chapter deals with the various fabrication techniques adapted for applying polymer films on substrates using the spin coating method, along with the different and unique properties the films have because of this processing technique. The chapter introduces the theory of the mechanisms involved in the spin coating process, followed by the different fabrication techniques which have been used and finally ending with the discussions on the uniqueness in the properties of the coated polymer films. This work will thus help in understanding the basic concepts of the spin coating technique along with the different methods that have been employed to perform the spin coating process and the resultant properties of the polymer films.

Mechanisms of the Process

The most commonly used polymer is polystyrene, polyvinyl chloride or polyvinyl butyral. This is dissolved in a solvent, like toluene or tetrahydrofuron. The deposition of the liquid polymer solution on a rotating horizontal disc, which remains at rest or rotates at a very low velocity during deposition, results in a uniform polymer film. The disc is then rotated at a high angular velocity in the range of 1000–4000 rpm during the coating process. A shearing force is experienced by the liquid due to the adhesive forces between the liquid and the substrate and the centrifugal force experienced by the rotating polymer solution. Evaporation is an intricate part of the entire process contributing to the reduction in the film thickness and also in the increase of the thickness uniformity [10]. The process of film thinning gradually slows down due to reduction in the film thickness and finally stops due to the excessive increase in the viscosity of the polymer solution. Defects in the coating process, such as skin formation, can be observed if the convective flow of the liquid does not stop during the formation of the final layer [51] and also due to low diffusivity of the solvent and high viscosity of the polymer solution [52]. Figure 10.2 illustrates the functioning of the spin coating process with all the stages involved.

Many parameters affect the effective and efficient performance of the spin coating process. These parameters are dependent on different aspects, such as the

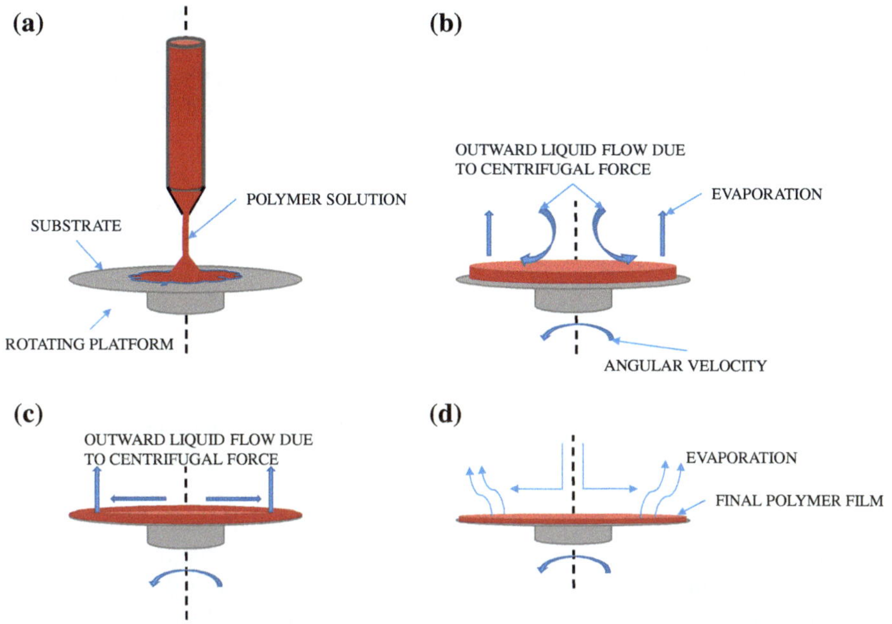

Fig. 10.2 Illustration of the process of spin coating where, **a** denotes the deposition stage, **b** illustrates the spin-up stage, **c** demonstrates the spin-off stage and **d** elucidates the final evaporation stage

rotational parameters and the properties of the solution. The morphology of the film is also an important aspect to be considered during the process. The different theories related to the modelling of the spin coating process, the morphological dependencies of the polymer film and other important factors are discussed in this part.

Modelling of the Process

The involvement of different mechanisms makes the modelling of the spin coating process a complex and difficult task. The complexity of the spin coating process has resulted in many approximations and assumptions for understanding the process via theoretical and analytical modelling. Emslie et al. [53] were the pioneers in theoretically describing the process. The Newtonian behaviour, that is, the linear relationship between the shear rate and shear stress, of the liquid solution, was assumed in the process. Acrivos et al. [54] proposed an analytical expression for spin coating of non-Newtonian fluids, following the power law. The other parameters involved in the process include Coriolis forces, temporal and spatial gradients of the concentrations, gravitational force, vertical diffusivity and viscosity of the solution

[23, 27, 55–58]. Many experimental studies have also been conducted in order to infer the correlation between the film thickness and the process and liquid parameters [59–69]. It has been observed that the angular velocity, concentration of the polymer solution, viscosity of the solution and rate of solvent evaporation, solvent diffusivity and solvent volatility considerably affect the final thickness of the polymer film. The parameters, such as the amount of polymer solution deposited, the rate of deposition, the initial angular velocity and the time required for spinning, were not found to influence the final film thickness considerably [10].

Critical Parameters

The most important parameters governing the quality of the polymer film are the angular velocity of the coater, the concentration and the viscosity of the polymer solution. The angular velocity of the spin-coater and the concentration of the solution directly affect the final film thickness [23, 57]. The viscosity of the polymer solution is, in turn, directly related to its concentration. Thus, the three parameters are inter-related and have pronounced effects on the thickness of the final polymer film. A general relationship exists between the angular velocity and the film thickness depending on the concentration and thus the viscosity of the polymer solution [10]. The relationship can be stated as

$$h = k\omega^{\alpha} \tag{10.4}$$

where ω is the angular velocity of the spin-coater, h is the thickness of the film and α and k are the constant exponents derived empirically depending on the physical aspects of the polymer solution, the substrate, the solvent and the interaction between the polymer solution and the substrate.

The value of α has been found to be around -0.5 for different polymer solutions from experiments [23, 27, 60, 61, 65, 67]. Lawrence [57] proposed that the deviation of the derived α is dependent on the process and is found to be more negative (<-0.5) when the coating ceases before the completion of the radial flow, resulting in the formation of thick films at lower speeds. On the other hand, if the angular velocity is high and the concentration of the polymer solution is low, the derived exponent (α) becomes less negative (>-0.5) [10].

The value of k is directly dependent on the viscosity of the polymer and can be expressed as:

$$k = k'\eta_0^{\beta} \tag{10.5}$$

where, k' is a proportionality constant derived from experiments, η_0 is the viscosity of the initial solution and β is a constant exponent and has a value in the range of 0.29–0.39 in case of polymer solutions [59, 60, 67, 68]. Equations (10.4) and (10.5) were studied by Spangler et al. [67] and they inferred that the values of the exponents α and β are not dependent on the polymer solution; whereas, the constant of proportionality, k', is completely dependent on the solution.

Solution-Dependent Parameters

The polymer solution used for spin coating consists of the polymer and a solvent that is selected according to the application. In the recent times, the use of volatile solvents has gained popularity in several applications, for example, in both electrical [70] and solar cells [71, 72]. Keith et al. [73] were the first to use volatile solvents. These advancements have led to the dependency of the output of the spin coating process (film quality) on the solvent evaporation and volatility, the solute diffusivity and also the surface topology of the final film, which is primarily affected by the solvent.

Solvent Evaporation and Volatility

The effect of solvent evaporation has been considered in many studies, which makes the theoretical models more consistent with the experimental results [23, 27, 55–58]. The evaporation of the solvent of the polymer solution alters the rheological properties of the solution, especially for those solutions which have a high latent heat of evaporation. Evaporation also helps in reducing the temperature due to the effective cooling effect inducing shear thinning and elasticity on the polymer film [10].

Chen [59] amended Eqs. (10.4) and (10.5) in order to incorporate the effect of solvent evaporation on the spin coating process. This incorporated the effects of mass and momentum transfer due to solvent evaporation for quantifying the thickness of the film with respect to the angular velocity, solvent evaporation and viscosity of the solution. The evolved equation can be expressed as

$$h = k'\left(\frac{E\lambda}{C_P}\right)^{\gamma}\eta_0^{\beta}\omega^{\alpha} \tag{10.6}$$

where, C_P is the heat capacity of the solvent, E is the average rate of evaporation and λ is the latent heat of evaporation. α, β and γ are the exponents in Eq. (10.6) with the typical values being, $\alpha = -0.5, \beta = 0.36$ and $\gamma = 0.6$.

The effect of solvent volatility on the final film thickness is relatively straight forward. Solvents with high volatility produce thicker polymer films when compared to those with low volatility, for a given initial viscosity (η_0) and concentration of the polymer [59, 67].

Diffusivity of the Solute

Lawrence [57] first proposed an expression which considered the diffusivity of the solute, which was also found to significantly affect the result. However, the lack of data, on the diffusivity of the solute, limits the application of the equation to some extent. The equation proposed by Lawrence, by again extending Eqs. (10.4) and (10.5), can be stated as:

$$h = k'' C_0 (\eta_0 \Delta_0)^{\beta'} \omega^{\alpha} \tag{10.7}$$

where, C_0 is the initial concentration of the polymer, Δ_0 is the diffusivity of the solute and k'' is a number, of the unit order, which changes with the change in the viscosity and the diffusivity of the solution. The value of the exponent, α, was derived to be −0.5 and the exponent of the diffusivity part, β', was 0.25. According to Eq. (10.7), the viscosity of the initial solution (η_0) and the solute diffusivity increase with an increase in C_0, thus inferring that, increasing C_0 will increase the thickness of the film.

Lawrence [57] assumed the concentration of the solution to be constant, a limitation that was addressed by Bornside et al. [23]. The authors did not make assumptions on the solute concentration behaviour along the depth of the film. Formation of a solid skin on the top surface of the film was observed, a phenomenon which can result in a defective film. This can be minimised or prevented by saturating the atmosphere immediately above the solution with solvent vapour, that is, by using both high and low volatile solvents. Although the work by Bornside et al. [23] was not analogous to the ones by Lawrence [57] and other authors, one key finding was that the temperature has negligible variations during the spin coating process.

Effect of Solvent on the Topology of the Film Surface

Studies have confirmed that the solvent volatility and the polymer–solvent interaction influence the surface topography of the final polymer film formed by spin coating. Spangler et al. [67] considered the extent of coiling, the polymer chain experienced, in the solution and inferred that for an effective solvent, the interaction between the solute(polymer) and the solvent should be more favourable than the interaction between themselves (solvent/solvent or polymer/polymer). The authors also observed that uniform polymer films were produced only for good and less volatile solvents and even experimentally derived the coefficients and exponents according to Mark-Houwink-Sukarada [10]. This exponent relates the viscosity t, the average weight of the polymer and the intrinsic viscosity of the polymer solution and was found to be 0.8 for solvents in the good category.

Graessley [74], before the work of Spangler et al. [67], proposed an expression which approximately considered the concentration of entanglement. The proposed formula can be expressed as:

$$C_e = \frac{\rho M_C}{M} \tag{10.8}$$

where, C_e represents the concentration of entanglement in the solution, M_C is that critical molecular weight at which the entanglement of the polymer in the melt phase is strong enough to affect the viscosity significantly, M is the polymer molecular weight and ρ is the bulk density of the polymer. Other studies, for example that by Lai [65], have also found the influence of solvent on the final appearance of the film, which is in agreement with the results by Spangler et al. [67].

Thin Film Morphology

The synergy and compatibility among the different factors during a spin coating process is of crucial importance for generating uniform polymer films with controlled thickness. The main factors to be considered are the interaction between the polymer and the substrate and that between the polymer and the surrounding air. The other influencing factors include the transition temperature of the film, the molecular weight and its effect on the phase separation, the film thickness and aggregate formations and their control.

Interfacial Interactions

The two main interfaces present during a spin coating process are the polymer–substrate and the polymer–air interfaces. The compatibility between the substrate and the polymer greatly influences the strength of interaction between the substrate and the polymer chain segments in the vicinity of the substrate and thus directly affects the properties of the final polymer film [64, 66, 75]. The mobility of the polymer chains is significantly greater on the free side of the film; although the dynamic effects are believed to further enhance the mobility not only on the surface, but also up to a certain depth below it [76–78]. Thus, localised layer properties, possessed by the thin films, are quite likely to be present at the polymer and substrate interface and at the free surface exposed to air. The average properties of a polymer film are affected mostly due to the interfacial interactions for thin films because of the greater surface-to-volume ratio [10].

Transition Temperature

The glass transition temperature, (T_g), is a very important property and has been extensively studied due to its immense significance [79–90]. The dependence of the glass transition temperature on the film thickness varies with the compatibility between the polymer and the substrate and can thus increase or decrease depending on it [84, 85, 88–90]. An increase in the mobility of polymer chains in the solution, during the coating process, results in reduced glass transition temperature of the film; whereas, the more compatibility of the substrate with the polymer will result in hindering the movement of the polymer chains and thus, will subsequently increase the glass transition temperature of the film [10]. Experiments on anisotropic and isotropic polymer films have shown that the glass transition temperature for the anisotropic films were much higher [91], thus, emphasising on the fact that the interaction between the polymer and the substrate has a pronounced effect on the T_g of the polymer film.

Kim et al. [64] proposed a model which can be used to estimate the glass transition temperature. The model can be stated as:

$$T_g(h) = T_{g,bulk}\frac{h(2k''' + h)}{(\sigma + h)^2} \tag{10.9}$$

where, $T_g(h)$ represents the glass transition temperature when the thickness of the film is h, $T_{g,bulk}$ represents the glass transition temperature of the bulk polymer, k''' is a constant representing the extent of influence of the polymer–substrate interaction on the average glass transition temperature and σ represents the statistical length of the polymer chain segment. This model is entirely based on multi-layered structures, each layer having a different glass transition temperature. Polymer solutions having strong interacting substrates were modelled by Long and Lequeux [75] considering the domains of both fast and slow dynamics. The model includes a critical density term which distinguishes between the fast dynamics at high densities and the slow dynamics at low densities.

Influence of Molecular Weight

The molecular weight of the constituent polymer and the solvent influences two important phenomena, namely, the phase separation and the film thickness. Incompatibility between copolymers (A and B) results in the repulsion between the constituent polymers and when this repulsion becomes strong enough, phase separation can be observed, that is, rich micro-domains of one type (for example A) is formed [92]. The state of the phase, as inferred by Leibler [92], is dependent on the volume fraction of the dominating component (A), the product of the number of monomers (N) in the polymer and the Flory parameters, χ [93]. The relationship can be expressed as:

$$\chi = \frac{\varepsilon_{AB} - 0.5(\varepsilon_{AA} + \varepsilon_{BB})}{k^{*}T} \tag{10.10}$$

where k^* represents the Boltzmann constant, T represents the temperature of the copolymer and ε is the energy of interaction between the two monomers of the individual polymers and also the copolymer (AB, AA and BB). Observation by De Gennes [76] has proved that the critical point for phase separation corresponds to that when $\chi N = 2$. The system was observed to undergo phase separation when the value of χN was in the range from 10 to 15. The value of the Flory parameter is always nearly equal to 0.1 and thus, the phase separation should not occur for copolymers of low molecular weight, that is, when the number of monomers (N) is typically less than 100. The size of the micro-domain is limited by the kinetics on the ordering of the domains signifying that the kinetics slow down for polymers with high molecular weights. Since χ is not dependent on the molecular weight of the polymer, an increase in the value of N results in more phase separation. Although it has also been noticed that the micro-domains that are normal to the surface require the polymer chains to have a coordinated motion, which is slower for higher values of N. Hence, it can be inferred that phase separation is dependent on the molecular weight of the polymer solution and the low molecular weight restricts phase separation. Even with very high values of N, the molecular weight will increase, resulting in the film being in a non-equilibrium state and preventing the monomers from forming ordered domains.

Studies have been conducted on the effects of molecular weight on the final polymer film thickness by Spangler et al. [67]. In order to understand the effect, a new term, called the 'dispersity' of the polymer was introduced, which is the ratio of the weight average molecular weight to the number average molecular weight (M_w/M_n). The experimental results concluded that the final thickness of the film was not dependent on the aforementioned average molecular weights, but on the assumed viscosity average molecular weight M_v. The expression of this average molecular weight can be given as:

$$M_v = \left(\sum w_i M_i^{\delta}\right)^{1/\delta} \tag{10.11}$$

where w_i is the weight fraction of ith component, M_i represents the molecular weight of ith component and δ represents the exponent of Mark-Hauwink-Sukrada [10, 67]. Thus, the type and distribution of molecular weight of the solvent influences M_v. However, there is still a lack of in understanding relating to which type of molecular weight influences the film thickness. Weill and Dechenaux [69] concluded from their observations that the film thickness is dependent on the value of M_w.

Other Factors Influencing the Spin Coating Process

There are many other parameters which can affect the functioning and efficiency of the spin-coater in forming the final polymer film. Attraction forces in the short range may result in the formation of aggregates in a polymer solution and has a profound influence on the photonic and electronic properties of the polymer film [40]. With the increase in concentration, the forces between the polymer chains become stronger resulting in entanglement of the chains to form loose aggregates. A further rise in concentration results in the formation of strong aggregates. Shear thinning is another effect which influences the polymer films. Acrivos et al. [54] inferred, from their analysis, that the phenomenon of shear thinning is more prominent on the edges of the spin coating disc when compared to the interior parts; although the result was not analogous to the experimental results.

The effect of magnetic field in the transverse direction, with respect to the geometry of the body, for a coater with non-uniform rotation was studied numerically by Dandapat and Layek [94]. Finite-difference method was adopted for the study and they concluded that thinner films were produced for non-uniform rotations when an initial impulsive rotation is followed by a high angular velocity. The likelihood of skin formation was reduced by this method. The presence of a transverse magnetic field is supposed to stabilize the liquid flow and in turn suppress the non-uniformities that may form on the polymer film surface.

Fabrication and Properties of the Films

Spin coating has many advantages over other techniques, as mentioned before, which include reproducibility, uniformity, control over the final film thickness and comparatively low cost [95]. These advantages have led to wide usage of this process for flat substrates which has also been extended to optical devices. In order to effectively coat a substrate using spin coating, the fabrication of the polymer films is of utmost importance, and it is also required to understand the properties of the film for suitably applying them in the concerned field.

Experimental Parameters

The factors affecting the experimental process have already been detailed in the aforementioned parts of this chapter. The key aspects required to obtain an optimised final polymer film include a number of parameters. The substrate should be at rest or should be rotated at a very slow speed during the process of polymer deposition. The final thickness of the polymer can directly be controlled by the angular velocity of the spin-coater and also by the concentration of the polymer solution. An increase in the polymer concentration results in thicker films; where as a high angular velocity leads to thinner films of the polymer. The thickness of the polymer film usually varies in the range from 10 nm to 5 μm [95], although special fabrication techniques for unique applications can make the films even more thinner (around 100 nm) [10]. The volatility of the solvent of the polymer solution has also a key effect, since high volatility may result in a cooling effect and subsequent non-uniformities in thickness or defects in the polymer solution. It has also been observed that a high volatility of the solvent results in thick polymer films, when compared to low volatile solvents, for the same initial angular speed and concentration.

The compatibility between the solvent and the polymer is also an important aspect. Greater compatibility between the two produces uniform polymer films with less defects and variability in the final surface. Thus, it is quite useful to choose a solvent which is chemically compatible with the polymer. Some volatility of the solvent is also beneficial for preventing the undesirable skinning effect. Evaporation of the volatile solvent will result in the saturation of the surrounding atmosphere, and thus help in eliminating the formation of the solid skin during the coating process. Natural parameters, such temperature, relative humidity and airflow velocity, can also have pronounced effect on the final polymer film. Hence, determination of a set of operational conditions for specified criteria of a polymer film is of crucial importance for designing an efficient spin coating process [10].

Morphology of the Polymer Film

The morphology of polymer films is a vast area and extensive research has been conducted on the same. Li et al. [96] studied the effect of molecular weight on the morphology of polymer films. An immiscible mixture of Polymethyl methacrylate (PMMA) and Polystyrene (PS) polymer in toluene was spin coated to produce a film thickness of 70–90 nm. Varying the molecular weight of PS, from 2.9 to 129 kg/mol, the morphology of the film was effectively controlled. Pham and Green [66] studied the effect of morphology of polymer films on the glass transition temperature, T_g. Solutions of toluene and varying molecular weight of immiscible tetramethyl bisphenol-A polycarbonate was used for forming the polymer films. It was found that the glass transition temperature decreased with reduction in the film thickness.

The morphological effects of the homopolymers and sub-mono-layers were studied by Norrman et al. [97] with poly vinyl chloride (PVC) and PMMA solutions, spin coated on substrates made of silicon. They found that all the monomers had a distinct break in between, which resulted from the more favourable distribution of the polymer in the lateral plane, than in the normal plane, at high concentrations.

Many other significant studies have been conducted by several researchers, Petri [87], Kanaya et al. [81] and Anselmo et al. [98, 99] in understanding the morphology of the films and its correlation with the properties. It can be inferred that the morphology of the polymer films is the most critical part for controlling and fabricating them according to the required properties.

Fabrication Processes and Applications

The process of spin coating has been applied for coating substances for more than the past five decades. There are many fabrication methods that can be employed to obtain the desired coating or film quality. An attempt to summarise all the available work on spin coating fabrication processes is outside the scope of this chapter and hence, only the important fabrication methods along with key examples have been summarised in this part of the chapter. The contents presented in this chapter will mainly focus on the different fabrication techniques related to the spin coating process for various applications and the resultant properties of the final polymer film.

Microelectronics

Spin coating has been used in the microelectronics field for many decades and has found many applications in this field. One of the most important applications of spin coating can be found in field-effect transistors (organic FETs), which contain an organic semi-conducting polymer layer. A significant factor of a field effect transistor is the field-effect mobility, which is greatly influenced by the morphology of the semi-conducting film [10]. The mobility, for efficient working of the transistor, should be around the threshold value of 0.1 $cm^2/V/s$. Optimising this mobility has been a key aspect of many studies and it has been found that Pentacene films show the highest mobility in the range of 1–3 $cm^2/V/s$ [36]. The degree of order of the amorphous solvent of the polymer solution has been observed to impede the phenomenon of charge transport. Moreover, it is of utmost importance to have a controlled molecular ordering of the film constituent. Polymers have excellent mechanical properties and are very good in film forming, although the only disadvantage being the difficulty in controlling the ordering of the molecules [43].

The most commonly used and studied is the p-type semiconductor poly 3-hexylthiophene (P3HT) polymer that has a unique property of self-organization of molecules. Optimising the choice of solvent has also been proved to be a tedious work and thus, many studies have been conducted on the same. Experiments were conducted by Sirringhaus et al. [43] on a hydrophobic substrate and a 70–100 nm thick polymer film was spin-coated and also drop casted at 2000 rpm. It was observed that the amount of charge mobility was more in the case of drop casted polymer film. Other studies on the same P3HT polymer were conducted by Krebs et al. [100], who found that the absorption of the vacuum ultraviolet ray was less intense in the case of the drop casted polymer films. This also proved that the drop casted polymer films were more ordered, because they are more sensitive and hence have less vacuum ultraviolet ray absorption capabilities. Thus, it can be inferred that the use of spin coating lowers the mobility of the polymer film when compared to the one prepared by drop casting, mainly because of the inability of the polymer in the former process to form ordered molecules when the available time, for the process is less. A systematic investigation was carried out by Chang et al. [33] to test the mobility of the P3HT polymer mixed with different solvents, such as chloroform, cyclohexyl benzene, thiophene, 1,2,4-trichlorobenzene and xylene. The mobility of the field effects was observed to be dependent greatly on the boiling point of the solvent, implying the solvents with higher boiling points (for example, 1,2,4-trichlorobenzene) had greater mobility compared to those with lower ones (for example, chloroform). Other applications of spin coating in the field of microelectronics include organic light emitting diode and photovoltaic cells. Burroughes et al. [32] were the pioneers in the fabrication of the light emitting diodes of the organic type. A detailed explanation on the fabrication for other microelectronic applications and their respective properties are given in the study conducted by Norrman et al. [10].

Protective Coatings

Protection against harmful UV rays, humidity, scratch and corrosion is extremely crucial in industries and thus, protective coatings have always been used for the same purpose. The bulk of the application has been found in protecting hard discs, digital versatile discs (DVD) and compact discs (CD). Development of small components has resulted in the reduction of the thickness of the protective layers as well. The mostly used polymer layer for the protection purpose is made of polycarbonate. The spin coating process has been mostly applied because of its low cost and relatively less residual error [10].

A significant investigation was carried out by Chang et al. [9] where a non-ionic fluorine containing a surfactant and a copolymer, named polyether-siloxane, was used as the protective layer. The layer was applied using the spin coating process and was found to be slip resistant and improved the lubrication properties. The goal of the study was to obtain a protective layer which would improve the protective properties, especially against static electricity and scratching. The polymer solution was thus made from a UV-curable resin of oligomer, like trifunctional aliphatic urethane acrylate. Conductive indium tin oxide nano-particles, in the size range of 3–50 nm, were added to the solution as well. Satisfactory coatings of thickness about 100 ± 2 μm were obtained which allowed focusing of the light on the surface conveniently. It was observed that the surface resistivity was inversely proportional to the number of conductive particles. The transmittance loss due to the protective layer was also in the acceptable limit of about 2 % and the possibility of fabricating protective polymer layers for high definition-DVDs was also demonstrated in the study [9].

Optics

Optical coating is a very important field and the spin coating process has found wide applications in it. Since spin coating can be used even on a curved surface, the application has been immensely varied, such as spectacles. Other significant applications include light emitting diodes, emissive displays integrated optical circuits and optical sensors. Mirrors also form an intricate part of the optics industry. Komikado et al. [101] illustrated the possibility of fabricating multi-layered mirrors through spin coating of organic polymers. The optical thickness of individual layers was controllable and thus, the desired wavelength could be reached by modifying the concentration of the polymer and the angular velocity of the spin-coater. It was observed that an increase in the number of layers would increase the reflectivity of the mirror. The reflectivity attained by the mirror made by spin coating was more than 99 % and thus had immense applications in the optical and also in the electrical industries.

The use of spin-coated polymers was limited to the anti-reflection field, due to the lack of materials having comparatively lower refractive indices. Walheim et al. [48] demonstrated a superior anti-reflection nano-porous polymer-based film. Two immiscible polymers (PMMA and PS) were mixed in tetra-hydrofuron and the solution was spin coated on a substrate made of glass. This process resulted in the formation of the individual polymer domains and it was observed that the PS was selectively dissolved to form a nano-porous film. It was also found that increasing the volume ratio of the pores would result in minimising the refractive index of the material, a phenomenon which is highly desired in the anti-reflection coatings of broadband.

Membranes

Porous membranes have gained significant importance in the last few decades and thus, the membranes of uniform pore sizes have been the aim of many studies. Block copolymers are generally used to produce such membranes because of their unique ability of phase separation to form ordered and periodic arrays of pores. The membrane properties can be controlled by controlling the density of the pores, size of the pores and pore size distribution. The interfacial interactions have been found to be controllable for films with a low thickness and thus can influence the domains easily.

The effect of molecular weight on nano-porous films was studied by Xu et al. [50] with di-block copolymers made of PS and PMMA. The volume fractions of both the polymers were kept constant at 0.3 and the molecular weight was varied in the range of 42–295 kg/mol to form a variety of polymer films. The solvent used in this regard was toluene, and two different substrates were used for the spin coating process. The authors observed that the size of the micro-domain was controllable to some extent with a change in the molecular weight. It was also noted that an increase in the molecular weight increased the domain size of the cylindrical micro-domains. Yan et al. [49] also used PS mixed in toluene solution to produce a polymer membrane with mono-dispersed cylindrical pores. A chromium coated glass plate was placed over the PS film and heated at 120 °C to produce an ensemble of pillars under a voltage of 200 V. It was observed that the pore size was controllable by adjusting the conditions responsible for the formations of the pillars, that is, the PS film thickness and the electric field.

Another significant work was carried out by Le Roux and Paul [102] who used 4-methyl-1-pentene (PMP) for spin coating on a micro-porous support of polysulfone (PSF). The PMP layer thickness was controlled by the angular velocity of the spin-coater, the number of layers selectively coated and the concentration of the coating solution. Polymer films without an intermediate layer of Polydimethylsiloxane (PDMS) required two or more layers of selective polymers in order to bridge the gap of the pores and to make the skin defect-less. It was also observed that for thinner selective films, PSF matrix has better selectivity and is superior to the PDMS ones.

Anti-reflection (AR) Coatings

Anti-reflection (AR) coatings have important applications, because optical reflection is always prevalent when light travels through a boundary of two media, having different refraction indices. Thus, achieving low reflection is crucially important, and there are two potential ways of doing this. These include adding an inhomogeneous layer or fabricating a multilayered structure which will create interference with light [26]. The first type of multi-layered AR coating was developed in the 1960s and composed of TiO_2/SiO_2-TiO_2-SiO_2 three-layered substrate. Seo et al. [103] reported the fabrication of anti-static AGAR of two layers using spin coating. The film consisted of high indexed silicon alkoxides and metal alkoxides. Polymers have been extensively used for spin coating multi-layered AR surfaces on small substrates for applications in the optics industry.

Debsikdar [104] reported the formation of an inhomogeneous AR coating by utilising inherent aging characteristics of the sol-gel process. This resulted in coating the substrate with nano-sized particles which grow gradually with the process and thus form a film with varying particle sizes. Many other works have been carried out for preparing anti-reflection coating polymer films for the past five decades [105–107].

Other Applications

Spin coating has found applications in many other fields. One of them is the production of ALPO type thin film dielectric coatings. The ALPO system consists of $Al_2O_{3-3x}(PO_4)_{2x}$ and has superior performances with respect to both the solid phases and the precursor chemistry [108]. Spin coating has been extensively used to apply dielectric coating on substrates uniformly with high accuracy and low cost [95]. Application of the spin coating in liquid silicon materials for thin film transistors (TFT) is also an emerging field. The silicon-based liquid precursor developed by Shimoda et al. [109] and Tanaka et al. [110] were spin coated to form high quality silicon films enabling the transistor to operate at a mobility as high as 108 cm^2/Vs [95]. Another application of spin coating is found in the doped silicon films which can be of both n-type and p-type. Tanaka et al. [111] fabricated the n-type doped silicon film using a liquid precursor based on polysilane and coated the device using spin coating.

A field that has a vast application of the spin coating technique is the sensor technology. Sensors consist of thin films which help in its applications. Eaton [45] developed a calorimetric oxygen sensor based on a thin polymer film of ethyl cellulose and in the presence of fructose. The polymer film functioned as an encapsulating agent and in the presence of oxygen it turns blue, indicating the presence of oxygen in the vicinity. There are many other types of sensors based on polymer films as demonstrated in [44, 46, 47]. Another useful area of application of spin coating is solar cells. Solar cells are made of multiple layers and each layer needs to be produced individually, but should be of the same device geometry and dimensions. Thus, spin coating is widely used for its fabrication due to the advantages of uniformity, low cost and simple preparation of the ink or polymer solution for the layer; although it is often accompanied with wastage of ink and the final film having less minimum thickness, when compared to other processing techniques [38].

Summary

Spin coating as a polymer coating process has been established for more than six decades with the same basic functioning and processing system and tools. In spite of this, the spin coating process has evolved over time due to in-depth fundamental understanding of the process and the requirements from emerging applications. A wide range of unique and novel applications in various industries have been utilising this technique, due to the convenience offered by the advanced tools for spin coating with the ability to produce resultant fine resolution coatings. The emerging field of nano-technology has also aided in the continued and extended growth of the spin coating process, which is a well-established technology for producing nano-scaled objects. Studies have shown that this method has been applied to various applications, each having unique fabrication characteristics and application specific properties. In the view of these studies, this chapter described the important parameters which influence the efficiency of the spin coating process along with their effects on the different fabrication techniques employed to process the films and the unique properties possessed by the films.

The parameters, such as the angular velocity of a spin-coater, molecular weight of the constituents of the polymer solution, the viscosity and concentration of the solution and the solvent properties were found to greatly influence the final thickness and quality of the polymer film. The angular velocity of the spin-coater is a critical aspect and was found to aid in the formation of thinner and uniform polymer films. Similarly, the concentration and viscosity of the polymer also had a pronounced effect on the film, which was found to be thinner for low viscosity and low concentration of the initial solution. The molecular weight was also found to affect the size of the micro-domains during membrane formation. Even, the final topology of the film was found to be highly dependent on the solvent of the polymer solution. Volatility of the solvent is a unique feature that controls the evaporation rate. A very high volatility results in faster solidification and hence lower uniformity in the film thickness. A very low volatility may reduce evaporation and lead to the formation of a solid skin. Other factors were also found to somewhat affect the final quality of the polymer film and these influenced the different fabrication techniques of the spin-coated polymer films significantly. The properties of the final polymer film were observed to be dependent on the variation of these influencing parameters. A detailed description of the fabrication processes, different applications and the resulting properties of the polymer film is given in this work.

Spin coating has many future prospects. For example, the spin coating process can be extended to produce conductive polymer films by adding metal fillers to the polymer solution. The metal fillers aid in enhancing the properties of electrically conductive polymer composite films with improved magnetic receptibility, heat conductivity, heat capacity and radiation shielding [112]. Incorporation of metal fillers in the polymer film has been successfully performed in the macro-scale. Thus, transforming the same into nano-scale can be extremely beneficial and has considerable scopes.

References

1. Hall DB, Underhill P, Torkelson JM (1998) Spin coating of thin and ultrathin polymer films. Polym Eng Sci 38(12):2039–2045
2. Toolan DT, Howse JR (2013) Development of in situ studies of spin coated polymer films. J Mater Chem C 1(4):603–616
3. Mellbring O, Kihlman Øiseth S, Krozer A, Lausmaa J, Hjertberg T (2001) Spin coating and characterization of thin high-density polyethylene films. Macromolecules 34(21):7496–7503. doi:10.1021/ma000094x
4. Prest WM, Luca DJ (1980) The alignment of polymers during the solvent-coating process. J Appl Phys 51(10):5170–5174. doi:10.1063/1.327464

5. Despotopoulou M, Frank C, Miller R, Rabolt J (1996) Kinetics of chain organization in ultrathin poly (di-n-hexylsilane) films. Macromolecules 29(18):5797–5804
6. Forrest J, Dalnoki-Veress K, Dutcher J (1997) Interface and chain confinement effects on the glass transition temperature of thin polymer films. Phys Rev E 56(5):5705
7. Bartczak Z, Argon A, Cohen R, Weinberg M (1999) Toughness mechanism in semi-crystalline polymer blends: I. High-density polyethylene toughened with rubbers. Polymer 40(9):2331–2346
8. Muratoglu O, Argon A, Cohen R, Weinberg M (1995) Toughening mechanism of rubber-modified polyamides. Polymer 36(5):921–930
9. Chang D, Yoon D, Ro M, Hwang I, Park I, Shin D (2003) Synthesis and characteristics of protective coating on thin cover layer for high density-digital versatile disc. Jpn J Appl Phys 42(2S):754
10. Norrman K, Ghanbari-Siahkali A, Larsen N (2005) 6 Studies of spin-coated polymer films. Ann Rep Sect "C" (Phys Chem) 101:174–201
11. Callewaert M, Gohy J-F, Dupont-Gillain CC, Boulangé-Petermann L, Rouxhet PG (2005) Surface morphology and wetting properties of surfaces coated with an amphiphilic diblock copolymer. Surf Sci 575(1):125–135
12. Oliveira AR, Zarbin AJ (2005) Um procedimento simples e barato para a construção de um equipamento "dip-coating" para deposição de filmes em laboratório. Quim Nova 28(1):141–144
13. Willey RR (2002) Practical design and production of optical thin films. CRC Press, Boca Raton
14. Beers KL, Douglas JF, Amis EJ, Karim A (2003) Combinatorial measurements of crystallization growth rate and morphology in thin films of isotactic polystyrene. Langmuir 19(9):3935–3940
15. Belleville P, Bonnin C, Priotton J-J (2000) Room-temperature mirror preparation using sol-gel chemistry and laminar-flow coating technique. J Sol-Gel Sci Technol 19(1–3):223–226
16. Ichiki M, Zhang L, Yang Z, Ikehara T, Maeda R (2004) Thin film formation on non-planar surface with use of spray coating fabrication. Microsyst Technol 10(5):360–363
17. Ju D-Y, Ji V, Gassot H (2004) Computer predictions of thermo-mechanical behavior and residual stresses in spray coating process. In: Journal de Physique IV (Proceedings). EDP sciences, pp 381–388
18. Mostaghimi J, Chandra S, Ghafouri-Azar R, Dolatabadi A (2003) Modeling thermal spray coating processes: a powerful tool in design and optimization. Surf Coat Technol 163:1–11
19. Tucker RC (2002) Thermal spray coatings: broad and growing applications. Int J Powder Metall 38(7):45–53
20. Singh RK, Narayan J (1990) Pulsed-laser evaporation technique for deposition of thin films: physics and theoretical model. Phys Rev B 41(13):8843
21. Yasuda H (2012) Plasma polymerization. Academic press, New York
22. Minko S, Patil S, Datsyuk V, Simon F, Eichhorn K-J, Motornov M, Usov D, Tokarev I, Stamm M (2002) Synthesis of adaptive polymer brushes via "grafting to" approach from melt. Langmuir 18(1):289–296
23. Bornside D, Macosko C, Scriven L-E (1987) On the modeling of spin coating. J Imaging Technol 13(4):122–130
24. Schubert D, Dunkel T (2003) Spin coating from a molecular point of view: its concentration regimes, influence of molar mass and distribution. Mat Res Innovat 7(5):314–321. doi:10.1007/s10019-003-0270-2
25. Scriven L (1988) Physics and applications of dip coating and spin coating. In: MRS proceedings. Cambridge University Press, Cambridge, p 717
26. Chen D (2001) Anti-reflection (AR) coatings made by sol–gel processes: a review. Sol Energy Mater Sol Cells 68(3):313–336
27. Meyerhofer D (1978) Characteristics of resist films produced by spinning. J Appl Phys 49 (7):3993–3997
28. Walker P, Thompson J (1922) Some physical properties of paints. In: Proceedings of the American Society of testing materials, vol 464, p 100

29. Apperloo JJ, Janssen R, Nielsen MM, Bechgaard K (2000) Doping in solution as an order-inducing tool prior to film formation of regio-irregular polyalkylthiophenes. Adv Mater 12(21):1594–1597
30. Arias A, Corcoran N, Banach M, Friend R, MacKenzie J, Huck W (2002) Vertically segregated polymer-blend photovoltaic thin-film structures through surface-mediated solution processing. Appl Phys Lett 80(10):1695–1697
31. Arias A, MacKenzie J, Stevenson R, Halls J, Inbasekaran M, Woo E, Richards D, Friend R (2001) Photovoltaic performance and morphology of polyfluorene blends: a combined microscopic and photovoltaic investigation. Macromolecules 34(17):6005–6013
32. Burroughes J, Bradley D, Brown A, Marks R, Friend R, Burn PL, Holmes AB (1990) Nature (London) 347:539–541. Burroughes539347Nature (London)
33. Chang J-F, Sun B, Breiby DW, Nielsen MM, Sölling TI, Giles M, McCulloch I, Sirringhaus H (2004) Enhanced mobility of poly (3-hexylthiophene) transistors by spin-coating from high-boiling-point solvents. Chem Mater 16(23):4772–4776
34. Corcoran N, Arias A, Kim J, MacKenzie J, Friend R (2003) Increased efficiency in vertically segregated thin-film conjugated polymer blends for light-emitting diodes. Appl Phys Lett 82 (2):299–301
35. Geens W, Shaheen SE, Wessling B, Brabec CJ, Poortmans J, Sariciftci NS (2002) Dependence of field-effect hole mobility of PPV-based polymer films on the spin-casting solvent. Org Electron 3(3):105–110
36. Klauk H, Halik M, Zschieschang U, Schmid G, Radlik W, Weber W (2002) High-mobility polymer gate dielectric pentacene thin film transistors. J Appl Phys 92(9):5259–5263
37. Kline RJ, McGehee MD, Kadnikova EN, Liu J, Frechet JM (2003) Controlling the field-effect mobility of regioregular polythiophene by changing the molecular weight. Adv Mater 15(18):1519–1522
38. Krebs FC (2009) Fabrication and processing of polymer solar cells: a review of printing and coating techniques. Sol Energy Mater Sol Cells 93(4):394–412. doi:10.1016/j.solmat.2008.10.004
39. Salleo A, Chabinyc M, Yang M, Street R (2002) Polymer thin-film transistors with chemically modified dielectric interfaces. Appl Phys Lett 81(23):4383–4385
40. Shi Y, Liu J, Yang Y (2000) Device performance and polymer morphology in polymer light emitting diodes: the control of thin film morphology and device quantum efficiency. J Appl Phys 87(9):4254–4263
41. Sirringhaus H, Brown P, Friend R, Nielsen M, Bechgaard K, Langeveld-Voss B, Spiering AJH, Janssen RAJ, Meijer EW, Herwig P, de Leeuw DM (1999) Nature 401:1038
42. Sirringhaus H, Brown P, Friend R, Nielsen MM, Bechgaard K, Langeveld-Voss B, Spiering A, Janssen R, Meijer E (2000) Microstructure–mobility correlation in self-organised, conjugated polymer field-effect transistors. Synth Met 111:129–132
43. Sirringhaus H, Tessler N, Friend RH (1998) Integrated optoelectronic devices based on conjugated polymers. Science 280(5370):1741–1744
44. Douglas P, Eaton K (2002) Response characteristics of thin film oxygen sensors, Pt and Pd octaethylporphyrins in polymer films. Sens Actuators B: Chem 82(2):200–208
45. Eaton K (2002) A novel colorimetric oxygen sensor: dye redox chemistry in a thin polymer film. Sens Actuators B: Chem 85(1):42–51
46. Mirkhalaf F, Schiffrin D (2000) Metal-ion sensing by surface plasmon resonance on film electrodes. J Electroanal Chem 484(2):182–188
47. Penza M, Anisimkin V (1999) Surface acoustic wave humidity sensor using polyvinyl-alcohol film. Sens Actuators, A 76(1):162–166
48. Walheim S, Schäffer E, Mlynek J, Steiner U (1999) Nanophase-separated polymer films as high-performance antireflection coatings. Science 283(5401):520–522
49. Yan X, Liu G, Dickey M, Willson CG (2004) Preparation of porous polymer membranes using nano-or micro-pillar arrays as templates. Polymer 45(25):8469–8474
50. Xu T, Kim H-C, DeRouchey J, Seney C, Levesque C, Martin P, Stafford C, Russell T (2001) The influence of molecular weight on nanoporous polymer films. Polymer 42(21):9091–9095

51. Bornside DE, Macosko CW, Scriven LE (1989) Spin coating: one-dimensional model. J Appl Phys 66(11):5185–5193. doi:10.1063/1.343754
52. Münch A, Please CP, Wagner B (2011) Spin coating of an evaporating polymer solution. Phys Fluids (1994-present) 23(10):102101
53. Emslie AG, Bonner FT, Peck LG (1958) Flow of a viscous liquid on a rotating disk. J Appl Phys 29(5):858–862. doi:10.1063/1.1723300
54. Acrivos A, Shah MJ, Petersen EE (1960) On the flow of a non-newtonian liquid on a rotating disk. J Appl Phys 31(6):963–968. doi:10.1063/1.1735785
55. Flack WW, Soong DS, Bell AT, Hess DW (1984) A mathematical model for spin coating of polymer resists. J Appl Phys 56(4):1199–1206
56. Jenekhe SA (1984) Effects of solvent mass transfer on flow of polymer solutions on a flat rotating disk. Ind Eng Chem Fundam 23(4):425–432
57. Lawrence C (1988) The mechanics of spin coating of polymer films. Phys Fluids (1958–1988) 31(10):2786–2795
58. Sukanek PC (1985) Spin coating. J Imaging Technol 11(4):184–190
59. Chen B (1983) Investigation of the solvent-evaporation effect on spin coating of thin films. Polym Eng Sci 23(7):399–403
60. Daughton W, Givens F (1982) An investigation of the thickness variation of spun-on thin films commonly associated with the semiconductor industry. J Electrochem Soc 129(1):173–179
61. Givens F, Daughton W (1979) On the uniformity of thin films: a new technique applied to polyimides. J Electrochem Soc 126(2):269–272
62. Higgins BG (1986) Film flow on a rotating disk. Phys Fluids 29(11):3522
63. Jenekhe SA (1983) The rheology and spin coating of polyimide solutions. Polym Eng Sci 23(15):830–834
64. Kim JH, Jang J, Zin W-C (2001) Thickness dependence of the glass transition temperature in thin polymer films. Langmuir 17(9):2703–2710
65. Lai JH (1979) An investigation of spin coating of electron resists. Polym Eng Sci 19(15):1117–1121
66. Pham JQ, Green PF (2002) The glass transition of thin film polymer/polymer blends: interfacial interactions and confinement. J Chem Phys 116(13):5801–5806
67. Spangler LL, Torkelson JM, Royal JS (1990) Influence of solvent and molecular weight on thickness and surface topography of spin-coated polymer films. Polym Eng Sci 30(11): 644–653
68. Washo B (1977) Rheology and modeling of the spin coating process. IBM J Res Dev 21(2):190–198
69. Weill A, Dechenaux E (1988) The spin-coating process mechanism related to polymer solution properties. Polym Eng Sci 28(15):945–948
70. Chang C-C, Pai C-L, Chen W-C, Jenekhe SA (2005) Spin coating of conjugated polymers for electronic and optoelectronic applications. Thin Solid Films 479(1–2):254–260. doi:10.1016/j.tsf.2004.12.013
71. Jukes PC, Heriot SY, Sharp JS, Jones RA (2005) Time-resolved light scattering studies of phase separation in thin film semiconducting polymer blends during spin-coating. Macromolecules 38(6):2030–2032
72. Kim J-S, Ho PK, Murphy CE, Friend RH (2004) Phase separation in polyfluorene-based conjugated polymer blends: lateral and vertical analysis of blend spin-cast thin films. Macromolecules 37(8):2861–2871
73. Keith F, Taylor J, Chong J (1958) Heat and mass transfer from a rotating disk. DTIC document
74. Graessley WW (1980) Polymer chain dimensions and the dependence of viscoelastic properties on concentration, molecular weight and solvent power. Polymer 21(3):258–262
75. Long D, Lequeux F (2001) Heterogeneous dynamics at the glass transition in van der Waals liquids, in the bulk and in thin films. Eur Phys J E 4(3):371–387
76. De Gennes P (2000) Glass transitions in thin polymer films. Eur Phys J E 2(3):201–205

77. Wallace WE, Fischer DA, Efimenko K, Wu W-L, Genzer J (2001) Polymer chain relaxation: surface outpaces bulk. Macromolecules 34(15):5081–5082
78. Xie L, DeMaggio G, Frieze W, DeVries J, Gidley D, Hristov H, Yee A (1995) Positronium formation as a probe of polymer surfaces and thin films. Phys Rev Lett 74(24):4947
79. Forrest J, Dalnoki-Veress K, Dutcher J, Rowat A, Stevens JR (1995) Brillouin light scattering determination of the glass transition in thin, freely-standing poly (styrene) films. In: MRS proceedings. Cambridge University Press, Cambridge, p 131
80. Forrest J, Dalnoki-Veress K, Stevens J, Dutcher J (1996) Effect of free surfaces on the glass transition temperature of thin polymer films. Phys Rev Lett 77(10):2002
81. Kanaya T, Miyazaki T, Watanabe H, Nishida K, Yamano H, Tasaki S, Bucknall D (2003) Annealing effects on thickness of polystyrene thin films as studied by neutron reflectivity. Polymer 44(14):3769–3773
82. Keddie J, Jones R (1995) Glass transition behavior in ultra-thin polystyrene films. Isr J Chem 35(1):21–26
83. Kleideiter G, Prucker O, Bock H, Frank CW, Lechner MD, Knoll W (1999) Polymer thin film properties as a function of temperature and pressure. In: Macromolecular symposia, vol 1. Wiley Online Library, pp 95–102
84. Lin EK, W-l Wu, Satija SK (1997) Polymer interdiffusion near an attractive solid substrate. Macromolecules 30(23):7224–7231
85. Mansfield KF, Theodorou DN (1991) Molecular dynamics simulation of a glassy polymer surface. Macromolecules 24(23):6283–6294
86. Orts WJ, van Zanten JH, W-l Wu, Satija SK (1993) Observation of temperature dependent thicknesses in ultrathin polystyrene films on silicon. Phys Rev Lett 71(6):867
87. Petri D (2002) Characterization of spin-coated polymer films. J Braz Chem Soc 13(5): 695–699
88. Reiter G (1994) Dewetting as a probe of polymer mobility in thin films. Macromolecules 27(11):3046–3052
89. Richard A (1994) Interface and surface effects on the glass-transition temperature in thin polymer films. Faraday Discuss 98:219–230
90. Wallace W, Tan NB, Wu W, Satija S (1998) Mass density of polystyrene thin films measured by twin neutron reflectivity. J Chem Phys 108(9):3798–3804
91. Forrest J (2002) A decade of dynamics in thin films of polystyrene: where are we now? Eur Phys J E: Soft Matter Biol Phys 8(2):261–266
92. Leibler L (1980) Theory of microphase separation in block copolymers. Macromolecules 13(6):1602–1617
93. Flory PJ (1953) Principles of polymer chemistry. Cornell University Press
94. Dandapat B, Layek G (1999) Spin coating in the presence of a transverse magnetic field and non-uniform rotation: a numerical study. J Phys D Appl Phys 32(19):2483
95. Mitzi D (2008) Solution processing of inorganic materials. Wiley, New York
96. Li X, Han Y, An L (2003) Surface morphology control of immiscible polymer-blend thin films. Polymer 44(26):8155–8165
97. Norrman K, Haugshøj K, Larsen N (2002) Lateral and vertical quantification of spin-coated polymer films on silicon by TOF-SIMS, XPS, and AFM. J Phys Chem B 106(51): 13114–13121
98. Anselmo AS, Dzwilewski A, Svensson K, Moons E (2013) Molecular orientation and composition at the surface of spin-coated polyfluorene: fullerene blend films. J Polym Sci, Part B: Polym Phys 51(3):176–182
99. Anselmo AS, Lindgren L, Rysz J, Bernasik A, Budkowski A, Andersson RM, Svensson K, van Stam J, Moons E (2011) Tuning the vertical phase separation in polyfluorene: fullerene blend films by polymer functionalization. Chem Mater 23(9):2295–2302
100. Krebs FC, Hoffmann SV, Jørgensen M (2003) Orientation effects in self-organized, highly conducting regioregular poly (3-hexylthiophene) determined by vacuum ultraviolet spectroscopy. Synth Met 138(3):471–474

101. Komikado T, Inoue A, Masuda K, Ando T, Umegaki S (2007) Multi-layered mirrors fabricated by spin-coating organic polymers. Thin Solid Films 515(7–8):3887–3892. doi:10.1016/j.tsf.2006.10.119
102. Le Roux JD, Paul DR (1992) Preparation of composite membranes by a spin coating process. J Membr Sci 74(3):233–252. doi:10.1016/0376-7388(92)80064-Q
103. Seo K-I, Jang D-S, Kim H-S, Jeong S-M (1996) Method for preparing anti-reflective coating for display devices. Google patents
104. Debsikdar J (1989) Broadband antireflective coating composition and method. Google Patents
105. Boström TK, Wäckelgård E, Westin G (2004) Anti-reflection coatings for solution-chemically derived nickel—Alumina solar absorbers. Sol Energy Mater Sol Cells 84(1–4):183–191. doi:10.1016/j.solmat.2003.12.015
106. Chao Y-C, Chen C-Y, Lin C-A, He J-H (2011) Light scattering by nanostructured anti-reflection coatings. Energy Environ Sci 4(9):3436–3441
107. Chen JY, Sun KW (2010) Enhancement of the light conversion efficiency of silicon solar cells by using nanoimprint anti-reflection layer. Sol Energy Mater Sol Cells 94(3):629–633. doi:10.1016/j.solmat.2009.11.028
108. Meyers ST, Anderson JT, Hong D, Hung CM, Wager JF, Keszler DA (2007) Solution-processed aluminum oxide phosphate thin-film dielectrics. Chem Mater 19(16):4023–4029
109. Shimoda T, Matsuki Y, Furusawa M, Aoki T, Yudasaka I, Tanaka H, Iwasawa H, Wang D, Miyasaka M, Takeuchi Y (2006) Solution-processed silicon films and transistors. Nature 440(7085):783–786
110. Tanaka H, Matsuki Y, Shimoda T, Iwasawa H, Aoki T, Yudasaka I, Wang D, Miyasaka M, Furusawa M (2006) Solution-processed silicon films and transistors using novel liquid silicon materials. Digest Tech Papers AM-FPD 6:27–30
111. Tanaka H, Iwasawa H, Wang D, Toyoda N, Aoki T, Yudasaka I, Matsuki Y, Shimoda T, Furusawa M (2007) Spin-on n-type silicon films using phosphorous-doped polysilanes. Jpn J Appl Phys 46(10L):L886
112. Gul VE (1996) Structure and properties of conducting polymer composites, vol 8. VSP

Part IV
Application Opportunities of Nano-size Polymers

Chapter 11
Electrospinning—Commercial Applications, Challenges and Opportunities

Bhuvana Kannan, Hansol Cha and Iain C. Hosie

Introduction

The high surface area to volume ratio makes nanofibres the ideal candidate for various applications where high porosity and high surface area is desirable. Although several methods such as melt spinning [1, 2], chemical vapour deposition [3, 4], self-assembly [5, 6], gas jet [7, 8] and nanolithographic techniques [9, 10] are used to produce nanofibre, electrospinning is the most widely used technique. It has been recognized as an indispensable technique for generating nanoscale fibres by the research community for more than a decade due to its low cost, operational simplicity and high production rate.

B. Kannan (✉) · H. Cha · I.C. Hosie
Revolution Fibres Ltd., 9A Corban Ave., Henderson, Auckland 0612, New Zealand
e-mail: bhuvana@revolutionfibres.com

H. Cha
e-mail: hansol@revolutionfibres.com

I.C. Hosie
e-mail: iain@revolutionfibres.com

B. Kannan
School of Chemical Sciences, University of Auckland, Auckland 1142, New Zealand

H. Cha
Department of Chemical and Materials Engineering, University of Auckland, Auckland 1142, New Zealand

S. Fakirov (ed.), *Nano-size Polymers*,
DOI 10.1007/978-3-319-39715-3_11

The physical phenomenon of electrospinning (polymer solution affected by electrical charge to form a fibre from a cone-shaped droplet) was known for over hundred years. Its discovery is credited to Sir William Gilbert (1628) who showed that a piece of charged amber could deform a droplet of water into a cone shape [11]. But the initial study investigating the mechanism regulating the formation of fibre was conducted only during early 1960s by Sir Geoffery Taylor [12]. In his experiments he defined a phenomenon known as a "Taylor cone" using a mathematical model to define the process whereby a viscous polymer droplet elongates into a cone to form fibre under an electric field. After the initial studies by Taylor many researchers explored different ways to manipulate the Taylor cone process. Although Cooley [13], Morton [14] and Formhals [15] from USA filed patents on electrospinning in the early 1930s none of them could be used for actual industrial applications.

The tremendous effort of I.V. Petryanov-Sokolov and collaborators led to first industrial facility for electrospinning fibrous material in the late 1930s for gas masks. The electrospun fibres were named after Petryanov as Petryanov fibres (PF) and showed excellent filtration properties such as high breathability, aerosol capturing efficiency and long shelf life. But due to security reasons and military secrets the products were confined to USSR [16]. The first commercial sub-micron nanofibre products were introduced to market in USA by Donaldson co., in 1980s [17], by Dupont in 1995 [17, 18] and Reneker and co-workers in 1995 [17, 19, 20].

Figure 11.1a clearly shows that from the late 1990s to early 2000s, a massive movement within the "electrospinning" research community and industry burgeoned. From the year 2000, there was an approximate 20 % increase annually in the number of journal articles published around the world which utilized the concept of "electrospinning" within their research schemes (SciFinder® Retrieved on 14 April 2015, keywords "electrospinning" references were found containing the concept "electrospinning").

A survey of officially published documents related with electrospinning in different subject areas are listed in Fig. 11.2a, whereas those publications were categorized into country of origin are shown in Fig. 11.2b. These graphs and data clearly demonstrate that "electrospinning" technique has regained the attention all over the world probably due in part to a swelling interest in nanotechnology.

Figure 11.1b shows the number of patents against year of publication. Every year there are about 500 new patents published for novel applications utilizing nanofibres fabricated via different methods including electrospinning. (SciFinder® Retrieved on 14th of April 2015, Keywords "nanofiber" references were found containing the concept "nanofiber").

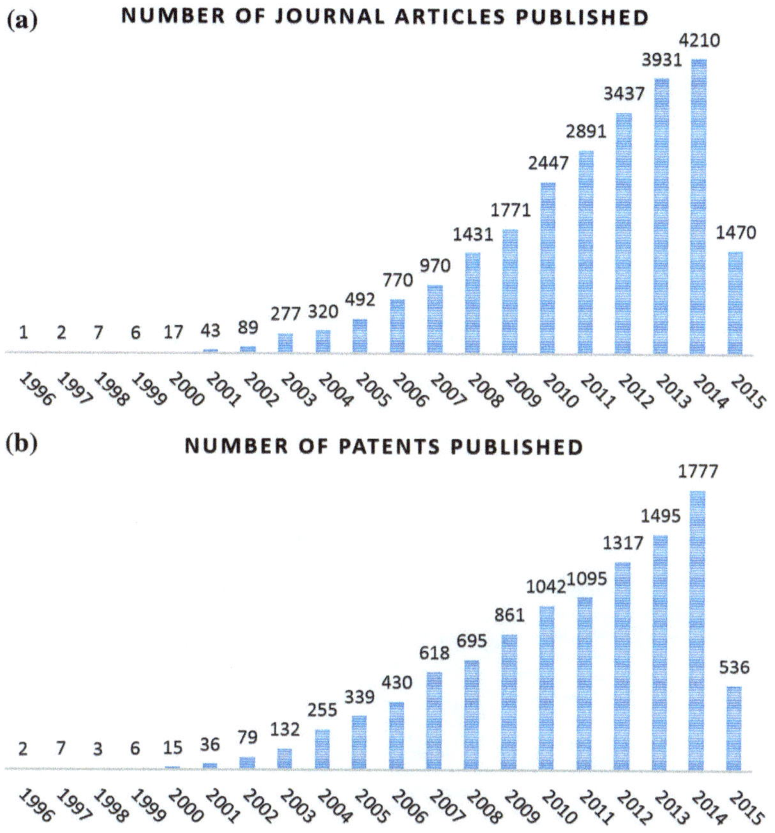

Fig. 11.1 Number of **a** Journal articles between 1995 and 2015 under the concept "electrospinning" (Data generated from SciFinder® on 14th of April 2015. Keywords "electrospinning") and **b** Patents published between 1995 and 2015 under the concept "nanofibre"

Electrospinning Process

Taylor Cone Formation

The working principle of electrospinning has been reported in many instances, elsewhere [21–24]. The fundamental working mechanism of electrospinning uses an electrostatically repulsive force and an electric field between a capillary tube, needle or any surface that could form Taylor cone and a metal or any conducting screen under high voltage. Many researchers studied the influence of electric voltage on Taylor cone for stable electrospinning processes [25–28] and categorized various morphologies of the electrospun jet based on different flow rates at 15 kV into five electrospinning regimes as (a) dripping, (b) jet with drops, (c) jet with large, (d) minimal jet fluctuations and (e) narrow, variable jet (as shown in the

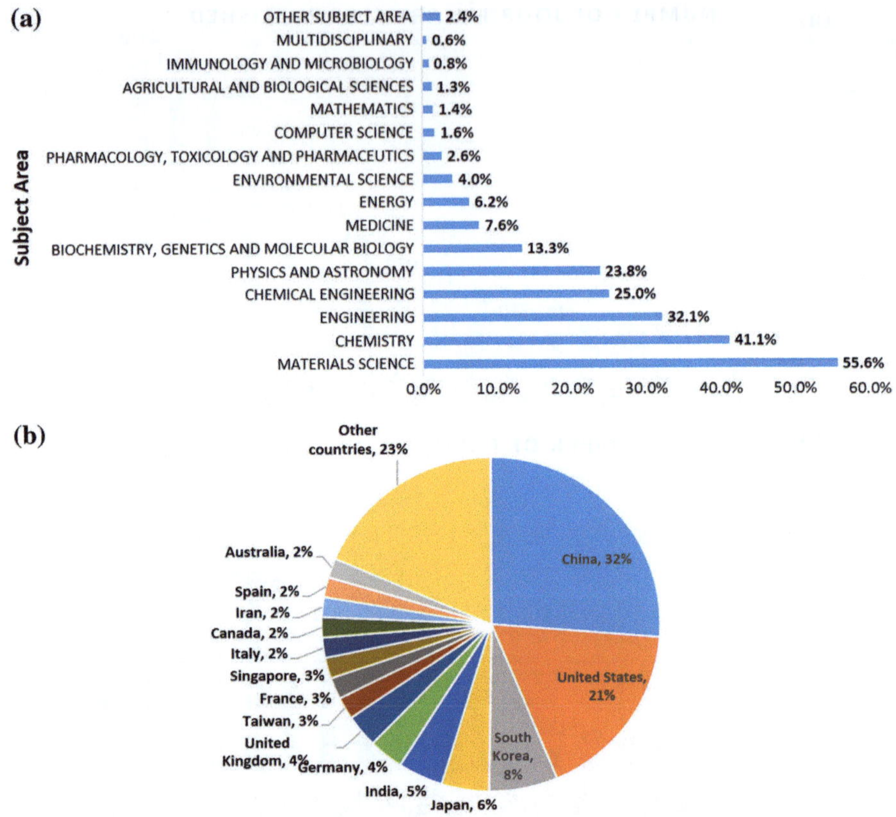

Fig. 11.2 **a** The top 15 subject areas that use electrospinning or nanofibres and **b** the total number of officially published document from 1970 to 2015 by categorizing them into countries of origin. (Retrieved on 14th of April 2015, Scopus. Keywords: "electrospinning" OR "electrospun" OR "electrostatic spinning" OR "nanofiber" OR "nanofibre" OR "nanofibrous" OR all)

Fig. 11.3) [27, 28]. Indeed, an in-depth study on Taylor cone morphologies was carried out at the tip of the needle at different flow rates and field strengths using an optical camera by Barua et al. [27] and Cai et al. [28]. Altering the flow rates significantly affected the Taylor cone morphology and the spinning jet. Cai and Gevelber have shown that out of these five electrospinning regimes a minimum jet fluctuation (regime 4) in the volume of the droplet is required to maintain a stable electrospinning process for an extended period of time [28] (Fig. 11.3). Renekaer et al. in 2008 also found that the rate at which solution flows into the Taylor cone must be equal to the rate at which fluid is carried away by the jet to achieve a stable electrospinning process [25].

While various studies have been carried out to study the profile of various morphologies of the electrospun jet it is important to deliberate the bending

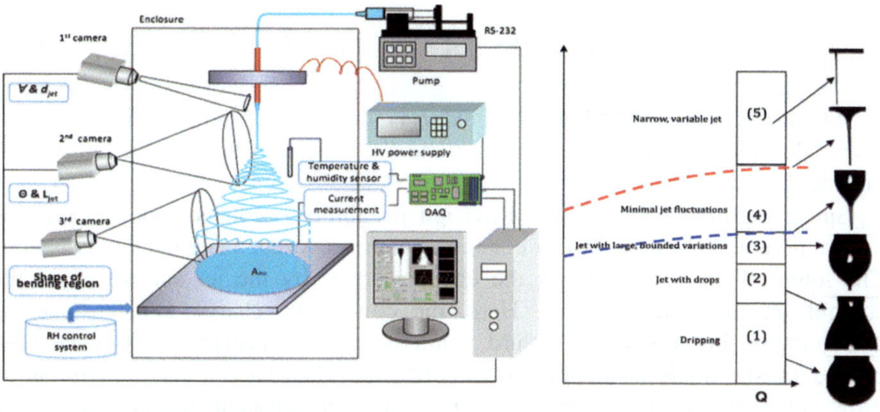

Fig. 11.3 Schematic representation of the electrospinning equipment (*left*) and the five electrospinning regimes and the corresponding Taylor cone formations [28]

instabilities suffered by the electrospun fibres which are responsible for fibre production and reduction of fibre diameter.

Jetting Mechanism of Viscoelastic Material

In order to upscale the electrospinning technology to an industrial level, the electrodes tips should be carefully designed to produce continuous fibre over time at a controlled deposition rate and thickness as the tip diameter affects the morphology of the Taylor cone. Subsequently, an in-depth understanding of the viscoelastic materials are also equally important which are affected (in the case of electrospinning) by complex physical perturbations causing a nonlinear bending instability leading to the formation of nanofibres. The solvent contributing to the bending instability of the viscoelastic polymeric solution under an electrically charged environment is responsible for enormously strong stretching and formation of nanofibres [29]. Reneker et al. demonstrated the rapid bending instability or whipping instability causes the long waveforms of perturbations of a liquid column driven by the lateral electric force and the aerodynamic interaction and inertia [30]. After him, Shin et al. [31, 32] and Hohman et al. [33] used mathematical expositions and asymptotic analyses to explain the instabilities of the jetting mechanism. They predicted three different instabilities associated with electrospinning. An axisymmetric instability better known as Rayleigh plateau instability, which causes beads when the surface tension surpasses viscoelastic forces at high electric field which is given by:

$$(\varepsilon - \varepsilon')E_{\infty}^2 + 4\frac{\pi^2\sigma^2}{\varepsilon'} = h2\pi\gamma/h \tag{11.1}$$

Where E_{∞} is the applied electric field, h is the radius of the jet, σ is the surface charge density, γ is the surface tension, ε and ε' are the dielectric constant inside and outside the jet, respectively.

But at high electric field, the Rayleigh instability becomes irrelevant and the critical potential for static Taylor cone formation is exceeded and the jetting begins. The longitudinal stress acts on the charge carried by the jet and stabilizes the straight jet for some distance. Here, the conductive charge transfer (causing axisymmetric instability and a third non-axisymmetric instability) dominates the charge transfer mode inside the Taylor cone [34]. Then later, a lateral perturbation grows in response to the repulsive forces between adjacent elements of charge carried by the jet to force the formation of larger loops (spinning) resulting in nanofibres [35]. Here, the conductive charge gives way to convective charge as the jet moves away from the Taylor cone towards the collector. This implies that the sum of conductive and convective charges acts on electrospun jet influencing the fibre formation. This also indicates that the total electric field near the tip is approximately the same as the applied external field and it declines as the jet moves from the needle tip to the collector surface [34]. This physical phenomenon of electrospinning solution depends on the direction of the spinneret and type of spinneret method that is being executed.

Spinneret

As discussed earlier the electrically driven non-axisymmetric bending instability and aerodynamic interactions become prevalent as the jet elongates and approaches the collector due to the reduction of total electric force near the collector. For instance, in an upward needleless electrospinning set-up these forces become negligible and forms more porous and loosely bound polymer fibre when compared to downward electrospinning set-up due to inertia [36]. Therefore, while implementing the lab scale technique to industrial scale machines all the electrospinning parameters including the spinneret extrusion process need to be decided and adjusted accordingly to avoid poor performances in terms of quality, reproducibility and accuracy in the production rate. For example, a needleless electrospinning system is of high advantage to a manufacturer as it uses no capillaries and avoids the risk of blockages forming during production and thus less labour intensive. But needle-based electrospinning offers options of different extrusion processes to produce continuous production of single fibre such as solution spinning, melt spinning and emulsion spinning. In a manufacturing point of view needle-based spinneret is of high importance as it offers unique advantages for melted polymer to electrospin with high throughput rate under a solvent-free environment.

Conventional solution electrospinning and emulsion electrospinning can be performed with a needle-based system to produce polymers exhibiting high melting points and have the capability to encapsulate functional materials. But the potential problems associated with needle-based electrospinning technique such as blockages in the needle (due to precipitation of the polymer at the tip) and spacing between the needles can inhibit the capability of continuous nanofibre production. Nevertheless with intelligent needle design these difficulties can be sidestepped. For example, BioInicia [37] produces various types of nanofibres at pilot and mass production scales with its needle-based electrospinning machine called FluidNatek®.

To avoid the risk of blockages associated with needle-based electrospinning and to allow the fibre production at higher rates Elmarco [38], an electrospinning company developed by the Technical University of Liberec has chosen an alternative "free-surface" method that uses no capillaries but a partially submerged rotating drum or wire electrode in a polymer solution. This design leads to larger fibre production with no mess due to absence of capillaries. But due to large surface area of volatile solutions exposed to free space the solution could evaporate (over time) before the fibre formation which can lead to uneven size distribution of nanofibres [39].

An alternate design that uses a needleless spinneret to overcome the disadvantages of both needle-based and free-surface electrospinning, whilst controlling the Taylor cone formation is credited to Revolution Fibres Ltd [40], who scaled up this process to achieve continuous production of nanofibres from less than 100 nm to sub-micron range with more than 30 varieties of polymers. Elmarco developed similar techniques with coated wires as spinneret and recently Stellenbosch University has developed a regenerating bubble method [41] as an alternative to needle-based electrospinning.

A major disadvantage of these methods, is that due to the different spinneret design it is difficult to replicate the solutions developed at research laboratories with needle-based electrospinning set-ups. Therefore major modifications in solution chemistry are required before an optimum solution and processing conditions can be achieved from lab to industrial scale.

Solution Electrospinning

More than 50 polymers have been successfully electrospun into fibres by researchers and most of these polymers were essentially dissolved in solvents before electrospinning. The polymer solution then passes through the charged needle (with few exceptions) and hits the collector substrate as solidified fibre with the solvent evaporating during the process. The solution electrospinning method can be classified into two categories based on the collectors as (a) dry or (b) wet. In the dry electrospinning process the solvent evaporation and fibre solidification are achieved simultaneously by blowing a hot air gun, pressurized gas or by using an extraction fan system. Nevertheless, the solvent composition plays a vital role in dry electrospinning technique. To overcome rapid volatilization of solvents during the electrospinning process high pressurized gases are used. For example, electrospinning of Poly (vinyl pyrrolidone) (PVP) nanofibres were conducted under pressurized carbon dioxide to reduce the viscosity of the polymer [42]. High vapour pressure solvent systems need extraction speeds at a minimal rate and vice versa for low vapour pressure solvents. Gou et al. modelled a two-dimensional analysis of the dry spinning process to predict the solidification behaviour and chain orientation for polymer fibres [43]. This incorporates both viscous and viscoelastic contributions along with the radial and axial variations in the concentration and temperature.

The wet electrospinning process is a variant of dry spinning process where the modification is based on the usage of a liquid collector. In "wet spinning", the polymer solution is extruded into a viscous coagulation bath that acts a liquid collector which is non-solvent for the polymer. The polymer solidifies in the bath as a precipitate due to the phase separation between the solvent and non-solvent system. This process was favoured for producing nanofibres for applications like tissue engineering where fluffy and three dimensional (3D) sponge—like fibres are desirable for cell scaffolds. A three-dimensional structure produced from poly(caprolactone) (PCL) by the wet electrospinning into ethanol/distilled water mixture was produced by Kostakova et al. [44]. This method was adapted by Xanofi, a commercial electrospinning company [45] and demonstrated continuous production of micro and nanofibres at rates exceeding 60 g min^{-1} for application in life sciences.

Melt Electrospinning and Melt Blowing

One major drawback of solution electrospinning is the use of volatile solvents which is not only difficult to extrude but also pose environmental concerns and health and safety concerns for researchers and machine operators. The melt-blown technique has become a process of choice for the production of microfibers in which high-velocity heated air blows a molten polymer resin from an extruder die tip/nozzle on to a conveyor to draw fibres without use of a solvent. This method was first developed in the 1950s and used at commercial scale by Exxon in the 1970s. Since then this technology attracted many synthetic textile manufacturers including Vose, 3M, Kimberley-Clark, Cummings and Johns Manville. This method could achieve fast production rates in the order 3 m/min for 3–4 μm diameter fibres. But later, by reducing the internal diameter of the nozzle and/or by increasing the number of orifices per nozzle, sub-micron fibres are produced. However due to the complex machine set-up and energy intensive process, has resulted in uses of melt blown nanofibre. Hills, Inc., [46] uses a bi-/multi-component fibre extrusion process called 'islands in the sea' (a conjugated nozzle melt blown technique) to produce nanofibres. In this example, one of the soluble polymer is dissolved in a selected solvent and a second polymer is insoluble. Then the bi-component fluids are extruded into fibres within a single filament followed by subsequent dissolution of the soluble polymer, leaving only the insoluble polymer in the form microfibres.

Melt electrospinning is different to melt blowing as it uses centrifugal forces to draw fibres, rather than pressurized air [35, 47]. Moreover, unlike typical electrospinning that uses electrostatic forces to draw the fibres from the tip, the melt electrospinning process is based on mechanical drawing of polymer melts. Moreover, melt electrospinning requires cooling of the polymeric jet, while solution electrospinning relies on evaporation of the solvent to produce fibres [48]. Larrondo and Manley [49–51] were the first to electrospin molten polymer (polypropylene). Although melt electrospinning has huge potential to replace solution electrospinning its usage is limited due to their inability to spin fibres less than sub-micrometre. The fibres produced by melt electrospinning are at least one order of magnitude thicker than the fibres produced by the solution electrospinning. There are two crucial factors that affect the formation and the morphology of the fibres in melt electrospinning, these are low surface charge density of the polymer melt and the solidification process of polymer melt.

As aforementioned in electrospinning, the large surface charge densities of the polymer solution due to the presence of solvent causes bending instabilities that are responsible for the reductions of the jet diameter leading to formation of nanofibres. But in melt electrospinning—a non-solvent system experiences little or no bending instabilities in the jet due to the absence of surface charge density. Suppression of these bending stabilities will greatly affect the stretching of the polymer jet leading to the formation of fibres with diameter in the sub-micron to micron range. Moreover, the characteristics of the polymer itself affect the process more than in solvent electrospinning. For instance, the tendency of polymer to solidify as it flows greatly hinders the formation of polymer jet, thus resulting in microfibres.

Nevertheless, coaxial melt electrospinning is heavily advantageous if two polymers with different melting points can create core–sheath nanofibres which can be incorporated for phase change applications [1, 52]. By understanding the factors affecting the spinning of nanofibres via melt electrospinning recently Liu et al. designed patented electrospinning apparatus to extrude ultrafine nanofibres [53–55]. Electrospunra produces micro- and nanofibres using melt electrospinning process [56].

A small variation in melt electrospinning was applied to discover two new techniques called Forcespinning [57, 58] and Rotary jet spinning [59]. Recently Forcespinning technique is gaining substantial attention, it was developed at the University of Texas and commercialized by FibeRio Technology Corporation. This technique employs centrifugal forces to drive polymer melt through the single or multiple orifices in a rotating spinneret. Rotary jet spinning which also employs the same principle was first developed at the Harvard School of Engineering and Applied Sciences. These centrifugal techniques have attracted more attention recently as they not only enable the production of insoluble thermoplastic polymer but also produce fibres under sub-micron range.

Emulsion Electrospinning

Emulsion electrospinning is a promising alternative to solution and melt electrospinning as it overcomes the restrictions of solution and melt electrospinning [60]. Although the emulsion interpolymerization technique was invented in the 1940s there has been only little development in this technique [61, 62]. Emulsion electrospinning was first patented by DuPont in 1956 to produce composite or matrix fibres composed of mixture of two or more polymers where one of which is in the form of immiscible microscopic droplets, dispersed into a continuous liquid phase solution [63–66]. Materials that are insoluble or non-melting are generally finely grounded and mixed with an electrospinnable polymer solution in the presence of catalyst(s) and emulsifiers to produce core–sheath nanofibres via emulsion electrospinning. For example, hydrophobic PVC polymer emulsion was dispersed in unsaturated hydrophilic PVOH in the presence of catalyst and emulsifier to produce flame-retardant matrix fibres such as Cordelan® [67, 68].

In addition, emulsion electrospinning has been identified as an ideal best method to encapsulate drugs/therapeutics in an emulsified polymer mixture as an encapsulated core–sheath fibres. The as-spun fibres have markedly ease the release of drugs [69–74]. Water/oil based emulsions are preferred to carry bioactives in order to overcome the challenge of incorporation and sustained release of the hydrophilic bioactive molecules (such drugs, therapeutics, etc.) from a hydrophobic polymer nanofibres [72, 73].

Seema Agarwal et al., a clear advocate of emulsion and suspension electrospinning, coined the technique as "green electrospinning" as it uses no hazardous solvent and is a promising option to clean and safe electrospinning [60]. The main differences between the three major electrospinning techniques are summarized in Table 11.1.

Commercial Electrospinning Companies

Despite the efforts taken in modifying and redesigning the basic electrospinning set-up in regards to spinneret and the spinning techniques, the challenges in upscaling electrospinning are still a significant barrier. Leading materials companies attempting to address the challenges of productivity, profitability and environmental responsibility and have created new systems, adapted from conventional electrospinning, to mass produce nanofibre at higher rates. Although some of these companies have patented their technique most of them are protected as trade secrets.

So far, companies such as Espin technologies (USA) and Finetex inc (Korea), produce nanofibres for specific applications such as air filtration. New innovative structuring of nanofibres are developed by industries for its application in other domains beyond the traditional air filtration application.

Table 11.1 The main difference between the three major electrospinning techniques

Parameters	Solution electrospinning	Melt electrospinning	Emulsion electrospinning
Analyte solution	Polymer dissolved in volatile solvent	Polymer melt with no solvent	Two or more polymer in emulsion
Viscosity of the solution	Highly viscous solution but low comparative to polymer melt	Highly viscous melt	Low viscous solution
Dominant force	Electrostatic force	Mechanical drawing of melt	Electrostatic force
Source	High voltage	High voltage and heating filament	High voltage
Fibre diameter	Sub-micron–nanometre	Micron–sub-micron	Sub-micron–nanometre
Environment	Controlled environment required for high ventilation	Ventilation required for most of the polymer melts due to corona discharge that produces endocrine disruptors	Controlled environment required for high ventilation
Major drawbacks	Use of high volatile solvents	Complexity of the experimental set-up where two or more heating filament is required for continues melt of polymers Comparatively high energy consumption Generally produce micro fibres	Complexity of the experimental set up where spinning parameters, interfacial tension, and viscoelasticity of the two polymers must be precisely controlled Need to use one or more catalyst and emulsifiers to produce neatly shaped fibres

Start-up companies like The Electrospinning Company (UK) and Nanofibre solutions (USA) have been using nanofibres for bio medical applications. Whereas few companies including Revolution fibres Ltd (New Zealand) and Fibre Rio Technology Corp (USA) are seeking to procure the unique advantage of electrospun polymer nanofibres for the wide range of versatile applications. Most of these companies specifically customize the fibre producing manufacturing unit according to their end users.

Table 11.2 is a list of selected companies, where their core business is around production of nanofibres, customization of nanofibres or production of electrospinning machine. Companies like Donaldson, Samsung, LG, Toray, Mitsubishi Chemicals and Boeing do have R&D programmes or have the ability to make nanofibres but they were intentionally left out of the list as the nanofibre production within their business very specialized to their own needs, and are not major participants in the core sector.

Table 11.2 Summary of list of electrospinning companies

Company name	HQ country	Website	Business*	Selected commercial products	Electrospinning technique	Related electrospinning technique Patent	Machine
Elmarco	Chezh Republic	http://www.elmarco.com/	MM, PT, CS	Nanospider™ (lab to industrial scale)	Needleless	DE10136255 A1, WO2012013167 A2, WO2008106903 A2, WO2009049566 A2	Own design
Revolution Fibres Ltd	New Zealand	http://www.revolutionfibres.com/	PM, PT, CS	Seta™ (air filtration), Phnoix™ (acoustics), Xantu. Layr™ (composite reinforcement), ActiVLayr™ (skin care and health), nanodream™ (bedding)	Sonic needleless electrospinning	Unknown/trade secret	Own design
Finetex EnE, Inc	South Korea	http://ftene.com/	PM, CS	Technoweb™ series X, V, R, Pro and Filtrepro (air and liquid filtration), Technoweb™ C series (cosmetics), Finemask® (facial mask), Nexture™ (technical textiles)	Needle	US7980838	Own design
Xanofi	USA	http://xanofi.com/	PM, CS	XanoLoft™ & XanoRail™ (air filtration), QuietMat™ (acoustics), XanoMatrix™ (cell culture), XanoCapture™ (protein purification)	XanoShear™—shear force spinning	WO2015023943 A1	Own design

(continued)

Table 11.2 (continued)

Company name	HQ country	Website	Business*	Selected commercial products	Electrospinning technique	Related electrospinning technique Patent	Machine
Nanofiber Solutions	USA	http://www.nanofibersolutions.com/	PM, CS	NanoAligned ™ (aligned nanofibre culture substrate), NanoECM™ (randomly oriented nanofibre culture substrate)	Unknown/trade secret	WO2010042647 A2	Unknown/trade secret
Electrospinning Company	UK	http://www.electrospinning.co.uk/	PM, CS	MIMETIX® (aligned or randomly oriented nanofibre culture substrate)	Needle	Unknown/trade secret	IME technologies
eSpin Technologies	USA	http://www.espintechnologies.com/	PM, CS	Exceed® and nWeb™ filters (air filtration), SIMWyPES® (high-performance wipes), technical textiles, nanofibre culture substrate, filtratraion media	Unknown/trade secret	Unknown/trade secret	Own design
Stellenbosch Nanofiber Company	South Africa	http://www.sncfibers.com/	PM, CS	Various biodegradable or synthetic polymer nanofibre sheets	SNC BEST™ free-surface ball electrospinning	US20140302245 A1, US8522520 B2,	Own design
SNS Nano Fiber Tehcnology, LLC	USA	http://www.snsnano.com/	PM, CS	NANOSAN® (super absorbant pad)	Needle, coaxial	US20090093585 A1	Own design
Pardam nanotechnology	Czech Republic	http://pardam.cz/	PM, CS	NnF CERAM® (inorganic nanofibrous powders), NnF MBRANE® (Polymer Nanofibrous membranes, PUR, PA6, PAN, PVB, PCL)	ForceSpinning®, Needleless	FibreRio's & Elmacro's Machines	FibreRio, Elmarco

(continued)

Table 11.2 (continued)

Company name	HQ country	Website	Business*	Selected commercial products	Electrospinning technique	Related electrospinning technique Patent	Machine
MEMPRO® Materials	USA	http://mempro.com/	PM, CS	Inorganic nanofbres, polymeric nanofibres, catalysts	PreciseFibre™	Unknown/trade Secret	Own design
Nicast Ltd	Israel	http://www.nicast.com/	PM, CS	AVflo™ vascular access graft, NovaMesh™ ventral hernia mesh	Needleless (drum) or Needle	US7112293 B2	Own design
Nanofiber Separations LLC	USA	http://www.nanofiberseparations.com/	PM, CS	Functional separation membranes	Unknown/trade secret	WO2014126575 A9	Unknown/trade secret
NANO-FM B. V.	Netherlands	http://www.nano-fm.nl/	PM, CS	Cell scaffold	Unknown/trade secret	Unknown/trade secret	Unknown/trade secret
ANF Technology (NAFEN)	USA	http://anftechnology.com/	PM, CS	NAFEN Alumina Nano Fibres	Chemical reaction	US20130192517 A1	Own design
nanoScaffold Technologies, LLC	USA	http://nanoscaffoldtech.com/index.php	PM, CS	Cell scaffold	Unknown/trade secret	Unknown/trade secret	Unknown/trade secret
Arsenal Medical	USA	http://www.arsenalmedical.com/	PM, CS	AxioCore™ fibre technology (cell integration and tissue regeneration)	Needle	US20120193836 A1	Own design
BioSurfaces Inc.	USA	http://www.biosurfaces.us/	PM, CS	Cell scaffold, medical article	Needle	US8771582 B2	Own design
Soft Material and Technologies S.r.l.	Italy	http://www.smtnano.com/	PM, CS	Cell scaffold, customized nanofibre sheets	Unknown/trade secret	Unknown/trade secret	Unknown/trade secret

(continued)

Table 11.2 (continued)

Company name	HQ country	Website	Business*	Selected commercial products	Electrospinning technique	Related electrospinning technique Patent	Machine
nanomyp	Spain	http://nanomyp.com/	PM, CS	Tiss® Series (nanofibre tissues)	NanoMyP® (needle)	Unknown/trade secret	Own disegn
Redspeare & Co., LLP (Respilon®)	Czech Republic	http://www.respilon.com/en/	PM, CS	ReSpimask® (nanofibre mask), RespiPro® (certified nanofibre mask), Nanofibre Filter, Respilon Night Care® (bedding), Respilon AIR® (window and door insect barrier and air filter screen)	Needleless, Elmarco's Machine	Elmarco's Machine	Elmarco
Welcron Co., Ltd	South Korea	http://eng.welcron.com/	PM	Technical textiles, wound dressings and medical textiles, heat exchanging element	Melt blown	WO2012081819 A1	Own design
Amofiber	South Korea	http://www.amofiber.co.kr/	PM	Amotex® (technical textiles), nanofibre membranes for energy storage, fluid filters	Air assisted needle	WO2010143914 A2	Own design
Nano109	Turkey	http://www.nano109.com	PM	Fineweb™ Series F & H (air filtration)	Unknown/trade secret	Unknown/trade secret	Unknown
BMCbio Co., Ltd	South Korea	http://www.bmcbio.com	PM	MEDIWIZ Natural Mask Pack, Wipes	Unknown/trade secret	KR101379249000	Own design
Wooree Nanophil Co. Ltd	South Korea	http://www.wooreenanophil.co.kr/	PM	Battery separator, texhnical textiles, high efficiency filtration media	Needle	US20140087013 A1	Own design

(continued)

Table 11.2 (continued)

Company name	HQ country	Website	Business*	Selected commercial products	Electrospinning technique	Related electrospinning technique Patent	Machine
Nanofiber veterinary	USA	http://nanofiberveterinary.com/	PM	NanoLigV™ (tendons and ligaments), NanoCareV™ (wound care), NanoMeshV™ (surgical mesh), NanoBoneV™ (bone grafts and stents), NanoVesslV™ (vascular)	Unknown/trade secret	Unknown/trade Secret	Unknown/trade secret
Hirose Paper MFG Co. Ltd	Japan	http://www.hirose-paper-mfg.co.jp/english/index.html	PM	A range of composite nanofibre non-woven textiles	Needleless	Unknown/trade secret	Unknown/trade secret
Neotherix Ltd	UK	http://www.neotherix.com/technology.php	PM	Neotherix® scaffold	Unknown/trade secret	US20120015331 A1	Unknown/trade secret
FiberTrap	USA	http://www.fibertrap.com/	PM	Insect/bug barrier for bedding and textiles	Unknown/trade secret	Unknown/trade secret	Unknown/trade secret
NanoSpun Technologies	Israel	http://www.nanospuntech.com/	PM	Water filtration, fuel filters, biomedical	Needle, coaxial	WO2014006621 A1	Own design
polyremedy inc	USA	http://www.polyremedy.com/	PM	HEALSMART® smart wound dressing	Air jet assisted	US7105058 B1	Own design
Vaporsens	USA	http://www.vaporsens.com/	PM	VSENS (nanofibre chemical Sensor)	Unknown/trade secret	Unknown/trade secret	Unknown/trade secret
Fuence	Japan	http://www.fuence.co.jp/en	MM,CS	Esprayer Series (lab to industrial Scale)	ESD Technology (air jet assisted needle)	WO2008044737 A1	Own design

(continued)

Table 11.2 (continued)

Company name	HQ country	Website	Business*	Selected commercial products	Electrospinning technique	Related electrospinning technique Patent	Machine
SPUR a.s.	Czech Republic	http://www.spur-nanotechnologies.cz/	MM, PM, CS	SpurTex (air filter media), SPIN Line (nanafibre production line)	Needleless	WO2010118708 A2	Own design
TOPTEC	South Korea	http://www.toptec.co.kr/	MM, PM, CS	Production Line, factory set-up, battery separator	Needle	WO2012111930 A3, WO2012128473 A2	Own design
BioInicia	Spain	http://www.bioinicia.com/	MM, CS	Fluidnatek® (lab to industrial scale machine)	eStretching™ (multicomponent coaxial electrospinning), Needle, Electrospraying	Unknown/trade secret	Own design
Yflow S.D	Spain	http://www.yflow.com/	MM, CS	Yflow® coaxial electrospinning machine (lab to industrial scale)	Coaxial, needle, electrospraying	US20100203315 A1, WO2005089042 A3, US7794634 B2	Own design
Inovenso	Turkey	http://www.inovenso.com/	MM, CS	Nanospinner series and NE series (lab to industrial scale)	Needle, coaxial	Unknown/trade secret	Own design
HILLS Inc	USA	http://hillsinc.net/	MM, CS	Fibre and textile machinery expert, technical textiles	Melt blowing	US20080023888 A1	Own design
Electrospunra PTE LTD	Singapore	http://www.electrospunra.com/	MM, CS	Electrospunra ES210 (Lab scale electrospinning Machine), Melt Electrospinning Device, Polymeric nanofibres	Needle, melt electrospinning	Unknown/trade secret	Own design
ANSTCO	Iran	http://anstco.com/english/indexen.html	MM, CS	ANSTCO-NF Series (lab to industrial scale), peripherals	Needle	Unknown/trade secret	Own design

(continued)

Table 11.2 (continued)

Company name	HQ country	Website	Business*	Selected commercial products	Electrospinning technique	Related electrospinning technique Patent	Machine
Profector Life Science (Spraybase®)	USA	http://www.spraybase.com/	MM, CS	Lab scale electrospinning, electrospraying, melt electrospinning solutions	Needle	Unknown/trade secret	Own design
SKE Advanced Therapies	Italy	http://www.ske.it/	MM, CS	Lab scale electrospinning machines, cell scaffold	Needle	Unknown/trade secret	Own design
Fnm Co., (Fanavaran Nano-Meghyas Research & Development Co.)	Iran	http://en.fnm.ir/	MM, CS	Electrolis-Dip®, lab scale to industrial electrospinning machines	Needle, Needleless, Melt Electrospinning	Unknown/trade secret	Own design
FiberRio® Technology Corp.	USA	http://fiberiotech.com/	MM	Fibrelab® & Fibre Engine® (lab to Industrial Scale)	ForceSpinning®	WO2012109215 A3, WO2012109240 A3, WO2012109242 A3, US8647540 B2, US20140035177 A1, US20140042651 A1	Own design
MECC Co., Ltd	Japan	http://www.mecc.co.jp/en/	MM	NF series, nanon and snan (lab scale)	Needle, coaxial	Unknown/trade secret	Own design
Shenzhen Tong Li Tech Co., Ltd	China	http://www.electro-spinning.com/	MM	NEU series & peripherals	Needle	Unknown/trade secret	Own design
Contipro Biotech s.r.o (4SPIN®)	Czech Republic	https://www.4spin.info/	MM	4SPIN® series with lots of spinneret and collector options	Needle, coaxial, air assited needle, spinneret, coaxial	US8721313 B2, US8727756 B2, WO2014015843 A1	Own design

(continued)

Table 11.2 (continued)

Company name	HQ country	Website	Business*	Selected commercial products	Electrospinning technique	Related electrospinning technique Patent	Machine
nanonc	South Korea	http://www.nanonc.co.kr/	MM	eS-robot® (lab scale machines)	Air assisted needle, spinneret, coaxial, eSpray®	Unknown/trade secret	Own design
eSpin NanoTech Pvt Ltd	India	http://www.espinnanotech.com/	MM	Super ES Series (lab scale electrospinning machine)	Needle, coaxial	Unknown/trade secret	Own design
Linari Engineering S.r.l	Italy	http://www.linaribiomedical.com/index.php/en/	MM	RT series (lab scale electrospinning machine)	Needle, coaxial	Unknown/trade secret	Own design
Electrospinz	New Zealand	http://www.electrospinz.co.nz/	MM	ES series (lab scale electrospinning machine)	Needle	Unknown/trade secret	Own design
Erich Huber GmbH	Germany	http://www.ehuber.de/index.php	MM	E-Spintronic (lab scale electrospinning machines)	Needle	Unknown/Trade secret	Own design
Technical Fine Co., Ltd	Japan	http://www.technicalfine.com	CS	Nanofibre customization service (Fuence's business partner)	ESD technology (air jet assisted needle)	Fuence's Machine	Fuence

*PM Product manufacturer (produces pure nanofibre membranes for various applications). *PT* Platform technology provider such as consultation market research, business analysis, new material development and product development, prototyping, trialling pilot production, designing and installing, purpose-built nanofibre manufacturing machines, designing and implementing post-processing equipment and quality assurance. *CS* Customization service provider either for nanofibre products for electrospinning machines or both. *MM* Machine manufacturer

Application of Nanofibres

The level of interest of creating small, powerful and efficient materials and commercialzing them have increased exponentially over time due to their infinite potential and endless application in many fields [39]. Several properties of nanofibres make them suitable for a wide range of applications. Fields adapting nanofibres in recent years are depicted in the Fig. 11.4.

Commercially Available Nanofibre Products

Filtration Industry

A clean and safe environment has become a priority to everyone in the world and air pollution is one of the top environmental concerns of our time. On 18 December 2013, the European Union adopted "*The Clean Air Policy Package*" to reduce environmental concerns and to improve the well-being of people. The filtration industry has been at the forefront of this matter making products for filtering polluted air and water to the state where people can breathe and drink safely. The filtration industry was the first to exploit the uses of nanofibre in the product as highly efficient filtration media.

By incorporating nanofibres into particulate filter media, companies were able to get higher particulate filtration efficiency for particulate sizes less than 2.5 μm with almost negligible effect on pressure drop or often less pressure drop across the filter. This is because nanofibre layer is non-woven and has controllable nanometre range pore size that is distributed evenly. Furthermore, each individual nanofibres has van der Waals forces which attract charged particles towards themselves. Table 11.3 shows the list of popular companies that have successfully commercialized air filter media and products.

High efficiency liquid filtration is also an area where nanofibres are being increasingly used. Nanofibres membranes are extremely porous hence offering reduced pressure drop across the filter and hence less energy required to pump liquid through the filter membrane. The disadvantage of nanofibres in liquid filtration is the strength of the membrane itself when subdued to high pressure pumps to push the liquid through the membranes. As liquid filtration requires more maintenance, skilled knowledge and experience the nanofibre liquid filtration industry itself is largely dominated by specialized companies and established filtration companies around the world. Table 11.4 shows some of the liquid nanofibre filters currently available on the market.

Another area of filtration that is growing fast and wide is "smart" filters. These filters are designed specifically to do special functions additional to filtering air particulates or liquid. These functions can involve targeting certain gas particles, microbes, chemicals, heavy metals, viruses and other pollutants or substances. The

Fig. 11.4 Application of electrospun nanofibres in different fields

Table 11.3 List of world leading companies that commercialize nanofibre for air filter media

Company	Filter brand
AMSOIL	Ea Air Filters [75]
Clarcor	Protura® [76]
Donaldson	Ultraweb® [77]
DHA Filter	Duraweb™ [78]
eSpin	Exceed® [79]
Finetex EnE, Inc	Technoweb™ [80]
Hollingsworth & Vose	Nanoweb® filter media [81]
MANN + HUMMEL	Micrograde NF filter [82]
Revolution Fibres	Seta™ [40]
Vokes Air	Synsafe REVO® [83]

Table 11.4 List of companies that commercialize nanofibres for liquid filter media

Company	Filter brand
Donaldson	Tetratex® [84]
DuPont	Hybrid Membrane Technology [85]
Finetex Ene Inc	Technoweb™ Filtrepro [80]
Hollingsworth & Vose	Nanoweb® filter media [81]
Pardem	NnF MBRANE® [86]
SPUR Nanotechnologies	SpurTEX FM [87]

"smart" filters are normally produced by adding functional substances within the filter media nanofibres or by coating surfaces of filter media. Nanofibres are usually customized by adding functional additives such as silver, platinum particles or bioactives such as Manuka extract. These additives are bounded within nanofibres and distribute themselves evenly within nanofibre membrane of the filter media. Revolution Fibres Ltd. (New Zealand) developed Manuka extract incorporated residential air filtration brand Seta™ which approximately 150,000 of New Zealand homes are breathing through. This range of filters possesses nanofibre layer where each nanofibre is enriched with anti-microbial properties of Manuka extract along with high filtration efficiency at a comparatively low pressure drop (Fig. 11.5).

Acoustics Industry

With an increasing global population, the awareness of personal space and privacy has also dramatically escalated. Sound pollution is a growing concern in many parts of the world. Sound is normally controlled by combining appropriate acoustic materials (foams, rubber, metal blocks, fabrics, etc.) and locations where these materials are installed. This works by orienting sound waves to desired angles and movement. However, as personal space becomes more limited and devices get

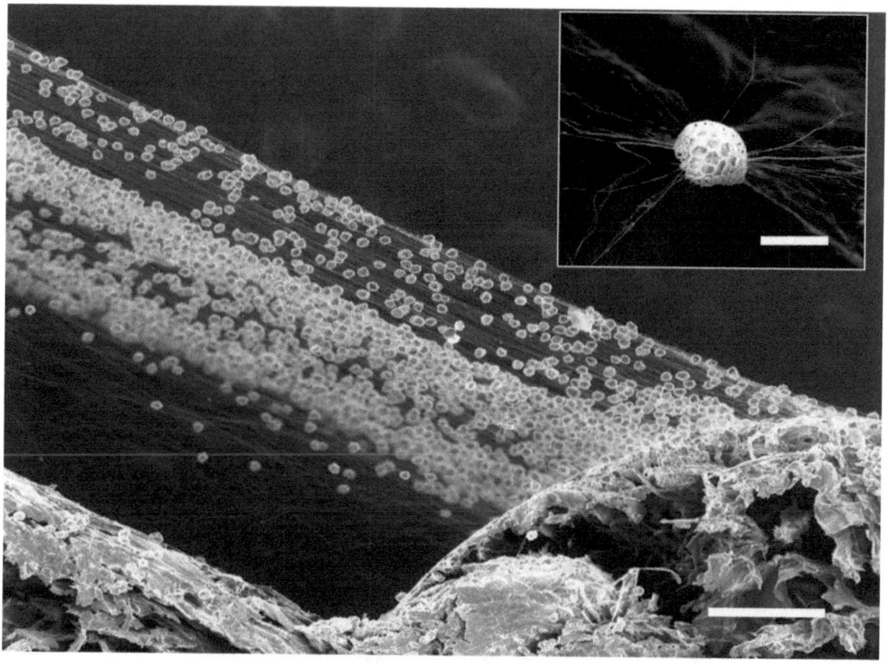

Fig. 11.5 SEM image (scale bar 100 μm) of "smart" Seta™ disk nanofibre filter with pollens and dust trapped within the fibre. *Inset* zoomed SEM image with scale bar 20 μm

smaller, thinner and lighter the conventional materials cannot be used in some applications. Sound controlling is crucial to protect people from unwanted and harmful noise but concurrently improve quality of sound performance. Personal electronics companies like LG and Denon have already incorporated nanofibre acoustic membranes into their headphones and speakers [88, 89]. The trend of nanofibre acoustic membrane is used in high-end premium equipment. Amogreentech Ltd (South Korea) have also developed patented acoustic nanofibre membrane and is collaborating with a Japanese company to make high-performing speaker system.

Revolution Fibres' nanofibre acoustic products Phonix™ can be tailored and customized to target a range of frequency levels. Its innovative acoustic nanofibre membrane Phonix™ was able to achieve improved sound absorption by in combination with IQ Commercial's furniture foam. Figure 11.6a shows the sound absorption efficiency of nanofibre at high frequencies. The two companies collaborated successfully and launched the Return Focus Pod for sound absorbing office work stations for improved client privacy and efficient working desk for office users (Fig. 11.6b).

(a)

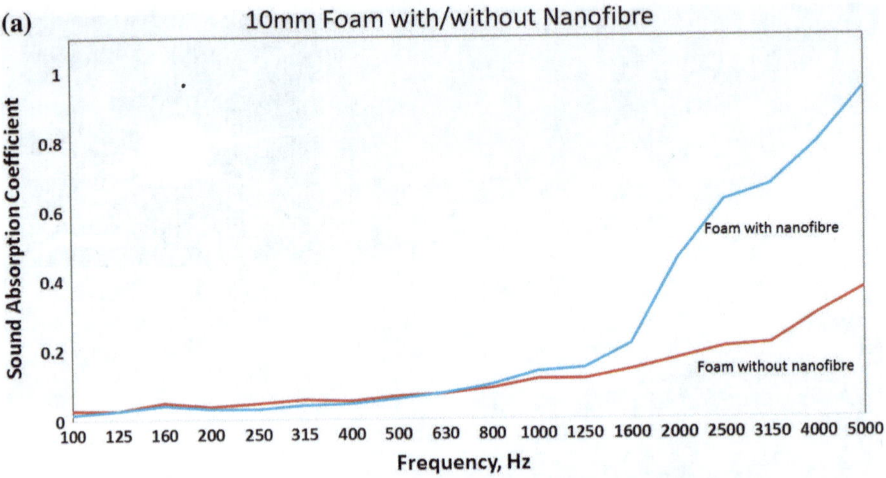

(b)

Fig. 11.6 **a** The graph showing sound absorption efficiency of foam with and without nanofibre from low to high frequency, **b** The commercially available acoustic foam office furniture with "Phonix"

Skin Care and Biomedical Industry

The morphology and characteristics of nanofibres have similar morphology to human body tissues. It is believed that electrospun nanofibres closely resemble the human body tissue extracellular matrix (ECM). Nanofibres can be spun with bio-compatible polymers and Food and Drug Administration (FDA) approved polymers which may be directly implanted within a human body. Nanofibres can also contain

Table 11.5 List of companies that commercialize nanofibres for biomedical industries

Company	Brand
Arsenal Medical	AxioCore™ [90]
Electrospinning Company	MIMETIX® [91]
Nanofiber Solutions	NanoAligned™ [92]
Xanofi	XanoMatrix™ [93]

active ingredients or cells to act as a drug delivery mechanism or kill certain bacteria and viruses.

One of many fast growing applications for nanofibres in biomedical space is tissue engineering. It is a multidisciplinary field that has been governed by generic concept of combining cells and scaffold in the design and fabrication of tissues and organs. Artificial scaffolds can be generated that are usable for tissue engineering and skin regenerative medicine due to their ability to mimic the nanoscale properties of certain fibrous components of the native extracellular matrix in tissues. Nanofibres are currently being investigated and researched for tissue regeneration and growth. Some companies have already commercialized nanofibre cell scaffold and can be purchased through online ordering. The following table shows the some of the companies which have nanofibre cell scaffold as their major nanofibre products (Table 11.5).

Nanofibres in wound care systems are also proving to be the wonder material for fast and effective treatment method. Nanofibre's large surface area and highly permeable characteristics make them suitable for use in healing wounds. Nanofiber Veterinary (USA) has recently discovered remarkable performance of functional nanofibre wound care system through in situ trials on wounded animals [94]. Nicast (Israel) has developed novel electrospun self-sealing nanofibrous CE certified vascular access graft AVflo™. This product has recently gone through clinical studies and is actively being distributed worldwide. Its amazing property makes the nanofibre graft to stimulate hemostasis when punched with a needle for dialysis but leaves no puncture on the graft.

The cosmetics and skin performance sector is another application where nanofibres are emerging, offering instantaneous delivery nutrient mechanism to the skin. In this mechanism, the nanofibre membrane acts similar to wound care but instead the nanofibres are usually water soluble or biodegradable and contain healthy nutrients for the skin. Revolution Fibres Ltd (New Zealand) has developed nutrient infused marine collagen nanofibres called ActiVLayr™. This unique water soluble collagen nanofibres act as a carrier to deliver natural nutrients to the skin to improve performance of the skin. The active content can be modified and customized in order to target certain skin benefits [40].

Composite Materials Industry

Composite materials is the use of two or more components to build a structure, and can date back to early as where people built houses with clay and straw. In some applications reinforced plastics are used due to its desirable weight to strength ratio, light weight, durability and relatively low maintenance cost for high resistance against corrosion. Furthermore, these reinforced plastics can be carefully engineered to have required mechanical properties. The reinforced plastics industry, especially carbon fibre reinforced plastic industry has always been advancing with automotive, aerospace, defence technology, wind turbine and sporting equipment industries.

Reinforced plastic composite parts are manufactured using reinforcing materials—usually fibres of glass, plastics or carbon and thermosetting polymer resin. Often mechanical failure occurs at the brittle polymer resin due to the resin content's relatively poor resistance against compression, impact, fracture and delamination. In order to increase physical and mechanical properties of the resin system for the fibre reinforced plastic parts, often toughening particles are used to toughen the resin system.

However, when particle enriched polymer resin is used, because particles are free to flow during the curing process, there is uneven distribution of particles and possibly agglomeration within the polymer matrix that can propose unexpected mechanical failure or can results in random mechanical properties within the plastic parts. Sometimes, microfibre interlayer is used to toughen the system but this can introduce voids within the resin system, unnecessary increase in volume and mass.

Nanofibres veils can be used in between carbon fibre laminate to significantly improve the following mechanical properties [95] of carbon fibre reinforced plastic materials.

- Mode 1 Interlaminar fracture toughness (Crack Opening)
- Mode 2 Interlaminar fracture toughness (Crack Sliding)
- Compression After Impact (CAI)
- Delamination Resistance
- Fatigue resistance (repetitive strain)

Additionally, due to small mass and volume of the nanofibre veils, it introduces almost negligible amount of addition of mass and volume, which is critical in thin-ply application where high-tech reinforced plastic parts are required. It also showed some promising improvement in resistance against fatigue and vibration properties of plastic parts in service.

XantuLayr™ is the first commericially available nanofibre veil that is currently being used by carbon fibre reinforced plastic industry. This innovative product is manufactured by Revolution Fibres Ltd (New Zealand) and was first implemented by Kilwell (New Zealand) carbon fibre fishing rod with XantuLayr™. This porous nanofibre membrane does not hinder resin flow in between carbon fibre laminate during curing process and is compatible with any resin system. XantuLayr™

Fig. 11.7 The photograph of Kilwell's fishing Xantu rod

doubled Kilwell's fishing rod breaking load. The fishing rod is building its reputation as a premium fishing rod and it is getting exported to around the world [95] (Fig. 11.7).

Sensing and Electronics Industry

Semiconducting nanofibres have been extensively explored in recent years as potential substrate/components for nanoscale flexible electronics [96], energy harvesting [97], storage and conversion [98] and biological, optical [99], chemical and electrochemical sensing [100, 101]. As discussed else where the large surface-area-to volume ratio, large stacking density, tailored pore structures of nanofibres are highly advantageous particularly for sensors as transducers.

Whether it is chemical, optical or electrochemical sensing, nanofibres as a substrate has been proved to improve the selectivity and specificity towards the target analyte [102]. For example, by combining the advantage of the redox active conducting polymer (CP) and the high surface area of the nanofibre, Vaporsens [103], created a portable gas detector (which was originally developed at the University of Utah) based on electrochemical mechanism for detecting trace amounts of chemicals with greater sensitivity and accuracy.

The main sensing material/transducer consists of building blocks of conducting polymer (based on perylene core molecule) coated on nanofibres. Because the core molecule is the CP they were directly coated onto interdigitated electrodes to form an electrical circuit. Due to the inherent redox nature of CP [104, 105] the transducer is sensitive especially to redox active explosives, pesticides and other chemicals including amines, nitros, peroxides, ammonia, phenols and phosphines. Another similar electrochemical devices based on nanofibres were created by Nottingham Trent University and Nano products ltd (UK).

Like chemical sensing devices the biomedical diagnostics devices also requires a sensitive and more stable sensing material surface that allows it to react readily with the target analyte. Such sensing material should also discriminate the impurities and interact specifically to target molecule to avoid false positive results. Recently,

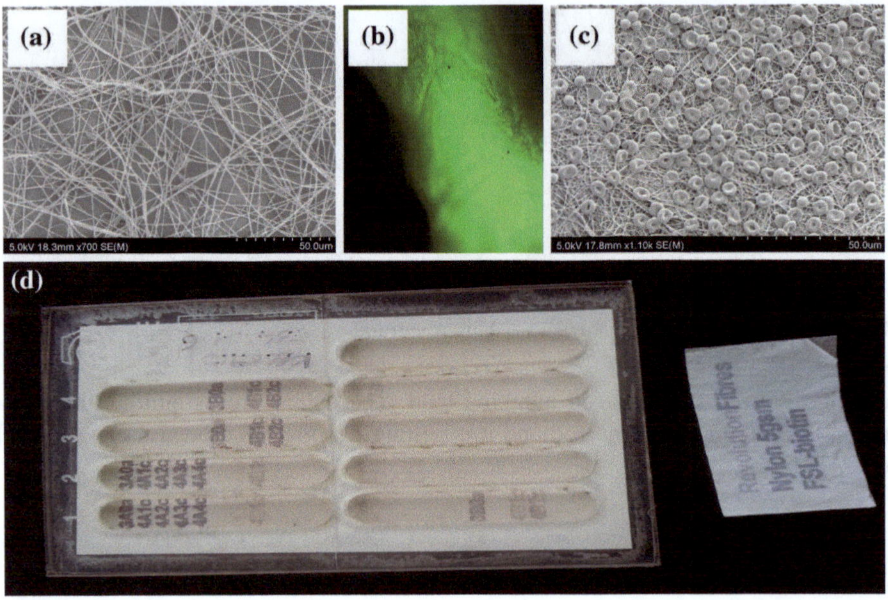

Fig. 11.8 Image showing the **a** SEM image of nanofibre electrospun on wax paper **b** Nanofibre with KODE™ constructs (FSL-biotin) with fluorescent marker showing the biomolecules are homogeneously present throughout the fibre, **c** SEM image of nanofibre with the KODE™ constructs and **d** KODE™ constructs inkjet printed on nanofibres and reacted with dyed immunological assay to stain the presence of FSL

KODE Biotech Ltd. (New Zealand) collaborated with the Revolution Fibres and developed a patented technology platform to create Nano-textile immunosensor. Here the sensing material consists of nanofibres incorporated with the patented KODE™ molecule, a Function-Spacer-Lipid (FSL). The KODE constructs that are readily dispersible in water coated on the nanofibre textiles either pre or post spinning. Figure 11.8 shows the Koded Nano-textiles (PCT/NZ2014/050015), inkjet printed with KODE FSL-biotin and subsequently reacted in an immunological assay to stain for the presence of FSL. The initial trails of KODE Nano-textiles exhibited outstanding sensitivity and selectivity towards the target analyte when compared to the previous technology that uses different substrates such as paper and micro fibres. The created KODE Nano-textiles holds huge promise in application including but not limited to anti-microbial medical textiles, tissue engineering scaffolds, drug delivery medium, diagnostic assays and lab-on-a-chip devices.

Apart from sensing market, nanofibres are also envisaged to bridge the gap between existing supercapacitors and lithium ion batteries as electrode material (conducting nanofibres) and separators (non-conducting nanofibres) to form a new generation ultracapacitors with more power capacity flexible electronics [106, 107].

Recently Teijin Techno Products Limited introduced its commercially viable nanofibre based product Dream Weaver Gold™ as Li-ion battery separator. Teijin's technology uses aramid nanofibres on a porous substrates to create a porous separator substrate with high temperature stability—up to 300 °C which is about double than is offered by other commercially available separators. The main advantage of this technology is helps to reduce the risk of ignition of batteries at high temperatures and also holds the electrolyte effectively to maintain battery perform at low temperatures [108].

Conclusion

The developments of electrospinning technology have seen dramatic increase over the last decade. Despite the clear potential for nanofibres to be used in versatile applications, the uptake in terms of the number of commercialized products that utilize electrospun nanofibre remains very limited, with air filtration being the only sector in which they are used extensively. This is mainly due to “invisibility” of potential of the nanofibre in the technology market. Therefore it is essential for the nanofibre manufactures to showcase the full potential of nanofibres in various diverse sectors to the global market as early as possible by collaborating with research organizations and academics at early stage. But the academic research and the industrial manufactures always step aside from each other during early stage development and often end up collaborating at a point where the concepts are ready to scale up but need significant amount of rework/redesign and additional costs to upscale the technology. Early engagement of academic researchers with the commercial nanofibre manufactures would benefit both industry and academic and accelerate the manufacturing speed and reduce cost and delays.

Moreover the early collaboration with industries would help the university “spin-out” companies to work on designing the machine that would suit industrial setup as early as possible during their research to avoid spending huge resources in new infrastructures and machine developments. Or it is cheaper to subcontract the manufacturer at early stage to turn their inventions to commercially ready product.

So far most of the researches are based on nanofibre as non-woven mat. Other assemblies of nanofibres are expected to grow and developed for novel applications especially in biomedical sectors are nanofibres in the form of yarns and hydrogels. For example, nano yarns can be assembled into any form of fibrous structures through knitting or weaving process which can be useful in high performances clothing and smart textiles. Similarly nanofibre in the form of hydrogel are becoming very popular as they are transparent hydrophilic gel, they are useful in the fields such as cosmetics as preventive medicines and tissue engineering as scaffolding.

The nanofibre industry is still growing diversely and helping other industries like filtration industries by adding value to the product, there is no doubt nanofibres will emerge in more diverse shape and sizes and will find its application in variety of domains other than filtration sector ranging from cosmetics to catalysis.

References

1. McCann JT, Marquez M, Xia Y (2006) Melt Coaxial electrospinning: a versatile method for the encapsulation of solid materials and fabrication of phase change nanofibers. Nano Lett 6 (12):2868–2872. doi:10.1021/nl0620839
2. Huang T, Marshall LR, Armantrout JE, Yembrick S, Dunn WH, Oconnor JM, Mueller T, Avgousti M, Wetzel MD (2012) Production of nanofibers by melt spinning. In: Google Patents
3. Che G, Lakshmi BB, Martin CR, Fisher ER, Ruoff RS (1998) Chemical vapor deposition based synthesis of carbon nanotubes and nanofibers using a template method. Chem Mater 10(1):260–267. doi:10.1021/cm970412f
4. Yoon YJ, Baik HK (2001) Catalytic growth mechanism of carbon nanofibers through chemical vapor deposition. Diam Relat Mater 10(3–7):1214–1217. doi:10.1016/S0925-9635 (00)00585-9
5. Hartgerink JD, Beniash E, Stupp SI (2001) Self-assembly and mineralization of peptide-amphiphile nanofibers. Science 294(5547):1684–1688. doi:10.1126/science.1063187
6. Zhang S (2003) Fabrication of novel biomaterials through molecular self-assembly. Nat Biotech 21(10):1171–1178
7. Lin Y, Yao Y, Yang X, Wei N, Li X, Gong P, Li R, Wu D (2008) Preparation of poly(ether sulfone) nanofibers by gas-jet/electrospinning. J Appl Polym Sci 107(2):909–917. doi:10.1002/app.26445
8. Wang B, Yao Y, Peng J, Lin Y, Liu W, Luo Y, Xiang R, Li R, Wu D (2009) Preparation of poly(ester imide) ultrafine fibers by gas-jet/electrospinning. J Appl Polym Sci 114(2):883–891. doi:10.1002/app.30505
9. Hung AM, Stupp SI (2007) Simultaneous self-assembly, orientation, and patterning of Peptide–Amphiphile nanofibers by soft lithography. Nano Lett 7(5):1165–1171. doi:10.1021/nl062835z
10. Cui H, Kalinin SV, Yang X, Lowndes DH (2004) Growth of carbon nanofibers on tipless cantilevers for high resolution topography and magnetic force imaging. Nano Lett 4 (11):2157–2161. doi:10.1021/nl048740j
11. Gilbert W, De Magnete (1991) Translated 1893 from Latin to English by Paul Fleury Mottelay. New York
12. Taylor G (1964) Disintegration of Water Droplets in an Electric Field. Proceed Roy Soc A 280(1382):383
13. Cooley JF (1902) Apparatus for electrically dispersing fluids. In: Google Patents
14. Morton WJ (1902) Method of dispersing fluids. In: Google Patents
15. Anton F (1934) Process and apparatus for preparing artificial threads. In: Google Patents
16. Barhate RS, Ramakrishna S (2007) Nanofibrous filtering media: Filtration problems and solutions from tiny materials. J Membr Sci 296(1–2):1–8. doi:10.1016/j.memsci.2007.03.038
17. Persano L, Camposeo A, Tekmen C, Pisignano D (2013) Industrial upscaling of electrospinning and applications of polymer nanofibers: a review. Macromol Mater Eng 298(5):504–520. doi:10.1002/mame.201200290
18. Andrady AL (2008) Science and Technology of polymer nanofibers. Wiley, Hoboken
19. Reneker DH, Chun I (1996) Nanometre diameter fibres of polymer, produced by electrospinning. Nanotechnology 7(3):216
20. Srinivasan G, Reneker DH (1995) Structure and morphology of small diameter electrospun aramid fibers. Polym Int 36(2):195–201. doi:10.1002/pi.1995.210360210
21. Wang X, Drew C, Lee S-H, Senecal KJ, Kumar J, Samuelson LA (2002) Electrospun nanofibrous membranes for highly sensitive optical sensors. Nano Lett 2(11):1273–1275. doi:10.1021/nl020216u
22. Wang T, Kumar S (2006) Electrospinning of polyacrylonitrile nanofibers. J Appl Polym Sci 102(2):1023–1029. doi:10.1002/app.24123

23. Li D, Xia Y (2004) Electrospinning of nanofibers: reinventing the wheel? Adv Mater 16 (14):1151–1170. doi:10.1002/adma.200400719
24. Wang X, Niu H, Lin T, Wang X (2009) Needleless electrospinning of nanofibers with a conical wire coil. Polym Eng Sci 49(8):1582–1586. doi:10.1002/pen.21377
25. Reneker DH, Yarin AL (2008) Electrospinning jets and polymer nanofibers. Polymer 49 (10):2387–2425. doi:10.1016/j.polymer.2008.02.002
26. Frenot A, Chronakis IS (2003) Polymer nanofibers assembled by electrospinning. Curr Opin Colloid Interface Sci 8(1):64–75. doi:10.1016/S1359-0294(03)00004-9
27. Barua B, Saha MC (2015) Investigation on jet stability, fiber diameter, and tensile properties of electrospun polyacrylonitrile nanofibrous yarns. Journal of Applied Polymer Science, n/a-n/a (2015). doi:10.1002/app.41918
28. Cai Y, Gevelber M (2013) The effect of relative humidity and evaporation rate on electrospinning: fiber diameter and measurement for control implications. J Mater Sci 48 (22):7812–7826. doi:10.1007/s10853-013-7544-x
29. Yarin AL, Koombhongse S, Reneker DH (2001) Bending instability in electrospinning of nanofibers. J Appl Phys 89(5):3018–3026. doi:10.1063/1.1333035
30. Reneker DH, Yarin AL, Fong H, Koombhongse S (2000) Bending instability of electrically charged liquid jets of polymer solutions in electrospinning. J Appl Phys 87(9):4531–4547. doi:10.1063/1.373532
31. Shin YM, Hohman MM, Brenner MP, Rutledge GC (2001) Experimental characterization of electrospinning: the electrically forced jet and instabilities. Polymer 42(25):09955–09967. doi:10.1016/S0032-3861(01)00540-7
32. Shin YM, Hohman MM, Brenner MP, Rutledge GC (2001) Electrospinning: a whipping fluid jet generates submicron polymer fibers. Appl Phys Lett 78(8):1149–1151. doi:10.1063/1.1345798
33. Hohman MM, Shin M, Rutledge G, Brenner MP (2001) Electrospinning and electrically forced jets. II. Appl Phys Fluids (1994-present) 13(8):2221–2236 (2001). doi:http://dx.doi.org/10.1063/1.1384013
34. Yarin AL, Koombhongse S, Reneker DH (2001) Taylor cone and jetting from liquid droplets in electrospinning of nanofibers. J Appl Phys 90(9):4836–4846. doi:10.1063/1.1408260
35. Reneker DH, Yarin AL, Zussman E, Xu H (2007) Electrospinning of nanofibers from polymer solutions and melts. In: Hassan A, van Erik der G (eds) Adv Appl Mech 41:43–346 (Elsevier)
36. Sahay R, Thavasi V, Ramakrishna S (2011) Design modifications in electrospinning setup for advanced applications. J Nanomaterials 2011:17. doi:10.1155/2011/317673
37. BioInicia, http://www.bioinicia.com/. Accessed 2015
38. Elmarco N http://www.elmarco.com/. Accessed 2015
39. Hayes T, Hosie I (2015) Turning nanofibres into products: electrospinning from a manufacturer's perspective. In: Macagnano A, Zampetti E, Kny E (eds) Electrospinning for high performance sensors. NanoSci Technol 305–329. Springer International Publishing
40. Revolution Fibres Ltd http://www.revolutionfibres.com/ (2013–2015)
41. Pringle C Single bubble-electrospinning of polyvinyl alcohol and polyacrylonitrile. Stellenbosch University (2011–2012)
42. Wahyudiono, Murakami K, Machmudah S, Sasaki M, Goto M (2012) A dry process for polymer nano-microfibers prepared by electrospinning under pressurized CO_2. Jpn J Appl Phys 51(8S1):08HF07
43. Gou Z, McHugh AJ (2004) Two-dimensional modeling of dry spinning of polymer fibers. J Nonnewton Fluid Mech 118(2–3):121–136. doi:10.1016/j.jnnfm.2004.03.003
44. Kostakova MSE, Pokorny P, Lukas D, Kostakova E, Seps M, Pokorny P, Lukas D (2014) Express Polym Lett 8:554–564. doi:10.3144/expresspolymlett.2014.59
45. Xanofi, Xanoshear, XanoMatrix (http://www.xanofi.com/tech.html). Accessed 2014
46. i. Hills www.hillsinc.net/. Accessed 2015

47. Yuan X, Mak AFT, Kwok KW, Yung BKO, Yao K (2001) Characterization of poly(L-lactic acid) fibers produced by melt spinning. J Appl Polym Sci 81(1):251–260. doi:10.1002/app.1436
48. Dalton PD, Grafahrend D, Klinkhammer K, Klee D, Möller M (2007) Electrospinning of polymer melts: phenomenological observations. Polymer 48(23):6823–6833. doi:10.1016/j.polymer.2007.09.037
49. Larrondo L, R. St. John Manley (1981) Electrostatic fiber spinning from polymer melts. I. Experimental observations on fiber formation and properties. J Polym Sci: Polym Phys Ed 19(6):909–920. doi:10.1002/pol.1981.180190601
50. Larrondo L, R. St. John Manley (1981) Electrostatic fiber spinning from polymer melts. II. Examination of the flow field in an electrically driven jet. J Polym Sci: Polym Phys Ed 19 (6):921–932. doi:10.1002/pol.1981.180190602
51. Larrondo L, R. St. John Manley (1981) Electrostatic fiber spinning from polymer melts. III. Electrostatic deformation of a pendant drop of polymer melt. J Polym Sci: Polym Phys Ed 19 (6):933–940. doi:10.1002/pol.1981.180190603
52. Li F, Zhao Y, Wang S, Han D, Jiang L, Song Y (2009) Thermochromic core–shell nanofibers fabricated by melt coaxial electrospinning. J Appl Polym Sci 112(1):269–274. doi:10.1002/app.29384
53. Liu Y, Li X, Ramakrishna S (2014) Melt electrospinning in a parallel electric field. J Polym Sci, Part B: Polym Phys 52(14):946–952. doi:10.1002/polb.23511
54. Deng R, Liu Y, Ding Y, Xie P, Luo L, Yang W (2009) Melt electrospinning of low-density polyethylene having a low-melt flow index. J Appl Polym Sci 114(1):166–175. doi:10.1002/app.29864
55. Yang RJDWM, Liu Y, Ding YM A high-efficiency spinning head for electrospinning. China Patent ZL20082008287.3
56. Electrospunra. (http://www.electrospunra.com/index.html#page_1) Accessed 2014
57. FibeRio. http://fiberiotech.com/technology/how-it-works/. Accessed 2015
58. Sarkar K, Gomez C, Zambrano S, Ramirez M, de Hoyos E, Vasquez H, Lozano K (2010) Electrospinning to Forcespinning™. Mater Today 13(11):12–14. doi:10.1016/S1369-7021(10)70199-1
59. Badrossamay MR, McIlwee HA, Goss JA, Parker KK (2010) Nanofiber assembly by rotary jet-spinning. Nano Lett 10(6):2257–2261. doi:10.1021/nl101355x
60. Agarwal S, Greiner A (2011) On the way to clean and safe electrospinning—Green electrospinning: emulsion and suspension electrospinning. Polym Adv Technol 22(3):372–378. doi:10.1002/pat.1883
61. Yarin AL (2011) Coaxial electrospinning and emulsion electrospinning of core–shell fibers. Polym Adv Technol 22(3):310–317. doi:10.1002/pat.1781
62. Angeles M, Cheng H-L, Velankar SS (2008) Emulsion electrospinning: composite fibers from drop breakup during electrospinning. Polym Adv Technol 19(7):728–733. doi:10.1002/pat.1031
63. Hener W (1942) Interpolymers of vinyl sulphonic acid with another vinyl compound and aqueous emulsions thereof. In: Google Patents
64. Seizo O, Tadashi M, Hiroshi A, Ippei C (1963) Process for the production of fibers having polyvinyl chloride as the principal constituent and also containing polyvinyl alcohol. In: Google Patents
65. Arthur BL, Edwin JW (1956) Composition comprising a polyhalogenated ethylene polymer and viscose and process of shaping the same. In: Google Patents
66. Le Fevre Walter J, P SD (1963) Emulsion polymerization with amino alcohol esters as cationic comonomers. In: Google Patents
67. Matsuo K, Araki M, Matsugu T, Mikami T (1975) Emulsions useful in the preparation of heat resistant fibers and films. In: Google Patents
68. Tsuji T, Korematsu M (1975) Highly flame-retardant shaped articles and method for preparing the same. In: Google Patents

69. Camerlo A, Vebert-Nardin C, Rossi RM, Popa AM (2013) Fragrance encapsulation in polymeric matrices by emulsion electrospinning. Eur Polym J 49(12):3806–3813. doi:10.1016/j.eurpolymj.2013.08.028
70. Sinha-Ray S, Zhang Y, Placke D, Megaridis CM, Yarin AL (2010) Resins with "Nano-Raisins". Langmuir 26(12):10243–10249. doi:10.1021/la1004177
71. Qi P, Hu J, Wang Xu (2006) Encapsulation of drug reservoirs in fibers by emulsion electrospinning: morphology characterization and preliminary release assessment. Biomacromolecules 7(8):2327–2330. doi:10.1021/bm060264z
72. Gentsch R, Pippig F, Schmidt S, Cernoch P, Polleux J, Börner HG (2011) Single-step electrospinning to bioactive polymer nanofibers. Macromolecules 44(3):453–461. doi:10.1021/ma102847a
73. Choi S-H, Youn D-Y, Jo SM, Oh S-G, Kim I-D (2011) Micelle-Mediated synthesis of single-crystalline β(3C)-SiC fibers via emulsion electrospinning. ACS Appl Mater Interfaces 3(5):1385–1389. doi:10.1021/am200171v
74. Bazilevsky AV, Yarin AL, Megaridis CM (2007) Co-electrospinning of Core–Shell fibers using a single-nozzle technique. Langmuir 23(5):2311–2314. doi:10.1021/la063194q
75. AMSOIL Inc, AMSOIL Ea Air Filters https://www.amsoil.com/shop/by-product/filters-and-by-pass-systems/air/amsoil-ea-air-filters/. Accessed 29 Apr 2015
76. United Air Specialists Inc., Clarcor, PROTURA® Advanced nanofiber. http://www.uasinc.com/Products/Replacement-Filters/ProTura-Nanofiber-Technology. Accessed 29 Apr 2015
77. Donaldson Company Inc., Ultra-Web Media Technology. http://www2.donaldson.com/torit/en-us/pages/products/ultra-webmediatechnology.aspx. Accessed 29 Apr 2015
78. DHA Filter LLC, Duraweb™ Nanofiber Dust Cartridges. http://www.dhafilter.com/#!duraweb-nanofiber-dust-cartridges/c1poj. Accessed 29 Apr 2015
79. eSpin Technologies http://www.espintechnologies.com/. Accessed 29 March 2015
80. Finetex EnE Inc. http://ftene.com/. Accessed 1 Apr 2015
81. Vose H NanoWeb Nanofiber Technology. http://www.hollingsworth-vose.com/en/KnowledgeCenter/White-Papers/nanofiber/. Accessed 29 Apr 2015
82. Hummel M Small fibers–big effect. https://www.mann-hummel.com/en/corp/news/news/newsdetail/?tx_ttnews%5Btt_news%5D=191&cHash=bc13c54ba5b73e17fcd8f05cf8600a44. Accessed 29 Apr 2015
83. Air V, Revo II. http://www.vokesair.com/products/bag-filters-0/revo-ii. Accessed 29 Apr 2015
84. Donaldson Company Inc., Donaldson Membranes: Improving filter performance with Tetratex PTFE membrane filter media. http://www2.donaldson.com/tetratex/en-us/pages/home.aspx. Accessed 29 Apr 2015
85. Pont D, Hybrid Membrane Technology (HMT). http://www2.dupont.com/Separation_Solutions/en_US/tech_info/hmt/hmt.html. Accessed 29 Apr 2015
86. Pardam nanotechnology. http://pardam.cz/. Accessed 1 Apr 2015
87. SPUR a.s. http://www.spur-nanotechnologies.cz/. Accessed 5 Apr 2015
88. Denon. http://headphone.usa.denon.com/#slideCont1. Accessed 2015
89. LG http://www.lg.com/uk/press-release/smart-hi-fi-audio-wireless-multi-room-solution-from-lg. Accessed 2015
90. Arsenal Medical. http://www.arsenalmedical.com/. Accessed 10 Apr 2015
91. Electrospinning Company. http://www.electrospinning.co.uk/. Accessed 3 Apr 2015
92. Nanofiber Solutions. http://www.nanofibersolutions.com/. Accessed 3 Apr 2015
93. Xanofi. http://xanofi.com/. Accessed 2 April 2015
94. Veterinary N http://nanofiberveterinary.com/. Accessed 2015
95. Beckermann GW, Pickering KL (2015) Mode I and Mode II interlaminar fracture toughness of composite laminates interleaved with electrospun nanofibre veils. Compos A Appl Sci Manuf 72:11–21. doi:10.1016/j.compositesa.2015.01.028
96. Nogi M, Iwamoto S, Nakagaito AN, Yano H (2009) Optically transparent nanofiber paper. Adv Mater 21(16):1595–1598. doi:10.1002/adma.200803174

97. Chang J, Dommer M, Chang C, Lin L (2012) Piezoelectric nanofibers for energy scavenging applications. Nano Energy 1(3):356–371. doi:10.1016/j.nanoen.2012.02.003
98. Ji L, Zhang X (2009) Electrospun carbon nanofibers containing silicon particles as an energy-storage medium. Carbon 47(14):3219–3226. doi:10.1016/j.carbon.2009.07.039
99. Meng C, Xiao Y, Wang P, Zhang L, Liu Y, Tong L (2011) Quantum-dot-doped polymer nanofibers for optical sensing. Adv Mater 23(33):3770–3774. doi:10.1002/adma.201101392
100. Sawicka K, Gouma P, Simon S (2005) Electrospun biocomposite nanofibers for urea biosensing. Sens Actuators B: Chem 108(1–2):585–588. doi:10.1016/j.snb.2004.12.013
101. Virji S, Huang J, Kaner RB, Weiller BH (2004) Polyaniline Nanofiber gas sensors: examination of response mechanisms. Nano Lett 4(3):491–496. doi:10.1021/nl035122e
102. Su X, Ren J, Meng X, Ren X, Tang F (2013) A novel platform for enhanced biosensing based on the synergy effects of electrospun polymer nanofibers and graphene oxides. Analyst 138(5):1459–1466. doi:10.1039/C2AN36663K
103. Vaporsens. http://www.vaporsens.com/sensors/. Accessed 2015
104. Kannan B, Williams DE, Laslau C, Travas-Sejdic J (2012) A highly sensitive, label-free gene sensor based on a single conducting polymer nanowire. Biosens Bioelectron 35(1):258–264. doi:10.1016/j.bios.2012.02.058
105. Kannan B, Williams DE, Khoshmanesh K, Bowmaker GA, Travas-Sejdic J (2012) The electrochemical growth of conducting polymer "nanowires". J Electroanal Chem 669:82–89. doi:10.1016/j.jelechem.2012.01.022
106. Ra EJ, Raymundo-Piñero E, Lee YH, Béguin F (2009) High power supercapacitors using polyacrylonitrile-based carbon nanofiber paper. Carbon 47(13):2984–2992. doi:10.1016/j.carbon.2009.06.051
107. Wu Q, Xu Y, Yao Z, Liu A, Shi G (2010) Supercapacitors based on flexible graphene/polyaniline nanofiber composite films. ACS Nano 4(4):1963–1970. doi:10.1021/nn1000035
108. Teijin. http://www.teijin.com/news/2012/ebd120426_00.html. Accessed 2015

Chapter 12
Nanofibrillar Single Polymer Composites: Preparation and Mechanical Properties

Stoyko Fakirov

Introduction

The *concept of single polymer composites* was formulated some 40 years ago by Roger Porter and demonstrated in his publication with Capiati and Porter [1]. They used two types of samples of high-density polyethylene (HDPE) differing mostly in their melting temperatures. The preparation of this new material, called by them "one polymer composite" was possible due to the fact that aligned and extended chains provide thermodynamically more stable crystals, which thus will have higher melting points than conventionally crystallized melts.

During the last two decades, the interest of academia and industry in single polymer composites (SPCs) increased immensely due to the steady increasing adverse environmental impact of synthetic, petroleum-based polymers and their glass fiber-reinforced composites. In parallel, studies driven mostly by environmental concerns and aiming at replacement of the mineral reinforcing component in polymer composites resulted in creation of new types of composite materials. To them belong the *polymer–polymer composites* (PPCs) with natural fibers as reinforcement [2] or synthetic fibrous components as reinforcement [3].

Fundamental contributions to the field of single polymer composites were made by Ward et al. (e.g., [4–12]). They not only succeeded to prepare SPCs of large number of homopolymers, to develop new techniques for their manufacturing but they also patented [8] and commercialized products with SPC structure. Interesting contribution to SPCs was made also by J. Karger–Kocsis demonstrating that such composites can be prepared using the polymorphic phenomenon in polymers (e.g., [13]). And, finally, to the SPCs was recently added the family of *nanofibrillar*

S. Fakirov (✉)
Department of Mechanical Engineering, Centre for Advanced Composite Materials,
The University of Auckland, Private Bag 92019, Auckland 1142, New Zealand
e-mail: s.fakirov@auckland.ac.nz

S. Fakirov (ed.), *Nano-size Polymers*,
DOI 10.1007/978-3-319-39715-3_12

single polymer composites prepared from parallel aligned non-interconnected nanofibris (diameters between 50 and 250 nm) having superior mechanical properties (e.g. [14, 15]). Just representatives of this last group, the nanofibrillar SPCs, are the main subject of this chapter. Their outstanding mechanical properties will be considered in more detail after offering some general information on nomenclature of SPCs, on the methods of their preparation and their importance for the polymer science and technology.

Single Polymer Composites: Definitions, Nomenclature, Advantages, and Disadvantages

It should be noted that the designation of single polymer composites, as they are referred to in this chapter, has been an issue of some debate. SPCs have also been called: *one polymer composites*, *homocomposites*, *self-reinforced composites*, *one-phase composites*, *homogeneous composites*, or *all-polymer composites* [13–20]. Of course, the terminology is less important as long as it is used consistently. *Composite material* typically means a (hopefully) synergistic combination of two *chemically different* materials. Polymer–polymer composites are then such composites whose reinforcement and matrix belong to two chemically different materials. At the same time, the reinforcement and matrix in SPCs are *chemically identical*; indeed they come from the same original starting material. Therefore, the typical definition of composites must be expanded to cover SPCs.

The polymer–polymer composites as well as the single polymer composites are distinguished by another characteristic feature, namely the fact that the matrix is always an isotropic material. Dealing with the traditional composites usually it is not stressed on this detail, possibly because such a large class of composites as the glass fibers reinforced *thermosets* are always characterized by inherent isotropic matrix as well as the injection molded glass fiber-reinforced *thermoplastics,* while the situation with PPCs and SPCs is completely different. The starting material for their manufacturing is always highly oriented polymer (for SPCs) or polymer blend (for PPCs) and a processing step is needed when at higher temperature one of the blend components is converted from highly oriented into isotropic state. In this way, the highly oriented polymer blend is converted in a typical composite, i.e., isotropic matrix reinforced with stronger fibrous material. Without this isotropization step, we are supposed to consider the highly oriented polymer blends (e.g., textile yarn spun from polymer blends) as composite material, what hardly could be correct (and if yes which of the drawn components represents the matrix?).

The same holds for SPCs when at higher temperature an isotropic matrix is formed due to a surface partial melting. However, common to all composite types is that they consist of a *matrix* and *reinforcement*, which differ in their mechanical properties and a synergistic combination of the latter characterizes the final product.

Dealing with nomenclature issues, it seems worth to call again the attention to a rather wrong practice in the composite community: the use of term "phase" instead of "*component*". As it was discussed earlier [13–15], the term "*phase*" is very well defined in thermodynamics and is frequently used in polymer physics for describing the various phases in one-component systems. A good example in this respect is poly(vinylidene fluoride) (PVDF) exhibiting five crystalline polymorphic modifications (phases!) and one amorphous phase but PVDF is still a one-component system. The use of the term "phase" instead of "component" would require the definition of another term for describing the phases in sense of thermodynamics. Obviously, the misuse of the term "phase" is a remnant from the time when the colloid chemistry was formulated (second half of nineteenth century, i.e., long before the polymer science was defined). Since the polymer solutions, even the true ones, behaved as the colloid systems of low-molecular-weight substances, they were called "lyophilic colloids" and to them was applied the terminology typical for the colloid chemistry—"dispersed phase" [15]!

What about SPCs, which according to their definition are one-component systems but comprise two different (*but chemically identical*) materials as matrix and reinforcement? For such cases it was suggested [13] to use the term "*constituents*", which will mean chemically identical materials but differing in some properties, such as melting temperature, physical structure (polymorphic modifications), mechanical properties, and others [13]. This means that we could have basically two approaches for preparation of SPCs, namely, *one-constituent approach* when we start with only one constituent (usually the reinforcement) and during the processing we are supposed to create the second constituent (the matrix). In the case of *two-constituent approach,* one starts with two different but chemically identical materials as matrix and reinforcement. Accordingly, in the multicomponent composite systems each component has different chemical composition.

The above definitions have been suggested by Karger–Kocsis and Fakirov [13] and they will be followed strictly in the current text. This means that the commercially available and widely cited [17–20] as SPC type of materials based on the homopolymer polypropylene (PP) (as reinforcement) and the random copolymer of PP [usually with polyethylene (PE)] (as a matrix) do not belong to SPCs because the two components are chemically different, i.e., such composites belong to the category of polymer–polymer composites. The same holds for thermoplastic polyester copolymers reinforced with poly(ethylene terephthalate) (PET) or with liquid crystalline polyester. Details about these PPCs can be found in the review by Matabola et al. [17] as well as in a very recent review of Karger–Kocsis and Barany [20].

Serious contributions to the development and commercialization of this special type of polymer–polymer composites (they comprise two chemically different polymers, which belong to the same polymer family as stressed by Karger–Kocsis [20]) have been made by Peijs et al. (e.g. [21, 22]). It seems important to note that their commercial importance would hardly be negatively affected if they will be properly classified as polymer–polymer but not as single polymer composites. The argument that the two chemically different components belong to the same polymer family [20] could hardly be considered seriously—it only smears the boundary

Table 12.1 Comparison of single polymer composites with other polymer composites

Characteristic feature	Common composites (incl. PPCs)	Single polymer composites
Components	Two	One
Elements	Matrix + reinforcement	Matrix + reinforcement
Chemical composition	Different	Identical
Interface adhesion	Poor	Perfect
Matrix/Reinf. ratio (wt%)	70/30, 80/20	20/80
Preparation	Melt blending	Thermal treatment
Processing equipment	Wearing from minerals	No wearing
Mechanical performance	Good to excellent	Very good to superior
Environmental impact	Serious	No

between the two types of composites, the PPCs and the SPCs and makes the clear definition of single polymer composites hardly possible!

In Table 12.1 are summarized some of the characteristic features of single polymer composites and the rest of other composites involving polymers.

Among the common characteristics of these two groups of polymer composites, the most important seems to be the fact that in both cases the composites comprize the two basic elements, the matrix, and the reinforcement. At the same time, in the case of SPCs the two basic elements are of the same chemical composition and this fact has an extremely important consequence—the interfacial adhesion quality is ideal and thus leading to superior mechanical performance. This issue, and particularly the approaches for improvement of the adhesion between matrix and reinforcement, is discussed in more detail in the subsequent pages.

The major disadvantage of all techniques for manufacturing of single polymer composites is the very small processing window, typically only a few degrees Centigrade.

Importance of Single Polymer Composites for Science and Technology of Polymer Composites and Their Environmental Impact

The awareness about adverse environmental impacts of synthetic polymers is, possibly, the main problem in polymer science and technology. What is more, this concern is ascending because the use of synthetic polymers is increasing rather than decreasing. A good example in this respect is the usage of PET whose production has an annual growth of 10 %, mostly due to its excellent properties as packaging material for various products that include pressurized beverages, food articles, and medicines. For this reason in many countries legislations are introduced to control the amount of plastics used. For example, in the European Union it was planned

[23] not to allow after 2015 the use in the cars manufacturing plastics having more than 5 wt% incineration quota.

The well-motivated efforts for replacing glass and other inorganic fibers as reinforcements in polymer composites resulted in application of natural fibers as reinforcement [2]. The next step in this direction was the use of synthetic, petroleum-based polymers as reinforcement but prepared as textile fibers, micro- or nanofibrils [3]. Of course, the latter approach is not as advantageous as using natural fibers that are biodegradable and eco-friendly. At the same time, the synthetic polymer–polymer composites seem to be much more acceptable from the environmental point of view in contrast to their inorganic counterparts because they, being organic in nature, could be subjected to incineration.

An even better solution than polymer–polymer composites are the single polymer composites as they, by definition, can be melted down to a single polymer, which can then be used in the manufacture of SPC, used in a common composite, or any other application for that particular polymer.

Regardless of the fact that recycling is becoming increasingly important to consider, the primary concern for most composites is still achieving the desired mechanical performance. But a common problem with many polymer composites is that the strength of the reinforcement is not fully utilized as the matrix and reinforcement do not have perfect adhesion. The problem of adhesion arises, because typically the word composite describes a combination of matrix and reinforcement which are chemically different, and as a rule they are not inclined to bond together. In the case of PPCs, the matrix and the reinforcement are of course much closer chemically to each other than their mineral counterparts, and so easier to choose for good adhesion. For example, while in the PPCs both component are hydrophobic, in the glass fiber-reinforced polymer composites the matrix is hydrophobic and the reinforcement—hydrophilic.

Considering the importance of the interface adhesion quality to the mechanical performance of polymer composites, it should be noted that particularly favorable opportunities for extreme improvement the adhesion quality in the case of polymer–polymer composites are available if the two composite components are condensation polymers. This is because such polymers can undergo additional chemical interactions in their melts [24] but also in a solid state [25] at elevated temperatures. These reactions are the *additional condensation* and *transreactions* [26].

If the improvement of interfacial adhesion is the main target, but not deep chemical changes in the matrix of PPCs, a short thermal treatment during the processing of the polymer–polymer composites of condensation partners is quite enough for establishing of chemical bonds between the blend components or matrix and reinforcement in case of PPCs.

A completely different is the situation when dealing with SPCs—they can show almost perfect adhesion because the matrix and reinforcement are chemically identical, and therefore show no tendency to separate from each other, allowing the formation of strong chemical bonds between the two due to mutual diffusion. In addition to this mechanism, for the cases when SPCs are prepared from condensation polymers, the establishment of covalent chemical bonds during the thermal treatment of the precursor for SPC is possible via the reactions mentioned above. What is more, the chemical interactions could be further enhanced by means of *transreactions catalyst* (e.g., Sb_2O_3) as recently demonstrated for nylon 6 (PA-6) [27]. While the improvement of the tensile modulus is almost the same for the cases without and with catalyst, the tensile strength of SPC as compared with the isotropic PA-6 improves by 300–400 % (depending on the moisture content) for the samples without catalyst, and by 500–650 % when catalyst is used [27].

In author's opinion, as mentioned elsewhere [15], this potential for drastic improvement of the interfacial adhesion quality between matrix and reinforcement is not exploited enough by the composite community possibly because of lack of chemistry knowledge among the engineers involved in this type of research.

Another interesting approach for considerable improvement of the interfacial adhesion is offered by the phenomenon of *transcrystallization*. The formation of transcrystalline layers comprising PP spherolites on the surface of glass and carbon fibers reinforced PP and other polymer matrix composites is well known and widely studied case (e.g., [28]). Lamellar transcrystalline layers of PP or of PE (matrix) have been observed also on the surface of PET microfibrils (reinforcement) in the respective microfibrillar composites (MFCs) prepared from the PP/PET or PE/PET blend [29–32].

It can be assumed that the transcrystalline layers play an important role on the adhesion between the reinforcing elements and the matrix thus determining to a considerable extent the mechanical performance of the final composite material. It has to be noted here that Capiati and Porter [1] considered the formation of transcrystalline layers in their HDPE single polymer composites for explanation of their impressive mechanical properties.

What the aspect ratio concerns as a factor determining to the great extent the reinforcing effect in the case of PPCs and SPCs, it should be mentioned that the values of this ratio are of some couple of hundreds since the reinforcement represents micro- or nanofibrils with diameters in the nano- or micrometer range and a length of couple of hundreds micrometer [33].

Methods for Preparation of Single Polymer Composites

The preparation of SPCs explores some inherent properties of polymers. For the case of completely amorphous SPCs the peculiarities of the glass transition temperature (T_g) could be used. Though T_g of amorphous polymers does not depend on the molecular weight (MW) above a given threshold (that is usually surpassed in commercial grades), their temperature-induced softening does depend. The appearance of a leathery state is a clear manifestation of this effect. Similar to that, the orientation also affects the softening of amorphous polymers. Accordingly, MW and orientation effects may be exploited in preparation of amorphous SPCs.

The scenario is different in semi-crystalline polymers. They always comprise of crystallites with varying perfection and thus having different melting temperatures

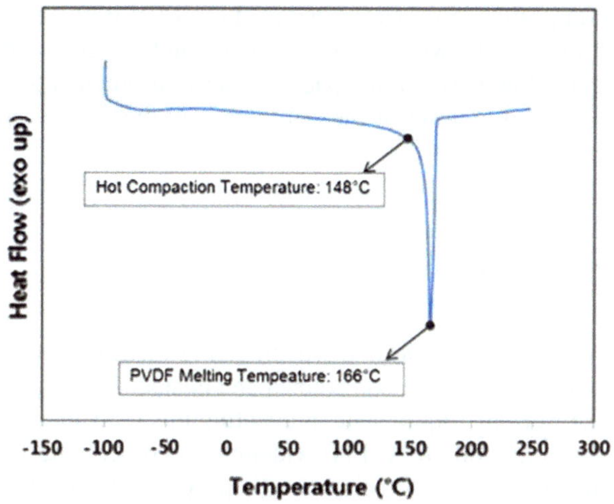

Fig. 12.1 DSC curve of PVDF demonstrating the melting and compression molding temperatures

(T_m). So, the related polymer has a melting interval instead of a sharp "melting point". During the SPC manufacturing the highly drawn fibers or tapes are "treated" at a temperature which corresponds to the T_m of the less perfect crystallites and represents the onset of the melting interval as demonstrated on the differential scanning calorimeter (DSC) curve for poly(vinylidene fluoride) in Fig. 12.1. In this way, due mostly to a partial surface melting, the aligned pressed fibers, tapes or filaments "stick" together by creating an isotropic "matrix" within the SPCs. A peculiarity of this approach is that the content of the highly oriented reinforcement may reach 80 wt% or more thus contrasting the common composites where the matrix strongly dominates.

The other inherent property of some semi-crystalline polymers is their strong tendency to crystallize in two or more *polymorphic modifications*. These modifications have different properties including T_m-values. A lower melting modification —for the role of the matrix—can be prepared using an appropriate nucleating agent, whereas the reinforcement should be a modification with higher T_m. Hot compression at a temperature between the T_m-values of the two polymorphic modifications can be used again to manufacture SPCs. It is also worth mentioning that this approach, mostly applied to PP, allows us preparation of SPC materials, the properties of which are close to those of traditional glass fiber-reinforced polymer composites.

Resin Infusion Method

This method, resembling to conventional composite manufacturing techniques, uses melt or powder impregnations. The reinforcing fibers are impregnated with a highly viscous polymer resin (similar to the production of glass and carbon fiber-reinforced composites via resin infusion). Low-MW polymers are often used to improve the flow rate. However, this is a slow and expensive process as the polymer matrix must flow between the packed fiber reinforcements. In addition, there is a high risk of partial melting of the reinforcement due to the required high melt impregnation temperature [34].

Overheating Method

Another technique is the overheating method whereby highly oriented, crystalline polymer fibers are constrained and heated. As the polymer chain must relax to reach the isotropic liquid state, preventing this relaxation is associated with an increase in T_m of around 10 °C [34] widening the processing window. Unfortunately, it is not always possible to constrain the polymer chains effectively if the molecular chains are folded, or the chain mobility is high.

Film Stacking Method

The film stacking method is a technique where the fiber reinforcement is sandwiched between films of the same polymer, and then hot pressing is applied to consolidate them together. Note that the melting of the matrix giving film should start at lower temperature than the reinforcement. This technique has been applied to a wide range of polymers, such as PET [35, 36], PP [37–40], PE [39, 41], and poly(lactic acid) (PLA) [42].

As many polymers are already available in fibers and films with a wide range of melting and other properties, the material selection is relatively straightforward, inexpensive, and materials can be selected to give a wide temperature window.

The tendency of some polymers to crystallize into two or more polymorphic modifications, with different mechanical properties and melting temperatures, can also be exploited to create SPCs. Karger–Kocsis patented [43] and explored [44–47] this method of creating polymorphic SPCs from PP, calling them *all-PP composites*. The matrix with a lower T_m than the reinforcement was prepared using β-phase nucleating agent, while the reinforcement was a PP fiber of α-modification. The same approach was applied to polyamide 6 (PA-6) [27] and polyamide 6.6 (PA-6.6) [48]. For the case of PA-6 impressive improvements in modulus (of 200 %) and ultimate tensile strength (of 300–400 %) were obtained over the isotropic matrix [27]. For PA-6.6 the modulus and tensile strength could be improved by 28 and 160 %, respectively, at only 20 wt% of the reinforcing constituent [48].

Co-extrusion Method

Co-extrusion is another technique to produce SPCs. The outer layer of the co-extruded and drawn tapes has a lower melting temperature than the core. The desired objects are prepared from various assemblies of the co-extruded tapes (woven, laminates of different lay-ups) by hot-pressing methods. The majority of the research has focused on a combination of a PP homopolymer for the core of the tape, and PP copolymer as the skin. After hot compaction the copolymer skin melts and becomes the isotropic matrix, bonding the still highly oriented homopolymer together (e.g., [21, 34, 49, 50]). Tapes are used as the reduced thickness leads to less crimping of the woven fabric, and therefore more efficient reinforcement. The specific strength and stiffness of the resulting composites are comparable to those of glass fiber-reinforced PP, however the former is fully recyclable [50, 51]. Another major advantage of this process is that when the co-extruded tapes are heated under high pressure in the mold, the movement of the polymer chains is constrained; and this causes the beneficial overheating effect. This, when combined with the use of different polymer grades, can extend the processing window to more than 30 °C [50]. Thanks to this large temperature interval, the commercially available co-extruded PP tapes are already converted into various products. Strictly speaking, however, this method does not lead to the creation of “true” SPCs because the reinforcement and the matrix are not completely identical in chemical composition as already discussed above. This is the major reason why only some selected works have been cited from the comprehensive studies of these co-extruded PP tapes by Peijs et al. [21], Alcock [49], Peijs [50, 51].

Hot-Compaction Method

Finally, there is the hot-compaction process which was patented [8] and pioneered by Ward and Hine (e.g. [4, 8, 10–12, 52, 53]) as a method of creating SPCs, while using only one constituent. SPC films are prepared by placing aligned polymer fibers or sheets under high pressure and at a critical temperature slightly below the T_m of the material [53]. The hot-compaction technique exploits an outlined above inherent property of polymers: the strong dependence of their T_m on their crystallization conditions (temperature, duration, pressure, etc.), as well as on their mechanical pretreatment (orientation). For this reason, due to the wide variation in the crystal perfection throughout the polymer, the melting takes place in a wide temperature range instead of at a point as low-MW crystals do (Fig. 12.1). By selecting a temperature on the lower end of this range, a small amount of the surface of the reinforcement melts and forms an isotropic matrix to bond the “residual” reinforcement together. The advantages of this approach are the following: (i) only a single starting material is required, (ii) the reinforcement volume fraction is very high, and (iii) the fibers are perfectly distributed in the matrix. The major disadvantage is the very small processing window, typically only a few degrees Centigrade. If the temperature is too high, too much of the reinforcement is transformed into the isotropic matrix. If it is too low there is not enough matrix material to properly bond the reinforcement together. This small processing window can be expanded using fibers with different draw ratios [54] or by exploiting some polymers variation in melting temperature with pressure to more precisely control the amount of reinforcement that is melted [55]. Hot-compaction has again been successfully performed on a wide variety of polymers: PE [4, 5, 36], PET [6, 36, 56], PP [10, 11, 52], PA-6.6 [57], and poly(ethylene naphthalate) (PEN) [58]. For reviews to the hot-compaction topic, the reader is addressed to Refs. [53, 59].

Methods for Preparation of Neat Polymer Nanofibrils and SPCs Thereof

As a matter of fact, the isolation of neat polymer nanofibrils takes place via the MFC concept.

There are three key requirements that must be satisfied when manufacturing a MFCs or NFCs. First, the involved polymers must have sufficient draw ability to allow the formation of reinforcing fibrils to occur; second, both components of the polymer blend must be able to be processed at a single temperature without the onset of degradation in either polymer; and third, the melting temperature of the reinforcing polymer must exceed that of the matrix polymer by at least 40 °C, to allow fibril retention during matrix consolidation. It has to be stressed that once these requirements have been satisfied, MFCs or NFCs can be manufactured using standard industrial polymer processing equipment as shown in Fig. 12.2.

The processing can be divided into three distinct steps, each vital to the successful creation of a micro- or nanofibrils-reinforced composites:

– *Mixing and extrusion*: The matrix and reinforcing polymers are dried and mixed, before being compounded and extruded. This forms an isotropic, continuous blend filament.

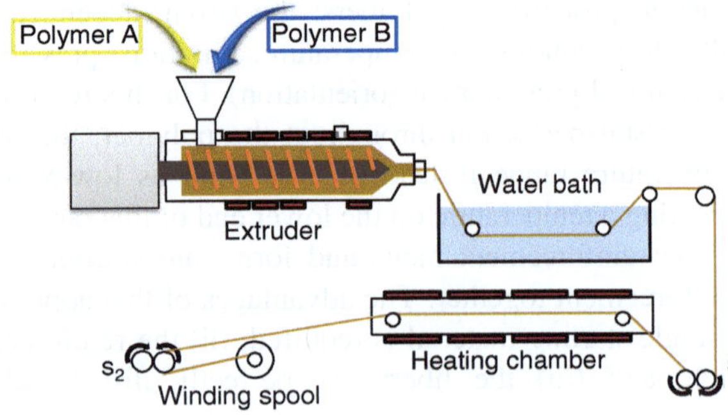

Fig. 12.2 Schematic setup for the manufacturing of polymer blends comprizing micro- or nanofibrils (s_1 and s_2 are the rotating velocities of the pairs of rolls, where $s_1 < s_2$, which causes drawing)

- *Drawing and fibrillation*: The blend filament is drawn through pairs of rollers (Fig. 12.2). This step creates highly oriented fibrils of the minor component with properties biased predominantly along a linear dimension or symmetry axis [60]. The drawing ratio is defined as the ratio of the linear speeds (s_2/s_1) of the two sets of rollers used to draw the filament and gives an indication as to the amount of alignment imparted to the blend. Next the filament is either collected on a spool (Fig. 12.2) or pelletized.
- *Matrix consolidation through thermal treatment*: The drawn filaments or pellets are formed into a composite at a processing temperature (T_{proc}) that lies between the melting temperatures of the constituent polymers. Control of this temperature is critical to the successful creation of MFCs as it ensures the formation of an isotropic matrix while still retaining the highly oriented reinforcing fibrils. If T_{proc} is too high the fibrils will melt and the reinforcing effect will be lost. Depending upon the post-processing method, the final composite structure can exhibit either quasi-isotropic or anisotropic tendencies depending on the production method.

The result of this process is a micro- or nanofibrils-reinforced composite material with mechanical properties superior to those of the plain matrix polymer [61]. It should be noted that variations in the MFC manufacturing process do exist.

Two different in situ fibril formation techniques can be employed to create the reinforcing fibrils: cold drawing of the solidified filament at a temperature below the melting temperature (T_m) of each blend constituent [61–75] or hot-stretching directly from the melt at a temperature above both polymers' glass transition temperatures [76–80]. Cold drawing generally results in a better molecular orientation.

To prepare neat fibrils from the drawn blend or from the respective polymer–polymer composite, one has to extract the second dominating blend component. This can be done using the modified Soxhlet extractor [66], which allows to perform the extraction of the matrix polymer (usually PP) with boiling organic solvent. This procedure could be performed immediately after the drawing or after the isotropization of the matrix polymer, i.e., after the preparation of the polymer–polymer composites (Fig. 12.3, Route A).

The final step in the preparation of the micro- or nanofibrillar SPCs is the hot compaction of the fibrils themselves (Fig. 12.3, Route B). For this purpose, the sample of neat fibrils was placed in a heating press. As mentioned already above, the compression molding temperature is of paramount importance in this case, which has been determined by DSC analysis (Fig. 12.1).

This approach has been applied to poly(butylene terephthalate) (PBT) for the preparation of nanofibrils. More specifically, PP and PBT were dry-mixed in a PP/PBT weight ratio of 70/30. Melt blending was performed using a Brabender DSE20 extruder with a 25 mm screw and *L*/*D* ratio of 40, at 250 °C and 5 rpm, using a 1.3 mm die. The extrudate was cooled using a water bath immediately after its exit from the die. This blending conditions result in a rather fine and homogeneous distribution of PBT particles in the PP matrix—particle sizes are typically

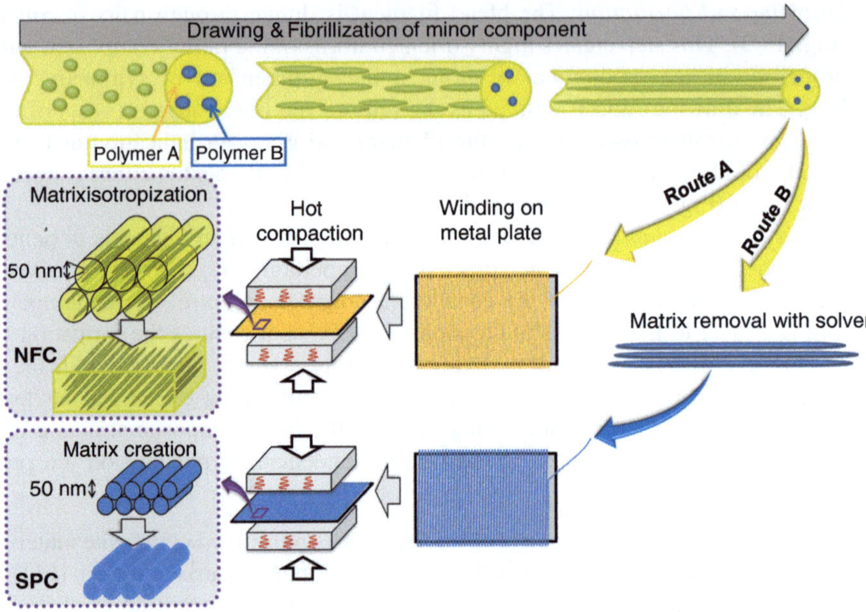

Fig. 12.3 Manufacturing of polymer–polymer nanofibrillar composites (Route A) and nanofibrillar single polymer composites (Route B) via the NFC concept

between 300 and 500 nm as can be concluded from the SEM micrograph on Fig. 12.4a. After cold drawing of the blend at 80 °C to a ratio of between 4.7 and 5 and removing of PP via extraction with selective solvent [66] it could be seen that the PBT component in the drawn PP/PBT blend exists in the form of extremely fine fibrils having diameters in the range of 100–150 nm (Fig. 12.4b). This observation confirms that the starting material for preparation of SPCs was indeed nano-sized.

The neat PBT nanofibrils were wound on to a metal plate and subjected to hot compaction at 215 °C, which is nearly 10 °C below their melting peak temperature, applying a pressure of 34 MPa for 5 min, and cooling the material to room temperature, keeping the pressure applied. This procedure is schematically shown in Fig. 12.3, Route B.

Figure 12.4c shows the surface of a one-constituent SPC after hot compaction. As can be seen, fibrils are visible on the surface of the specimen, demonstrating only partial melting of PBT nanofibrils during processing resulting in the formation of thin continuous film. Finally, Fig. 12.4d shows the cross section of the same sample after cryofracturing. A rather dense structure is formed of adhered to each other nanofibrils.

Another illustrative example of nanofibrillar SPCs is that of PVDF [81], shown in Fig. 12.5, where the PVDF nanofibrils were isolated from the nanofibrillar polymer–polymer composites but not from the drawn blend directly as in the previous case (Fig. 12.4). Basically, again the same above described protocol for isolation of neat nanofibrils was followed.

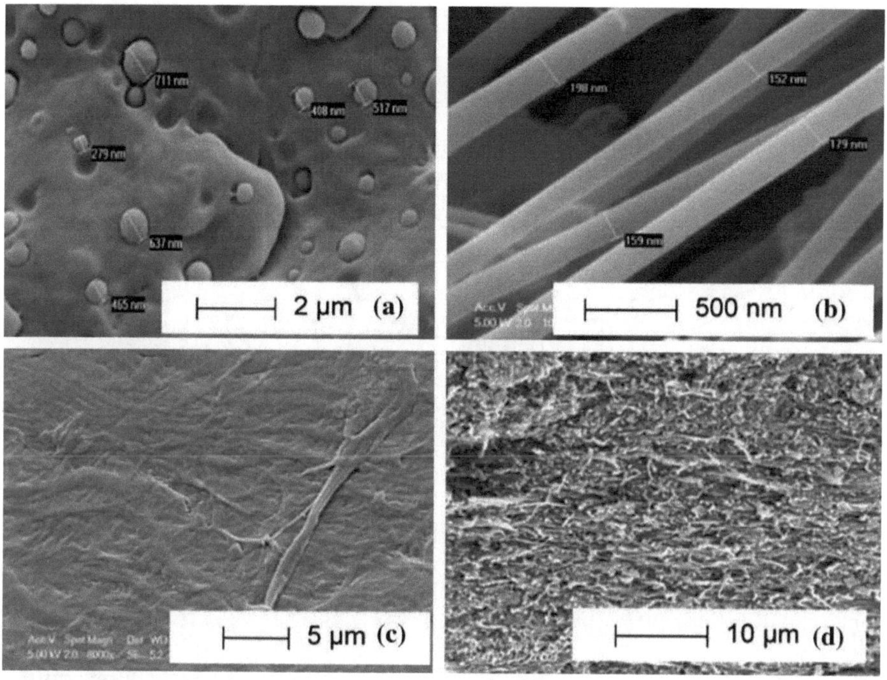

Fig. 12.4 SEM micrographs of: **a** PP/PBT (70:30 w/w) blend just after extrusion (before drawing), **b** neat PBT nanofibrils after drawing and removal of PP with selective solvent, **c** the surface of the hot compacted SPC of PBT nanofibrils, and **d** cryofractured cross-sectional area of the SPC film (sample c)

In more detail, PVDF was melt blended with linear low-density polyethylene (LLDPE), extruded and cold drawn, followed by winding on a metal plate and subjected to compression molding as schematically shown in Fig. 12.3, Route A. In Fig. 12.5a, one can see the cryofractured surface of the prepared polymer–polymer composites parallel to the draw direction, where the smooth nanofibrils are individually surrounded by the matrix material (LLDPE). After extraction with boiling xylene [66] of this nanofibrillar PPC one obtains the neat PVDF nanofibrils with dominating diameters of 200–300 nm (Fig. 12.5b).

The next step of preparation of the nanofibrillar SPC is the compression molding of the parallel aligned PVDF nanofibrils (Fig. 12.3, Route B). It was carried out at 148 °C (i.e., some 18 °C below the temperature of the complete melting, Fig. 12.1) and pressure of 80 MPa for 5 min. As a result, a thin film of 20 μm thickness was prepared with quite smooth surface (Figure, top left angle). Observations in SEM at higher magnification (Fig. 12.5d) revealed that the surface is formed of not completely molten nanofibrils, which adhere to each other due to a surface melting. The same consolidation process occurred not only on the surface, but also in the bulk of the film (Fig. 12.5c). And again this takes place with the preservation of the major part of nanofibrils as can be concluded from the SEM micrograph in Figs. 12.5c, d.

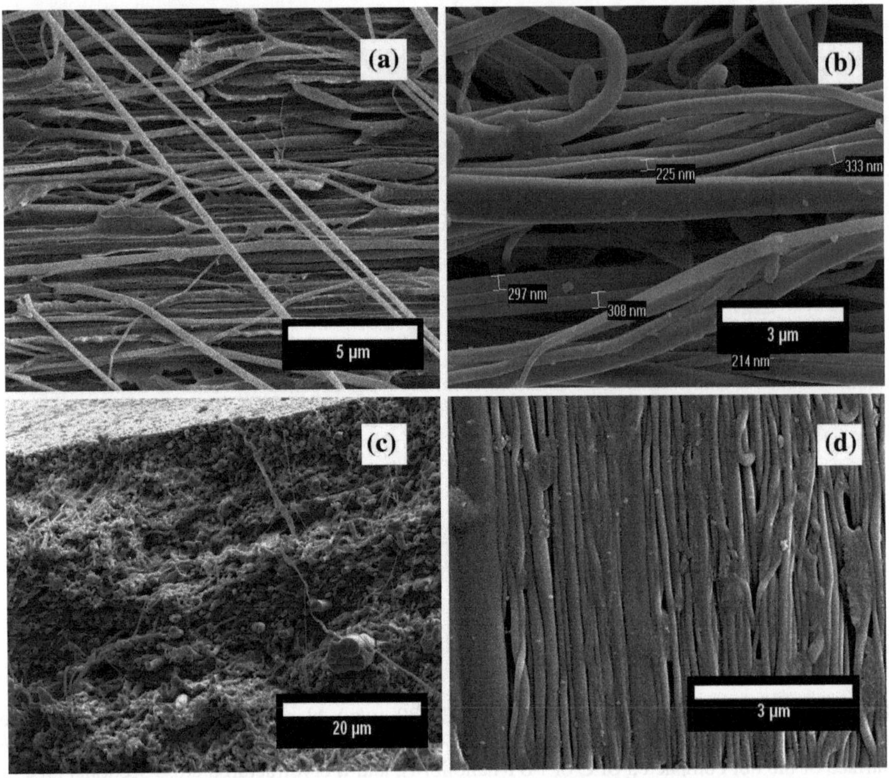

Fig. 12.5 SEM micrographs of the various stages of preparation of nanofibrillar SPCs: **a** cryofractured surface made parallel to the draw direction of nanofibrillar polymer–polymer composite based on LLDPE/PVDF highly drawn blend subjected to compression molding, **b** neat PVDF nanofibrils after extraction with selective solvent of LLDPE from the LLDPE/PVDF polymer–polymer composite (**a**), **c** cryofractured cross section of nanofibrillar single polymer composite prepared via compression molding of the parallel aligned neat PVDF nanofibrils (**b**), **d** surface of the nanofibrillar SPC shown in (**c**) demonstrating the adherence of PVDF nanofibrils due to surface premelting only

It is quite obvious that the thermal treatment at temperatures well below the melting temperature of the polymer results in creation of tiny amount of isotropic binder, which keeps together the dominating reinforcing nanofibrils.

In conclusion, it should be stressed that the described technique for the preparation of one-constituent nanofibrillar SPCs by hot compaction represents a new concept for manufacturing of single polymer nanocomposites. Instead of the usual blending with nano-size materials, the polymer itself is converted into nanofibrils, which are subjected to hot compaction causing surface melting. The obtained nanofibrillar SPC is characterized by superior mechanical properties (the tensile modulus and strength are improved up to 350 % compared to the isotropic film of the same polymer as it will be discussed in more detail in the subsequent section).

Therefore, the SPCs may challenge the respective glass fiber-reinforced polymers from the viewpoint of the mechanical property profile. It is also worth mentioning that the final nanocomposite is prepared using nano-size starting material only, a case which seems to be very rare if not unique.

Mechanical Performance of Nanofibrillar Single Polymer Composites

Before discussing the mechanical properties of the *true* nanofibrillar SPCs let remind those of the *nanofibrillar* polymer–polymer composites prepared according to the MFC concept [33, 60–69]. In Table 12.2 are summarized the improvements in tensile modulus, E, and tensile strength, σ, of three different nanofibrillar polymer–polymer composites as compared with isotropic matrix material (PP and LLDPE), as well as with PP/glass fibers (GF) composite. The improvements of tensile modulus are between 50 and 150 % and for the tensile strength—between 20 and 230 % (Table 12.2).

It seems interesting to remind here that according to Bousmina [84], only in exceptional circumstances can one observe an improvement greater than 30 % in the mechanical performance of nanocomposites prepared via blending the nano-size filler with the matrix. More specifically, Zhang et al. [85] thoroughly studied the case of PP/SiO_2 nanocomposites in which nanoparticles have been coated by various polymers to improve the interfacial adhesion. They reported mechanical properties (tensile modulus and tensile strength) only 20–25 % higher than those of the neat PP [85]. These conclusions are supported by the statement of Schaefer and Justice [86] that "*With the exception of reinforced elastomers, nanocomposites have not lived up to expectations. Although claims of modulus enhancement by factors of 10 exist, these claims are offset by measurements that show little or no improvement*" [86].

The superior mechanical performance of the nanofibrillar polymer–polymer composites, prepared according to the concept of converting instead of adding [15] (Table 12.2) originates from the very high aspect ratio of nanofibrils, their better adhesion then the mineral fillers to the matrix and mostly from the perfect

Table 12.2 Improvements of tensile modulus, E, and tensile strength, σ, of *nanofibrillar* polymer–polymer composites

#	Components	Composition (wt%)	E (%)	σ (%)	References
1	PP/PET	80/20	50	22	[64]
2	LLDPE/PVDF	70/30	165	230	[81]
3	PP/PBT	70/30	45	190	[83]
For comparison					
4	PP/GF	80/20	100	400	[33]

Comparison with isotropic matrix material, as well as with PP/GF composite distribution of nanofibrils in the matrix. As a matter of fact, each micro- or nanofibril is individually surrounded by matrix material with absolute lacking of any aggregate as demonstrated in Fig. 12.5a for the nanofibrillar polymer–polymer composite prepared from the LLDPE/PVDF blend. These favorable characteristics are the reason for an interesting observation during the tensile testing—the accuracy (error bar) for the polymer–polymer composites (Table 12.2, samples 1, 2 and 3) represents 5–10 % only of the measured value, while for the PP/GF (Table 12.2, sample #4)—it is between 50 and 100 % [33].

In Fig. 12.5 are demonstrated also other characteristic features of the nanofibrillar SPCs prepared according to the concept of converting instead of adding [15]. Due to the fact that the compacting of the aligned nanofibrils takes place at temperatures included in the melting interval but far below the complete melting temperature, the adherence of nanofibrils is due to partial surface melting. As a matter of fact, the nanofibrils preserve their identity as can be concluded from Fig. 12.5d, thus supporting the conclusion that complete melting is lacking.

The fact that the manufacturing steps of SPCs exclude complete melting of the polymer has immense importance for their mechanical performance. Contrasting all the other composite materials where the reinforcing component amounts typically 30 %, in the case of SPCs it is 70–80 % being highly drawn material of perfect molecular orientation. These two peculiarities of the reinforcing constituent, the amount and its perfect molecular orientation, together with the ideal adhesion between matrix and reinforcement result in the superior mechanical properties of single polymer composites.

In Table 12.3 are listed data regarding the reinforcing effect in tensile experiments of the *true* nanofibrillar (diameter of nanofibrils between 10 and 250 nm) SPCs, prepared via one- or two-constituent approaches. The comparison is done with isotropic film of the same polymer prepared via compression molding with a thickness similar to that of the SPC samples.

Table 12.3 Improvements of tensile modulus, E, and tensile strength, σ, of *true* nanofibrillar SPCs

#	Polymer	Constituent one/two	E (%)	σ (%)	References
1	PET	Two	Up to 100	Up to 140	[88]
2	PVDF	One	40	330	[81]
3	PET	One	350 (for E = 10.57 MPa)	300	[65]
4	PBT	Two	35	5	[83]
5	PBT	One	31	35	[83]
6	LLDPE	One	112	325	[90]
7	PP	One	70	440	[90]
For comparison					
8	PET/GF	Two	Up to 100–150		[56]

Comparison with isotropic matrix material, i.e., isotropic film of the same polymer

For the case when SPCs are prepared according to the two-constituent approach with nanofibrils as reinforcement, the improvements of the tensile modulus and the tensile strength are 100 and 140 %, respectively, (for PET, Table 12.3, sample #1) and much lower for PBT (Table 12.3, sample #4). Considering the differences in the mechanical performance of the SPCs prepared using the one- or two-constituent approach (Table 12.3, samples 2, 3, 5 and 1, 4, respectively), one has to take into account the drastic difference in the structure of the two types of materials. While the one-constituent approach results in a more or less homogeneous distribution of the reinforcing fibrils in the matrix, the two-constituent approach (Table 12.3, samples 1 and 4) leads to a typical layered structure characterized by a completely different mechanical behavior during the loading process. For example, the PET nanofibrillar SPCs prepared according to the one-constituent approach (Table 12.2, sample 3) are characterized by an average modulus value of 6.85 GPa and a maximum of 10.57 GPa thus surpassing all previously reported SPCs based on PET [13–19, 56, 65, 88, 89], and approaching even the stiffness of the glass fiber-reinforced PET composite (with an E modulus of 11 GPa for 40 wt% glass [90]). Quite similar is the situation with the tensile strength where improvements of at least 300 % are achieved in contrast to all other SPCs found in the literature [13–19, 56, 65, 88, 89], including the PET/glass fiber composite, for which the improvement is 100–150 % [90]. The tensile strength of the same type SPC based on PVDF, LLDPE and PP (Table 12.3, samples 2, 6 and 7) are also very impressive.

The improvements in the tensile mechanical properties of all SPCs so far studied [13–19, 56, 65, 88, 89], and particularly the superior mechanical properties of the new nanofibrillar SPCs prepared by hot compaction using only one constituent, are evident (Table 12.3). What could be the reason for this impressive mechanical performance? There are at least four reasons: (i) in the current SPCs the reinforcing "component" dominates strongly, and has much better mechanical properties when compared with the isotropic matrix of the same polymer, (ii) excellent adhesion between matrix and reinforcement because they have the same chemical composition, (iii) better orientation of the macromolecules in nanofibrils as compared to that in microfibrils and textile filaments of the same polymer, and (iv) in the test specimen of SPCs the nanofibrils are uniaxially aligned and the testing has been performed so far in the drawing direction only.

Another opportunity for further improvement of the properties (mostly the functional ones but not necessarily the mechanical performance) of the micro-/nanofibrillar single polymer composites is the possibility to reinforce the fibrils before their compacting [89, 91].

Very recently [92] again applying the concept of microfibrillar composites, microfibrils of poly(butylene terephthalate) loaded with 5 wt% multiwalled carbon nanotubes (MWCNTs) were prepared by extraction of PP from drawn PP/(PBT-CNT) blends and shaped into single polymer composites. The SPCs show superior mechanical performance with up to 300–600 % improvement in modulus and tensile strength as compared to neat isotropic PBT [92].

The reinforcing effect of MWCNTs on the microfibrillar SPC is quite well expressed. In addition, the microfibrillar SPCs as well as the single microfibrils have electrical volume conductivity of 8×10^{-3} S/cm demonstrating electrical conductive behavior. These results [92] indicate the possibility for preparation of micro-, and even nanofibrillar SPCs with new, very useful properties as for example electrical conductivity, i.e., to create and deal with functional single polymer composites.

Discussing the single polymer composites and comparing them with the polymer–polymer composites another system is worth of mentioning, the blend LLDPE/PVDF since the reinforcing element in the two types of composites is the same—the PVDF nanofibrils [81]. Tensile moduli, *E*, (chord moduli between 0.05 and 0.25 % strain) and stresses at break, σ, of the four samples, namely the isotropic films of LLDPE and of PVDF, the nanofibrillar LLDPE/PVDF composite, and the nanofibrillar SPCs based on PVDF are summarized in Table 12.4.

The tensile modulus of the neat isotropic LLDPE is extremely low, but it is significantly improved (by 165 %) after reinforcement with 30 wt% of PVDF nanofibrils, Table 12.4, samples 1 and 3). The improvement of the tensile modulus for the PVDF nanofibrillar SPCs as compared with the isotropic film of the same chemical composition is more modest—some 40 % only (Table 12.4, samples 4 and 2, respectively). Obviously, the reason for the observed difference in the improvement of the tensile moduli of the two composite types stems from the big difference in the tensile moduli of the starting neat materials—the isotropic PVDF has almost five times larger tensile modulus than that of the isotropic LLDPE (compare samples 1 and 2 in Table 12.4).

Quite different is the situation with the tensile stress as can be concluded from Table 12.4. While the tensile stress at break of the isotropic polymers is almost the same (Table 12.4, samples 1 and 2), their reinforced versions are characterized by drastic improvements (Table 12.4, samples 3 and 4). For the LLDPE/PVDF nanofibrillar polymer–polymer composite and the PVDF nanofibrillar SPC, the improvement of tensile stress is 230 and 330 %, respectively. These values, particularly those for the σ improvement in the case of the nanofibrillar SPCs, are in the range of results reported for other nanofibrillar SPCs [14], Table 12.3.

As a matter of fact, the described results (Table 12.4) help us in the best way to understand the important difference between PPCs and SPCs. In the two cases, we have the same reinforcing element—PVDF nanofibrils (Fig. 12.5b) dealing with PPCs the nanofibrils reinforce the LLDPE matrix (a two-component composite),

Table 12.4 Tensile modulus, *E*, and tensile stress at break, σ, of isotropic LLDPE, isotropic PVDF, *nanofibrillar* polymer–polymer composite of LLDPE/PVDF, and *nanofibrillar* single polymer composite based on PVDF, as well as the improvements as compared with the respective isotropic matrix [40]

#	Polymers	*E* (GPa)	Impr. (%)	σ (GPa)	Impr. (%)
1	LLDPE (isotropic)	0.29	–	28.7	–
2	PVDF (isotropic)	1.38	–	36.3	–
3	LLDPE/PVDF (nanofibrillar PPC)	0.68	165	87	230
4	PVDF (nanofibrillar SPC)	1.79	40	146.9	330

(Fig. 12.5a) and in the case of the SPCs the PVDF nanofibrils reinforce a matrix comprised of isotropic PVDF, i.e., the same polymer (one-component composite), Fig. 12.5c, d. These two types of composites having the same reinforcing element and different polymer matrixes, exhibit quite different mechanical behavior—the improvement of tensile stress is 230 % in the case of PPC and 330 % in the case of SPC (Table 12.4), (the comparizon in both cases is done with the respective isotropic matrix). Obviously, the main factor contributing to these excellent mechanical properties of SPCs is the perfect adhesion quality between the matrix and reinforcement due to the fact that they have the same chemical composition. The change of matrix from PVDF to LLDPE (in the current example), i.e., converting the SPC into PPC means to lose one third of the improvement in the tensile stress value (Table 12.4). This is the reason why we need a clear definition of single polymer composites as well as strictly following of it if we want to have polymer composites with perfect adhesion quality between matrix and reinforcement.

In conclusion, it should be stressed that the described technique for preparation of one-constituent nanofibrillar SPCs by hot compaction represents a new concept for manufacturing of polymer nanocomposites. Instead of the usual blending with nano-size materials, the polymer itself has to be converted into nanofibrils and subjected to hot compaction causing surface premelting, which results in nanofibrils' adhering, as shown in Fig. 12.5c, d. The obtained nanofibrillar SPCs are characterized by superior mechanical properties (in some cases the tensile modulus and strength are improved up to 450 %, Table 12.3, competing with glass fiber-reinforced PET. It is also worth to stress that the final nanocomposite is prepared using *nano-size material only*, a case which could be very rare if not unique.

Conclusions and Outlook

Speaking about SPCs, it should be mentioned that single polymer composites will remain in the focus of interest further on because of their advantages: (i) lack of matrix-reinforcement adhesion problem, (ii) high aspect ratio (always a fibrous constituent is involved), (iii) strongly dominating reinforcing constituent (up to 80 or 90 %), (iv) environmentally friendly (no mineral additives), (v) complete regeneration of SPCs, (vi) for the preparation of nanofibrillar SPCs a nano-size material only is used, and (vii) the nanofibrillar SPCs demonstrate extremely high mechanical properties.

This prediction is based on the fact that they are lightweight (their density is lower than most of the traditional composites), environmentally benign (especially due to their easy recycling via reprocessing in the melt), and they offer novel properties (related to nanoporosity, for example). For this reason, they enjoy a continuously increasing interest as demonstrated by the very recent reviews on this topic [13–15, 17–20]. To their basic disadvantages belongs the very narrow processing window—in many cases a couple of degrees Centigrade only.

And finally, in order to be able to make use of the unique properties of single polymer composites outlined above, we should not forget that they are one-component systems where the matrix and the reinforcement are of the same chemical composition. They should not be mixed up with the closely related to them polymer–polymer composites where the matrix and the reinforcement have different chemical composition [95, 96].

Acknowledgment The author would like to thank the Foundation for Research Science and Technology of New Zealand for the financial support (Grant No. UOAX 0406), as well as The University of Auckland, Department of Mechanical Engineering for hospitality, where this study has been completed.

References

1. Capiati NJ, Porter RS (1975) Concept of one polymer composites modeled with high-density of polyethylene. J Mater Sci 10(10):1671–1677
2. Fakirov S, Bhattacharya D (eds) (2007) Handbook of engineering biopolymers: homopolymers, blends and composites. Hanser Publisher, Munich
3. Bhattacharyya D, Fakirov S (eds) (2012) Synthetic polymer–polymer composites. Hanser Publisher, Munich
4. Hine PJ, Ward IM, Olley RH, Bassett DC (1992) The hot compaction of high modulus melt-spun polyethylene fibers. J Mater Sci 28(2):316–324
5. Kabeel MA, Bassett DC, Olley RH, Hine PJ, Ward IM (1994) Compaction of high-modulus melt-spun polyethylene fibers at temperatures above and below the optimum. J Mater Sci 29 (18):4694–4699
6. Rasburn J, Hine PJ, Ward IM, Olley RH, Bassett DC, Kabeel MA (1995) The hot compaction of polyethylene terephthalate. J Mater Sci 30(3):615–622
7. Yan R, Hine P, Ward I, Olley R, Basset D (1997) The hot compaction of SPECTRA gel-spun polyethylene fibre. J Mater Sci 32(18):4821–4832
8. Ward IM, Hine PJ, Norris K (1992) GB 225:3420
9. Jordan ND, Olley RH, Bassett DC, Hine R, Ward IM (2002) The development of morphology during hot compaction of tensylon high-modulus polyethylene tapes and woven clot. Polymer 43(12):3397–3404
10. Hine PJ, Ward IM, Jordan ND, Olley R, Bassett DC (2003) The hot compaction behaviour of woven oriented polypropylene fibres and tapes. I. Mechanical properties. Polymer 44 (4):1117–1131
11. Jordan ND, Basset DC, Olley RH, Hine PJ, Ward IM (2003) The hot compaction behaviour of woven oriented polypropylene fibres and tapes. II. Morphology of cloths before and after compaction. Polymer 44(4):1133–1143
12. Hine PJ, Astruc A, Ward IM (2004) Hot compaction of polyethylene naphthalate. J Appl Polym Sci 93(2):796–802
13. Karger-Kocsis J, Fakirov S (2012) Polymorphism- and stereoregularity-based single polymer composites. In: Bhattacharyya D, Fakirov S (eds) Synthetic polymer–polymer composites. Hanser Publisher, Munich, pp 673–698
14. Fakirov S (2013) Nano- and microfibrillar single-polymer composites: a review. Macromol Mater Eng 298(1):9–32
15. Fakirov S (2013) Nano-/microfibrillar polymer–polymer and single polymer composites: the converting instead of adding concept. Compos Sci Technol 89:211–225

16. Hine P, Broo V, Ward I (2005) The incorporation of carbon nanofibres to enhance the properties of self-reinforced, single polymer composites. Polymer 46(24):10936–10944
17. Matabola K, De Vries A, Moolman F, Luyt A (2009) Single polymer composites: a review. J Mater Sci 44(23):6213–6222
18. Kmetty A, Bárány T, Karger-Kocsis J (2010) Self-reinforced polymeric materials: a review. Prog Polym Sci 35(10):1288–1310
19. Gao C, Yu L, Liu H, Chen L (2012) Development of self-reinforced polymer composites. Prog Polym Sci 37(6):767–780
20. Karger-Kocsis J, Barany T (2014) Single-polymer composites (SPCs): status and future trends. Comp Sci Technol 92:77–94
21. Alcock B, Cabrera NO, Barkoula N-M, Loos L, Peijs T (2006) The mechanical properties of unidirectional all-polypropylene composites. Compos Part A: Appl Sci Manuf 37(5):716–726
22. Alcock B, Cabrera NO, Barkoula N-M, Peijs T (2006) Low velocity impact performance of recyclable all-polypropylene composites. Compos Sci Technol 66(11–12):1724–1737
23. Bismarck A, Misra S, Lamoe T, Mohanty TA, Misra M, Drzal LT (2005) Plant fibres as reinforcement for green composites. In: Mohanty TA, Misra M, Drzal LT (eds) Natural fibres, biopolymers, and biocomposites. CRC/Taylor & Francis, pp 37–108
24. Flory PJ (1953) Principles of polymer chemistry. Cornell University Press, Ithaca
25. Fakirov S (1990) Solid state reactions in linear polycondensates. In: Schultz JM, Fakirov S (eds) Solid state behavior of linear polyesters and polyamides. Prentice Hall, Englewood Cliffs (NJ), pp 1–74
26. Fakirov S (ed) (2008) Transreactions in condensation polymers. Wiley-VCH, Wenheim
27. Bhattacharyya D, Maitrot P, Fakirov S (2009) Polyamide 6 single polymer composites. Expr Polym Lett 3(8):525–532
28. Incardona S, Migliaresi C, Wagner HD, Gilbert AH, Marom G (1993) The mechanical role of the fiber matrix transcrystalline interphase in carbon-fiber reinforced J-polymer microcomposites. Compos Sci Technol 47(1):43–50
29. Evstatiev M, Fakirov S, Krasteva B, Friedrich K, Covas J, Cunha A (2002) Recycling of PET as polymer–polymer composites. Polym Eng Sci 42(4):826–835
30. Fakirov S, Kamo H, Evstatiev M, Friedrich K (2004) Microfibrillar reinforced composites from PET/LDPE blends: morphology and mechanical properties. J Macromol Sci B Phys B 43 (4):775–789
31. Friedrich K, Ueda E, Kamo H, Evstatiev M, Fakirov S, Krasteva B (2002) Direct electron microscopic observation of transcrystalline layers in microfibrillar reinforced polymer–polymer composites. J Mater Sci 37(20):4299–4305
32. Krumova M, Michler GH, Evstatiev M, Friedrich K, Stribeck N, Fakirov S (2005) Transcrystallization with reorientation of polypropylene in drawn PET/PP and PA66/PP blends. Part 2. Electron microscopic observations on the PET/PP blend. Prog Colloid Polym Sci 130:167–173
33. Fakirov S (2012) The concept of micro- or nanofibrils reinforced polymer–polymer composites. In: Bhattacharyya D, Fakirov S (eds) Synthetic polymer–polymer composites. Hanser Publisher, Munich, pp 353–400
34. Barkoula NM, Peijs T, Schimanski T, Loos J (2005) Processing of single polymer composites using the concept of constrained fibers. Polym Compos 26(1):114–120
35. Yao D, Li RH, Nagarajan P (2006) Single-polymer composites based on slowly crystallizing polymers. Polym Eng Sci 46(9):1223–1230
36. Hine PH, Ey RH, Ward IM (2008) The use of interleaved films for optimizing the production and properties of hot compacted, self-reinforced polymer composites. Compos Sci Technol 68 (6):1413–1421
37. Abraham TN, Wanjale SD, Barany T, Karger-Kocsis J (2009) Tensile mechanical and perforation impact behavior of all-PP composites containing random PP copolymer as matrix and stretched PP homopolymer as reinforcement: effect of beta nucleation of the matrix. Compos A 40(5):662–668

38. Barany T, Izer A, Czigany T (2006) On consolidation of self-reinforced polypropylene composites. Plast, Rubber Compos 35(9):375–379
39. Jenkins MJ, Hine PJ, Hay JN, Ward IM (2006) Mechanical and acoustic frequency responses in flat hot-compacted polyethylene and polypropylene panel. J Appl Polym Sci 99 (5):2789–2796
40. Sun X, Li H, Zhang X, Wang J, Wang D, Yan S (2006) Effect of fiber molecular weight on the interfacial morphology of iPP fiber/matrix single polymer composites. Macromolecules 39 (3):1087–1092
41. Mosleh M, Suh NP, Arinez J (1998) Manufacture and properties of a polyethylene homocomposite. Composites: Part A-Appl Sci Manuf 29(5–6):611–617
42. Li RH, Yao DG (2008) Preparation of single poly(lactic acid) composites. J Appl Polym Sci 107(5):2909–2916
43. Karger-Kocsis J (2007) DE 1023:7803
44. Abraham TN, Siengchin S, Karger-Kocsis J (2008) Dynamic mechanical thermal analysis of all-PP composites based on β and α polymorphic forms. J Mater Sci 43(10):3697–3703
45. Banik K, Karger-Kocsis J, Abraham T (2008) Flexural creep of all-polypropylene composites: model analysis. Polym Eng Sci 48(5):941–948
46. Romhany G, Barany T, Czigany T, Karger-Kocsis J (2007) Fracture and failure behavior of fabric-reinforced all-poly(propylene) composite (Curv (R)). Polym Adv Technol 18(2):90–96
47. Barany T, Izer A, Karger-Kocsis J (2009) Impact resistance of all-polypropylene composites composed of alpha and beta modifications. Polym Test 28(2):176–182
48. Duhovic M, Maitrot P, Fakirov S (2009) Polyamide 66 polymorphic single polymer composites. Open Macromol J 3(1):37–40
49. Alcock B, Barkoula NM, Reynolds C, Govaert L, Peijs T (2007) The effect of temperature and strain rate on the mechanical properties of highly oriented polypropylene tapes and all-polypropylene composites. Compos Sci Technol 67(10):2061–2070
50. Peijs T (2003) Composites for recyclability. Mater Today 6(4):30–35
51. Peijs T (2003) Greener plastic. Mater Perform 42(6):13-13
52. El Maaty MIA, Bassett DC, Olley RH, Hine PJ, Ward IM (1996) The hot compaction of polypropylene fibres. J Mater Sci 31(5):1157–1163
53. Ward IM, Hine PJ (2004) The science and technology of hot compaction. Polymer 45 (5):1413–1427
54. Izer A, Bárány T (2007) Hot consolidated all-PP composites from textile fabrics composed of isotactic PP filaments with different degrees of orientation. Expr Polym Lett 1(12):790–796
55. Rein DM, Vaykhansky L, Khalfin RL, Cohen Y (2002) Controlling the properties of single-polymer composites by surface melting of the reinforcing fibers. Polym Adv Technol 13(10–12):1046–1054
56. Hine PJ, Ward IM (2004) Hot compaction of woven poly(ethylene terephthalate) multifilaments. J Appl Polym Sci 91(4):2223–2233
57. Hine PJ, Ward IM (2006) Hot compaction of woven nylon 6,6 multifilaments. J Appl Polym Sci 101(2):991–997
58. Hine PJ, Astruc A, Ward IM (2004) Hot compaction of polyethylene naphthalate. J Appl Polym Sci 93(2):796–802
59. Hine PJ, Ward IM (2005). In: Michler GH, Baltá-Calleja FJ (eds) Mechanical properties of polymers based on nanostructure and morphology. CRC Press, Boca Raton, p 683
60. Ihm DW, Hiltner A, Baer E (1991) Microfiber systems: a review. In: High performance polymers, Hanser, Munich, pp 280–327
61. Sarkissova M, Harrats C, Groeninckx G, Thomas S (2004) Design and characterisation of microfibrillar reinforced composite materials based on PET/PA12 blends. Compos Part A: Appl Sci Manuf 35(4):489–499
62. Shields RJ, Bhattacharyya D, Fakirov S (2008) Fibrillar polymer-polymer composites: morphology, properties and applications. J Mater Sci 43(20):6758–6770

63. Fuchs C, Bhattacharyya D, Fakirov S (2006) Microfibril reinforced polymer-polymer composites: application of Tsai-Hill equation to PP/PET composites. Comp Sci Technol 66 (16):3161–3171
64. Fakirov S, Bhattacharyya D, Shields RJ (2008) Nanofibril reinforced composites from polymer blends. Coll Surf A: Physicochem Eng Aspects 313:2–8
65. Fakirov S, Duhovic M, Maitrot P, Bhattacharyya D (2010) From PET nanofibrils to nanofibrillar single-polymer composites. Macromol Mater Eng 295(6):515–518
66. Fakirov S (2006) Modified Soxhlet apparatus for high-temperature extraction. J Appl Polym Sci 102(2):2013–2014
67. Fakirov S, Evstatiev, Schultz JM (1993) Microfibrillar reinforced composite from binary and ternary blends of polyesters and nylon 6. Macromolecules 34(22):4479–4669
68. Fakirov S, Evstatiev M, Petrovich S (1993) Microfibrillar reinforced composites. Macromolecules 26(19):5219–5226
69. Fakirov S, Evstatiev M, Friedrich K (2000) From polymer blends to microfibrillar reinforced composites. In Paul DR, Bucknall CB (eds) Polymer blends. Vol. 2. Performance. Wiley, New York, pp 455–475
70. Fakirov S, Evstatiev M, Friedrich K (1995) Interfacial interactions in microfibrillar composites from condensation polymers. In: High technology composites in modern applications. University of Patras, Patras, pp 12–20
71. Evstatiev M, Nicolov N, Fakirov S (1996) Morphology of microfibrillar reinforced composites from polymer blends. Polymer 37(20):4455–4463
72. Evstatiev M, Fakirov S, Friedrich K (2000) Microfibrillar reinforced composite: another approach to polymer blends processing. In: Cunha A, Fakirov S (eds) Structure development during polymer processing. Kluwer Academic Publishers, Dordrecht, pp 311–325
73. Fakirov S, Evstatiev M, Friedrich K, Nanostructured polymer composites from polyester blends: structure–properties relationship. In: Fakirov S (ed) Handbook of thermoplastic polyesters. Wiley-VCH, Weinheim, pp 1093–1129
74. Evstatiev M, Fakirov S, Friedrich K (2005) Manufacturing and characterization of microfibrilla reinforced composites from polymer blends. In: Friedrich K, Fakirov S, Zhang Z (eds) Polymer composites: from nano- to macroscale. Springer, Boston, pp 149–167
75. Friedrich K, Evstatiev M, Fakirov S, Evstatiev O, Ishii M, Harrass M (2005) Microfibrillar reinforced composites from PET/PP blends: processing, morphology and mechanical properties. Comp Sci Technol 65:107–116
76. Li ZM, Huang CG, Yang W, Yang MB, Huang R (2004) Morphology dependent double yielding in injection molded polycarbonate/polyethylene blend. Macromol Mater Eng 289 (11):1004–1011
77. Li ZM, Lu A, Lu ZY, Shen KZ, Li LB, Yang MB (2005) In-situ microfibrillar PET/iPP blend via a slit die extrusion, hot stretching and quenching process: influences of PET concentration on morphology and crystallization of iPP at a fixed hot stretching ratio. J Macromol Sci Phys B44(2):203–216
78. Li ZM, Yang MB, Lu A, Feng JM, Huang R (2002) Tensile properties of poly(ethylene terephthalate) and polyethylene in-situ microfiber reinforced composite formed via slit die extrusion and hot stretching. Mater Lett 56(5):756–762
79. Li ZM, Yang MB, Xie BH, Feng JM, Huang R (2003) In-situ microfiber reinforced composite based on PET and PE via slit die extrusion and hot stretching: Influences of hot stretching ratio on morphology and tensile properties at a fixed composition. Polym Eng Sci 43(3):615–628
80. Li ZM, Li LB, Shen KZ, Yang MB, Huang R (2004) In-situ microfibrillar PET/iPP blend via slit die extrusion, hot stretching, and quenching: influence of hot stretch ratio on morphology, crystallization, and crystal structure of iPP at a fixed PET concentration. J Polym Sci, Part B: Polym Phys 42(22):4095–4106
81. Kim NK, Lin RJT, Bhattacharyya D, Fakirov S (2014) Nanofibrillar poly(vinylidene fluoride) single polymer and polymer–polymer composites: manufacturing and mechanical properties. J Macromol Sci Phys B53(7):1168–1181

82. Fakirov S, Evstatiev M (1994) Microfibrillar reinforced composites—new materials from polymer blends. Adv Mater 6(5):395–398
83. Panamoottil SM, Bhattacharyya D, Fakirov S (2013) Nanofibrillar polymer–polymer and single polymer composite involving poly(butylene terephthalate): preparation and mechanical properties. Polym-Plast Technol Eng 52(11):1106–1112
84. Bousmina M (2005) Fundamental insight into polymer nanocomposites. In: Abstract book 21st annual meet polymer processing society (PPS 21), Leipzig, Germany, 19–23 June 2005, pp 219–220
85. Zhang MQ, Rong MZ, Friedrich K (2005) Application of non-layered nanoparticles in polymer modification. In: Friedrich K, Fakirov S, Zhang Z (eds) Polymer composites: from nano- to macro-scale. Springer, Heidelberg, pp 25–44
86. Schaefer DW (2007) Justice RS how nano are nanocomposites? Macromolecules 40 (24):8501–8517
87. Friedrich K, Fakirov S, Zhang Z (eds) (2005) Polymer composites: from nano- to macro-scale. Springer, New York
88. McCardle R, Bhattacharyya D, Fakirov S (2012) Effect of reinforcement orientation on the mechanical properties of microfibrillar PP/PET and PET single-polymer composites. Macromol Mater Eng 297(7):711–723
89. Fakirov S, Duhovic M, Maitrop P, Bhattacharyya D (2010) Nanofibrillar single polymer composites of poly(ethylene terephthalate). Macromol Mater Eng 295(6):515–518
90. Fakirov S, Bhattacharyya D, Panamoottil SM (2014) Converting of bulk polymers into nanosized materials with controlled nanomorphology. Int J Polym Mater Polym Biomater 63 (15):777–793
91. Hine PJ, Ward IM (2004) Hot compaction of woven poly(ethylene terephthalate) multifilaments. J Appl Polym Sci 91(4):2223–2233
92. Hine PJ, Ward IM, El Matty MIA, Olley RH, Bassett DC (2000) The hot compaction of 2-dimensional woven melt spun high modulus polyethylene fibres. J Mater Sci 35 (20):5091–5099
93. Lin RJT, Bhattacharyya D, Fakirov S (2010) Innovative manufacturing of carbon nanotubes-loaded fibrillar polymer composites. Int J Mod Phys B 24(15–16):2459–2465
94. Fakirov S, Rahman Md Z, Poetschke P, Bhattacharyya D (2014) Single polymer composites of poly(butylene terephthalate) microfibrils loaded with carbon nanotubes exhibiting electrical conductivity and improved mechanical properties. Macromol Mater Eng 299(7):799–806
95. Fakirov S, Composite materials—is the use of proper definitions important?. Materi Today 18(10):529–529
96. Fakirov S (2015) Is the use of correct terms and definitions important in creation of new materials? Expr Polym Lett 9(8):671-671

Chapter 13
Template-Assisted Approaches for Preparation of Nano-sized Polymer Structures

Stanislav Rangelov and Petar Petrov

Polymer Capsules via Template-Assisted Methods

General Remarks

The polymer self-assembly can be considered as a template-free approach to produce nanomaterials. In contrast, the *template-assisted approaches* make use of a sacrificial template, which is initially covered by a polymeric layer and afterwards removed thus yielding polymeric capsules or containers. The latter are structures, composed of a hollow core and a polymeric shell (membrane), that have shown potential as drug and vaccine carriers as well as in applications such as gene and protein delivery, nanoreactors, and artificial organelles.

The size, size distribution, shape, and morphology of the final structures prepared by these approaches are largely dependent on the template. Therefore, the proper selection of an appropriate template is of essential importance. Monodispersed and narrowly distributed particulate structures of submicron dimensions can be considered as ideal templates. In addition, they should be stable during the process of outer layer formation and exhibit ability of being removed at mild conditions thus not affecting the structure, properties and stability of the final capsules.

A variety of particulate structures have been used as templates for capsule formation. These can be inorganic (carbonates such as $CaCO_3$, $MnCO_3$ [1–3] and solid or porous silica nanoparticles [4, 5]), organic (melamine formaldehyde [6]) and polymeric (polystyrene latex [7], mesoglobules, prepared from thermo-responsive polymers [8–10]) materials, emulsion droplets [11–13], bubbles [14], and even red

S. Rangelov (✉) · P. Petrov
Institute of Polymers, Bulgarian Academy of Sciences,
Akad. G. Bonchev St. 103-A, 1113, Sofia, Bulgaria
e-mail: rangelov@polymer.bas.bg

S. Fakirov (ed.), *Nano-size Polymers*,
DOI 10.1007/978-3-319-39715-3_13

blood cells [15, 16]. In size, these substrates vary from a few nm to $\sim$100 μm. Some of them, e.g., polystyrene, mesoglobules, silica nanoparticles, are characterized with very narrow size distribution. Most of them are spherical but others can be prepared in varying aspect ratios. The methods for core removal largely depend on the nature of the particulate material. Typically, the methods involve treatment with aggressive reagents such as HCl, HF/NH_4F, NaClO, EDTA, organic solvents, and change of pH or temperature. The latter is the mildest and the least invasive approach applicable for removal of templates based on mesoglobules (see below).

Numerous materials, layering methods, and strategies have been reported for the fabrication of (thin) films on the sacrificial particulate substrates (templates) generally aiming at combining different properties in one system. The methods for depositing (multiple) layers can be broadly divided into *surface polymerization* and *layer-by-layer (LbL) assembly*. "Grafting from" and "grafting to" techniques as well as seeded radical polymerization have emerged as facile and flexible routes to prepare hollow polymer capsules with diverse properties and functionalities. These techniques allow for simultaneous (via copolymerization of a cross-linking monomer) or post-polymerization (via chemical, UV, and physical reactions) cross-linking of the layer, which provides an additional dimension to influence the properties of the final structures. Numerous interactions have been utilized for the LbL assembly. Electrostatic interactions [17], hydrogen bonding [18–20], and covalent bonding [21] are the three main strategies for alternate adsorption of materials on the surface of the sacrificial template but other interactions such as DNA hybridization [22–27], stereocomplexation [28, 29], hydrophobic [30], and host–guest [31–33] interactions have also been used for production of multilayer capsules (Fig. 13.1).

The polymers which have typically been used for construction of membranes are listed in Table 13.1. The group of synthetic polymers is the largest one; however, recent research is increasingly focused on using biomacromolecules such as polypeptides, proteins, DNA, and polysaccharides. The electrostatic interactions are the most frequently used interactions for fabrication of LbL polymer capsules. That is why the polyelectrolytes constitute a large group. The group of polymers capable of forming hydrogen-bonded layers is composed of hydrogen bonding donors such as poly(methacrylic acid), poly(L-glutamic acid), and acceptors, e.g., poly(ethylene glycol), poly(N-vinylpyrrolidone). Additional modification of the polymers used as well as presence of functional groups and addition of specific substances are generally needed for construction of stable and functional membranes.

Layer-by-Layer Assembly

LbL assembly refers to depositing multiple layers on planar as well as particle supports. Nano- and microcapsules can be obtained after removal of the sacrificial mainly spherical substrates onto which the multiple layers are deposited [6, 35, 36]. The multilayer structure of the resulting capsules enables the combination of

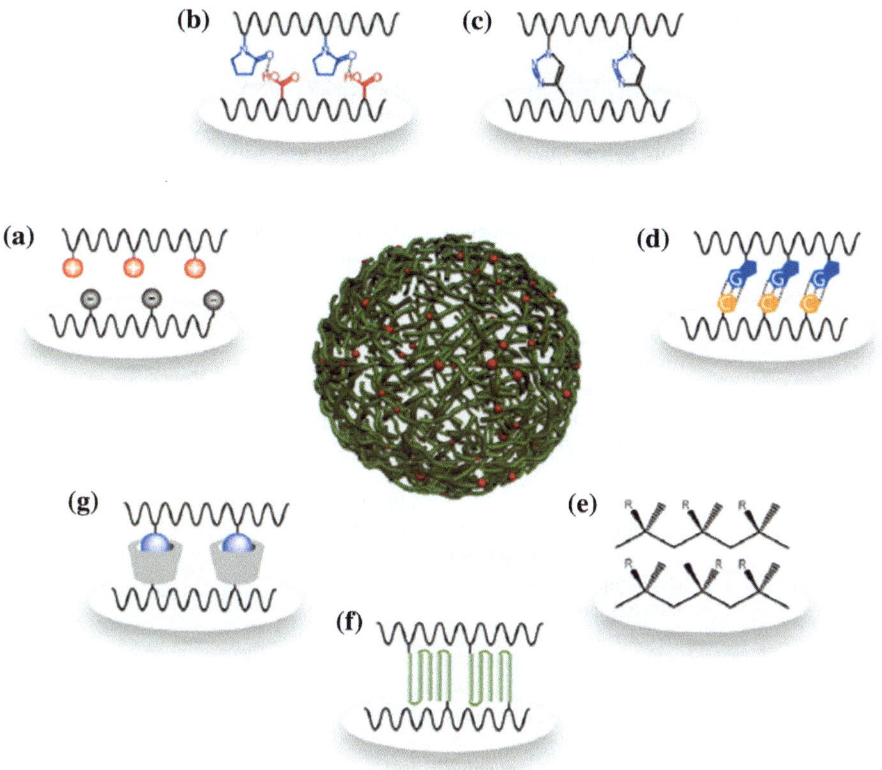

Fig. 13.1 Different polymer interactions for the assembly of multilayer capsules (*center*). Interactions clockwise from *left*: **a** electrostatic; **b** hydrogen bonding; **c** covalent bonding; **d** DNA hybridization; **e** stereocomplexation; **f** hydrophobic; and **g** host–guest interactions. Reproduced from [34] with permission from Elsevier

different properties. Precise control over the properties such as layer thickness, morphology, stimuli-responsiveness, and loading capacity can be achieved as well [37, 38].

Centrifugation, filtration, and electrophoretic assembly are the most frequently applied techniques for the fabrication of LbL capsules [39]. Schematic representation of these techniques is shown in Fig. 13.2. The centrifugation, (Fig. 13.2a), is employed to separate the free polymer from the coated particles. It suffers from the necessity to sediment the latter and redisperse them before depositing the subsequent layer. Problems related with agglomeration and difficulties with redispersion of the coated particles frequently arise. The membrane filtration (Fig. 13.2b) allows the particles to remain suspended thus lowering their tendency to agglomerate. Here, the template particles and polymer are suspended in a vessel for layer deposition; the free polymer is separated by applying a pressure differential across the filter membrane while adding a washing medium. In the electrophoretic polymer

Table 13.1 List of synthetic polymers, polypeptides, nucleotides, and polysaccharides typically used for fabrication of capsules

Interaction	Synthetic polymers	Polypeptides, DNA, proteins	Polysaccharides
Electrostatic	Poly(styrene sulfonate), polyallylamine hydrochloride, poly (diallyldimethyl ammonium chloride), polyethyleneimine, Nafion/Fe^{3+}, poly (4-vinylpyridine), poly (acrylic acid), poly (methacrylic acid), poly (hydroxypropyl methacrylamide), polyferrocenylsillane [sulfonate]	Poly(L-aspartic acid), poly(L-lysine) [imine formation], poly (L-arginine), albumin, protamine, silk proteins	Dextran [sulfate], chitosan [quaternazied, sulfate], hyaluronic acid, alginate
Hydrogen bonding	Poly (2-disiopropylamino-ethyl methacrylate) [alkyne], poly (N-vinylpyrrolidone) [alkyne], poly(diethylene glycol methacrylate-r-oligoethylene glycol methacrylate) [alkyne], poly(acrylic acid) [cysteamine, pyridinedithioethanolamine, alkene], poly(methacrylic acid) [cysteamine, pyridinedithioethanolamine, alkene]	Poly(L-glutamic acid) [alkyne, azide]	
Covalent	Poly(N-isopropylacryl amide) [alkyne, azide], poly (glycidyl methacrylate) [ring-opening], polyallylamine [imine formation]	Poly(L-glutamic acid) [alkyne, azide], poly (L-lysine) [carbodiimide chemistry]	Dextran [alkyne, azide], alginate [imine formation]
DNA hybridization	Poly(hydroxypropyl methacrylamide) [oligonucleotides]	Oligonucleotides	
Stereocomplexation	Poly(methyl methacrylate) [isotactic/syndiotactic]		
Hydrophobic		Silk proteins [physical cross-links, β-sheets]	
Host-guest	Polyallylamine [β-cyclodextrine], poly (acrylic acid) [azobenzene], poly(methacrylic acid) [azobenzene]	Poly(L-aspartic acid) [adamantine]	Dextran [carboxymethyl-, α-cyclodextrin, β-cyclodextrin]

The respective modification, chemistry and additional functional group/substance needed for formation of the membrane are given in parentheses

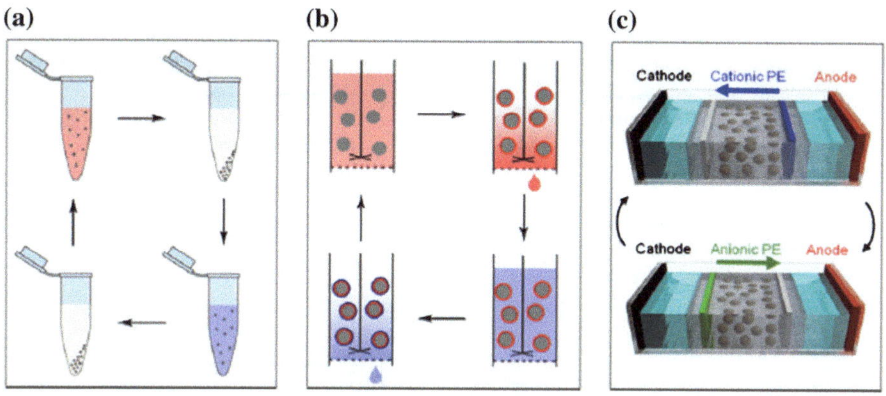

Fig. 13.2 Schematic representation of LbL assembly of polymer capsules via centrifugation (**a**), filtration (**b**), and electrophoresis (**c**). Reproduced from [39] with permission from John Wiley and Sons

assembly, the layer deposition is achieved by electrophoresis of polymers through agarose gel (Fig. 13.2c). By this technique, a diverse size range of particles can be layered. Each of those techniques has its own advantages and drawbacks. A common drawback of the LbL assembly is the necessity of multiple polymer adsorption steps, which is time- and material consuming.

The multilayer shells of the capsules are usually built-up by electrostatic deposition of oppositely charged polyelectrolytes, though other interactions have been utilized for the LbL assembly (see above, Fig. 13.1). These capsules (as referred to as PEM, polyelectrolyte multilayer, capsules) are synthesized under mild conditions using different materials (Fig. 13.3), which allows introducing multiple functionalities, e.g., enhanced colloidal stability via PEGylation, targeting to specific cells via functionalization with biologically active ligands, imparting magnetic properties via incorporation of magnetic nanoparticles, etc.

The typical protocol consists of subsequent alternate immersion of a solid substrate in aqueous solution of either positively or negatively charged polymers for 15–30 min to deposit a polymer layer on the substrate surface. The effects of pH and ionic strength are strongly pronounced, since they affect the conformation of the polyelectrolyte chains. A variety of particulate substrates have been used as templates (see above). Melamine formaldehyde colloidal particles were the first utilized to transfer the LbL technique from planar films to 3-D capsules [6], whereas $CaCO_3$ particles are mostly preferred for biological applications since they can be completely removed at mild conditions without affecting the properties of the entrapped molecules [2, 41]. The choice of the template for the preparation of polyelectrolyte capsules is particularly important depending on the field of application. Besides the above-mentioned division of the nanoparticulate templates into inorganic, organic, and polymeric materials, Parakhonskiy et al. [42] consider two important classes—particulate templates with smooth and porous surface. Melamine formaldehyde,

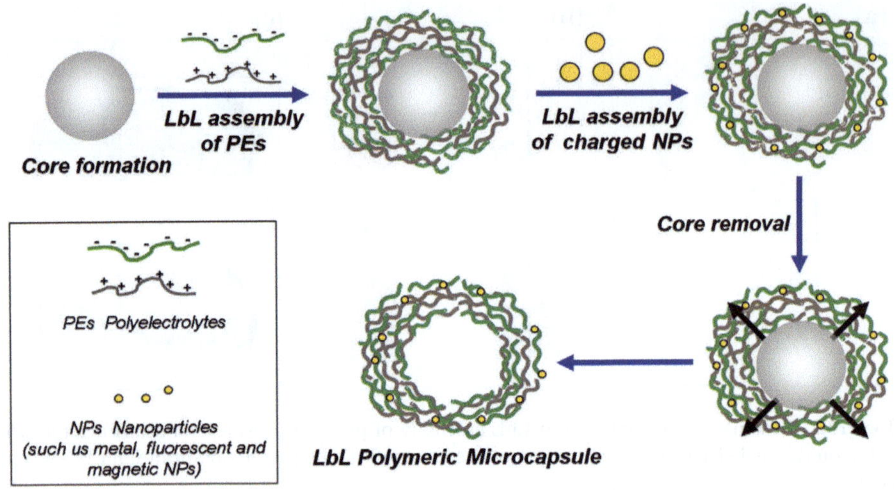

Fig. 13.3 Schematic representation of capsules prepared by LbL adsorption of oppositely charged polyelectrolytes onto spherical charged spheres, followed by dissolution of sacrificial templates. During LbL coating, various charged elements (such as metal, magnetic or fluorescent nanoparticles) can be incorporated within the multilayers to add functionality to the capsules. Reproduced from [40] with permission from Elsevier

silica, and polystyrene latex are examples for particles with smooth surface. Such particles typically need deposition of many polyelectrolyte layers and are characterized with thin shells. The interior of the capsules obtained from smooth templates is typically clean and transparent and they look like empty balloons [42]. Porous particles encompass, for example, mesoporous silica, and calcium and manganese carbonates. The obtained capsules after core dissolution look like sponges since during the LbL adsorption, the polyelectrolytes penetrate into the matrix, resulting in considerably thicker polymer shell. The pores can be utilized for loading of the capsules, which produces relatively large loading capacity. In contrast, the former capsules have a larger inner volume; however, the encapsulation and release of the cargo typically require application of external stimuli [42].

More sophisticated capsule systems, e.g., multicompartment hybrid assemblies, have recently been reported [43–47], which endow the systems with novel and advanced properties. Multicompartmentalized capsules can be constructed via (i) assembly from preformed different subcompartments; (ii) synthesis of particles on which capsules are assembled; (iii) a combination of synthesis and assembly from preformed subcompartments [48]. The main advantages of these sophisticated structures are, for example, simultaneous delivery of different molecules using one entity with protection from the outside environment as well as conduction of cascade reactions. Multicompartmentalization has been recently described as a suitable approach for *theranostics* because elements for diagnostics can be encapsulated in some subcompartments and drugs for therapy can be encapsulated in other subcompartments [49].

The geometry of the particles could drastically change their properties. Recent findings indicate that the shape could influence the cellular uptake, endothelial targeting in the vasculature, the rate of endocytosis and lysosomal transport within endothelial cells [50, 51]. Hence, capsules templated on specifically shaped particles would be beneficial for certain applications. Furthermore, the flexibility of the polyelectrolyte molecules allows reproducing of the template shape after removal of the solid core [52–54]. $CaCO_3$ particles of different shape (spherical, elliptic, and cubic) have been used for fabrication of polyelectrolyte capsules, which were found to replicate the shape of the initial particles [55]. The properties of microcapsule shells assembled on cubic $CdCO_3$ particles were compared to the same shells assembled on spherical silica [56].

The LbL assembly has not yet found industrial application in spite of its versatility. The main reason for this is the time-consuming multistep assembly procedure, which often is additionally complicated with particle agglomeration. As noted elsewhere, deposition of a film consisting of dozens of layers takes more than 10 h using a robotic dipping apparatus [57]. Nonetheless, recently the use of spray-assisted LbL assembly, both for planar films as well as for the construction of polymeric particles, has emerged as a possible strategy to reduce the deposition times per layer from tens of minutes to a few seconds [57].

Surface Polymerization

For the fabrication of polymeric nanocapsules using templates, the surface heterophase polymerization is an emerging technique. It is a facile and flexible route to construct a (cross-linked) shell around templates and produce different shape polymeric capsules that can be loaded with biologically active molecules either during the process of their formation or post-formation. Preparation of polymeric nanocapsules using *mesoglobules* as templates has been demonstrated in a series of recent studies [8–10]. *Mesoglobules* are typically monomodal and spherical nanoaggregates, prepared from thermo-responsive polymers. The latter are soluble in cold water and phase separate upon heating above the phase transition temperature or lower critical solution temperature (LCST). The mesoglobules can be prepared in a very narrow size distribution with dimensions ranging from ca. 50 to 400 nm. The size can be controlled via the initial concentration, heating protocol, molar mass, and presence of additives such as surfactants, osmolytes and ionic liquids. It is noteworthy that the process of mesoglobule formation is fully reversible: upon cooling below the LCST the mesoglobules disintegrate into individual macromolecules very frequently without any hysteresis. This is of the paramount importance for the present approach: if the mesoglobules are coated with a water-swollen cross-linked polymeric membrane, it is expected that the core would start to dissolve upon cooling down below the LCST of the thermo-responsive polymer, and the dissolved macromolecules would diffuse through the polymeric

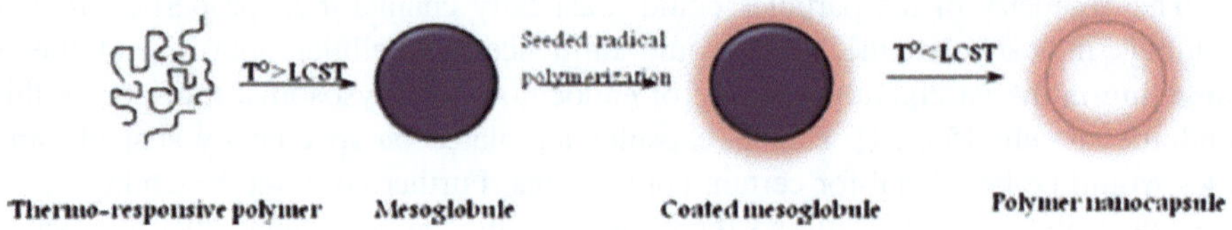

Fig. 13.4 Schematic presentation of the method for preparation of nanocapsules using mesoglobules as templates. Reproduced from [10] with permission from Elsevier

membrane, thus creating empty spaces in the interior of the particles. The approach is schematically presented in Fig. 13.4.

The most commonly studied thermo-responsive polymers are poly (N-isopropylacrylamide), poly(2-alkyl-2-oxazoline)s, poly(vinyl methyl ether), poly(N-vinyl caprolactam), and polymers of oligoethylene glycol (meth)acrylates. Each of them is characterized with specific LCST, mechanism of heat-driven phase transition, structure and properties of the resulting mesoglobules, as well as with varying extents of reversibility and reproducibility of the process of mesoglobule formation [58]. The most preferable characteristics of the thermo-responsive polymers and mesoglobules used as templates for nanocapsule preparation have been formulated below [58]:

- The aqueous colloidal dispersions of mesoglobules should be stable in time preferably in the absence of surfactants. The mesoglobules should be uniform in shape and size (preferably below 200 nm in diameter) and of narrow size distribution.
- The mesoglobules should be able to quickly dissolve upon cooling without any significant increase in volume due to swelling.
- Low-molar mass polymers (preferably below 10 kDa) with narrow dispersity and flexible chains without functional groups able to interact with the membrane would diffuse more easily through the membrane and, hence, more effectively be removed from the interior of the particles.
- The process of mesoglobule formation should not be hampered by interactions with molecules that are to be loaded in the nanocapsules.

The approach of using mesoglobules as templates provides an effective control over the size, dispersity, and morphology of the resulting nanocapsules. Beside this, the proper selection of monomers, cross-linking agents, and initiators are of crucial importance for the membrane formation and properties. The first attempt to construct an outer shell on the surface of mesoglobule was done in 2008 [8]. These authors utilized poly(N-isopropylacrylamide), PNIPAM, and mesoglobules on the surface of which an outer shell was formed by radical copolymerization of 2-hydroxyethyl methacrylate (HEMA) and poly(ethylene glycol) dimethacrylate (PEG-DMA) as a cross-linking agent. The selection of the comonomers was based on the expectations that the–CH_2CH_2OH group from PHEMA and the PEG moiety from PEG–DMA would prevent flocculation of the particles. The polymerization was initiated by the hydrophilic radical initiator potassium persulfate and was carried out in aqueous solution at 60–80 °C, which is well above the LCST of the core-forming PNIPAM. The PNIPAM chains were partially released from the interior via dialysis at low (below the LCST) temperatures thus resulting in semi-hollow particles.

In another contribution, mesoglobules from hydroxyl end-functionalized poly (2-isopropyl-2-oxazoline)s (PiPOZ-OH) of molecular weights in the 3600–8900 g mol^{-1} range have been used as templates [13]. They were coated with a thermosensitive cross-linked shell formed via seeded radical copolymerization of N-isopropylacrylamide (NIPAM) and N,N′-bis-methylene acrylamide to produce core-shell nanoparticles, which were subsequently subjected to extensive dialysis below the LCSTs of both the core-forming PiPOZ-OH and shell-forming PNIPAM to remove the core. The use of a core-forming polymer of low-molecular weight (<8900 g mol^{-1}), narrow dispersity (<1.15), and relatively low T_g (52–68 °C) is beneficial as far as the effectiveness of the removal of the cores is concerned. The inherent immiscibility between PiPOZ-OH and PNIPAM as well as the specific raspberry-like structure of the resulting core-shell particles also contributed for enhancement of the core removal effectiveness [9].

Other combinations of template-forming and membrane-forming polymers have been explored elsewhere [10]. In that study, polymers of the family of polymethacrylates with pendant oligo(ethylene glycol) chains were employed as templates for preparation of polymeric nanocapsules, whereas the membranes were based on PNIPAM and PHEMA. The coated particles displayed different morphology and local ordering, which was attributed to the different compatibility of the polymers: the membrane, composed of cross-linked PNIPAM was raspberry-like and uneven, in contrast to the homogeneous, smooth, and uniform in thickness membrane of PHEMA. Following the extensive dialysis against water at low temperatures resulting in template removal, nanocapsules were obtained. They also displayed distinctive differences in morphology, whereas the membrane based on PNIPAM was prone to bend and form kinks and, accordingly, the capsules collapsed, the one based on PHEMA was self-supported because of its constant thickness and homogeneity and the capsules retained the spherical shape. A very strong advantage of the approach is the possibility to incorporate in the membrane specific monomers and cross-linking agents, which can impart specific functions and properties such as biodegradability, targeting, and sensitivity to certain stimuli and presence of specific substances.

Emulsion droplets can also serve as soft templates to produce capsules. A range of cross-linking chemistries on the interface of oil-in-water emulsion droplets, whereby introducing various polymer functionalities in the resulting capsules has been employed. These are ring-opening metathesis polymerization [59], interfacial step-growth polymerization [60], metal coordination polymerization also termed miniemulsion periphery polymerization [61], surface-initiated miniemulsion ATRP [62, 63], and inverse miniemulsion periphery RAFT polymerization [64, 65].

Hollow capsules can be prepared using polymer brushes polymerized from particulate hard templates. The formation of polymer brushes proceeds via chain growth polymerization of monomer units from initiating groups that are anchored to the selected substrate (template). In aqueous solution the resulting capsules, obtained following core removal, are free-standing and self-supported if the grafted chains are hydrophobic. However, cross-linking of the brushes is needed when the capsules are dispersed in a good solvent for the grafted polymer. Various cross-linking strategies can be applied depending on the chemical nature of the brushes, end application of the capsules. and the desired properties. An elegant approach is using UV light as demonstrated for cross-linking of polystyrene-grafted silica particles [66], copolymerization of UV cross-linkable comonomers [67], and non-covalent cross-linking utilizing photoisomerization of spiropyran groups in brushes grafted from SiO_2 particles [68].

Additional functionalities can be incorporated into the capsules using small molecule cross-linkers such as diamines [69] and dithiols [70], which allows tuning of the capsule properties by varying the cross-linker size and architecture. Recently, Qiao and Caruso used macrocross-linkers, that is, preformed polymer chains, functionalized with multiple polymerizable groups, to produce cross-linked films in one step [71]. The technique is termed CAP (Continuous Assembly of Polymers) and, similar to "grafting from", proceeds via controlled chain growth polymerization of polymerizable species from the surface of the template particles as shown in Fig. 13.5.

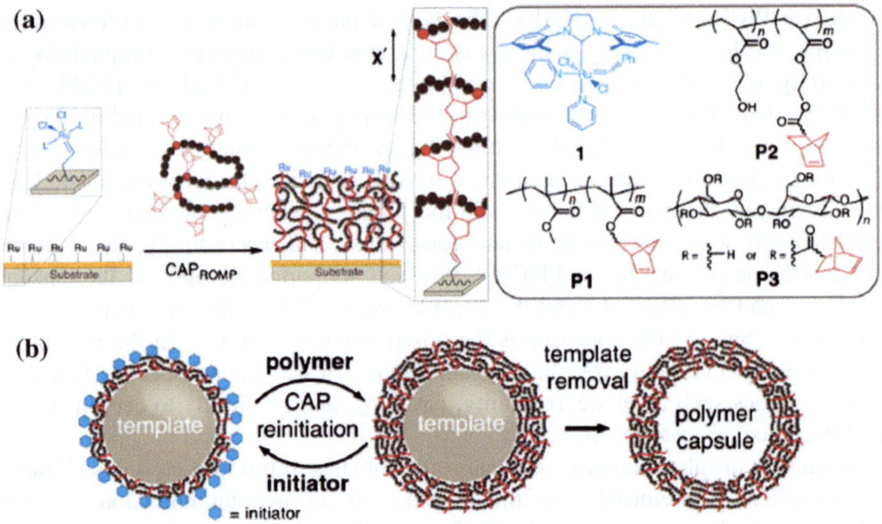

Fig. 13.5 **a** Schematic of film formation via the CAP$_{ROMP}$ process on a planar substrate. Surface initiator functionalization was achieved using catalyst 1 followed by the CAP$_{ROMP}$ reaction with either *P1*, *P2* or *P3* macrocross-linkers. *X′* represents the interlayer spacing. **b** Schematic of capsule formation via CAP$_{ROMP}$. Increased layer buildup was achieved by reinitiation of the polymer film with catalyst *1* followed by a CAP$_{ROMP}$ assembly step. This process was repeated until the desired film thickness was achieved followed by core removal. Reproduced from [71] with permission from John Wiley and Sons

Single-step deposition of a polymer onto sacrificial template is an alternative route to prepare polymeric nanocapsules. Silica particles with a solid core/mesoporous shell structure are particularly appropriate as templates, since they are able to entrap various materials via adsorption. Moreover, significant research efforts have been made to fabricate mesoporous silica particles of smaller (e.g., 100 nm) size, monodispersity and controlled size of mesopores. Wang and co-workers have recently reported fabrication of single-component capsules prepared via infiltration of various polymers in the mesoporous shell, followed by cross-linking of the polymer chains and removal of the template [72]. Polyelectrolyte such as polyallylamine hydrochloride, polypeptides (poly(L-lysine), poly(L-glutamic acid)), as well as polypeptide-drug conjugates have been used as a polymer component.

Biopolymer capsules have been prepared via non-covalent interactions between various biopolymers and bromoisobutyramide surface-functionalized templates [73]. These authors proposed that the adsorption of the biopolymer was achieved through non-covalent halogen bonding (Fig. 13.6). Enzymes, proteins, polypeptides, polynucleotides, and polysaccharides have been used for preparation of the capsules. Their mechanical stability has been improved by cross-linking with an amine reactive cross-linker [74] or via repeated refunctionalization of the deposited layer/biopolymer adsorption [75].

Through the use of a radial assembly technique, polyrotoxane capsules have been prepared [76]. Polyrotoxanes are supramolecular structures consisting of mechanically interlocked cyclic molecules threading a linear macromolecule. Cyclodextrines and poly(ethylene glycol) are the most typically used polyrotoxane materials due to their low cytotoxicity and commercial availability at low cost. Figure 13.7 presents the formation of polyrotoxane capsules by radial assembly [76]. PEG-functionalized gold nanoparticles were threaded with α-cyclodextrin. In order to prevent dethreading, the free PEG chain end was blocked with 2,4,6-trinnitrobenzene sulfonic acid. The resulting structures were stabilized by covalent cross-linking of the α-cyclodextrin toroids. Following etching, the gold nanoparticle templates were removed, which resulted in formation of polyrotoxane nanocapsules.

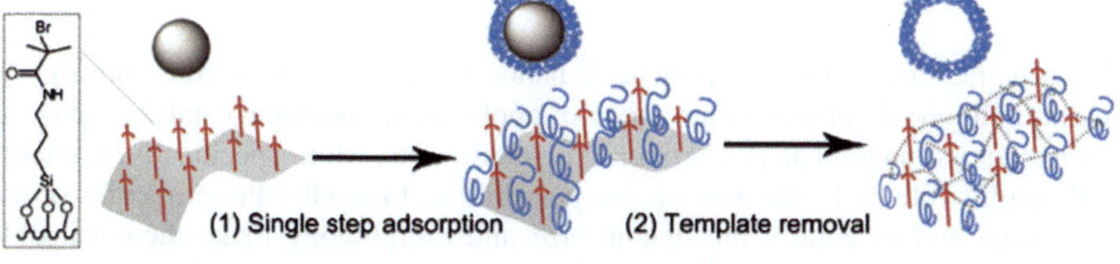

Fig. 13.6 Formation of free-standing biopolymer capsules via single-step adsorption of a biopolymer (enzymes, proteins, polypeptides, polynucleotides, polysaccharides) from aqueous solution onto bromoisobutyramide surface-modified silica templates followed by template removal. Reproduced from [73] with permission from John Wiley and Sons

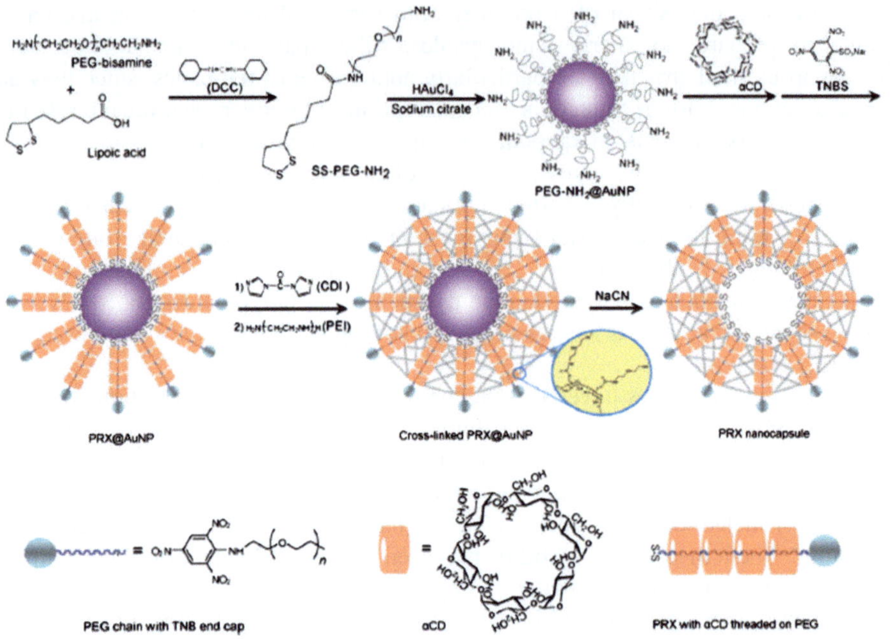

Fig. 13.7 Polyrotoxane nanocapsules prepared by radial assembly from PEG-functionalized gold nanoparticles. Cross-linking was achieved by activation of the α-cyclodextrin hydroxyl groups followed by reaction with polyethyleneimine. Reproduced from [76] with permission from John Wiley and Sons

Dam and Caruso [77] used preformed alkyne end-functionalized α-cyclodextrin/PEG polyrotoxanes, which were grafted to the surface of azido-functionalized silica particles. The polyrotoxane shell was cross-linked with a disulfide cross-linker. After removal of the templates via silica etching, capsules were obtained. Both, the disulfide moieties in the cross-linker and the blocking groups allowed degradation of the capsules to free PEG and α-cyclodextrin upon exposure to glutathione.

Biomedical Applications of Polymeric Capsules

The specific structure of the polymeric nanocapsules—a hollow core and a polymeric shell—holds promise for a range of applications including catalysis, sensing, microreactors, nanomedicine, artificial organelles, etc. More specifically, the large hydrophilic volume in the interior, the possibilities to modify the properties of the membrane and to control the overall size and morphology make them valuable materials with potential in delivery of drugs, vaccines, and genes. Considering the alternative techniques for preparation, the variety of templates, and the plethora of different polymers used to tune the properties of polymeric nanocapsules, it seems likely that the key requirements for effective drug-delivery systems such as low toxicity and optional degradability, high loading capacity, triggered release mechanisms, low immunogenicity, and targeting (directing) to designated sites are easily achievable. In addition, the nanocapsules can be precisely tailored by modulating the thickness and construction of the membrane and by modifying the latter with a number of active compounds such as metal nanoparticles, light-sensitive polymers, or functional dyes, specific properties can be endowed. Furthermore, different drugs can be combined within a single capsule to overcome, e.g., multidrug resistance in cancer therapy, which is also an important advantage [40].

Therapeutic molecules can be loaded by several routes, including: (i) preloading of the cargo molecules to the template followed by membrane formation and template removal; (ii) loading of therapeutics in preformed hollow capsules by temporarily altering membrane permeability, and (iii) drug incorporation into specific domains of the capsule. A range of external and biological stimuli such as pH, redox potential, ionic strength, solvent, temperature, light, ultrasound, magnetic field, enzymatic reactivity, etc., can initiate the release of payload molecules. Some of these stimuli create tiny pores in the membrane, which facilitate the diffusion. Others are able to fully disintegrate the membrane thus causing burst release of the cargo molecules. The PEGylation of polymeric nanocapsules is typically used to provide "stealth" properties and prevent nonspecific adsorption and cellular uptake, whereas attaching specific ligands may improve the effectiveness of drug accumulation and minimize harmful side effects.

Preloading of the cargo molecules to the template is used for molecules of high-molecular weight and dimensions. Encapsulation of the therapeutic enzyme L-asparaginase using porous particles like $CaCO_3$ into multilayered polyelectrolyte capsules is shown elsewhere [78]. The efficiency of loading strongly depends on the properties of the template matrix such as surface area, size of the pores, and surface charge [42]. An interesting approach is the method for DNA encapsulation via nanotemplates from cationic block copolymer micelles [79]. The micelles are based on polystyrene-b-poly(quaternized 2-vinylpyridine) diblock copolymer, which, upon addition of DNA, formed nano-sized polyplexes. The latter were coated by a cross-linked polymeric shell and afterwards conditions were found to simultaneously destroy the polyplex and disintegrate the micelles. The overall encapsulation procedure is shown in Fig. 13.8. The results indicated that the block copolymer was effectively removed thus leaving DNA macromolecules entrapped in polymeric nanocapsules.

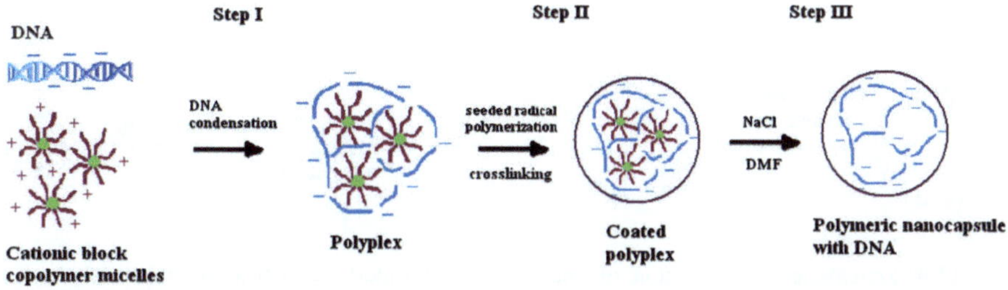

Fig. 13.8 Schematic representation of DNA encapsulation via nanotemplates from cationic block copolymer micelles. Reproduced from [79] with permission from Royal Society of Chemistry

Besides molecules of high-molecular weight, water-insoluble drugs can be encapsulated in polymeric capsules using the template-assisted approach. Wang et al. [80] encapsulated water-insoluble compounds (thiocoraline, paclitaxel) using mesoporous silica templates. After removal of the silica particles, the water-insoluble compounds agglomerated into clusters but were retained in the capsules. The drug-loaded capsules exhibited excellent colloidal stability and high potency of colorectal cancer cells in vitro with similar cytotoxicity to the free drug dissolved in organic solvent [80]. Hydrophobic drugs can be encapsulated also by preloading in emulsion droplets followed by shell formation via one-step interfacial polymerization [81] or polyelectrolyte LbL deposition on the liquid core [82].

Post-loading in preformed capsules is mainly but not exclusively applicable to low-molecular weight therapeutics. Tonq et al. [83] described high-efficient loading and release of water-soluble low-molecular drugs in/from capsules consisting of five bilayers of poly(diallyldimethyl ammonium chloride) and poly(styrene sulfonate) assembled on $CaCO_3$ particles. After removal of the latter, the resulting capsules showed temperature-dependent shrinkage accompanied by thickening of the capsule walls. Doxorubicin hydrochloride was effectively loaded (and released) using these features. Basically, the same system has been shown earlier to be effective in encapsulating larger molecules bearing different charges such as fluorescein isothiocyanate-labeled dextran (10 and 70 kDa) and poly(acrylic acid) (30 kDa) employing morphological changes in the capsules upon heating as depicted in Fig. 13.9 [84].

The capsules can be engineered to contain specific domains into which drugs/therapeutics can be incorporated. Hosta-Rigau et al. [85] have recently reported on polymer capsules containing liposomes. The antitumor hydrophobic peptide, thiocoraline, was entrapped within the lipid membrane of the liposomal subcompartments. Cell viability assays verified the activity of the trapped antitumor cargo. In a recent work, Boudou et al. engineered paclitaxel-loaded capsules by precomplexing paclitaxel with chemically modified derivative of hyaluronic acid containing hydrophobic nanocavities, and subsequent assembly with either poly(L-lysine) or quaternized chitosan as polycations (Fig. 13.10). The paclitaxel-loaded capsules were found to decrease the viability and proliferation of MDA MB 231 breast cancer cells [86].

An appropriate release mechanism must be selected depending on the desired application. The biological stimuli are dependent on intracellular mechanisms and

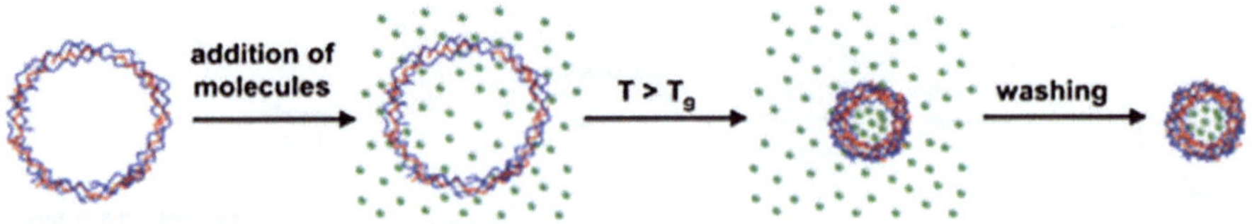

Fig. 13.9 Schematic representation of encapsulation of model compounds via morphological changes in the capsules upon heating. Reproduced from [84] with permission from John Wiley and Sons

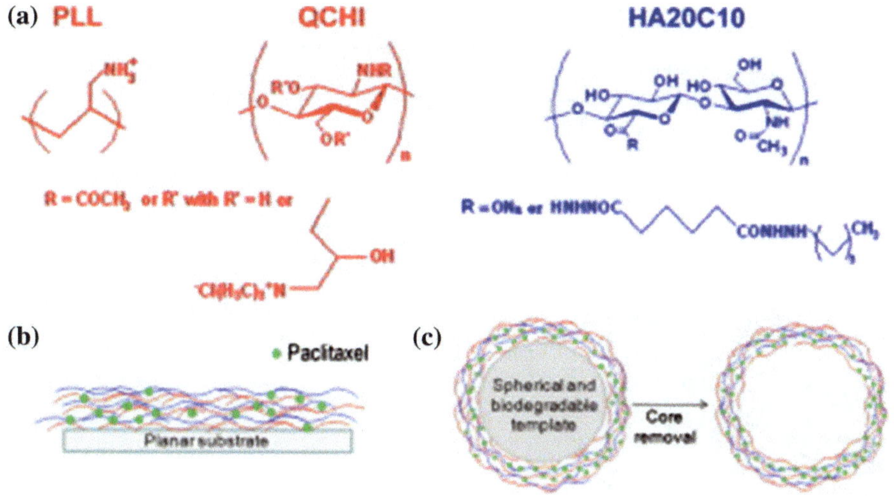

Fig. 13.10 Schematic presentation of loading of paclitaxel in the shell of polymeric capsules. Reproduced from [86] with permission from Elsevier

typically promote slow release. On the contrary, by on/off switching of applied external physical stimuli, opening of the capsules and, hence, release of their payloads, can be remotely controlled. The shell of the capsules prepared via the LbL technique from oppositely charged polyelectrolytes are intrinsically sensitive to pH. They typically exhibit pH-dependent shrinking/swelling and thus are interesting for intracellular delivery to target sites characterized by acidic environments such as tumor tissues (pH < 6.8). Acidic pH-responsive nanocapsules composed by weak and biodegradable polyelectrolytes chitosan and poly (gamma-glutamic acid) have recently been reported [87], whereas the capsules were stable at neutral and slightly alkaline pH, the encapsulated substances were released in response to acidic pH. An interesting example of release from the shell of capsules has recently been described elsewhere [88]. The capsules were prepared by self-assembly of polyaldehyde dextran-graft-adamantane and carboxymethyl dextran-β-cyclodextrin on $CaCO_3$ particles via host–guest interactions. The shell was loaded with doxorubicin conjugated with adamantane through hydrazone bonding. Adamantane was grafted to polyaldehyde dextran also through hydrazone bonding. At slightly acidic conditions the hydrazone bonds were cleaved, the capsules underwent destruction and doxorubicin was released (Fig. 13.11).

Application of external stimuli such as high-frequency magnetic field, ultrasound, microwave, and UV irradiation has been suggested as strategies to remotely affect the stability of the capsule membrane. Carrying a component, which exhibits different behavior when exposed to the specific stimulus, is an essential prerequisite to control the release of cargo molecules. The external stimulus typically affects the integrity of the capsule membrane, which may lead to destruction of the entire capsule or (reversible) formation of defects that would ultimately result in release of

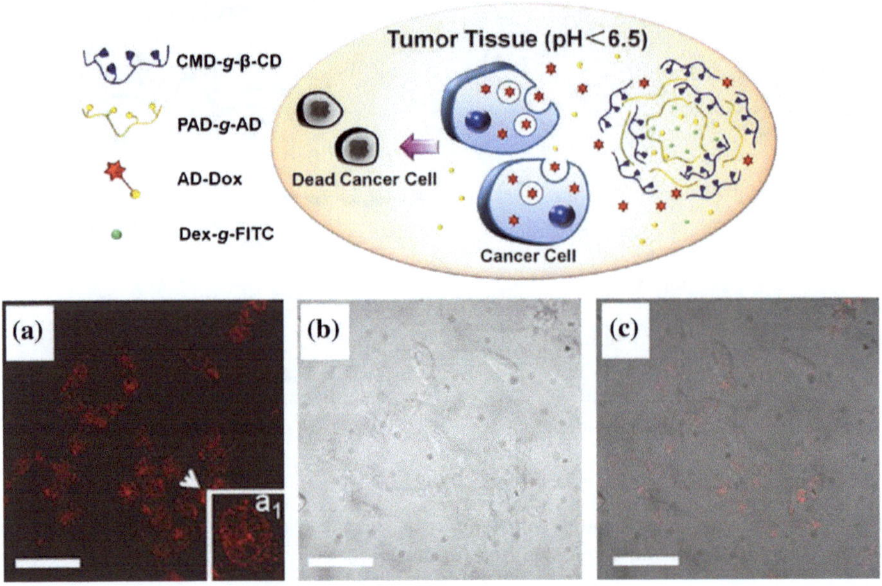

Fig. 13.11 *Top* Schematic diagram of pH-induced drug release from pH-cleavable drug-loaded supramolecular capsule. CMD-g-β-CD (carboxymethyl dextran-β-cyclodextrin), PAD-g-AD (polyaldehyde dextran-graft-adamantane), AD-Dox (adamantine-doxorubicin conjugate), Dex-g-FITC (fluorescein isothiocyanate-labeled dextran). *Bottom* Confocal microscope image of HeLa cells treated with AD-Dox loaded capsules at (**a–c**) pH 5.5: (*a1*) enlarged image; **a** Dox fluorescence in cells; **b** phase contrast image; **c** overlay fluorescence and phase contrast image. Scale bar 50 μm. Reproduced from [88] with permission from American Chemical Society

the contents. del Meracato et al. [89] have recently reported on polyelectrolyte capsules with Au nanoparticles embedded into their membranes and their behavior upon microwave irradiation. The modified capsules underwent more rapid damage compared to their unmodified analogs, which could be harnessed for drug-delivery purposes. Benzophenone-modified hollow shells prepared by alternate LbL adsorption of poly(allylamine hydrochloride) and poly(sodium 4-styrenesulfonate) on polystyrene particles have been reported by Park et al. [90]. The release kinetics of a model drug, rhodamine B, was investigated and it was found that the release of the model drug could be effectively controlled based on UV irradiation time. An essential drawback of UV-assisted approaches for in vivo applications are the damages to biological samples and the limited penetration of the UV light in living tissues. In that aspect, near infrared light (NIR) is more suitable for biological applications. The NIR-based approaches also require embedding of the membranes with metal nanoparticles to induce morphological changes upon irradiation with laser light in the NIR. NIR-adsorbing capsules with metal nanoparticles in the shells and fluorescently labeled polymers as cargo inside their cavities have been reported to internalize in living cells [91]. At moderate light intensities, the photo-induced heating of the metal nanoparticles lead to rupture of the capsule walls and release of the cargo in the cytosol. Carregal-Romero et al. [92] have recently demonstrated that capsules can be opened individually, which allowed for sequentially releasing the cargo from different capsules within a single cell and that proteins such as green fluorescent protein could be released without destruction by the local heating. Triggering of an enzymatic reaction upon sequential opening of capsules containing either the enzyme or the substrate is shown in Fig. 13.12 [93]. Furthermore, this technique allowed controlled release of mRNA from capsules, thereby resulting in synthesis of green fluorescent protein.

The enzyme-sensitive capsules, similarly to the pH-sensitive ones, are bioresponsive. Such systems are based on the enzymatic degradation of the capsule shells for the controlled and sustained release of the encapsulated molecules. Here, the release profile is controlled via the thickness of the membrane (the fewer the layers, the faster the degradation) or the nature of biodegradable polymers building the shell (incorporation of synthetic polyelectrolytes slows down degradation). Natural polymers such as chitosan, dextran, hyaluronic acid, as well as various polypeptides and synthetic polyesters containing degradable units have been used as building layers of the capsule membranes. The angiogenic cytokine basic fibroblast growth factor (bFGF) has been loaded in capsules based on dextran sulfate and chitosan through pH-controlled switching of the capsule permeability [94]. When placed in physiological media, competition with salts allowed a controlled release of the growth factor. The bFGF was directly and continuously

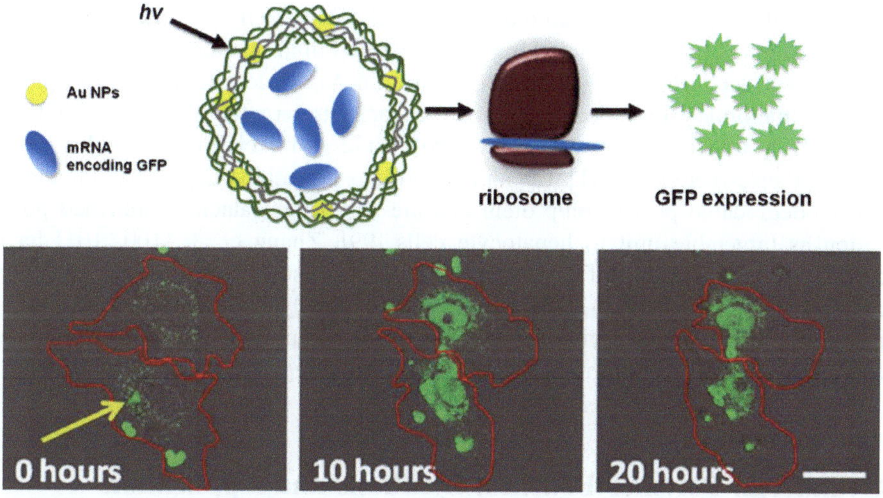

Fig. 13.12 *Top* Schematic light-opening of mRNA loaded capsules carrying Au nanoparticles into their shells. *Bottom* Confocal microscope images of capsules internalized by HeLa cells. Capsules were irradiated with a light pointer, as indicated by the *yellow arrow*. At time t = 0, mRNA is released into the cytosol where GFP expression started. At times t = 10 h and t = 20 h, the mean fluorescence intensity per cell significantly increases indicating successful expression of GFP in HeLa cells with an intensity maximum around 10 h after induced release. Scale bars 20 μm. Reproduced from [93] with permission from John Wiley and Sons

released from the capsules to the fibroblast cells without denaturation or diffusion of the bFGF. The bFGF-loaded capsules were able to prolong proliferation of L929 fibroblasts to 15 days, compared to proliferation of only 4 days of the free bFGF. A versatile method to encapsulate water-insoluble molecules in the shell of polysaccharide-based capsules (alkylated hyaluronic acid and chitosan) is described elsewhere [95]. The hydrophobic molecules can be delivered intracellularly in dendritic cells.

In order to maximize the therapeutic efficiency and minimize the side effects of a certain carrier, targeted delivery to the site of action is needed. Decorating the external surface of the carrier loaded with therapeutics with specific targeting ligands could direct the carrier to specific cells or tissues. Antibodies, antibody fragments, receptors, polypeptides, lipids, carbohydrates covalently, or non-covalently attached to the outer surface of the carrier have typically been used. A general approach for functionalization of capsules with antibodies using click chemistry is presented in Fig. 13.13 [96]. It was shown that the antibody-functionalized capsules specifically bound to colorectal cancer cells expressing the target antigen. In an earlier paper, capsules prepared by LbL deposition of poly(sodium 4-styrenesulfonate) and poly (allylamine hydrochloride) were coated with a humanized monoclonal antibody (huA33 mAB) [97]. The biofunctionalized capsules bound to the A33 antigen present on colorectal cancer cells. The targeting showed selective binding and internalization of the particles.

Polysaccharide capsules coated with lipid molecules have been shown not only to provide a sealed barrier to prevent the leakage of the encapsulated drug, but also can be used as a means to attach biological functionality to the capsule surface [98]. Thus, neoglycolipid incorporated into the supported lipid membrane displayed high efficiency in lectin binding studies (concanavalin A), while a folate-linked lipid in the lipid membrane provided a functional capacity to target cancer cells (MCF-7).

Finally, surface modification of microcapsules with carbohydrates has been described elsewhere [99–101]. Lin and co-workers assembled hepatocyte-targetable and biodegradable polycation/protein capsules by using galactose-branched polycation as targetable unit to hepatocyte cells [99]. Zhang et al. [100, 101] have described the fabrication of multilayers and microcapsules by alternate deposition of a cationic D-galactose-branched copolymers, poly(vinyl galactose ester-co-methacryloxyethyl trimethylammonium chloride) and poly(styrene sulfonate). The capsules exhibited recognition abilities with peanut agglutinin lectin.

Cylindrical Polymer Brushes: Synthesis and Application as Templates for Organic–Inorganic Nanostructures

Whereas the LbL assembly typically uses (inorganic) sacrificial templates to prepare capsules, the situation is somewhat opposite for another type of polymeric nanostructures, named cylindrical polymer brushes (CPBs). In the last decade, these

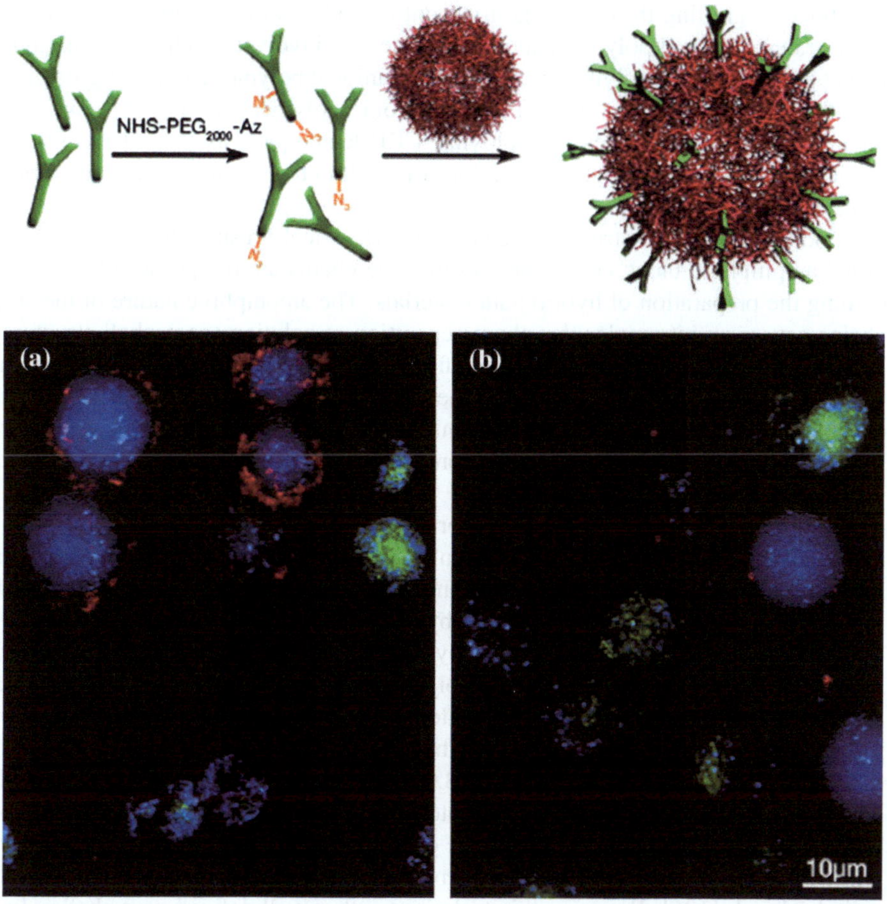

Fig. 13.13 Targeting of cancer cells using click-functionalized polymer capsules. *Top* Scheme showing Ab azide functionalization with the NHS–PEG2000–Az linker and subsequent capsule functionalization. *Bottom* Fluorescence microscopy images of LIM2405+ cells (*blue*) and LIM2405− cells (*green*) incubated with **a** huA33 mAb- or **b** IgG-functionalized capsules (*red*). Capsule/cell ratio of 50:1. Scale bar corresponds to both images. Reproduced from [96] with permission from American Chemical Society

unique polymeric nano-sized structures have received great attention due to their specific intrinsic properties. Particularly, many efforts have been focused on the use of cylindrical brushes as single molecular templates for the synthesis of inorganic nanoparticles or nanowires [102]. Cylindrical brushes (known also as molecular brushes or bottlebrushes) are a special type of graft copolymers in which multiple polymer chains are densely grafted to a linear polymer [103]. Most cylindrical polymer brushes synthesized so far possess anisotropic one-dimensional (1D) shape. Generally, there are three strategies for the synthesis of cylindrical brushes possessing densely grafted side chains covalently bonded to a linear backbone—“grafting through”, “grafting onto”, and “grafting from” [104]. Within each strategy, several polymerization techniques are involved, such as conventional and controlled radical polymerizations (CRP), anionic polymerization, ring-opening metathesis polymerization (ROMP), and cationic polymerization [105]. Particular attention is paid to syntheses employing CRP techniques, as these have proven particularly useful for the preparation of well-defined, functional cylindrical brushes.

Among a variety of structures, core-shell cylindrical brushes that contain functional amphiphilic block copolymers in the side chains are of special interest concerning the preparation of hybrid nanomaterials. The amphiphilic nature of the side chains causes an intramolecular phase separation in solution (core–shell structure) due to the dissimilar interactions of the different blocks with the solvent. By selecting a polymer carrying functional groups one can create a structure comprising a core, capable of coordinating metal ions, surrounded by a protective shell. Consequently, the core can be practically used as a nanoreactor to synthesize and accommodate 1D inorganic or hybrid nanostructures.

The synthesis of cylindrical polymer brushes with amphiphilic poly(acrylic acid)-block-poly(n-butyl acrylate) (PAA-b-PnBA) diblock copolymer side chains is shown in Fig. 13.14. The procedure includes several steps: (i) synthesis of a well-defined macroinitiator, PBIEM, by esterification of poly(2-hydroxyethyl methacrylate) (PHEMA), which was synthesized via ATRP of 2-hydroxyethyl methacrylate (HEMA) or anionic polymerization of silyl-protected HEMA; (ii) ATRP of t-butyl acrylate (tBA) initiated by the pendant α-bromoester groups of PBIEM, yielding cylindrical brushes with PtBA homopolymer side chains; (iii) sequential ATRP of n-butyl acrylate (nBA) forming the cylindrical brushes with diblock copolymer [poly(t-butyl acrylate)-block-poly(n-butyl acrylate) (PtBA-b-PnBA)] side chains; and (iv) hydrolysis of the PtBA block to produce the hydrophilic poly(acrylic acid) (PAA) block forming the core of an amphiphilic core–shell cylinder brush [106]. By using this technique, other well-defined core-shell cylindrical polymer brushes with polystyrene (PS), PS-b-PAA or PAA-b-PS, as side chains have been successfully synthesized.

Well-defined amphiphilic core-shell cylindrical brushes with PAA core and PnBA shell have been exploited as templates for the synthesis of wire-like

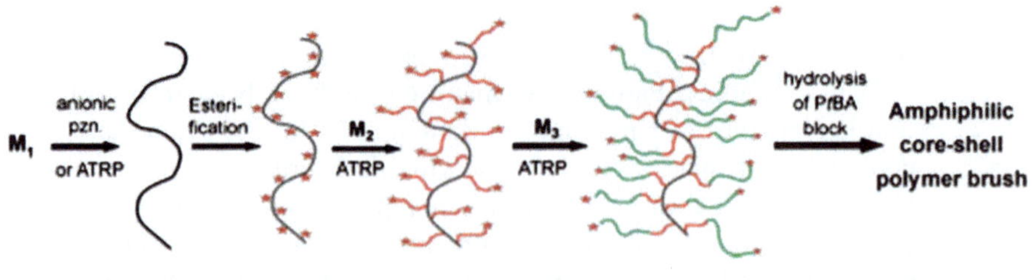

Fig. 13.14 Schematic route to a cylindrical polymer brush with amphiphilic diblock copolymer side chains (M1 = HEMA or silyl-protected HEMA; M2 = tBA; M3 = nBA; star = initiating site). Reproduced from [106] with permission from American Chemical Society

assemblies of cadmium sulfide (CdS) nanoparticles, utilizing the coordination of cadmium ions with carboxylate groups in the core of the brush [107]. The formation of CdS nanoparticles inside the polymer brush was achieved via reaction of the coordinated Cd^{2+} ions with H_2S. This route resulted in wire-like CdS nanoparticle assemblies of about 4–5 nm in diameter and about 170 nm long. Similarly, ultrafine magnetic nanoparticles were successfully synthesized within the PAA cores of these cylindrical brushes (Fig. 13.15) [108]. The polymer shell provided not only the stability of nanoparticles, but also the solubility of hybrid nanocylinders. The quantity and size of the fabricated nanoparticles were controlled by the limited number of coordinated iron ions inside the polymer core.

In another work, cadmium selenide (CdSe) nanowires were successfully in situ fabricated, utilizing amphiphilic core-shell (PAA-PnBA) cylindrical polymer brushes as well-defined single molecular templates [109]. The hydrophilic polymer brush core acted as the nanoreactor for generating and shaping CdSe nanoparticles into nanowires via the absorption of cadmium ions by carboxylate groups in the core and subsequent introduction of H_2Se gas. The formation of 170 nm-long CdSe nanowires proceeded simultaneously with the nucleation, growth and combination of CdSe nanoparticles. Later on, uniform one-dimensional titania hybrid nanotubes were prepared using core-shell-corona cylindrical polymer brushes as soft templates [110]. The CPBs consisted of a polymethacrylate backbone with densely grafted biodegradable poly(ε-caprolactone) as the core, poly(2-(dimethlamino)ethyl methacrylate) as the cationic shell, and poly(oligo(ethylene glycol) methyl ether methacrylate) (POEGMA) as the corona. The weak polyelectrolyte shell complexed an oppositely charged titania precursor, and then acted as a nanoreactor for the hydrolysis and condensation of the precursor, resulting in crystalline TiO_2. The POEGMA shell provides solubility in aqueous and organic solvents.

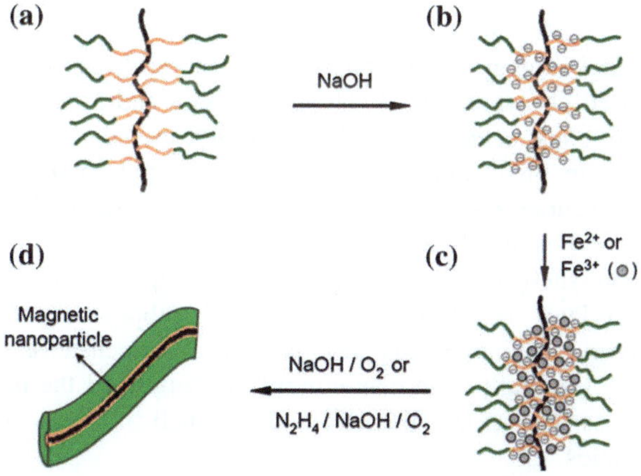

Fig. 13.15 Schematic illustration for the synthesis of a wire-like assembly of magnetic nanoparticles inside a cylindrical polymer brush: **a** polymer brush with PAA core and PnBA shell; **b** neutralized polymer brush with poly(sodium acrylate) core (Na^+ ions are not shown); **c** polychelate of the brush with Fe^{2+} or Fe^{3+} ions; and **d** hybrid nanocylinder of the brush and magnetic nanoparticles. Reproduced from [108] with permission from John Wiley and Sons

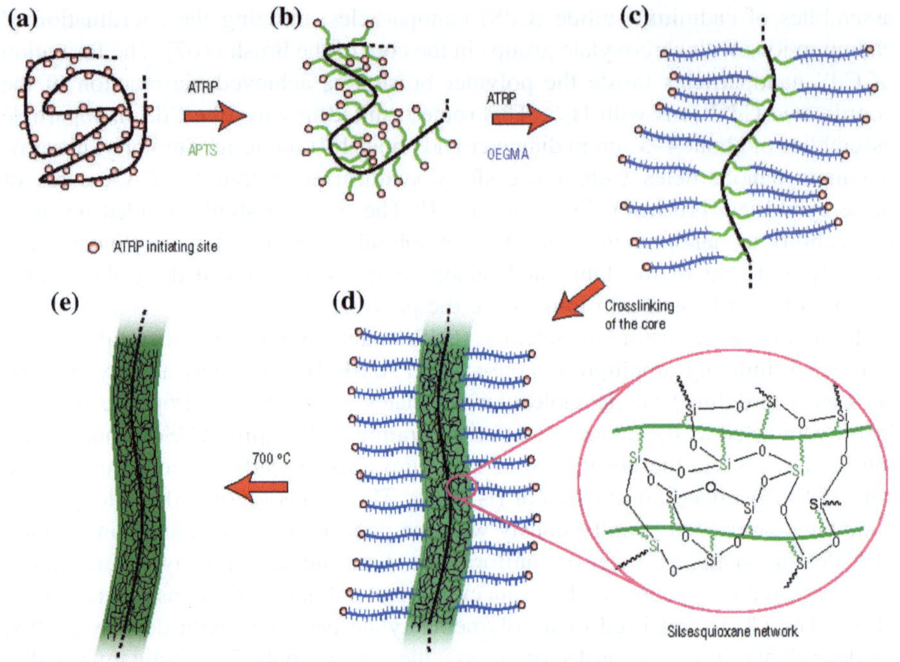

Fig. 13.16 Synthesis of soluble organo-silica and inorganic silica nanowires. **a** ATRP polyinitiator, poly(2-bromoisobutyryloxyethylmethacrylate) (PBIEM), with degree of polymerization (DP) = 3200; **b** CPB with side chains of 20 3-acryloylpropyltrimethoxysilane (APTS) units; **c** core–shell CPB with an additional 57 oligo(ethylene glycol) methacrylate (OEGMA) units; **d** soluble organo-silica hybrid nanowires with a cross-linked silsesquioxane network in the core; **e** inorganic silica nanowires after pyrolysis. Reproduced from [111] with permission from Nature Publishing Group

Another original strategy for the fabrication of hybrid nanostructures based on CPBs was reported elsewhere [111]. Thus, inorganic and hybrid silica nanowires were successfully constructed from CPBs in which the silica precursor was integrated into the core block simultaneously with the polymerization process via polymerizing a precursor-containing monomer (Fig. 13.16). Examples of organo-silica hybrid nanowires are shown in Fig. 13.17.

The use of core-shell-corona structured CPBs, comprising a poly(3-(trimethoxysilyl)propyl acrylate) (PAPTS) shell, allowed fabrication of water-soluble organo-silica hybrid nanotubes via base-catalyzed condensation of the PAPTS shell block [112]. Generally, one may control the lengths of the nanotubes by the length of CPBs, as well as the diameter and the shell thickness by adjusting the monomer feed ratios.

Alternatively, 1D brush-like templates for the synthesis of organic–inorganic hybrids have been fabricated by self-assembly of amphiphilic block copolymers in selective solvent and subsequent cross-linking. The cross-linking procedure is necessary to convert the block copolymer assemblies into permanent structures resistant to either disassembly or morphological transitions. These structures have the

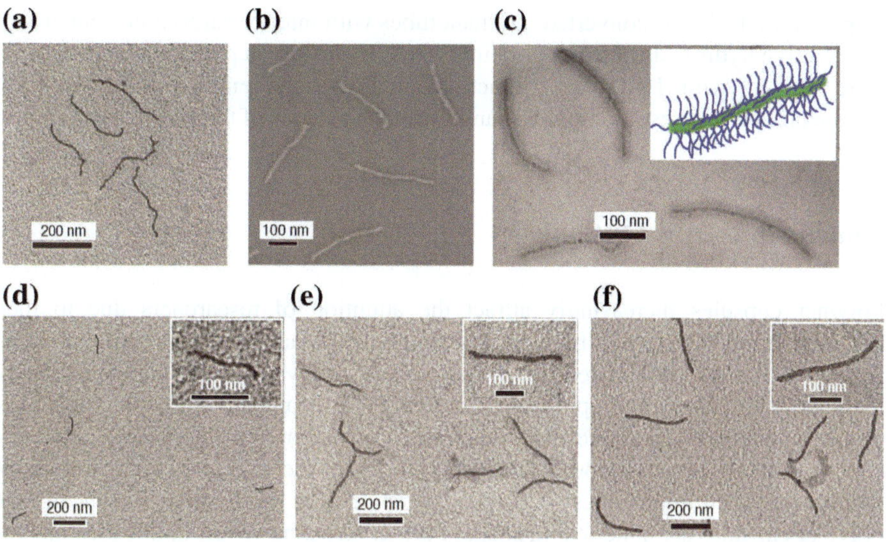

Fig. 13.17 Electron microscopy characterization of soluble organo-silica hybrid nanowires. **a** Non-stained TEM image of $[(SiO_{1.5})_{20}\text{-b-OEGMA}_{57}]_{3,200}$; **b** its SEM image; **c** its non-stained cryo-TEM image in water. **d–f** Are non-stained TEM images of $[(SiO_{1.5})_{21}\text{-b-OEGMA}_{58}]_{1,500}$, $[(SiO_{1.5})_{41}\text{-b-OEGMA}_{68}]_{3,200}$ and $[(SiO_{1.5})_{72}\text{-b-OEGMA}_{95}]_{3,200}$, respectively. Reproduced from [111] with permission from Nature Publishing Group

characteristic worm-like morphology at a similar size scale as CPBs. Liu and co-workers have prepared polymeric cylinders and nanotubes from poly(butyl methacrylate)-block-poly(2-cinnamoyloxyethyl methacrylate)-block-poly(tert-butyl acrylate) (PBMAb-PCEMA-b-PtBA) triblock copolymers [113]. The polymers, synthesized by living anionic polymerization, self-assembled in bulk into cylindrical aggregates with a PtBA core and a PCMEMA shell, embedded in a PBMA matrix (Fig. 13.18). The cylindrical morphology was locked by cross-linking the PCMEMA shell in the bulk via UV irradiation. Then, the cross-linked cylinders were

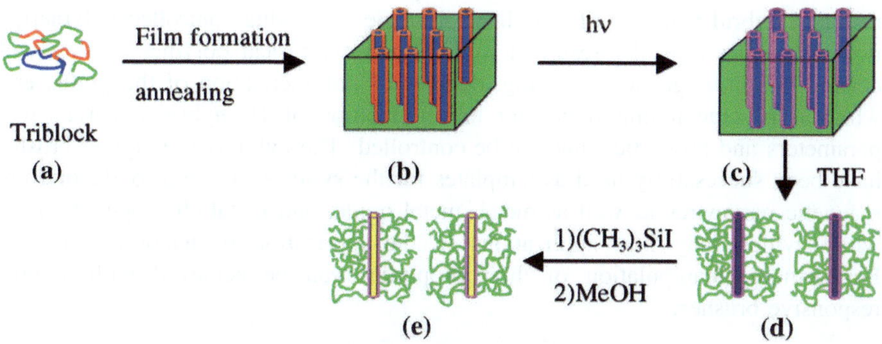

Fig. 13.18 Schematic pathway to the preparation of PBMA-b-PCEMA-b-PAA nanotubes. Reproduced from [113] with permission from American Chemical Society

dispersed in THF and converted into nanotubes with interior carboxylic acid groups by selective hydrolysis of PtBA core into PAA. In the last step, the nanotube interior were loaded with Fe_2O_3 by reduction of Fe^{2+} ions entered the PAA-lined nano-channels as a result of ion exchange with the protons of the acrylic acid groups.

Conclusions and Outlook

Polymer capsules increasingly attract the attention of researchers due to their potential applications as drug carriers, biomimetic nanoreactors, artificial organelles, multicompartment reactors, sensors, and catalysts. A plethora of different templates, polymers, and approaches can be used to prepare such structures. The proper selection of an appropriate template is of paramount importance, since the size, size distribution, geometry, and morphology of the structures, prepared by the template-assisted approaches are largely dependent on the template. Typically the polymer capsules are spherical in shape. Further steps, however, in this area include development of capsules of different aspect ratios, anisotropic and non-spherical capsules. Recent reports indicate a dramatic influence of particle geometry on their biological interactions and cellular uptake [114, 115]. Not less important are the conditions for template removal. They need to be tuned to the specificity of the material building the membrane and intended applications to avoid the influence on the polymer chemistry and properties, functionality of the cargo material, and effectiveness of the tethered receptors. Generally, the synthetic approaches used up to date allow imparting specific, e.g., *Stealth*, properties of the capsules, tethering particular receptors, decorating the capsule membrane with targeting ligands, endowing stimuli-responsiveness and/or biodegradability. All these advances expand our understanding of the importance of the polymeric nanocapsules and create new opportunities for their application.

The preparation and utilization of polymer brushes as single molecular templates for synthesis of hybrid nanomaterials have received special attention. When appropriately designed, the core of these templates can accommodate 1D inorganic nanostructures, whereas the side chains provide solubility and stability of the resulting hybrid materials. The utilized strategies involving controlled polymerization techniques typically result in well-defined cylindrical polymer brushes with desired length, degrees of grafting, degrees of polymerization of the grafts, etc., which is of crucial importance for the preparation of 1D hybrid materials with parameters and properties that can be controlled. The cylindrical polymer brushes have been successfully used as templates for the synthesis of semiconducting and magnetic nanowires as well as metal, metal oxide, and metal chalcogenide nanowires, which may find applications in the fabrication of nanoscale devices. Furthermore, manipulation of their properties can be achieved with stimuli-responsive brushes.

Acknowledgments The authors express gratitude to the EC project POLINNOVA.

References

1. Antipov AA, Shchukin D, Fedutik Y, Petrov AI, Sukhorukov GB, Möhwald H (2003) Carbonate microparticles for hollow polyelectrolyte capsules fabrication. Colloids Surf A 224:175–183
2. Volodkin DV, Larionova NI, Sukhorukov GB (2004) Protein encapsulation via porous $CaCO_3$ microparticles templating. Biomacromolecules 5:1962–1972
3. Gaponik N, Radtchenko IL, Gerstenberger MR, Fedutik YA, Sukhorukov GB, Rogach AL (2003) Labeling of biocompatible polymer microcapsules with near-infrared emitting nanocrystals. Nano Lett 3:369–372
4. Schuetz P, Caruso F (2003) Copper-assisted weak polyelectrolyte multilayer formation on microspheres and subsequent film crosslinking. Adv Funct Mater 13:929–937
5. Yu AM, Wang Y, Barlow E, Caruso F (2005) Mesoporous silica particles as templates for preparing enzyme-loaded biocompatible microcapsules. Adv Mater 17:1737–1741
6. Donath E, Sukhorukov GB, Caruso F, Davis SA, Möhwald H (1998) Novel hollow polymer shells by colloid-templated assembly of polyelectrolytes. Angew Chem Int Ed 37:2201–2205
7. Lee D, Rubner MF, Cohen RE (2005) Formation of nanoparticle-loaded microcapsules based on hydrogen-bonded multilayers. Chem Mater 17:1099–1105
8. Weda P, Trzebicka B, Dworak A, Tsvetanov CB (2008) Thermosensitive nanospheres of low density core—An approach to hollow nanoparticles. Polymer 49:1467–1474
9. Toncheva N, Tsvetanov C, Rangelov S, Trzebicka B, Dworak A (2013) Hydroxyl endfunctionalized poly(2-isopropyl oxazoline)s used as nano-sized colloidal templates for preparation of hollow polymeric nanocapsules. Polymer 54:5166–5173
10. Haladjova E, Rangelov S, Tsvetanov Ch, Simon P (2014) Preparation of polymeric nanocapsules via nano-sized poly(methoxydiethyleneglycol methacrylate) colloidal templates. Polymer 55:1621–1627
11. Tjipto E, Cadwell KD, Quinn JF, Johnston APR, Abbott NL, Caruso F (2006) Tailoring the interfaces between nematic liquid crystal emulsions and aqueous phases via layer-by-layer assembly. Nano Lett 6:2243–2248
12. Priest C, Quinn A, Postma A, Zelikin AN, Ralston J, Caruso F (2008) Microfluidic polymer multilayer adsorption on liquid crystal droplets for microcapsule synthesis. Lab Chip 8:2182–2187
13. Szczepanowicz K, Hoel HJ, Szyk-Warszynska L, Bielarinska E, Bouzga AM, Gaudernack G, Simon C, Warszynski P (2010) Formation of biocaompatible nanocapsules with emulsion core and PEGylated shell by polyemectrolyte multilayer adsorption. Langmuir 26:12592–12597
14. Shchukin DG, Kohler K, Möhwald H, Sukhorukov GB (2005) Gas-filled polyelectrolyte capsules. Angew Chem Int Ed 44:3310–3314
15. Donath E, Moya S, Neu B, Sukhorukov GB, Georgieva R, Voigt A et al (2002) Hollow polymer shells from biological templates: fabrication and potential applications. Chem Eur J 8:5481–5485
16. Shenoy DB, Antipov AA, Sukhorukov GB, Möhwald H (2003) Layer-by-layer engineering of biocompatible, decomposable core-shell structures. Biomacromolecules 4:265–272
17. Sukhorukov GB, Donath E, Davis S, Lichtenfeld H, Caruso F, Popov VI et al (1998) Stepwise polyelectrolyte assembly on particle surfaces: a novel approach to colloid design. Polym Adv Technol 9:759–767
18. Kozlovskaya V, Ok S, Sousa A, Libera M, Sukhishvili SA (2003) Hydrogen-bonded polymer capsules formed by layer-by-layer self-assembly. Macromolecules 36:8590–8592
19. Such GK, Johnston APR, Caruso F (2011) Engineered hydrogen-bonded polymer multilayers: from assembly to biomedical applications. Chem Soc Rev 40:19–29
20. Kharlampieva E, Kozlovskaya V, Sukhishvili SA (2009) Layer-by-layer hydrogen-bonded polymer films: from fundamentals to applications. Adv Mater 21:3053–3065
21. Bergbreiter DE, Liao K-S (2009) Covalent layer-by-layer assembly-an effective, forgiving way to construct functional robust ultrathin films and nanocomposites. Soft Matter 5:23–28

22. Johnston APR, Read ES, Caruso F (2005) DNA multilayer films on planar and colloidal supports: sequential assembly of like-charged polyelectrolytes. Nano Lett 5:953–956
23. Johnston APR, Mitomo H, Read ES, Caruso F (2006) Compositional and structural engineering of DNA multilayer films. Langmuir 22:3251–3258
24. Johnston APR, Lee L, Wang Y, Caruso F (2009) Controlled degradation of DNA capsules with engineered restriction-enzyme cut sites. Small 5:1418–1421
25. Johnston APR, Caruso F (2007) Exploiting the directionality of DNA: controlled shrinkage of engineered oligonucleotide capsules. Angew Chem Int Ed 46:2677–2680
26. Johnston APR, Caruso F (2008) Stabilization of DNA multilayer films through oligonucleotide crosslinking. Small 4:612–618
27. Cavalieri F, Ng SL, Mazzuca C, Jia ZF, Bulmus V, Davis TP et al (2011) Thinmultilayer films and microcapsules containing DNA quadruplex motifs. Small 7:101–111
28. Serizawa T, Hamada K, Akashi M (2004) Polymerization within a molecular-scale stereoregular template. Nature 429:52–55
29. Kida T, Mouri M, Akashi M (2006) Fabrication of hollow capsules composed of poly(methyl methacrylate) stereocomplex films. Angew Chem Int Ed 45:7534–7536
30. Shchepelina O, Drachuk I, Gupta MK, Lin J, Tsukruk VV (2011) Silk-on-silk layer-by-layer microcapsules. Adv Mater 23:4655–4660
31. Wang ZP, Feng ZQ, Gao CY (2008) Stepwise assembly of the same polyelectrolytes using host-guest interaction to obtain microcapsules with multiresponsive properties. Chem Mater 20:4194–4199
32. Xiao W, Chen W-H, Zhang J, Li C, Zhuo R-X, Zhang X-Z (2011) Design of a photoswitchable hollow microcapsular drug delivery system by using a supramolecular drugloading approach. J Phys Chem B 115:13796–13802
33. Li C, Luo G-F, Wang H-Y, Zhang J, Gong Y-H, Cheng S-X et al (2011) Host–guest assembly of pH-responsive degradable microcapsules with controlled drug release behavior. J Phys Chem C 115:17651–17659
34. Cui J, van Koeverden MP, Müllner M, Kempe K, Caruso F (2014) Emerging methods for the fabrication of polymer capsules. Adv Coll Interface Sci 207:14–31
35. Caruso F, Caruso RA, Molwald H (1998) Nanoengineering of inorganic and hybrod hollow spheres by colloidal templating. Science 282:1111–1114
36. Voigt A, Lichtenfeld H, Sukhorukov GB, Zastrow H, Donath E, Baumler H et al (1999) Membrane filtration for microencapsulation and microcappsule fabrication by layer-by-layer polyelectrolyte adsorption. Ind Eng Res 38:4037–4043
37. Esser-Kahn AP, Odom SA, Sottos NR, White SR, Moore JS (2011) Triggered release from polymer capsules. Macromolecules 44:5539–5553
38. Johnston APR, Such GK, Caruso F (2010) Triggering release of encapsulated cargo. Angew Chem Int Ed 49:2664–2666
39. Richardson II, Ejima H, Lorcher SI, Liang K, Senn P, Cui J et al (2013) Preparation of nano- and microcapsules by electrophoretic polymer assembly. Angew Chem Int Ed 52:6455–6458
40. Del Mercato LL, Ferraro MM, Baldassarre F, Mancarella S, Greco V, Rinaldi R, Leporatti S (2014) Biological applications of LbL multilayer capsules: from drug delivery to sensing. Adv Colloid Interface Sci 207:139–154
41. Volodkin D, von Klitzing R, Mohwald H (2010) Pure protein micrspheresby calcium carbonate templating. Angew Chem Int Ed 49:9258–9261
42. Parakhonskiy B, Yashchenok A, Konrad M, Skirtach A (2014) Colloidal micro- and nano-particles as templates for polyelectrolyte multilayer capsules. Adv Colloid Interface Sci 207:253–264
43. Boyer C, Zasadzinski JA (2007) Multiple lipid compartments slow vesicle contents release in lipases and serum. ACS Nano 1:176–182
44. Stadler B, Chandrawati R, Price AD, Chong SF, Breheney K, Postma A et al (2009) A microreactor with thousands of subcompartments: enzyme-loaded liposomes within polymer capsules. Angew Chem Int Ed Engl 48:4359–4362

45. Delcea M, Yashchenok A, Videnova K, Kreft O, Mohwald H, Skirtach AG (2010) Multicompartmental micro- and nanocapsules: hierarchy and applications in biosciences. Macromol Biosci 10:465–474
46. Chandrawati R, Caruso F (2012) Biomimetic liposome- and polymersome-based multicompartmentalized assemblies. Langmuir 28:13798–13807
47. Huang X, Voit B (2013) Progress on multi-compartment polymeric capsules. Polym Chem-UK 4:435–443
48. Sandre O, Moreaux L, Brochard-Wyart F (1999) Dynamics of transient pores in stretched vesicles. Proc Natl Acad Sci USA 96:10591–10596
49. Xiong R, Soenen SJ, Braeckmans K, Skirtach AG (2013) Towards theranostic multicompartment microcapsules: in-situ diagnostics and laser-induced treatment. Theranostics 3:141–151
50. Champion JA, Mitragotri S (2006) Role of target geometry in phagocytosis. Proc Natl Acad Sci USA 103:4930–4934
51. Muro S, Garnacho C, Champion JA, Leferovich J, Gajewski C, Schuchman EH et al (2008) Control of endothelial targeting and intracellular delivery of therapeutic enzymes by modulating the size and shape of ICAM-1-targeted carriers. Mol Ther 16:1450–1458
52. Shchepelina O, Kozlovskaya V, Kharlampieva E, Mao W, Alexeev A, Tsukruk VV (2010) Anisotropic micro- and nano-capsules. Macromol Rapid Commun 31:2041–2046
53. Kozlovskaya V, Higgins W, Chen J, Kharlampieva E (2011) Shape switching of hollow layer-by-layer hydrogel microcontainers. Chem Commun (Camb) 47:8352–8354
54. Kozlovskaya V, Yakovlev S, Libera M, Sukhishvili SA (2005) Surface priming and the selfassembly of hydrogen-bonded multilayer capsules and films. Macromolecules 38:4828–4836
55. Shchepelina O, Lisunova MO, Drachuk I, Tsukruk VV (2012) Morphology and properties of microcapsules with different core releases. Chem Mater 24:1245–1254
56. Yashchenok AM, Parakhonskiy BV, Donatan S, Kohler D, Skirtach AG, Möhwald H (2013) Polyelectrolyte multilayer microcapsules templated on spherical, elliptical and square calcium carbonate particles. J Mater Chem B 1:1223–1228
57. Dierendonck M, De Koker S, De Rycke R, Geest BD (2014) Just spray it—LbL assembly enters a new age. Soft Matter 10:804–807
58. Haladjova E, Toncheva-Moncheva N, Apostolova M, Trzebicka B, Dworak A, Petrov P, Dimitrov I, Rangelov S, Tsvetanov Ch (2014) Polymeric nanoparticles engineering: from temperature-responsive polymer mesoglobules to gene delivery systems. Biomacromolecules 15:4377–4395
59. Breitenkamp K, Emrick T (2003) Novel polymer capsules from amphiphilic graft copolymers and cross-metathesis. J Am Chem Soc 125:12070–12071
60. Roux R, Sallet L, Alcouffe P, Chambert S, Sintes-Zydowicz N, Fleury E et al (2012) Facile and rapid access to glyconanocapsules by CuAAC interfacial polyaddition in miniemulsion conditions. ACS Macro Lett 1:1074–1078
61. Ye S, Liu Y, Chen S, Liang S, McHale R, Ghasdian N et al (2011) Photoluminescent properties of Prussian blue (PB) nanoshells and polypyrrole (PPy)/PB core/shell nanoparticles prepared via miniemulsion (periphery) polymerization. Chem Commun 47:6831–6833
62. Li W, Matyjaszewski K, Albrecht K, Möller M (2009) Reactive surfactants for polymeric nanocapsules via interfacially confined miniemulsion ATRP. Macromolecules 42:8228–8233
63. Li W, Yoon JA, Matyjaszewski K (2010) Dual-reactive surfactant used for synthesis of functional nanocapsules in miniemulsion. J Am Chem Soc 132:7823–7825
64. Utama RH, Guo Y, Zetterlund PB, Stenzel MH (2012) Synthesis of hollow polymeric nanoparticles for protein delivery via inverse miniemulsion periphery RAFT polymerization. Chem Commun 48:11103–11105
65. Utama RH, Stenzel MH, Zetterlund PB (2013) Inverse miniemulsion periphery RAFT polymerization: a convenient route to hollow polymeric nanoparticles with an aqueous core. Macromolecules 46:2118–2127

66. Fu GD, Shang Z, Hong L, Kang ET, Neoh KG (2005) Preparation of cross-linked polystyrene hollow nanospheres via surface-initiated atom transfer radical polymerizations. Macromolecules 38:7867–7871
67. Huang X, Appelhans D, Formanek P, Simon F, Voit B (2011) Synthesis of well-defined photo-cross-linked polymeric nanocapsules by surface-initiated RAFT polymerization. Macromolecules 44:8351–8360
68. Achilleos DS, Hatton TA, Vamvakaki M (2012) Light-regulated supramolecular engineering of polymeric nanocapsules. J Am Chem Soc 134:5726–5729
69. Blomberg S, Ostberg S, Harth E, Bosman AW, Van Horn B, Hawker CJ (2002) Production of crosslinked, hollow nanoparticles by surface-initiated living free-radical polymerization. J Polym Sci, Part A: Polym Phys 40:1309–1320
70. Huang X, Appelhans D, Formanek P, Simon F, Voit B (2012) Tailored synthesis of intelligent polymer nanocapsules: an investigation of controlled permeability and pH dependent degradability. ACS Nano 6:9718–9726
71. Goh TK, Guntari SN, Ochs CJ, Blencowe A, Mertz D, Connal LA et al (2011) Nanoengineered films via surface-confined continuous assembly of polymers. Small 7:2863–2867
72. Wang Y, Price AD, Caruso F (2009) Nanoporous colloids: building blocks for a new generation of structured materials. J Mater Chem 19:6451–6464
73. Mertz D, Wang Y, Goh TK, Blencowe A, Caruso F (2011) Bromoisobutyramide as an intermolecular surface binder for the preparation of free-standing biopolymer assemblies. Adv Mater 23:5668–5673
74. Mertz D, Wu H, Wong JS, Cui J, Tan P, Alles R (2012) Ultrathin, bioreposnsive and drug-functionalized protein capsules. J Mater Chem 22:21434–21442
75. Mertz D, Cui J, Yan Y, Devlin G, Chaubaroux C, Dochter A (2012) Protein capsules assembled via isobutyramide grafts: sequential growth, biofunctionalization, and cellular uptake. ACS Nano 6:7584–7594
76. Wu Y-L, Li J (2009) Synthesis of supramolecular nanocapsules based on threading of multiple cyclodextrinsover polymers on gold nanoparticles. Angew Chem Int Ed 48:3842–3845
77. Dam HH, Caruso F (2012) Modular click assembly of degradable capsules using polyrotoxanes. ACS Nano 6:4686–4693
78. Karamitros CS, Yashchenok AM, Möhwald H, Skirtach AG, Konrad M (2013) Preserving catalytic activity and enhancing biochemical stability of the therapeutic enzyme asparaginase by biocompatible multilayered polyelectrolyte microcapsules. Biomacromolecules 14:4398–4406
79. Haladjova E, Rangelov S, Tsvetanov Ch, Pispas S (2012) DNA encapsulation via nanotemplates from cationic block copolymer micelles. Soft Matter 8:2884–2889
80. Wang Y, Yan Y, Cui J, Hosta-Rigau L, Heath JK, Nice EC, Caruso F (2010) Encapsulation of water-insoluble drugs in polymer capsules prepared using mesoporous silica templates for intracellular drug delivery. Adv Mater 22:4293–4297
81. Cui J, Wang Y, Postma A, Hao J, Hosta-Rigau L, Caruso F (2010) Monodisperse polymer capsules: tailoring size, shell thickness and hydrophobic cargo loading via emulsion templating. Adv Funct Mater 20:1625–1631
82. Grigoriev DO, Bukreeva T, Mohwald H, Shchuin DG (2008) New method for fabrication of loaded micro- and nanocontainers: emulsion encapsulation by polyelectrolyte layer-by-layer deposition on the liquid core. Langmuir 24:999–1004
83. Tonq W, She S, Xie L, Gao Ch (2011) High efficient loading and controlled release of low-molecular-weight drugs by combinantion of spontaneous deposition and heat-induced shrinkage of multilayer capsules. Soft Matter 7:8258–8265
84. Koehler K, Sukhorukov GB (2007) Heat treatment of polyelectrolute multilayer capsules: a versatile method for encapsulation. Adv Funct Mater 17:2053–2061
85. Hosta-Rigau L, Staedler B, Yan Y, Nice EC, Heath JK, Alberico F, Caruso F (2010) Capsosomes with multilayered subcompartments: assembly and loading with hydrophobic cargo. Adv Funct Mater 20:59–66

86. Boudou T, Kharkar P, Jing J, Guillot R, Pignot-Paintrand I, Auzely-Velty R, Picart C (2012) Polyelectrolyte multilayer nanoshells with hydrophobic nanodomains for delivery of paclitaxel. J Contr Rel 159:403–412
87. Imoto T, Kida T, Matsusaki M, Akashi M (2010) Preparation and unique pH-responsive properties of novel biodegradable nanocapsules composed of poly(gamma-glutamic acid) and chitosan as weak polyelectrolytes. Macromol Biosci 10:271–277
88. Luo GF, Xu XD, Zhang J, Yang J, Gong YH, Lei Q (2012) Encapsulation of an adamantine-doxorubicin prodrug in pH-responsive polysaccharide capsules for controlled release. ACS Appl Mater Interfaces 4:5317–5324
89. del Meracato LL, Gonazalez E, Abbasi A, Parak WJ, Puntes V (2011) Synthesis and evaluation of gold nanoparticle-modified polyelectrolyte capsules under microwave irradiation for remotely controlled release for cargo. J Mater Chem 21:11468–11471
90. Park MK, Deng S, Advincula RC (2005) Sustained release control via photo-cross-linking of polyelectrolyte layer-by-layer hollow capsules. Langmuir 21:5272–5277
91. Munoz Javier A, del Pino P, Bedard MF, Ho D, Skirtach AG, Sukhorukov GB, Plank C, Parak WJ (2008) Photoactivated release of cargo from the cavity of polyelectrolyte capsules to the cytosol of cells. Langmuir 24:12517–12520
92. Carregal-Romero S, Ochs M, Rivera-Gil P, Ganas C, Pavlov AM, Sukhorukov GB, Parak WJ (2012) NIR-light triggered delivery of macromolecules into the cytosol. J Contr Rel 159:120–127
93. Ochs M, Carregal-Romero S, Rejman J, Breackmans K, De Smedt SC, Parak WJ (2013) Light-addressable capsules as caged compound matrix fro controlled triggering of cytosolic reactions. Angew Chem Int Ed 52:695–699
94. Itoh Y, Matsusaki M, Kida T, Akashi M (2008) Locally controlled release of basic fibroblast growth factor from multilayered capsules. Biomacromolecules 9:2202–2206
95. Cui D, Jing J, Boudou T, Pignot-Paintrand I, De Koker S, De Geest BG, Picart C, Auzely-Velty R (2011) Hydrophobic shell loading of biopolyelectrolyte capsules. Adv Mater 23:H200–H204
96. Kamphius MM, Johnston AP, Such GK, Dam HH, Evans RA, Scott AM et al (2010) Targeting of cancer cells using click-functionalized polymer capsules. J Am Chem Soc 132:15881–15883
97. Cortez C, Tomaskovic-Crook E, Johnston APR, Radt B, Cody SH, Scott AM, Nice EC, Heath JK, Caruso F (2006) Targeting and uptake of multilayered particles to colorectal cancer cells. Adv Mater 18:1998–2003
98. Qi W, Wang A, Yang Y, Du M, Bouchu MN, Boulanger P, Li J (2010) The lectin binding and targetable cellular uptake of lipid coated polysaccharide microcapsules. J Mater Chem 20:2121–2127
99. Zhang F, Liu I, Wu Q, Lin X (2009) Design and in vitro biodegradation of novel hepatocyte-targetable (galactose polycation/hemoglobin) multilayers and microcapsules. Macromol Chem Phys 210:1052–1060
100. Zhang F, Wu Q, Chen Z, Li X, Jiang X, Lin X (2006) Bioactive galactose-branched polyelectrolyte multilayers and microcapsules: self-assembly, characterization, and biospecific lectin adsorption. Langmuir 22:8458–8464
101. Zhang F, Wu Q, Chen Z, Zhang M, Lin X (2008) Hepatic-targeting microcapsules construction by self-assembly of bioactive galactose-branched polyelectrolyte for controlled drug release system. J Colloid Int Sci 317:477–484
102. Lee H, Pietrasik J, Sheiko SS, Matyjaszewski K (2010) Stimuli-responsive molecular brushes. Prog Polym Sci 35:24–44
103. Yuan J, Müller AHE, Matyjaszewski K, Sheiko SS (2012) Molecular brushes. In: Müller AHE, Wooley KL (eds) Polymer science: a comprehensive reference, vol 6. Macromolecular architectures and soft nano-objects. Elsevier, Amsterdam
104. Zhang M, Müller AHE (2005) Cylindrical polymer brushes. J Polym Sci A Polym Chem 43:3461–3481

105. Sheiko SS, Sumerlin B, Matyjaszewski K (2008) Cylindrical molecular brushes: synthesis, characterization, and properties. Prog Polym Sci 33:759–785
106. Cheng G, Böker A, Zhang M, Krausch G, Müller AHE (2001) Amphiphilic cylindrical core-shell brushes via a "grafting from" process using ATRP. Macromolecules 34:6883–6888
107. Zhang M, Drechsler M, Müller AHE (2004) Template-controlled synthesis of wire-like cadmium sulfide nanoparticle assemblies within core-shell cylindrical polymer brushes. Chem Mater 16:537–543
108. Zhang M, Estournes C, Bietsch W, Müller AHE (2004) Superparamagnetic hybrid nanocylinders. Adv Funct Mater 14:871–882
109. Yuan J, Drechsler M, Xu Y, Zhang M, Müller AHE (2008) Cadmium selenide nanowires within core-shell cylindrical polymer brushes: synthesis, characterization and the double-loading process. Polymer 49:1547–1554
110. Müllner M, Lunkenbein T, Schieder M, Gröschel AH, Miyajima N, Förtsch M, Breu J, Caruso F, Müller AHE (2012) Template-directed mild synthesis of anatase hybrid nanotubes within cylindrical core-shell-corona polymer brushes. Macromolecules 45:6981–6988
111. Yuan J, Xu Y, Walther A, Bolisetty S, Schumacher M, Schmalz H, Ballauff M, Muller AHE (2008) Water-soluble organo-silica hybrid nanowires. Nat Mater 7:718–722
112. Müllner M, Yuan J, Weiss S, Walther A, Förtsch M, Drechsler M, Müller AHE (2010) Water-soluble organo-silica hybrid nanotubes template by cylindrical polymer brushes. J Am Chem Soc 132:16587–16592
113. Yan X, Liu F, Li Z, Liu G (2001) Poly(acrylic acid)-lined nanotubes of poly(butyl methacrylate)-block-poly(2-cinnamoyloxyethyl methacrylate). Macromolecules 34:9112– 9116
114. Best JP, Yan Y, Caruso F (2012) The role of particle geometry and mechanics in the biological domain. Adv Healthc Mater 1:35–47
115. Shimoni O, Yan Y, Wang Y, Caruso F (2012) Shape-dependent cellular processing of polyelectrolyte capsules. ACS Nano 7:522–530

Index

S. Fakirov (ed.), *Nano-size Polymers*,
DOI 10.1007/978-3-319-39715-3

MIX
Papier aus verantwortungsvollen Quellen
Paper from responsible sources
FSC® C105338

If you have any concerns about our products,
you can contact us on
ProductSafety@springernature.com

In case Publisher is established outside the EU,
the EU authorized representative is:
Springer Nature Customer Service Center GmbH
Europaplatz 3, 69115 Heidelberg, Germany

Printed by Libri Plureos GmbH
in Hamburg, Germany